This item must be returned or renewed by the last date shown below. The loan period may be shortened if it is reserved by another reader. A fine will be due if it is not returned on time.

DATE OF RETURN

30/SEPT /97.

23 OCT 2002

10 JAN 2000

3 JUN 2000

1 1 JUN 2004

0 2 MAY 2000

2 7 APR 2001

7/1/02
4/3/02
29/4/02

UNIVERSITY COLLEGE LONDON
Gower Street London WC1E 6BT

LF4D

rlo276.24.2.97

Dynamic Biological
Organization

Dynamic Biological Organization

Fundamentals as Applied to Cellular Systems

M.A. Aon

Associate Professor of Biological Chemistry
Research Career Investigator from CONICET
Argentina

and

S. Cortassa

Assistant Professor of Biological Chemistry
Research Career Investigator from CONICET
Argentina

CHAPMAN & HALL

London · Weinheim · New York · Tokyo · Melbourne · Madras

Published by Chapman & Hall, 2–6 Boundary Row, London SE1 8HN, UK

Chapman & Hall, 2–6 Boundary Row, London SE1 8HN, UK

Chapman & Hall GmbH, Pappelallee 3, 69469 Weinheim, Germany

Chapman & Hall USA, 115 Fifth Avenue, New York, NY 10003, USA

Chapman & Hall Japan, ITP-Japan, Kyowa Building, 3F, 2-2-1 Hirakawacho, Chiyoda-ku, Tokyo 102, Japan

Chapman & Hall Australia, 102 Dodds Street, South Melbourne, Victoria 3205, Australia

Chapman & Hall India, R. Seshadri, 32 Second Main Road, CIT East, Madras 600 035, India

First edition 1997

© 1997 M.A. Aon and S. Cortassa

Typeset in 10/12 pt Palatino by Saxon Graphics Ltd, Derby
Printed in Great Britain by T.J. International, Padstow, Cornwall

ISBN 0 412 79890 5

A catalogue record for this book is available from the British Library

Library of Congress Catalog Card Number: 96–86752

∞ Printed on permanent acid-free text paper, manufactured in accordance with ANSI/NISO Z39.48-1992 and ANSI/NISO Z39.48-1984 (Permanence of Paper).

Dedicated to the memory of Ranwel Caputto.

Contents

Foreword

Science is the meeting place
of two kinds of poetry:
The poetry of thought
and the poetry of action.

G. Agostino Da Silva

Classical physics   : organized simplicity
Statistical mechanics : chaotic complexity
Biology   : organized complexity
W. Weaver (1948) Science and complexity. *American Scientist*, **36**, 536–544

This volume is a fascinating account of the living organism as a dynamic system; its arrival has been anticipated by a flurry of research publications on the excellent experimental and theoretical work of the two young authors in recent years. Miguel Aon and Sonia Cortassa have travelled widely to work in some of the leading laboratories of the day. In researching their own studies they have assembled a formidable information base, and here we have the quintessence of their accumulating insights.

It may have seemed to some that the last decade has been a quiet one for bioenergetics, at least by comparison with the 'golden age' of the 1960s and 1970s. Those decades saw the elucidation of mitochondrial electron transport and oxidative phosphorylation, and the final *coup de grace* epitomized in the chemiosmotic theory and its predictions by Peter Mitchell. Metabolic control theory was formalized by Kacser and Burns and by Heinrich and Rapoport, building on the earlier ideas of Chance and Higgins, Sel'kov and Reich, and D.E. Atkinson. All these influences are clearly traceable antecedents to the enthusiasm of the present authors.

The enormous successes of reductionism in the years since the discovery of the usefulness of restriction enzymes have led to the fine dissection in great detail of the read-only memory of life. The grand vistas of biological form and function can so easily be lost. Now, with the latest technology, it becomes even more feasible to have rapid and sensitive non-invasive glimpses of complex processes in real time in ways that just a few years ago would have seemed like dreams of science fiction. Enormous strides in biophysics, as always, have depended on developments in opto-electronics, cryogenics, imaging and data acquisition techniques. The new possibilities for direct studies of living systems extend the existing techniques of mathematics; these are also exciting days for those interested in applying new analytical methods to the problems of nonlinear complex systems, and thereby to biological problems.

Aon and Cortassa have major achievements themselves in these areas and their up-to-date knowledge of a number of merging fields has stood them in good stead. It is at the interfaces that the new biology is being created. Biology is the science of the twenty-first century: this volume projects us towards the new millennium in a most engaging and exciting way.

David Lloyd, Cardiff
30 June 1996

Preface

At present our difficulties in understanding the functional organization of living systems is not in our already powerful technologies, but in the lack of a coherent and integrated approach to interpreting and working with biological complexity. As Goodwin (1986) noted:

> When we have failed to discover the logic of embryogenesis in a specific part, the genome with its genetic program, we can turn to a systematic exploration of the proposition that is the characteristic dynamic order of the whole process that defines the unique characteristics of organismic form and transformation.

That is precisely the attempt we pursue in our book and the main reason for writing it. We are convinced that the time is ripe for such a challenge.

R. Mayr (1982) wrote that although it is difficult to define scientific progress, in biological sciences it may be characterized 'by the gradual but decisive development of new concepts'. In this vein, we develop the concept of **dynamic organization** and analyse within its perspective several biological systems at different levels of organization.

The basic idea of the concept of dynamic organization is that function, understood as the spatio-temporal coherence of events, results from the intrinsic dynamics of the processes taking place in living systems. Our life continues because of blood flowing through our vessels, the air stream we breath 13 times a minute, the flux of food that we ingest and the flux of waste we discard, the wave of ionic fluxes through the axons of our neurons enabling us to communicate with our environment. In short, life is a flux of matter, energy and information empowering the emergence of unsuspected properties, of beauty and consciousness.

Our approach is transdisciplinary – beyond but through disciplines whose emphasis is on the theoretical connections between different fields. This is a very important component of scientific advance, as each field then reinforces the strength of the other (Halliday, 1988). The transdisciplinary approach suits our quest for understanding biological wholeness at the interface of biochemistry, genetics, physiology, thermodynamics, bioenergetics, kinetics and biomathematics. Mathematical modelling assists us throughout our approach to corroborate, predict, interpret and analyse biological complexity.

Most of the concept of dynamic organization owes its birth to a novel integration of previously established facts and theories with a strict biological sense. Accordingly, the whole perspective displayed in this book is nourished by the modern scientific paradigms which emerge from spatio-temporal self-organization and dynamics, chaos and fractals. The concept of dynamic organization, which is founded on a general biothermokinetic approach, is based on non-equilibrium thermodynamics, kinetics of nonlinear dynamic systems, metabolic control analysis and methods that integrate the known biochemistry, bioenergetics and physiology of cells. Dynamic organization is applied to the analysis of metabolic regulation and organization of growing, proliferating or differentiating organisms, prokaryotic or eukaryotic, either normal or malignant. Drawing heavily on the immense effort of many other researchers, biological function is analysed and integrated. According to the perspective we have adopted here, functional organization of the living organism takes into account its complexity; namely, the existence of several levels of organization, for the comprehension and detection of which we provide analytical quantitative tools. It is this attempt to include complexity in our approach that gives our theoretical concepts a biological status beyond the tools provided by different disciplines. Last but not least, we make an explicit effort to translate all those concepts in the realm of biology, focusing on concrete examples from our own work or from that of other authors, of whose work we offer a new reading with the help of mathematical modelling.

An old, albeit robust, question underlies the whole book: is biology reducible to physics and chemistry? A worthwhile answer came from F. Jacob, quoted by Peacocke (1983): 'Biology can neither be reduced to physics, nor do without it'. We do not want to be naively either reductionistic or holistic with respect to biology. In other words, we will not exert a naive reductionism to explain biology in terms of physics and chemistry but will use physics, chemistry, physiology and mathematics to understand living cells. We are ready to accept that there is no more than molecules and atoms in a cell (on the assumption that there is no such a thing as a 'vital force') and that these molecules are able to self-organize following the laws of thermodynamics and kinetics. We equally recognize that chemical processes at the cellular level – for example, the sequence in which the bases are assembled in DNA – are a function and property of the whole organism.

We live in a splendid epoch since we have at hand powerful technologies and quantitative methods as well as the computational resources to describe living systems. We lack a sound and integrative approach that leads us to the heart of the organizational principles of life. Our book primarily intends to contribute in that direction.

We especially thank Professor D. Lloyd (University of Wales) for his encouragement and interest in this work, which has benefited from collaborations with many colleagues through discussions, suggestions and conceptual and experimental enlightening over the years. Its maturing was completed during our stays at the Universidad Nacional de Córdoba, Argentina (1980–1986), the Université de Technologie de Compiègne, France (1987–1989), and the University of Amsterdam, The Netherlands (1990–1991). It was finally written at the Universidad Nacional de Tucumán, Argentina (1992–1995).

For the seminal scientific problems that form the starting point of this work, we are very much indebted to the professional and scientific richness at the Departmento de Química Biológica of the Facultad de Ciencias Químicas de la Universidad Nacional de Córdoba, Argentina, at the beginning of the 1980s. Some of those colleagues became teachers and friends forever. We especially acknowledge B. Maggio and C. Landa, whose intellectual guidance at the beginning of our scientific careers played a key role that helped us to 'bifurcate' toward a new approach of biological systems. Thanks are also given to the late R. Caputto and to J.A. Curtino, H.J.F. Maccioni and C. Argaraña. We are particularly indebted to Walter Monsberger (Centro de Cálculo, Universidad Nacional de Córdoba), by whom we were introduced to modelling and simulation of dynamic systems.

Our ideas were enriched and sharpened through numerous works as well as stimulating and fruitful discussions at the Université de Technologie de Compiègne with D. Thomas, J. Breton, S. Canu, J.P. Kernevez, P.F. Villon, J.F. Hervagault and J.P. Loza; at the Université Libre de Bruxelles with P. Borckmans, G. Dewel, A. Goldbeter and M. Kauffman; at the University of Amsterdam with K. van Dam, J.A. Berden, P. Postma, R. van Driel and N. Stuurman; at the University of Delft (The Netherlands) with W.A. Scheffers and V.P. van Dijken; at the Vrije University (Amsterdam) with A. Stouthamer; at the Netherlands Cancer Institute (Amsterdam) with H.V. Westerhoff, R. Welch, D. Kahn and H.

Daams; at the Max-Planck-Institut für Ernahrungsphysiologie (Dortmund) with B. Hess, S.C. Muller and M. Markus; and at the Instituto de Investigación Médica 'Mercedes y Martín Ferreyra' (Córdoba, Argentina) with A. Cáceres.

Special thanks are given for productive scientific collaboration with C. Rabouille (Imperial Cancer Research, London), N. Verdoni (Chemap, France), M. Goldberg (Université La Rochelle, France) and C. Briasco (Genetics Institute, Massachusetts) during our time at the Université de Technologie de Compiègne; E. van Spronsen (University of Amsterdam); and M. Manes, M.C. Manca de Nadra, J.C. Aon, M.E. Mónaco, P.A. Valdecantos, V.A. Rapisarda and H.D. Genta (Universidad Nacional de Tucumán).

Gratefully acknowledged are the assistance in AUTO installation and helpful advice from E. Doedle (Department of Computer Science, Concordia University, Canada), R. Belleman (Netherlands Cancer Institute, Amsterdam) and the staff of the Centro de Cálculo, Universidad Nacional de Tucumán; and in image analysis from W. Rasband and B. Sheriff (NIH, Bethesda); thanks are also due to the National Biomedical Simulation Resource (Duke University, Durham, USA) for providing us with information about SCoP. We gratefully acknowledge Professor Alice Schegel for translating the Borges and Saramago quotations.

We are indebted for fellowships to Ecôle Pratique des Hautes Etudes, Centre International d'Etudiants et Stagiaires (CIES) and the Ministère des Affaires Etrangères, France, during the period 1987–1989. Also acknowledged are the Commission of the European Communities and the Netherlands Organization for the Advancement of Pure Research (NWO) for financial support during 1989–1990 and 1990–1991, respectively.

We thank all the authors and publishers who kindly gave us permission to reproduce various figures and tables. Our thanks are also due to Rachel Young and Nigel Balmforth, Commissioning Editors in Life Sciences at Chapman & Hall, who handled the preliminary versions of this project, and to Kim Worham, Production Editor, and Valerie Porter, thanks to whom the style of this book was further improved.

In Argentina (1992–1995), our work has been supported by Fundación Antorchas and Centro de Investigaciones de la Universidad Nacional de Tucumán, to whom we are especially indebted.

> Creía en infinitas series de tiempos, en una red creciente y vertiginosa de tiempos divergentes, convergentes y paralelos. Esa trama de tiempos que se aproximan, se bifurcan, se cortan o que secularmente se ignoran, abarca todas las posibilidades.
>
> J.L. Borges (1941) El jardín de senderos que se bifurcan, *Obras Completas*, Emecé Editores

I believed in an infinite series of times, in a spinning and expanding net of divergent times, converging and parallel. This weave of times that approach one another, bifurcate, intersect, or that over the ages are unaware of each other, includes all possibilities.

It was inevitable that the 'timeless' conception of classical physics would clash with the metaphysical conceptions of the Western world. It is not by accident that the entire history of philosophy from Kant through Whitehead was either an attempt to eliminate this difficulty through the introduction of another reality ... or a new mode of description in which time and freedom, rather than determinism, would play a fundamental role. Be that as it may, time and change are essential in problems of biology and in sociocultural evolution.

I. Prigogine (1980) *From Being to Becoming*, W.H. Freeman and Co.

... the time has come to abandon the view of enzymes as isolated, albeit sophistically regulated, machines which produce some products. One has to go a step further and try to understand the dynamic properties of the complete integrated biochemical network in order to understand the functions of the living cell.

J.W. Stucki (1978) *Prog. Biophys. Molec. Biol.*, **33**, 99–187.

It is evident that the living organism has a time-structure which is hierarchically organized and is at least as complex as its spatial structure. Present-day biochemistry to a large extent ignores this time-structure ...

D. Lloyd, R.K. Poole and S.W. Edwards (1982) *The Cell Division Cycle. Temporal organization and control of cellular growth and reproduction*, Academic Press

... when we come to discuss complexity and reduction of biological complexity, there are many different levels in a biological organism that might be chosen as the reference level of distinguishing 'parts' (... molecule, organelle, cell, organ, organism, population of organisms, etc.). ... Although most biochemists and molecular biologists would have a predilection to choose the molecular level there is really no 'fundamental' level of reality that has an obvious priority ...

A.R. Peacocke (1983) *The Physical Chemistry of Biological Organization*, Clarendon Press

Fue ayer, y es lo mismo que si dijéramos, Fue hace mil años, el tiempo no es una cuerda que se pueda medir nudo a nudo, el tiempo es una superficie oblícua y ondulante que solo la memoria es capaz de hacer mover y aproximar.

J. Saramago (1992) *El evangelio según Jesucristo*, Ed. Seix Barral

It was yesterday, we might even say it was a thousand of years ago, time is not a rope that can be measured knot by knot; time is an oblique and undulating surface which only memory can call forth and approach.

PART ONE
Dynamic Organization

General concepts 1

This chapter aims to substantiate principles and fundamental concepts emerging from the thermodynamic and kinetic analysis of living cells. A search is made to recognize those analytic elements in biological systems, and specific mechanisms operating in the dynamics of processes at different levels of biological organization. We will attempt to show that organization is the result of the dynamics of processes occurring in living cells. Consequently, the starting point is to get in touch with the main concepts of thermodynamics and kinetics that are going to be utilized throughout the book.

Our hope is that a deeper knowledge of the dynamics at cellular level will lead to a more profound understanding of biological organization and function than one emerging from an exclusive knowledge of (macro)molecular constituents of living organisms.

From the concept of energy dissipation as a fundamental property of living systems we focus on the ability of (sub)cellular processes to self-organize. The emphasis is on the description and application of the tools of kinetics and thermodynamics that throw light on the coherent behaviour shown by living cells.

1.1 INTRODUCTION

A main concept related to cell function is that of organization in time and space. Dynamics is a widely used word in the scientific literature. For instance, it has been used to describe the growth and shrinking of a microtubule population in a steady state condition of tubulin polymerization, a phenomenon called 'dynamic instability' (Mitchison and Kirschner, 1984a,b; Kirschner and Mitchison, 1986). A different meaning is adopted when reference is made to dynamics in the context of mathematical dynamic systems or 'dynamical systems' of deterministic equations (Abraham and Shaw, 1987; Aon *et al.*, 1991; Cortassa *et al.*, 1991). In this latter sense and throughout this work, dynamics will be used to mean time-dependent (transient) or homeodynamic, time-independent (i.e. steady states, balanced growth) behaviour of biological processes occurring at different levels of organization (Figure 1.1).

To introduce the importance of dynamics for biological organization let us quote D'Arcy Thompson (1980). Discussing the shapes generated by flowing liquids, he said that 'the things which we see in the cell are less important than the actions which we recognize in the cell'.

These 'actions' result from the dynamic coherent behaviour of living systems. We regard the dynamics of cellular function as the outcome of flows of:

- energy, mainly studied with the tools provided by non-equilibrium thermodynamics;
- matter, studied through (bio)chemical, kinetic and thermodynamic methods.

Nowadays, biological organization is regarded from three different preconceptions (Harrison, 1987):

- The **structural preconception (self-assembly)** uses static geometry in order to account for the shape of the whole from fitting together small parts. This is the preconception underlying the thoughts of molecular biologists and it has been successful for explaining shape and form in viruses.
- The **equilibrium theory** attributes form to the minimization of free energy (approach to equilibrium).
- The **kinetic theory** envisages pattern and form as being generated by movement away from equilibrium, explainable in terms of rates of chemical reactions and transport processes. This is known as the kinetic preconception.

In general, the biological scientific community has been reluctant to adopt the kinetic preconception (Harrison, 1987). This may be due to the fact that organization in cells and intracellular dynamics are elusive subjects, because of the complexity of the spatio-temporal organization of molecular and cellular processes. A living cell appears to be an ensemble of levels of organization whose interactions give rise to processes (properties) of different natures – chemical, electrical, mechanical – such as solute transport, enzyme activity, protein synthesis, electron transport, cell interaction and gene expression (Aon and Cortassa, 1993).

Organization may result from:

- externally applied constraints;
- self-organization (occurring concomitantly with dissipation of matter and energy); or
- self-assembly in structures emerging in systems driven toward thermodynamic equilibrium, such as membranes (self-assembly of lipids and proteins) or viruses.

Almost every biological process shows aspects able to be addressed from the kinetic preconception, the equilibrium and the self-assembly approach.

To achieve progress in understanding the dynamics of living organisms we have to deal with concepts emerging from two large disciplines that provide tools to deal with fluxes and energy. These disciplines are thermodynamics and kinetics, from which we will present just some of the theoretical developments that will help us in our journey through the dynamic organization of cellular systems.

1.2 BIOLOGICAL SYSTEMS ARE IRREVERSIBLE BECAUSE OF THEIR CONTINUOUS FREE-ENERGY DISSIPATION

To support steady state operation, biological systems must dissipate energy and be continuously fed with matter (Figure 1.1b). Non-equilibrium steady states differ from equilibrium ones by exhibiting steady but non-zero fluxes, i.e. there has to be a continuous supply of substrate along with a continuous removal of product. By definition, this is impossible in an isolated system where the only possible steady state is equilibrium (zero net flux). These concepts are better exemplified by the 'sink analogue', for an open system (Figure 1.1b) or an isolated one (Figure 1.1a).

Non-equilibrium systems at steady state depend on the rates at which energy or matter, or both, are fed into the system along with the dependence of reaction rates on metabolite concentrations.

1.3 THERMODYNAMICS OF IRREVERSIBLE PROCESSES

For isolated systems, the **second law of thermodynamics** (Carnot–Clausius principle) states that the entropy which depends on the macroscopic state of the system increases irreversibly. In terms of statistical thermodynamics, the second law is expressed through the statistical law of evolution toward the most probable state which corresponds to the state of maximal disorder (Prigogine and Nicolis, 1971).

The entropy prediction function may be split into two contributing parts:

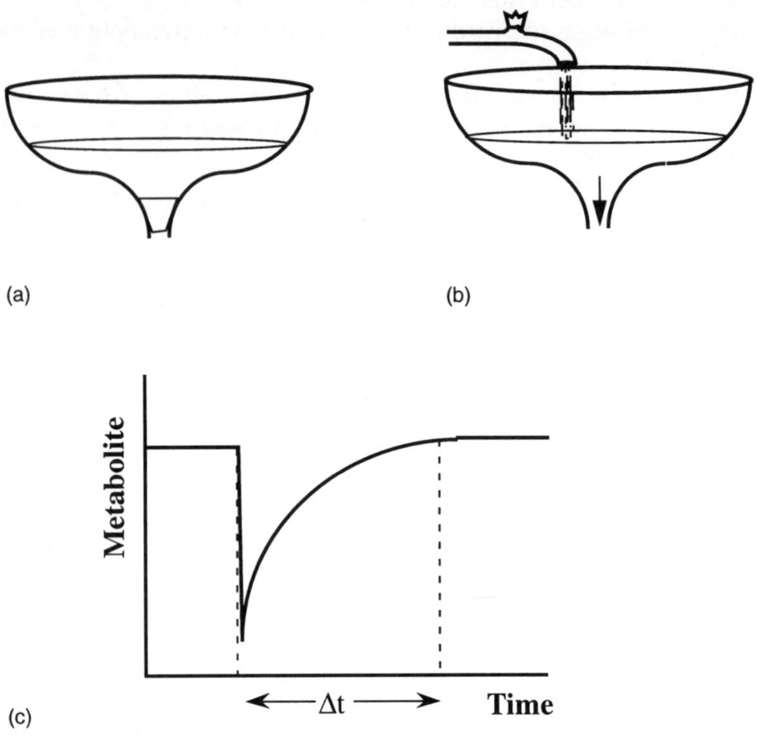

Figure 1.1 'Sink analogue' showing the steady state (time-independent) operation in open systems (b), as opposed to equilibrium ones in closed systems (a), or the transient (time-dependent) behaviour exhibited by the level of, say, a metabolite during Δt, in open systems initially at steady state (c).

$$dS = d_i S + d_e S \tag{1.1}$$

where $d_i S$ is the entropy production from irreversible processes inside the system which always increases; $d_e S$ is the entropy flow from the surroundings whose contribution is null in an isolated system. Then:

$$dS \geq 0 \tag{1.2}$$

In an isolated system, entropy always increases irreversibly. In biological systems $d_e S$ may be negative, so that the system may decrease its entropy because statistical fluctuations may favour the organized state. Instability of an initial state may give rise to more organized states, and in such cases fluctuations are amplified. The stabilization of an organized state after statistical fluctuations may occur because of the existence of nonlinear kinetic laws. The expression 'order through fluctuations' was coined because of that stabilization of ordered states upon random fluctuation (Nicolis and Prigogine, 1977).

Thermodynamically speaking, biological systems (a cell, for instance) are open systems, i.e. they continuously exchange energy and matter with their surroundings, and stay far from equilibrium due to that continuous flux exchange. But at what distance from equilibrium are they?

Classical thermodynamics has been extended to irreversible processes by Onsager (1931a,b), Katchalsky and Spangler (1968), Glansdorff and Prigogine (1971) and Nicolis and Prigogine (1977). The **near-equilibrium region**, within which most of the treatments of non-equilibrium thermodynamics (NET) have been dealing, is defined as the region where all the forces deviate much less than RT (= 0.597 kcal mol^{-1} at 25°C) from equilibrium. Because of this limitation imposed on the variations in the forces, the flow–force relations could be approximated by linear (proportional) ones as in equation 1.6. The variation of y with x is linear when $y = \alpha x + \beta$, and proportional when $\beta = 0$ (Westerhoff *et al.*, 1982; Cortassa *et al.*, 1991). Thermodynamics of irreversible processes has been extensively developed for situations in which the flows and rates of the processes are linear functions of the forces (temperature, concentration gradients, chemical potentials or chemical gradients). Such linear, non-equilibrium processes can, in fact, lead to the formation of configurations of lower entropy and higher order, so that non-equilibrium can be a source of order. In open systems, further away from equilibrium (though not necessarily in the nonlinear range of fluxes and forces), new spatio-temporal order can emerge and be stable since matter and energy are flowing to maintain them. These spatio-temporally organized forms are called **dissipative structures**, which are radically different from equilibrium structures in the sense that they are maintained by the continuous flow of matter and energy (Nicolis and Prigogine, 1977).

1.3.1 EVOLUTION AND STABILITY CRITERIA OF IRREVERSIBLE PROCESSES

Prigogine derived the criteria of stability for systems in the linear domain, i.e. where the laws applicable to the near equilibrium domain are still valid, and for the nonlinear domain where the forces, A (A = $-\Delta$G), are rather large to meet the condition:

$$\frac{A}{RT} \ll 1 \tag{1.3}$$

The calculation of the entropy, S, production per unit time, t, and volume, V, allows definition of the dissipation function, σ, as:

$$\frac{d_i S}{dt} = \int dV \sigma \tag{1.4}$$

which written in a bilinear form renders the expression:

$$\sigma = \sum_i J_i X_i \tag{1.5}$$

The dissipation function, σ, is the sum over all irreversible processes, i, of the product of the flows, J_i, of those processes, times the forces, X_i, driving those flows. In the case of a chemical reaction, the flux is the reaction rate; for transport reactions it is the flux of the substance through a membrane driven by the difference in chemical potential between two regions of the space; or for the specific case of an **electron motive force** (EMF) the flux is the current intensity driven by the EMF. Each term in the summation stands for a degree of freedom of the system (Caplan and Essig, 1983).

As always, $\sigma \geq 0$ when an irreversible process occurs; then in the dissipation function both factors – the flow and its conjugated force – will have the same sign. This means that no substance will be transported spontaneously against the chemical potential difference of that substance existing between both sides of the membrane (Caplan and Essig, 1983; Montero and Morán, 1992).

In case some terms of equation 1.5 are negative, the balance should be positive anyway. Under those conditions, if fluxes with different signs with respect to those of the conjugated forces appear, then there will be other coupled processes which will balance positively the total dissipation (σ). For processes to be coupled to each other, they should occur nearby in space interacting through some mechanism. This latter point is treated extensively in Chapters 7 and 12.

According to non-equilibrium thermodynamics (NET), in order to evaluate the relationship between forces and flows it is necessary to take advantage of the so-called phenomenological equations which, for the case of three independent processes, will take the form:

$$
\begin{aligned}
J_1 &= L_{11}X_1 + L_{12}X_2 + L_{13}X_3 \\
J_2 &= L_{21}X_1 + L_{22}X_2 + L_{23}X_3 \\
J_3 &= L_{31}X_1 + L_{32}X_2 + L_{33}X_3
\end{aligned}
\tag{1.6}
$$

These three equations show that each flow J_i is linearly driven by its 'conjugated' force X_i through the 'straight' coefficient L_{ii}. In turn, flow J_i can also be affected by any other force of the system X_j through the coupling or cross-coefficient L_{ij}. Equations 1.6 hold for the near-equilibrium domain where forces are small compared with the thermal forces, a condition expressed in equation 1.3. The linear domain of flow–force relations of NET, i.e. the near-equilibrium domain, is in fact an extrapolation of the equilibrium domain. In the near-equilibrium domain, it can be demonstrated that the matrix of coefficients is symmetrical, which is expressed by Onsager reciprocity relations (Onsager, 1931a,b), so that:

$$L_{ij} = L_{ji} \qquad (1.7)$$

The symmetry property means that in the near-equilibrium domain the phenomenological cross-coefficients are identical (Onsager, 1931a,b), which cannot be guaranteed outside the near-equilibrium domain. Symmetry and linearity define the limits of applicability of the near-equilibrium thermodynamic approach. In the so-called near-equilibrium domain and according to Onsager (1931a,b) the flow–force relations are symmetrical and proportional. The phenomenological coefficients are not constant since they implicitly contain information about the kinetic parameters determining the dynamic behaviour of the system (section 1.4).

For the steady state in the linear domain, Glansdorff and Prigogine (1971) have proved the theorem of minimum entropy production. According to this theorem, σ is a minimum in a non-equilibrium steady state (Figure 1.2b). This theorem provides an evolution criterion for the direction which will be followed by any physico-chemical system toward a non-equilibrium steady state.

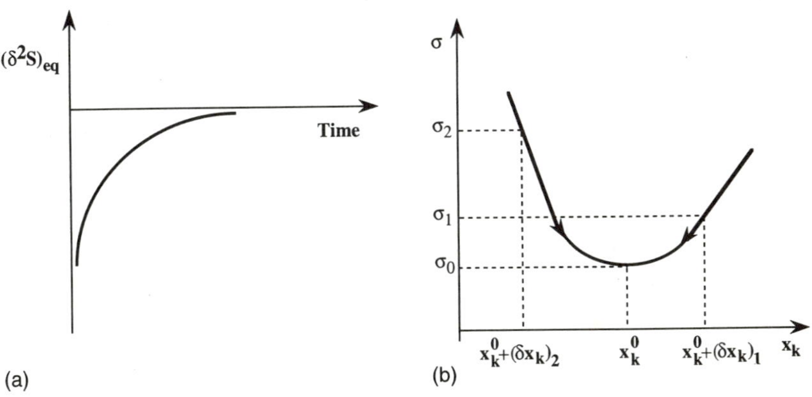

Figure 1.2 Stability criteria of non-equilibrium thermodynamics. (a) Time evolution of second-order excess entropy $(\delta^2 S)_{eq}$ around equilibrium; (b) illustration of the theorem of minimum entropy production. X_k^0 = steady state value of variables X_k; δX_k = perturbation from steady state. (Redrawn from Nicolis and Prigogine, 1977, quoted by Peacocke, 1983.)

Analytically, it may be expressed in the form:

$$\int dV \frac{dS}{dt} \geq 0 \qquad (1.8)$$

the terms of which are defined as in equation 1.4. The overall rate of entropy production should always be positive for the system to evolve toward an asymptotically stable steady state from any state close to it.

A careful consideration of the properties, namely linearity and symmetry, of the near equilibrium domain, which is an extrapolation of the equilibrium regime, allows the discarding of the appearance of self-organization.

A more general formulation is the stability criterion which states that:

$$\delta^2 S < 0 \qquad (1.9)$$

A small but finite change in entropy, with respect to that of the steady state, should be negative for a steady state to be stable (Figure 1.2a). If the criterion expressed by equation 1.9 is not fulfilled, the system may (but not necessarily will) become unstable and undergo a break in symmetry giving rise to self-organization.

1.3.2 INSTABILITIES, SELF-ORGANIZATION AND SYMMETRY-BREAKING

When the stability criteria expressed by equations 1.8 or 1.9 are not fulfilled, the system may undergo an instability that leads to the stabilization of a more organized state (i.e. self-organization). These states are maintained by energy and matter dissipation; therefore they are also known as **dissipative structures** (Peacocke, 1983). Dissipative structures may be characterized by higher dissipations. Peacocke quotes the Brussels group postulation that symmetry-breaking will lead a system even further from equilibrium, thus increasing its possibility of undergoing further instabilities to form new structures with increased dissipation (Peacocke, 1983). The presumption that dissipation further increases from bifurcation points was tested in a kinetic model of glycolysis (Appendix 1A). The thermodynamic analysis *a posteriori* of this model showed that the dissipation function behaved differently according to the bifurcation parameter under study (section 1.5.2 and Figure 1.12a,b). The analysis of thermodynamic functions demonstrates that not always should a self-organized state exhibit higher dissipation with respect to homogeneous ones. According to the model analysed in Appendix 1A, this may be highly dependent on the way the self-organized state is achieved (section 1.5.2).

Chapter 2 develops the idea that the successive symmetry-breaking steps arising may be interpreted as the dynamic emergence of new levels of organization.

1.3.3 DISSIPATION IN BIOLOGICAL SYSTEMS

Dissipation being the summation of the product of the flux times a force (equation 1.5), what does the statement that biological systems are dissipative mean? It means that biological systems sustain fluxes of matter and energy interconverting the nature of the forces (e.g. radiant, chemical, electrostatic) driving such fluxes. At the cellular level, for instance, free energy stores are exchanged essentially between the chemical potentials, electrochemical gradients and conformational energy of macromolecules.

In bacterial metabolism, the energy supplied by catabolic substrates is translated into phosphorylation, redox and transmembrane (electro)chemical gradients. Chemical potentials, such as phosphorylation, and redox potentials and electrochemical gradients such as those composing the proton motive force (ΔpH, $\Delta\psi$), or of neutral or charged solutes across the plasma membrane, drive most of the transport processes through prokaryotic or eukaryotic membranes. Those exchanges in the nature of potentials and gradients are usually mediated by chains of electron transport and/or ATPases. The former exchange redox into phosphorylation potentials and electrochemical gradients, e.g. the proton motive force ($\Delta\mu_{H^+}$), whose dissipation drives an ATPase in the sense of ATP synthesis or hydrolysis depending upon the values of the $\Delta\mu_{H^+}$ components, or the physico-chemical conditions of the intracellular milieu, or the cellular environment concerning, say, proton concentration. For instance, under anaerobic conditions bacterial cells perform increased glycolysis whose functioning increases intracellular proton concentrations (Maloney, 1977; Ten Brink and Konings, 1982; Verdoni *et al.*, 1990). Under those conditions, ATPases extrude protons with ATP consumption, changing a phosphorylation potential into the dissipation of a chemical gradient of protons. In this way, cells perform an essential biological function, i.e. to maintain an intracellular pH compatible with other cellular processes (Busa and Nuccitelli, 1984; Poolman *et al.*, 1985; Verdoni *et al.*, 1990).

Metabolite gradients also represent a form of energy which can be used in certain cases as metabolic energy. The efflux of end products of metabolism (organic acids) may occur via specific transport proteins in symport with protons, and this process can lead to the generation of a proton motive force ($\Delta\mu_{H^+}$) which subsequently can drive energy-demanding processes. In this way, energy recycling can occur (Konings, 1985). The excretion of metabolic end products (organic acids) in facultative anaerobes or 'strict' aerobes in the absence of electron acceptors (Verdoni *et al.*, 1990) is an alternative mechanism, other than ATP hydrolysis, of creating an electrochemical gradient (Chapter 12).

Within this conceptual framework, several questions arise. What is the nature of free energy intermediates sensed by the cell? How is it (are they) transduced to regulate growth? We will attempt to answer these questions in several chapters dealing with the organization of cellular metabolism in the context of cellular growth, proliferation or differentiation.

1.4 KINETICS: TOOLS AND SYSTEMS OF DETERMINISTIC EQUATIONS

Now we turn to the discipline that studies dynamic processes, emphasizing the rates at which processes occur rather than the initial and final states, as thermodynamics does.

1.4.1 NONLINEAR BEHAVIOUR

It must be remembered that the essential mathematical difference between linear and nonlinear systems represented by linear and nonlinear mathematical equations is given by the applicability of the **superposition principle**. The latter states that any two solutions of a linear equation can be added together to form a new solution; this is not true in systems of nonlinear equations (Hirsch and Smale, 1974). It is crucial to understand that the functioning of biological systems must be considered *in toto*. One cannot, in principle, break the problem into small subproblems.

Three main features characterize and distinguish nonlinear from linear behaviour (Campbell, 1989):

- irregular motion implying qualitative, non-smooth or regular, changes in behaviour when a control parameter exceeds a threshold value;
- abrupt changes in the response of a system to small changes in the parameters;
- a remarkable order exhibited by spatial patterns as stable and highly coherent localized structures instead of fading away after a while following, say, a pulse, as in linear systems.

A main precondition for a system to exhibit nonlinear behaviour (e.g. multiple steady states, oscillations, chaos) is to be ruled by nonlinear mechanisms. **Feedback** is a prominent biological one and many biological systems have built-in feedback control. Basically, feedback is when the product of one step in a reaction sequence has an effect on other reaction steps in the sequence. Kinetically, biological systems are nonlinear because of their multiple interactions, e.g. feedbacks (substrate inhibition), feed-forwards (product activation), cross-activation or cross-inhibition (Murray, 1989). **Autocatalysis** is a biologically ubiquitous nonlinear process whereby a chemical activates its own production, as follows:

$$A + X \underset{k_{-1}}{\overset{k_1}{\rightleftharpoons}} 2X \tag{1.10}$$

where a molecule of X combines with one of A to form two molecules of X. Processes of this kind are extensively treated along with specific examples in Chapters 2 and 5. Nonlinearities confer on biological systems the ability to bifurcate toward new steady states or attractors giving rise to self-organization, i.e. the appearance of spatio-temporal coherence (section 1.4.2).

1.4.2 MATHEMATICAL DYNAMIC SYSTEMS

Simply called dynamic systems (Abraham and Shaw, 1987), these consist of deterministic equations. Deterministic equations provide one of the most accurate mathematical descriptions of dynamic processes in biological systems. They include **ordinary differential equations** (ODEs), **partial differential equations** (PDEs) and **finite difference equations,** e.g. the logistic equation. We strongly encourage biologists to consult Abraham and Shaw (1984a,b, 1987) who explain mathematical ideas visually along with essential concepts of dynamics. The essential mathematical and dynamic concepts which will be utilized throughout this book are as follows.

Computer simulation and numerical computations are essential tools in the study and qualitative prediction of the evolution of dynamic systems (Stucki, 1978; Savageau, 1985; Smarr, 1985). ODEs are employed to describe spatially homogeneous systems mathematically; PDEs deal with systems where there are at least two independent variables: time and one spatial dimension. Although this book emphasizes mathematical formalisms dealing with deterministic equations, other mathematical approaches such as those of cellular automata have been very useful in understanding and describing behaviours peculiar to biological phenomena such as pattern formation (Ermentrout and Edelstein-

Keshet, 1993). The Boolean on/off idealization (Kauffman, 1989), neural networks based on the MacCullogh-Pitts approach of two possible states such as firing at the axon (one) or no firing (zero) (Aleksander and Morton, 1990), and information processing in molecular networks formalized by Boolean algebra (Lahoz-Beltra *et al.*, 1993) have provided manageable mathematical approaches to the functional behaviour of complex interacting networks.

ODEs describe the temporal evolution of variables (called **state variables**) as a function of other variables with which they interact, and of constants which are called **parameters** (Pratt, 1974; Hirsch and Smale, 1974). The slowly changing variables of a model are often incorporated as parameters; these change slowly enough to be approximated as constants, at least for the temporal window of the process under consideration. Formally, ODEs are described by:

$$\frac{dx}{dt} = F(k, x) \qquad (1.11)$$

where x stands for either a state variable or a vector representing a set of state variables and k a set of parameters that can be controlled. F(k,x) is a system of nonlinear functions of the dynamic variables.

A point in parametric space reflects a specific combination of values of all parameters of the system – values typically held constant in time that describe the behaviour of, say, a chemical system (Kauffman, 1989; Eubank and Farmer, 1989). The behaviour of the state variable x may change according to parameter values. Therefore, all behavioural possibilities of a system are contained in the parameter space or control space which represents in dynamic bifurcation theory (DBT) the set of all possible values for parameters (Kubicek and Marek, 1983; Abraham, 1987).

The representation of the evolution of a variable as a function of time is a **time series**. Figure 1.3a shows a time series in which the y-axis represents the **state space**. The state space is an n-dimensional space representing the set of all possible states of the system, n being the number of variables in the system. The state of a dynamic system (equation 1.11) on a state space of the state variable X is represented by a vector (with n-components) assigned to each point.

A starting point in the state space is called the **initial state**, and the curve it describes is the **trajectory**. Figure 1.3b shows a trajectory when two variables are plotted in the plane, i.e. a two-dimensional state space, that describes what is called a **phase space**; its coordinates are time-dependent variables that characterize the state of the system – for example, metabolite concentrations. Each point carries the label

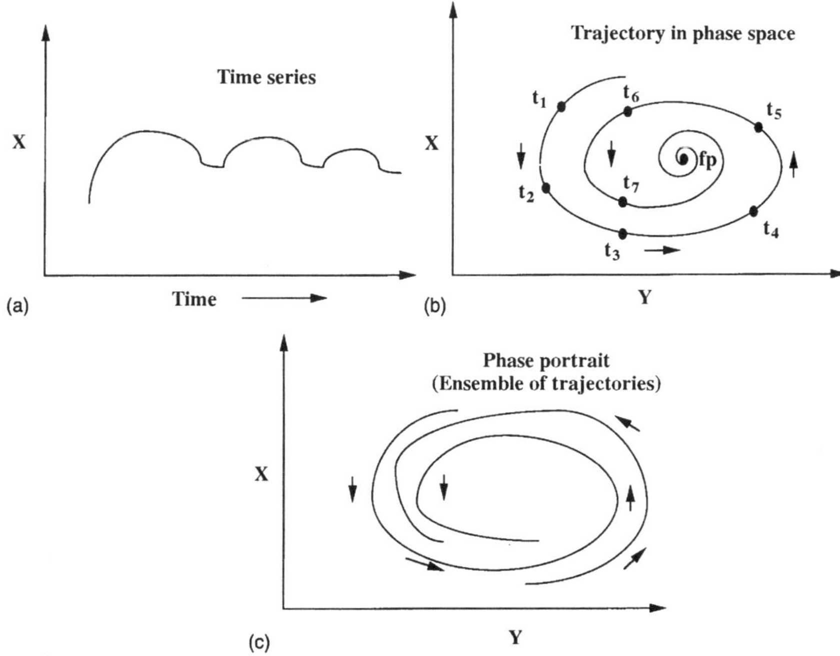

Figure 1.3 Basic concepts of dynamic systems. (a) Temporal evolution of the state variable X or time series. (b) A two-dimensional state space or phase space. Arrows indicate direction of trajectories or of the space of states when this is pictured as the flow of a fluid that tends toward an attractor of the type of a fixed point (fp). (In Figure 1.14, a phase space analysis is applied to glycolysis.) (c) An ensemble of trajectories – in fact several of that shown in (b) – giving the phase portrait of the dynamic system. The different trajectories may be followed by a dynamic system departing from distinct initial states after a perturbation. X, Y = state variables; $t_1...t_n$ = different times of observation.

corresponding to the time of observation (e.g. t_1, t_2). The phase space filled with trajectories is called the **phase portrait** of the dynamic system. The space of states may be imagined to flow as a fluid around itself, i.e. the **flow** of the dynamic system. A phase portrait of the dynamic system is represented in Figure 1.3b,c. Conceptually, the phase space represents the simultaneous temporal evolution of two state variables (X and Y in Figure 1.3b,c). Imagine that the time-axis of Figure 1.3a is perpendicular to the page in Figure 1.3b.

Figure 1.4 shows the results of a concrete biological example. Phase space analysis had been applied to different state variables of a bioelectromechanochemical (**Bemchem**) model of plant cell growth (Aon and Cortassa, 1989; see Chapter 12 for a more extensive discussion). This technique proved useful since complex systems (i.e. several nonlinearly interrelated state variables) exhibit systemic behaviour, which implies that variables with different relaxation times will optimize each other successively from a temporal point of view. Thus, in a series of sequential events moving in different time scales, the fast variables will reach their stationary state before any significant change in the slower processes has been observed (Reich and Sel'kov, 1981). In Figure 1.4a, it appears that there is no close relationship between the variables representing cell elongation and polysaccharide synthesis for the first part of the curve, i.e. a significant amount of carbohydrate was synthesized yet cell elongation was not sizeable. In other words, cell elongation and polysaccharide synthesis exhibit two different regions of behaviour in the phase space, corresponding to different moments of the system's evolution or two stages of the same phenomenon (Chapter 12).

The time scale for a significant variation of the transmembrane electric potential, $\Delta\psi$, is milliseconds–seconds; however, the membrane strain behaved even faster than $\Delta\psi$ and turgor pressure (Figure 1.4b,c). On the basis of these results, we suggested that electromechanical events are earlier processes affecting and conditioning plasma membrane's future behaviour, i.e. self-electrophoresis of plasmalemma proteins, H^+ pumps and permeability changes, during plant cell growth (Aon and Cortassa, 1989).

A fundamental idea from dynamic bifurcation theory is that of **parameter space**. A point in parameter space reflects a specific combination of values of all parameters of the dynamic system. As parameters change slowly, the trajectories typically change slowly, as do the attractors, i.e. the basin portrait (section 1.4.3) of the entire system is altered. But for particular changes of the parameters abrupt modifications in the behaviour of dynamic systems occur, i.e. **bifurcations**. These are sudden, dramatic changes in trajectories and attractors in the behaviour of dynamic systems for particular changes of the parameters (Nicolis and Prigogine, 1977; Segel, 1980, 1984; Aon *et al.*, 1991; Cortassa *et al.*, 1991). Upon bifurcations triggered by particular environmental conditions, or change of their internal parametric range, biological systems evolve toward other attractors of their dynamic state space (Hirsch and Smale, 1974; Segel, 1980,1984; Abraham and Shaw, 1987; Kauffman, 1989; Aon *et al.*, 1991; Aon and Cortassa, 1993; Cortassa and Aon, 1994b).

1.4.3 ATTRACTORS

Once a system is perturbed by, say, taking it out of a certain environment and dropping it into another one, there is an initial transient before its

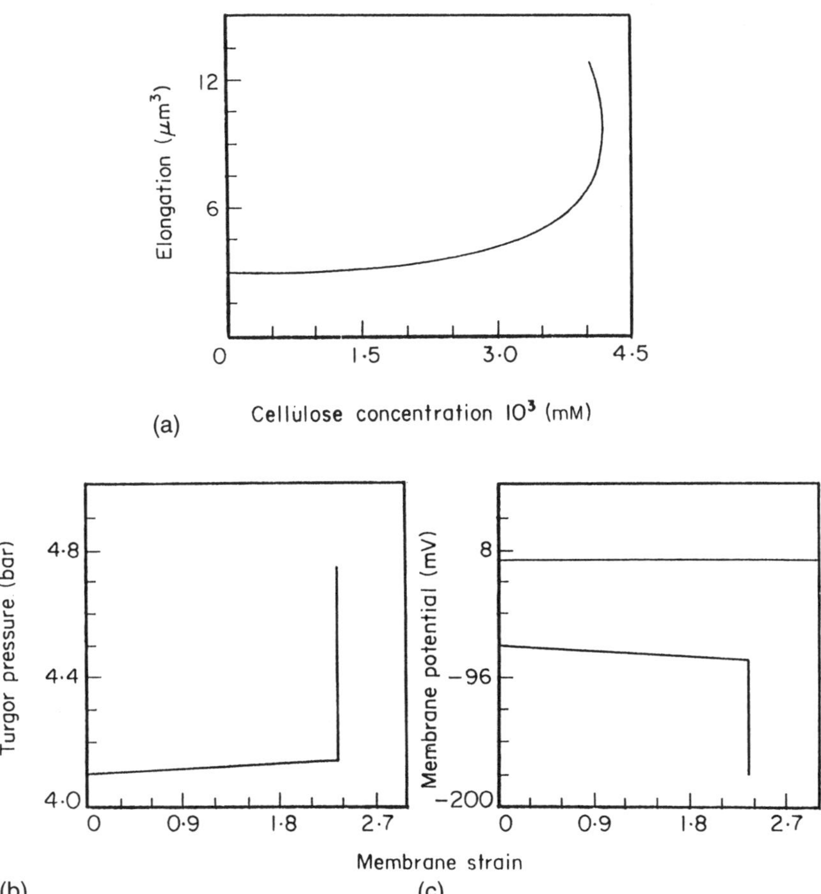

Figure 1.4 Phase space analysis as applied to a bioelectromechanochemical model (Bemchem). In (a)–(c), phase space analysis is applied to describe the relationship existing between several state variables in the Bemchem model (Chapter 12) of plant cell growth. In the Bemchem model, 14 ODEs (representing the same number of state variables) were numerically integrated simultaneously (Aon and Cortassa, 1989). The following state variables are plotted here: (a) cell elongation versus cellulose concentration and (b) turgor pressure and (c) membrane potential versus membrane strain (see text). (Reproduced from Aon and Cortassa, 1989, with permission of Academic Press.)

motion settles or approaches (asymptotically) a restricted region of the phase space characterized by a steady state (Figure 1.5). That restricted region is called an **attractor**, because the trajectories define a finite region of the state space and distant points are drawn toward it under successive iterations. The simplest attractor is a fixed point in phase space which is the only possible attractor for linear dynamic systems. In Figure 1.3, the initial transient is a damped oscillation (Figure 1.3a) which after damping will tend to a fixed point (Figure 1.3b), following a trajectory as represented in Figure 1.7a.

Since various initial conditions are 'attracted', the set of points that are so attracted is called the **basin of attraction** (Figure 1.5). An attractive point has a two-dimensional basin of attraction (Figure 1.5a). The basin is the inset of an attractor whose probability is the relative area or volume of its basin. The set of basins of attraction is often called the **basin portrait** of the dynamic system (Segel, 1980; Abraham and Shaw, 1984a,b).

Nonlinear dynamic systems have many more possibilities than fixed points. For instance, their asymptotic behaviour may settle on a periodic, time-dependent, oscillating state. Such periodic attractors are called **limit cycles**. Chaotic, **strange attractors** are also exhibited by dynamic systems (Abraham and Shaw, 1984a,b). They are called strange because they are infinitely intricate on every scale we might choose to examine. According to Mandelbrot's terminology, a strange attractor is a fractal and has a fractal dimension, D, of ≈ 1.3 (Mandelbrot, 1982; Shinbrot *et al.*, 1993). This exotic dynamic behaviour called **chaos**, which is conspicuously represented by strange attractors (equally so named because of their bizarre and unanticipated properties), underlies **deterministic chaos** in dissipative systems (Campbell, 1989). Deterministic chaos responds to the motion on a strange attractor which exhibits most of the properties associated with random functions, although the equations of motion are fully deterministic. The random behaviour arises intrinsically from the nonlinear mechanisms built into the dynamic system (see Chapter 3 for an extensive treatment of chaos).

Usually, biological systems are dynamic systems with multiple attractors, each with its own basin of attraction. Due to the existence of multiple basins of attraction, different initial conditions may lead to different types of long-term behaviour (Campbell, 1989; Kauffman, 1989; Haken, 1991).

1.4.4 STABILITY ANALYSIS

An important topic in the kinetic analysis of dynamic systems is the stability of the solutions of ODEs representing that particular system. From the viewpoint of stability, the important points in the state space

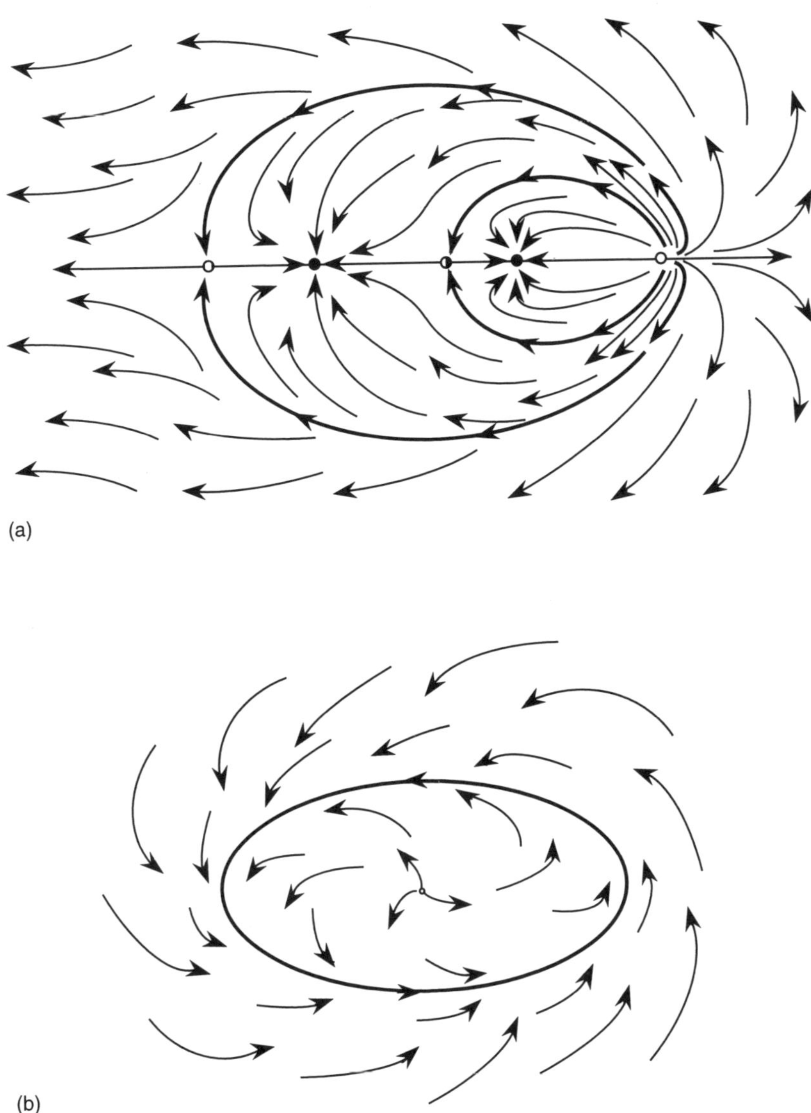

Figure 1.5 Basins of attraction of: (a) fixed points (filled circles) exhibiting saddle points (half-filled circles) in the limits between the basins of the two fixed points, indicated by bold arrows; (b) a limit cycle. Open circles represent unstable steady states; filled circles represent stable steady states. (See also Figure 1.7.)

are the critical (singular) points where the variables do not change in time. Thus, the critical points correspond to the steady states of the system that may be either stable or unstable. When a dissipative dynamic system is forced to do something that it would never do on its own, there are essentially two possibilities: either the motion is unstable and flies off to infinity, or the motion is bounded (stable) and approaches an attractor (Hirsch and Smale, 1974; Abraham and Shaw, 1987). Thus, the existence of an attractor is related to the stability of a family of trajectories.

From a biological point of view, unstable steady states cannot be experimentally visualized because for that state the system dynamics will depart toward a stable one. Graphically (Figure 1.6), the ball (the state of a system) will not stay on top of the convex curve; its motion will describe a trajectory toward a concave curve in which it will approach a stable state or attractor. However, although the ball may attain different steady states, the path followed may be different along with the fact that some of the states may be inherently unstable. For instance, to reach the valley (i.e. a stable steady state) on the right of Figure 1.6, the ball may take a direct path which may imply a transient (time-dependent) behaviour, i.e. the shorter arrow through the small dome, or a longer path before reaching the valley, i.e. the longer arrow. Figure 1.6 captures the essential meaning of the evolution of the dynamics of a complex system, e.g. a subcellular or cellular process, toward a stable behaviour. Such a view illustrates the 'epigenetic landscape' of Waddington (1947; see also Harold, 1990) in which a system undergoing development evolves toward valleys through domes or channels. All those states evolving in time may be more or less unstable to allow the developing system to reach the adult or differentiated organism (e.g. Chapter 10).

Those who are interested in the mathematical details of stability analysis may find a practical introduction in the review by Stucki (1978) or a more involved treatment in Hirsch and Smale (1974). We present here just the type of stability that ODE or PDE systems may exhibit. Mathematically, the stability of a steady state is given by the **eigenvalues**, which correspond to the roots of the **Jacobian matrix** of the linearized equations (around a critical or singular point) obtained from the system of ODEs (Hirsch and Smale, 1974; Stucki, 1978). The roots are complex numbers and the stability is given by the sign of the real part and whether or not there is an imaginary part. The eigenvalues measure the exponential behaviour of the trajectories of a dynamic system in the neighbourhood of a steady state. The logs of the eigenvalues are called the **Lyapunov exponents**. These measure the sensitive dependence on initial conditions which is one of the main features characterizing chaotic dynamics.

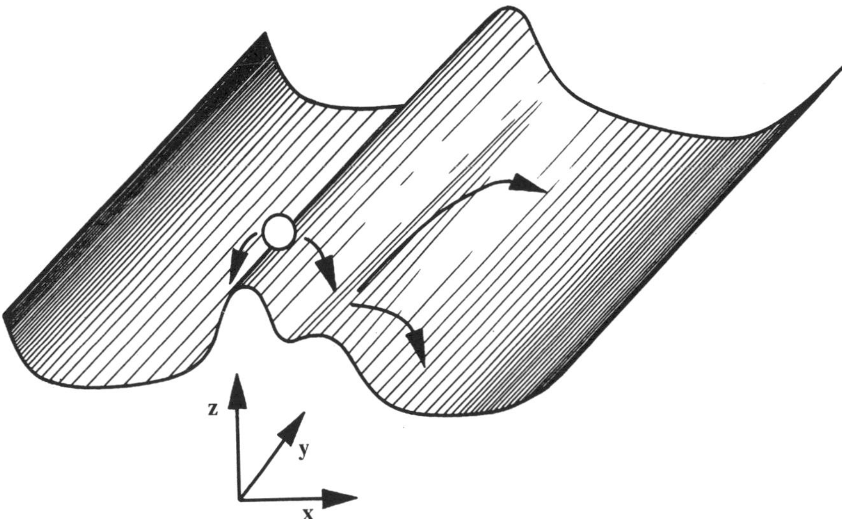

Figure 1.6 A graphical metaphor of the evolution of a dynamic system through succesesive states or attractors. The ball's position represents the state of a biological system, which in this particular case is given by the three state variables depicted in the x, y and z axes. The dynamics of the system is shown by the arrows, indicating the trajectory followed by a biological system (e.g. a cell, an enzymatic reaction, a secretory process). Biological systems are dynamic systems with multiple attractors. The ball in the apex of the convex surface is in an unstable position; thus, it will roll down toward concave surfaces (stable states or attractors). It must be stressed that situations like the ball on the top of the crest cannot be visualized experimentally. The paths followed by the ball (e.g. a developing system) may be different (i.e. to the valleys at the left or the right of the crest) and, supposedly, the fate of the biological system (e.g. spore or stalk cells in slime mould). Different parametric combinations (a state variable depends on parameter which have constant, time-independent values) will prompt the biological system to follow different paths to attain the same (the ball rolling down the valley on the right of the graph through two different excursions given by the long or short arrows) or different attractors (the valleys on the right or left of the ball).

Figure 1.7 illustrates different types of stability in a two-dimensional state space. When a thermodynamic analysis of a particular steady state is also possible, it is found that stability criteria (equations 1.8 and 1.9) will certainly fail in cases of unstable steady states (Figure 1.7b,d,e) or limit cycles (Figure 1.7f). For the case where the steady state solutions correspond to a stable focus or a stable node (Figure 1.7a and c, respectively) the stability criteria may or may not be fulfilled since the thermodynamic criteria of stability are a necessary though not sufficient condition.

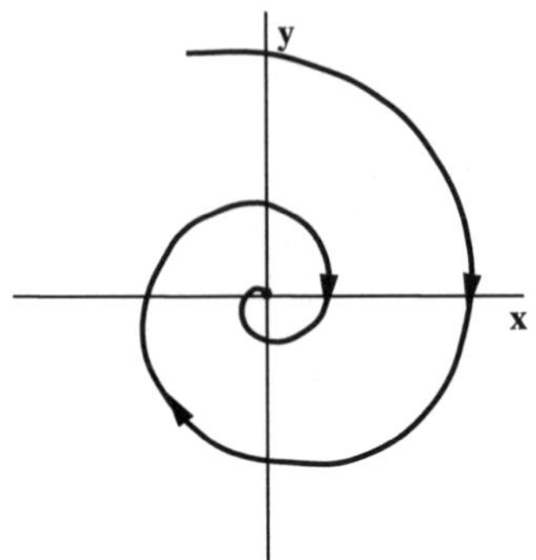

(a) **stable focus**

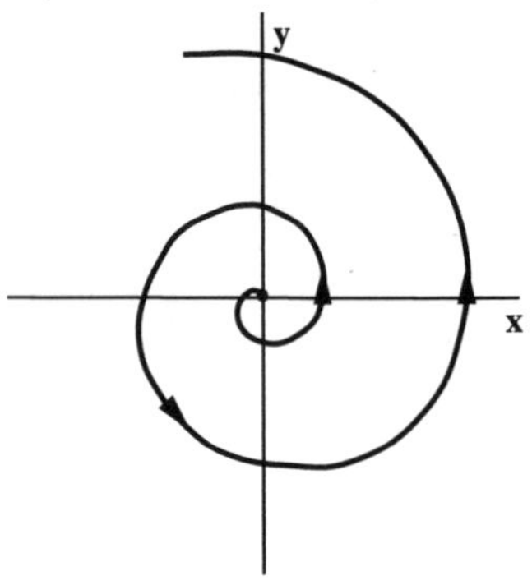

(b) **unstable focus**

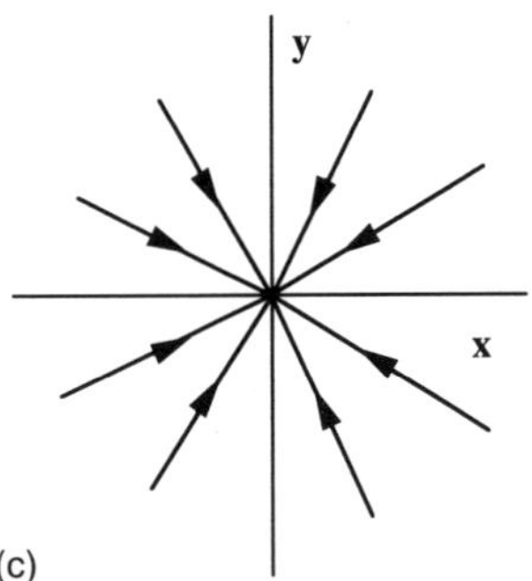

(c)

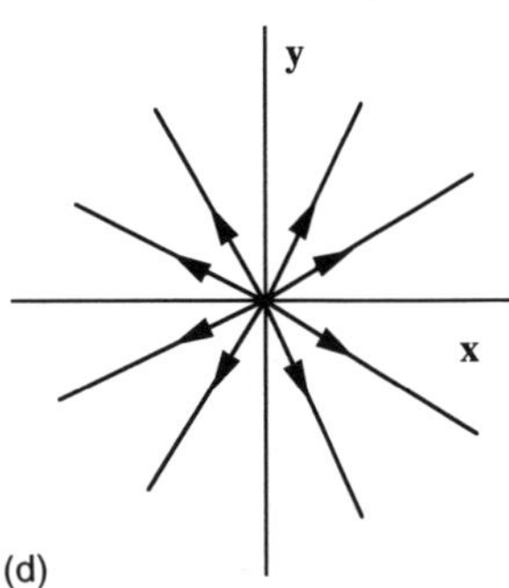

(d)

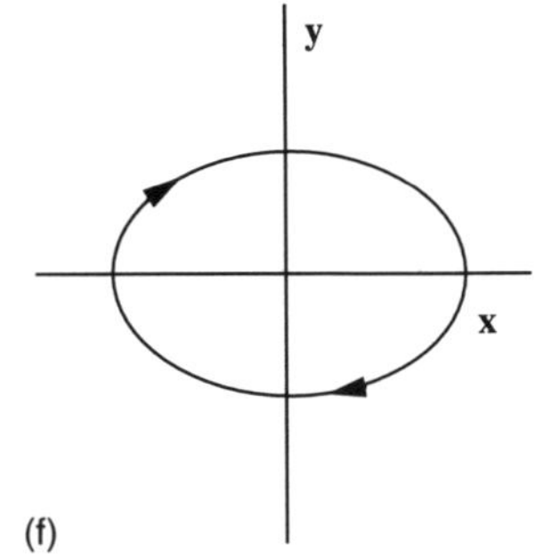

(e)

(f)

(a) The biological implications of stability in an example of two coupled subcellular systems

The system of ODEs discussed in detail in section 6.3.1 exhibits **bistability**, i.e. two possible steady states for a single parameter set. The two branches of (stable) steady states are represented as surfaces (shadowed) since two parameters are simultaneously varying (Figure 1.8). Figure 1.8a shows that as the total amount of MTP (C_o) increased, the system became unstable at lower K_{Pol} values (rate of polymerization), i.e. that at higher amounts of total microtubular protein (MTP), lower rates of polymerization are sufficient for the system to exhibit new functional properties. This means that the pyruvate kinase (PK) activity may 'jump' from the branch of low stable steady state values to the other branch of higher steady states (Figure 6.11). The branch of lower concentrations of phosphoenolpyruvate (PEP) reflects a high PK activity and, vice versa, the branch of high [PEP].

Figure 1.8b shows that as the total amount of C_o increased, the amount of polymerized MTP (C_p) also increased. As expected, at the same time as the amount of polymerized MTP increases, the activity of PK (reflected by the steady state concentration of PEP) is also augmented. In other words, the more spatially 'structured' the system becomes (i.e. a greater amount of polymerized MTP, C_p), the higher the enzymatic activity.

Combining the interpretation of Figure 1.8, which represents a true biochemical situation, with the metaphor depicted in Figure 1.6, the ball in the latter symbolizes the dynamic behaviour of state variables such as PEP and C_p. On the one hand, near the values of K_{Pol} and C_o for which the system bifurcates, the dynamics of the system is like that of the ball in the apex in Figure 1.6. On the other hand, when the system dynamics reaches a stable steady state (a valley) it remains there either reversibly or irreversibly (Chapter 6 explains this point).

This phenomenology describes the essential features of the concept of **dynamic organization**, developed in Chapter 2. The basic idea of this concept is that function in cellular systems, understood as the spatio-temporal coherence of events, results from the intrinsic, autonomous

Figure 1.7 (See facing page) Types of stability in a two-dimensional state space. Point attractors (a,c), point repellers (b,d) and a saddle point (e) or limit cycle (f) are plotted. Mathematically, the stability of a steady state is given by the eigenvalues, which measure the exponential behaviour of the trajectories of the dynamic system in the neighbourhood of a steady state. The trajectories in the plane x–y are described by the following eigenvalues: (a) complex conjugates with negative real parts; (b) complex conjugates with positive real parts; (c) real negative number; (d) real positive number; (e) one with a real positive part and the other real negative; (f) pure imaginary, with a zero real part.

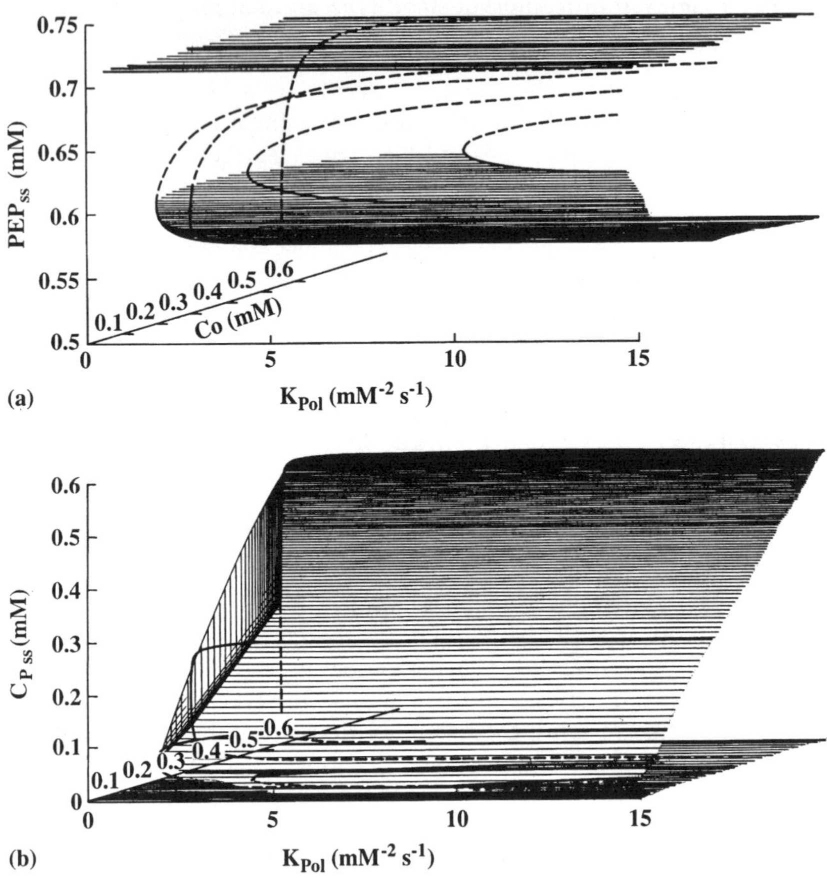

Figure 1.8 A three-dimensional representation of the qualitative dynamic properties of two coupled dynamic systems. The study of the qualitative dynamic behaviour of a model based on experimental data (Cortassa *et al.*, 1994a; Chapters 6 and 8), which takes into account the dynamics of microtubular protein (MTP) assembly–disassembly coupled to the kinetics of pyruvate kinase (PK) in the presence of polymerized or depolymerized MTP, was performed with the program AUTO (Doedle, 1986). The stability properties as well as the different types of steady state (ss) exhibited by the model are represented as a function of two bifurcation parameters of physiological relevance, such as the rate of polymerization of MTP (K$_{Pol}$) and the total amount of MTP species (C$_o$) present in the system. The steady state values of two representative state variables, each belonging to one of the two dynamic systems, are plotted: (a) phosphoenolpyruvate (PEP), the substrate of PK; (b) the polymerized microtubular protein, Cp. Plain lines indicate branches of stable steady states; dashed lines correspond to unstable ones. Shaded areas represent the domain (or 'surface') of stable behaviour of the system separated by a 'gap' of instability.

dynamics of the processes taking place in different biological phenomena, i.e. growth, proliferation, metabolic transitions and cellular differentiation.

A corollary of this idea is that the phenotype (e.g. cell form or transformation) is the result of the spatio-temporal deployment of the genetic programme in a constant and reciprocal interaction with the intra- or extracellular environment whose dynamic graphical view is given by the epigenetic landscape (Figure 1.6) (Waddington, 1947). The displacement of the system's dynamics (the ball running through valleys along crests of different height) occurs in a multidimensional parametric space, which implies that the cell affects the balance of more than one parameter simultaneously.

Figure 1.8 shows an example of a two-dimensional parameter space. The cellular system is able to 'move' both parameters (i.e. K_{Pol} and C_0) simultaneously, influencing in this way the route of the ball either transiently or at a stable steady state. The rate of MTP polymerization may be affected by intracellular levels of Ca^{2+}, phosphoinositides (Janmey, 1994) pH (Suprenant, 1991) or amounts of MAPs (Mitchison and Kirschner, 1984a,b) whereas C_0, the total amount of MTP, may be subjected to regulation by gene expression (Ben-Ze'ev, 1986; Cleveland, 1988).

To illustrate this point, Figure 1.6 shows that the ball may roll down toward the valley at the right by either two ways, each of them depending on, say, the fine tuning of the parameters in the x- and y-axes under the particular physiological condition given by the intra- or extracellular environment. When the instantaneous state of the system (depicted by the ball) reaches the bottom of the valley, it will be in a stable steady state, i.e. following a perturbation the system returns to the same original position.

In biological systems a network of interactions is established between dynamic subsystems which are coupled through common intermediates or effectors such as hormones or second messengers. Under those conditions, a variable in one dynamic subsystem due to its quick relaxation following a perturbation may behave as a control parameter of another dynamic subsystem. This is the main basis of the concept of **dynamic coupling**, developed in Chapter 7 and extensively in Chapter 12.

The concept of dynamic coupling may offer an explanation to a phenomenon in which a system shifts between attractors (or 'quasiattractors': Haken, 1991) all the time following perturbations. This is especially relevant for the physiology of cells or organisms. In particular, cells are dynamic systems in which attractors are of obvious importance since they represent their asymptotic long-term behaviour (Kauffman, 1989). In a fixed environment, and with a fixed set of parameter values, the system has a set of basins of attraction and attractors representing the integrated dynamic behaviour of the variables in the system as they mutually influence one another.

Very similar ideas have been applied by Kauffman (1989) to adaptation and selection processes. An environmental parameter can induce generalized cellular responses that lead them to stop or resume division, or choose between divergent developmental paths (Chapters 9 and 10). Environmental perturbations might 'tune' attractors so that at least a subset of variables are confined within some limits; attractors would not drift widely around state space. An extreme case for a dynamic system is represented by high-dimensional chaotic attractors because dynamics will exhibit complex flow sensitive to initial conditions and therefore any subset of 'essential' variables would hold within narrow limits (Chapter 3).

Within the above conceptual framework, Kauffman (1989) conjectured that dynamic systems which exhibit small attractors are more likely to behave homeostatically with respect to those with large attractors. Cellular adaptation will consist of a dynamic system to keep small attractors with bounded essential variables.

1.4.5 BIFURCATION DIAGRAMS

It has been shown above that stability analysis is the study of the qualitative dynamic behaviour of any system in terms of the stability of the steady states as a function of a bifurcation parameter (Figure 1.8). The systematic analysis of the solutions of systems of ODEs like equation 1.11 as a function of a bifurcation parameter allows the construction of bifurcation diagrams. The parameter values for which the stability properties change (e.g. from stable to unstable steady states) are called **bifurcation points**. For instance, a **Hopf bifurcation** indicates a change in the dynamic behaviour of a system from a region of stable steady states (attained after oscillatory transients or damped oscillations) to a region of sustained oscillations (limit cycles). Looking at the dependence of the solutions of equation 1.11 on parameters, it can be seen that regions of the parameter space k may be found where the solutions are 'simple', steady state or periodic, whereas in other parametric regions 'erratic' or chaotic solutions also appear. The borderline between the two regions of the parametric space is referred to as the **borderline of chaos** (Procaccia, 1988) (Chapter 3).

1.5 BIOLOGICAL DYNAMIC SYSTEMS FROM THE PERSPECTIVE OF THERMODYNAMICS AND KINETICS

The kinetic and thermodynamic descriptions of (sub)cellular processes are not mutually exclusive. Rather, they are largely complementary, reciprocally illuminating each other (Westerhoff and van Dam, 1987; Cortassa *et al.*, 1991, 1994b). The following describes the thermodynamics of a concrete biological example.

Mitochondrial oxidative phosphorylation has been studied largely through a thermodynamic approach at the steady state. Under those experimental conditions, both input and output forces (thermodynamic potentials) were fixed by the experimenter. Respiration may be viewed as an energy converter, a sophisticated chemical machine in which the energy stored as redox equivalents is used to move protons uphill in their chemical potential difference across the mitochondrial inner membrane, creating a proton motive force which in turn will drive the synthesis of ATP (Mitchell, 1961). Hence, the respiratory chain acts as a device to couple the oxidation of redox equivalents to the synthesis of ATP.

For processes in which energy conversion is achieved, thermodynamic efficiency has been defined as the ratio of the dissipation of the energy-consuming processes (output) to the dissipation of the energy-producing (input) processes (Stucki, 1982; Westerhoff and van Dam, 1987):

$$\eta = \frac{J_o X_o}{J_i X_i} \tag{1.12}$$

where the subscripts i and o refer to the input and output processes, respectively.

The efficiency of processes of free-energy transduction whose modelling assumes linear relations between fluxes and the driving forces produces a typical diagram when plotted as a function of the force ratio. Figure 1.9 shows different degrees of coupling, q, which for a system with two degrees of freedom is defined by the relation:

$$q = \sqrt{\frac{L_{12} L_{21}}{L_{11} L_{22}}} \tag{1.13}$$

The points where the curves reach the x-axis at low and high ratios of forces $(\Delta G_o / \Delta G_i)$ are known as level flow and static head states, respectively (Caplan and Essig, 1983). **Level flow** indicates that the efficiency is null when the output force is null even for a large flux ratio. A membrane uncoupled by a protonophore allows us to explain in a biological sense the concept of level flow. According to this example, level flow occurs when there is a high rate of electronic transport which drives a high rate of H^+ pumping $(J_o/J_i =$ flux ratio) without generating a $\Delta \mu_{H^+}$ $(\Delta G_o =$ output force). On the other hand, the system is at **static head** when it may generate a large output force so that the relation cannot further increase because the output flow is null. This is the case of inhibited ATP synthase in the presence of a functional electronic transport. Under such conditions, the system stops pumping protons because the $\Delta\mu_{H^+}$ is so large (ΔG_o = output force) that the free energy released during electron transport is not enough to pump protons against the electrochemical gradient, since the latter is not dissipated through ATP synthesis (J_o = output flow).

There is an optimum relationship of output to input force in terms of efficiency that depends globally on the nature of the coupled processes, and particularly on the value of the kinetic parameters contained implicitly in the phenomenological coefficients (section 1.5.3).

1.5.1 DYNAMIC COUPLING IN (SUB)CELLULAR PROCESSES

A flux is a rate of change of a process. Mathematically, a flux (amount or concentration per unit of time) may be defined in its general form by ODEs:

$$J_i: \frac{dx}{dt} = f(k, x, y) \tag{1.14}$$

where J_i is the flux, x and y are state variables and k stand for parameters determining the behaviour of the state variables (note that equation 1.14 is similar to 1.11).

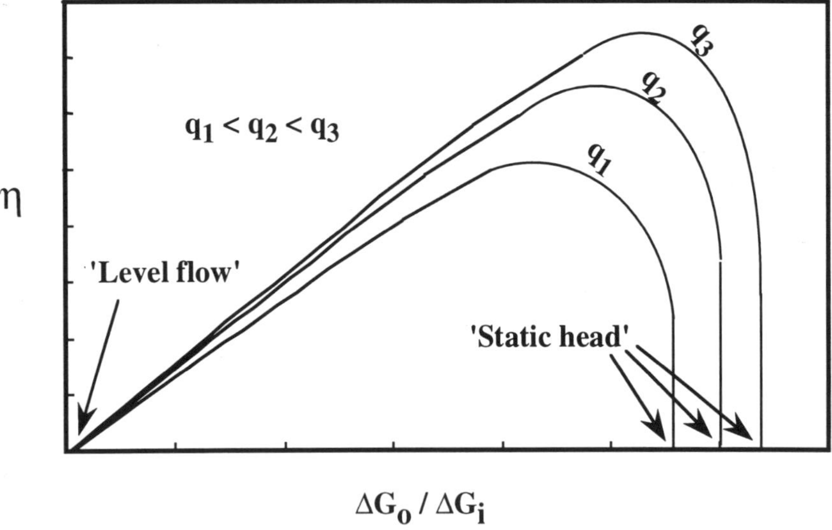

Figure 1.9 Dependence of thermodynamic efficiency, η, on the force ratio, $\Delta G_o/\Delta G_i$, at various degrees of coupling, q. The figure shows the general shape of the dependence of η on $\Delta G_o/\Delta G_i$ and q, which may be valid for either linear or nonlinear energy-transducing systems. It also shows the static head and level flow functioning of energy-transducing systems. In oxidative phosphorylation, the force ratio is given by the phosphate potential over the redox potential difference (see text).

The parametric space of a biological system represents its full behavioural possibilities. In principle, every dynamic behaviour will be significant, i.e. operative, depending upon the values of bifurcation parameters, which in turn will depend upon the physiological condition of the cell or the organism. Given two dynamically autonomous coupled processes, a time-dependent change of a state variable in one of the processes that affects the intermediate that couples those processes may influence the steady state behaviour of the second process. From the point of view of dynamic coupling, the relaxation time of the dynamics of coupled processes (e.g. electron transport, ATPase, glycolytic), becomes relevant (Aon and Cortassa, 1993). From a mathematical point of view, the relaxation time is reflected by the eigenvalues of the dynamic system; i.e. the relaxation time is given by the inverse of the eigenvalues of the equations that describe the dynamics of both subsystems.

Conceptually, when a process relaxes faster with respect to its dynamically coupled one, the common intermediate behaving as a state variable for the faster process will exhibit a nearly constant steady state value comporting itself as a parameter for the slower process (Higgins, 1965; Reich and Sel'kov, 1981; Abraham and Shaw, 1987).

Coupling can be defined broadly as how two processes are related through a common intermediate. If we consider two related processes as two subsystems of a larger system that contains them, a variable in one of the subsystems (with a low relaxation time) may be viewed as a parameter for the other. In such cases, that variable will be called the **bifurcation variable** or **self-regulated bifurcation parameter**. Such a concept is exemplified in the interactive coupling between glycolysis and mitochondrial processes in yeast (Chapter 7). We have observed changes in the dynamic properties of glycolysis by an increase in the cytoplasmic ATP (common intermediate) load by oxidative phosphorylation, i.e. transition from a stable branch of asymptotic steady states through a Hopf bifurcation to an unstable branch of oscillatory steady states (Aon *et al.*, 1991). In oscillating populations of yeast, we showed that individual cells influenced the glycolytic dynamics of other cells by affecting each other's dynamic redox balance (Aon *et al.*, 1992). Apparently, depending upon the intracellular balance between oxidative and reductive (ethanol catabolism) fluxes of NADH that was mediated by the extracellular concentration of ethanol, highly dense yeast cell populations were able to remain synchronous for longer periods (Chapter 12).

The concept of dynamic coupling may be useful in combined experimental and modelling approaches. The relaxation time towards perturbations of two or more coupled processes allows the quantitation of the relative rates at which they take place when interacting. In this way, it may be possible to discern a temporal causality, i.e. which processes may regulate others according to the temporal hierarchy of their occurrence (Cortassa and Aon, 1994b; also Chapters 7 and 12).

1.5.2 THERMODYNAMICS *A POSTERIORI*

An evaluation of thermodynamic potentials as a function of time was performed in a system driven by rate equations in which the state variables that are chemical concentrations appear explicitly. Essentially, **thermodynamics *a posteriori*** is an evaluation of thermodynamic potentials and efficiency from the dynamics of a biological system described by the kinetic equations. It is *a posteriori* because we do not impose certain flow–force relations as in the *a priori* approach but let the system attain them on its own. The *a posteriori* approach overcomes, to some extent, a limitation of the *a priori* one, in the sense that in the latter a system may only be studied at asymptotically stable steady states. Besides, the system should exhibit continuity between successive steady states when a thermodynamic potential is systematically varied. That is, whenever bifurcations in the dynamic behaviour occur toward unstable steady states, the *a priori* approach can no longer be applied.

Thermodynamic studies *a posteriori* in a pure stoichiometric model of anaerobic glycolysis (Sel'kov, 1975; Cortassa *et al.*, 1990a; Aon and Cortassa, 1991) were performed under conditions that simulate physiological oscillations experimentally observed at the cellular level (Chapter 3). Such an analysis has been performed by other authors (Ross and Schell, 1987; Ross *et al.*, 1988) and ourselves (Cortassa *et al.*, 1990a; Aon and Cortassa, 1991; Aon *et al.*, 1996b).

The model of anaerobic glycolysis is described by two consecutive reactions that convert a substrate such as glucose, through a phosphorylated intermediate, into an end product such as lactic acid or ethanol. The reactions were considered irreversible and only stoichiometric relationships were taken into account (Figure 1.10). The mathematical formulation of the model is presented in Appendix 1A.

Steady state and oscillatory behaviour were obtained through variations in the substrate input rate and in the energy load. The thermodynamic analysis performed *a posteriori*, based on the numerical integration of the kinetic equations, revealed the presence of a region where the relations between flows and forces are linear, albeit non-proportional (the intercept with the y-axis) (Figure 1.11). This behaviour corresponds to a region where the system exhibits an asymptotic temporal evolution of their variables to a steady state value. The relation between the flux and the force deviates from linearity when the system approaches the region where it exhibits sustained oscillations (the shaded

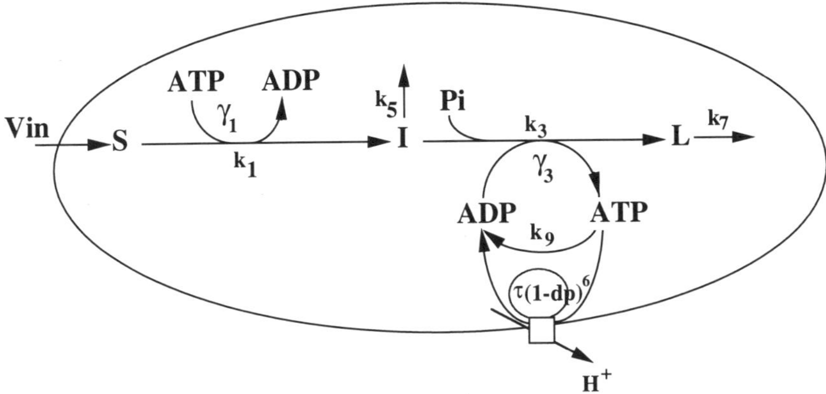

Figure 1.10 Scheme of a stoichiometric model of anaerobic glycolysis. (Reproduced from Cortassa, Aon and Thomas, 1990, by permission of Elsevier Science.)

regions in Figure 1.11). It should be pointed out that the force variations were achieved through changes in physiologically meaningful kinetic parameters, namely the rate at which a substrate enters a cell (V_{in}) and the energy demand (d_p) exerted by the non-glycolytic ATP-consuming processes.

The dissipation increased with the substrate input rate, V_{in}, being maximal at static head (Figure 1.12a). A very different profile was obtained for a systematic variation of the energy demand, d_p. In the latter case, thermodynamic efficiency decreased when approaching the Hopf bifurcation point from which sustained oscillations characterized the dynamic behaviour of the system (Figure 1.12b). The dissipation function increased monotonically with the energy demand, being higher in the oscillatory regime than in the domain characterized by asymptotic steady states. In the case of the kinetic parameter related to the rate of substrate uptake by the cell, the less substrate enters a cell, the less it will dissipate (Figure 1.12a). On the contrary, the higher the energetic demands (represented by the kinetic parameter d_p), the higher is the dissipation (Figure 1.12b).

The thermodynamic efficiency exhibited by the system also increased with the substrate input and was significantly larger in the region of asymptotic steady states with respect to the oscillatory domain (Figure 1.11). The typical curve exhibited by the thermodynamic efficiency as a function of the force ratio was obtained with the largest force ratios corresponding to the region where asymptotic behaviour is displayed, i.e. high rates of substrate input. On the other hand, the dissipation function

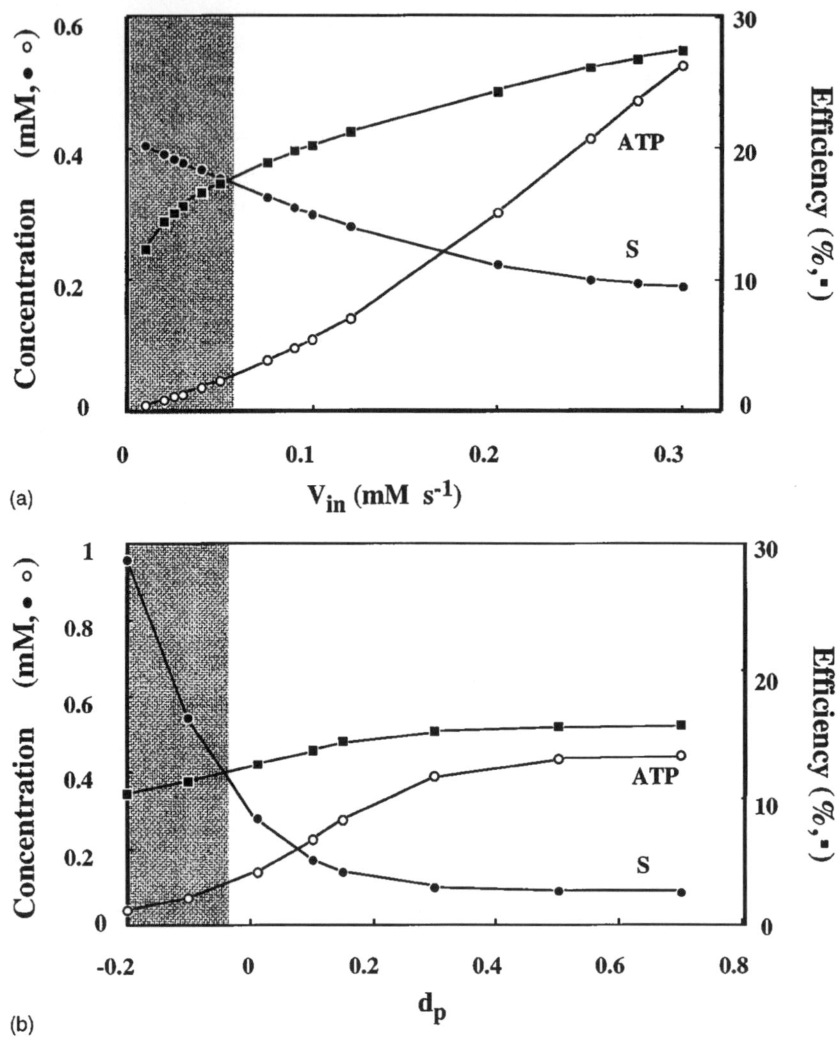

Figure 1.11 Steady state behaviour of substrate (S), ATP and thermodynamic efficiency under variable substrate input rate (V_{in}) or energy load (d_p). The parameter values used to perform the simulations were: V_{in} = 0.12 mM s^{-1}; τ = 0.83; k_1 = 0.3 mM^{-1} s^{-1}; k_3 (mM^{-2} s^{-1}) = k_5 (s^{-1}) = k_7 (s^{-1}) = 0.1; k_9 = 0.05 s^{-1}. The kinetic parameters of the proton pump were K_M = 2 mM; V_M = 0.5 mM s^{-1}. The total nucleotide (C_A) and phosphate pool (P_i) were both 10 mM. The shaded regions indicate the range of parameter values for which sustained oscillations have been observed. In all cases the ATP concentration plotted is one tenth of the actual value. (Reproduced from Cortassa, Aon and Thomas, 1990, with permission by Elsevier Science.)

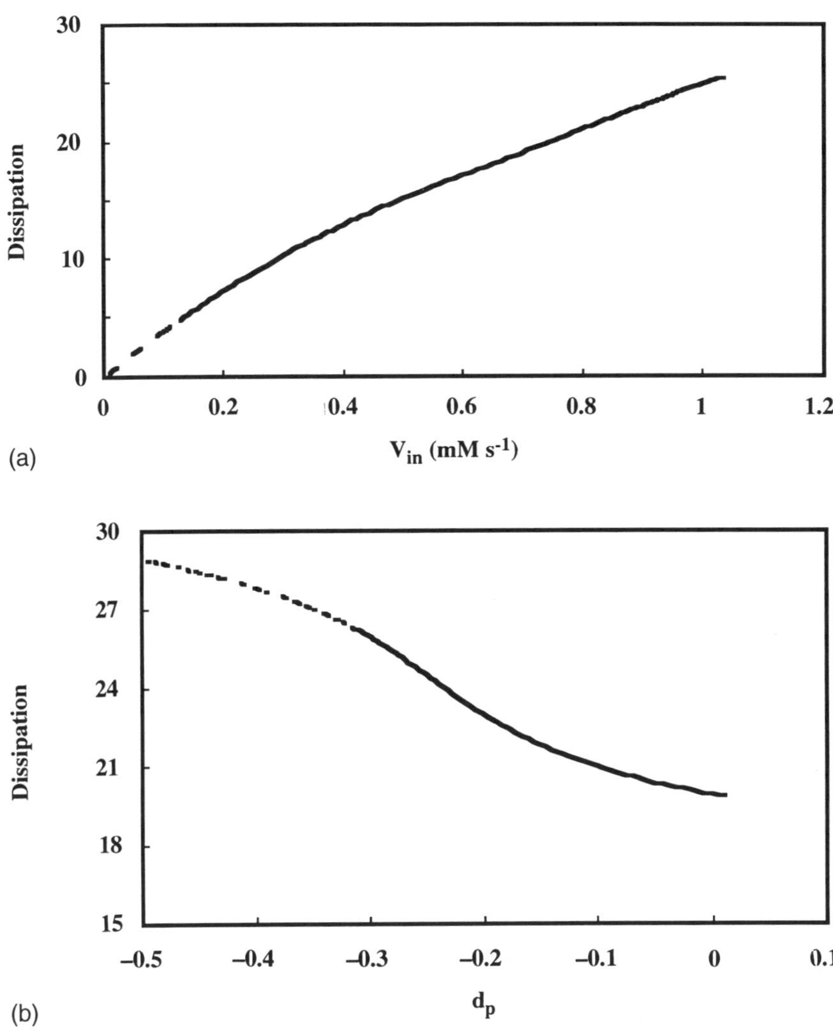

Figure 1.12 Varying dissipations of the glycolytic model as a function of (a) the rate of substrate uptake by the cell, V_{in}, or (b) the energy load, d_p, imposed by ATP-consuming active transport. Continuous lines represent stable asymptotic steady states; dashed lines represent oscillatory regimes. The ODEs system (1.21–1.24) in Appendix 1A was numerically integrated with the SCoP package (Duke University, Durham) and the stability analysis was performed with the AUTO package (Appendix 1A). The thermodynamic functions described by equations 1.21–1.24 were automatically computed as a function of the steady values of the state variables.

increased with growing slope with the force ratio. In other words, the dissipation function is minimal in the oscillatory domain (Figure 1.11b). It could be demonstrated that during the oscillatory regime, elicited by decreasing the substrate input rate, the average efficiency averaged over one cycle in the limit cycle behaviour was higher than the efficiency of the unstable asymptotic steady state under the same parametric conditions. This increased efficiency with respect to that of the unstable non-oscillatory state is achieved through a temporal shift between the occurrence of a maximum in the forces with respect to the maximum in fluxes. This result suggests that the oscillatory regime could be an energetically advantageous strategy for microorganisms under stress conditions such as food shortage (Cortassa *et al.*, 1990a). However, the latter appears to depend on the topology of the pathway (Chapter 8).

A similar result was obtained by Termonia and Ross (1981) when analysing a metabolic system consisting of glycolysis and other ATP-consuming processes. The system exhibited autonomous oscillations, and the thermodynamic efficiency during oscillatory behaviour was higher than that of the (unstable) asymptotic steady state. Ross *et al.* (1988) analysed the dynamics of a proton-translocating pump that showed parametric regions where the forced oscillatory behaviour resulted in a more efficient performance of the energy-transducing device. However, they also found parametric domains where the oscillatory regime implied a larger dissipation and a lower efficiency than that of the steady regime.

1.5.3 SELF-ORGANIZATION WITH LINEAR FLOW–FORCE RELATIONS IN AUTOCATALYTIC SYSTEMS THAT ARE FAR FROM THERMODYNAMIC EQUILIBRIUM

The analysis in the previous section showed that a system with an autocatalytic feedback loop may exhibit regions where the flow–force relations are linear (not shown). A natural question emerges: how far from equilibrium should a system be to exhibit self-organization? Do dissipative structures arise in the region where the flow–force relations are still linear? This is an important problem in view of the concept of dynamic organization developed in Chapter 2 (see also Figures 1.6 and 1.8). On the one hand, it had been suggested that linear flow–force relationships may still hold in biological systems far from thermodynamic equilibrium (Stucki, 1982; Berry *et al.*, 1987). On the other hand, bifurcations giving rise to nonlinear dynamics (e.g. bistability, oscillations, chaos) that phenomenologically characterize self-organization are assumed to occur when flow–force relations are nonlinear (Nicolis and Prigogine, 1977; Peacocke, 1983). The problem may be formulated as follows. If it is assumed that nonlinear phenomena happen only in regions of the thermodynamic regime with nonlinear flow–force relations, the evidence that shows linear flow–force relations for a wide range of physiological performances in cells then suggests that self-organization is a marginal, physiologically irrelevant phenomenon. This is clearly at odds with, say, the existence of rhythms as a fundamental property of biological organization at all levels of organization and complexity (Chapter 3). Let us fully address this problem.

We asked the question whether an extended NET – mosaic non-equilibrium thermodynamics (MNET) (Westerhoff and Van Dam, 1987) – is able to describe self-organization, at least qualitatively (Cortassa *et al.*, 1991, 1994b). Essentially, the approach focuses on the expressions for the fluxes where each metabolite concentration in the classical, kinetic rate equations was substituted for the exponential function of its chemical potential with respect to a reference state (Appendix 1B). These exponential functions were expanded around the steady state as a Taylor series. From the resulting nonlinear relations between rates and chemical potentials, only the zero- and first-order (linear) terms were taken into account.

Rather than analysing an abstract general scheme with little bearing on actual metabolism, the glycolytic pathway, which has been shown to exhibit oscillations under certain conditions (section 1.5.2), was considered. At relatively high values of the substrate input rate, V_{in}, and low energy loads (i.e. small values of k_p), the system exhibited a virtually monotonous evolution towards the steady state. At high energy load (high k_p), sustained oscillations were observed in all metabolite concentrations (Figure 1.13). This indicated the presence of a Hopf bifurcation point in the parameter space of the model, as could be confirmed by stability analysis of the linearized system (Cortassa *et al.*, 1991).

1.5.4 STABILITY ANALYSIS OF THE THERMODYNAMIC GLYCOLYTIC MODEL

Figure 1.13 indicates the parameter values for which locally (close to the reference state) monotonous asymptotically stable behaviour (grey region), damped oscillations (diagonally hatched region) or unstable spiralling (vertically hatched region) are expected. The stability of the glycolytic system was characterized by the eigenvalues which are the roots of the Jacobian matrix from the linearized system around a steady state. The thick line separating diagonally and vertically hatched regions corresponds to the Hopf bifurcation points, characterized by a pair of

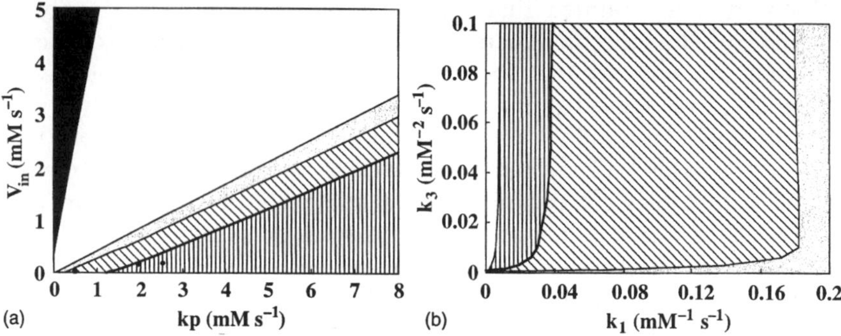

Figure 1.13 Bifurcation behaviour of the glycolysis model for different sets of parameters. The stability analysis of the system as a function of the parameters was performed after transformation of variables from concentrations to chemical potentials and linearization of the corresponding differential equations (Appendix 1B). The eigenvalues of the corresponding Jacobian matrix were used as the stability criteria. The different shadings indicate distinct qualitative behaviour as reflected by different kinds of eigenvalues. In the black region, the reference concentrations would take negative values; therefore this zone lacks physical meaning. The vertically hatched region corresponds to sets of parameters for which the system exhibited complex eigenvalues with positive real parts (the third eigenvalue was always real and negative); the reference state was unstable, i.e. states infinitesimally close to it evolve away from it in a spiralling fashion. In the parameter region depicted by diagonal hatching the system had two complex eigenvalues with negative real parts. Points around the reference state here spiralled back to it. The heavy line between the two hatched areas indicates the Hopf bifurcation points; here two eigenvalues were purely imaginary conjugates. In the shaded region, all eigenvalues were real and negative, giving rise to non-oscillatory relaxation to the reference state. In the white region, all eigenvalues were real and some were positive; states slightly deviating from the reference state develop monotonously away from it. In (a), k_1 and k_3 were kept at 0.3 mM^{-1} s^{-1} and 0.1 mM^{-2} s^{-1}, respectively; in (b), V_{in} = 0.3 mM s^{-1}; k_P = 1.0 mM s^{-1}. The values of the constants, C_A, P_t and K_M, were 10 mM, 10 mM and 2 mM, respectively. (Reproduced from *Biophysical Journal*, 1991, **60**, 794–803, with permission by The Rockefeller University Press.)

conjugate imaginary eigenvalues (Hopf, 1942; Hirsch and Smale, 1974; Segel, 1980). In the vertically hatched region, the eigenvalues of the Jacobian matrix are complex with positive real parts, indicating that the linear thermodynamic system was unstable close to the reference state. In that region, the system did evolve towards stable limit cycles (sustained oscillations).

An equivalent model where autocatalysis has been eliminated was subjected to stability analysis. Not a single region with complex roots could be found when the linearized equations were solved numerically for a broad parametric range. This demonstrated that the autocatalytic feedback through ATP was the nonlinearity involved in the appearance of oscillations (Cortassa *et al.*, 1990a, 1991).

1.5.5 THE MEANINGS OF LINEARITY FROM THERMODYNAMIC AND KINETIC VIEWPOINTS

It is timely to distinguish the various meanings of the word **linearity** as currently used in the literature. Any dependence of a rate law on the concentration of a compound to a power different from 1 produces nonlinear kinetics. Any positive or negative feedback or feed-forward loop transforms the kinetic structure of a system to a nonlinear one.

With respect to linearity in the sense of NET, the relations between the flows and the thermodynamic forces may be expanded into a Taylor series around any reference point. If only a small range of variation in the forces is considered, the second and higher order terms of the Taylor series may be neglected and the first order term may be considered as a good approximation to describe the flux. This type of NET linearity, which will be called **approximative linearity**, is strictly dependent on variations in the forces being much smaller than RT. Translation of kinetic rate equations into NET rate equations (Rottenberg, 1973; Van der Meer *et al.*, 1980; Rothschild *et al.*, 1980) has revealed that there also exist values of the forces around which there is so-called **extended linearity**. Thus, the second term of the Taylor series needs not be neglected because it is zero by itself.

The kinetic equations of the individual steps that served as the basis for the flow–force relations were kinetically linear with respect to each metabolite except for the non-glycolytic ATP-consuming reaction. However, there is an autocatalytic feedback loop arising from the fact that the number of ATP molecules produced in the second reaction exceeds the number consumed in the first step (Cortassa *et al.*, 1990a, 1991, 1994b). This autocatalytic feature of glycolysis provides the kinetic nonlinearity that in the present model was crucial for the appearance of self-organized behaviour. Indeed, the type of autocatalysis described in this work was shown to coexist with linear flow–force correlations. However, in section 1.5.4 it was noted that the oscillations appeared in regions of nonlinear flow–force relations. In the limit cycles appearing further from the Hopf bifurcation points the forces became large compared with RT. Thus, the quantitative aspects of the dynamics differed from those described with a kinetic model.

1.5.6 A COMPARISON BETWEEN KINETIC AND THERMODYNAMIC DESCRIPTIONS

The question of whether linearity in the sense of NET would preclude the occurrence of oscillations in a system with an autocatalytic kinetic structure was investigated by expanding the dependences of the fluxes on the forces linearly. That is to say, the flow–force relationships had been forced to be linear from a thermodynamic point of view.

The starting point of the thermodynamic formulation was the kinetic model, which was translated into a nonlinear NET model and then linearized. As a consequence of the linearization, the NET model used here is only an approximation of the original kinetic model, whenever states other than the reference state are considered (Figure 1.14). This has no implications for the local stability properties around steady states (stable or unstable), albeit it certainly does have for the dynamics of trajectories evolving towards stable steady states. We demonstrate that both descriptions resemble each other as the concentrations of reactants approach the corresponding concentrations in the reference state. The kinetic and NET descriptions will end up in the same stable steady state (Figure 1.14a,b).

Around unstable steady states, the fact that the linear NET model is only an approximation of the linear kinetic model has more conspicuous implications for the simulated dynamic behaviour. When the system evolves to a limit cycle, its linear NET description differs from that offered by the corresponding kinetic description. However, for bifurcation parameter values close to the Hopf bifurcation point, the (linear) thermodynamic description is still quantitatively similar to the kinetic description (Figure 1.14c,d). In the parametric region of sustained oscillations, the amplitude of limit cycles increases as the bifurcation parameter is taken further away from the Hopf bifurcation point, allowing the chemical potentials of metabolites to deviate strongly from the steady state values. In those cases, the MNET model is unable to predict, quantitatively, the time dependence of metabolite concentrations. The linear thermodynamic description could, however, predict the type of dynamic behaviour in the entire parametric space of the kinetic model. Therefore, the linear approximation has primarily a qualitative value when applied at unstable steady states.

The conclusion that linearity *per se* is not in conflict with self-organized behaviour does not contradict the demonstration that self-organization cannot be observed close to equilibrium: it is the symmetry property of the near-equilibrium domain, and not the linearity often attached to it, that precludes self-organized behaviour. The dynamic behaviour of the system, i.e. whether the system will undergo a bifurcation and change from asymptotic behaviour to oscillatory steady states, is not determined

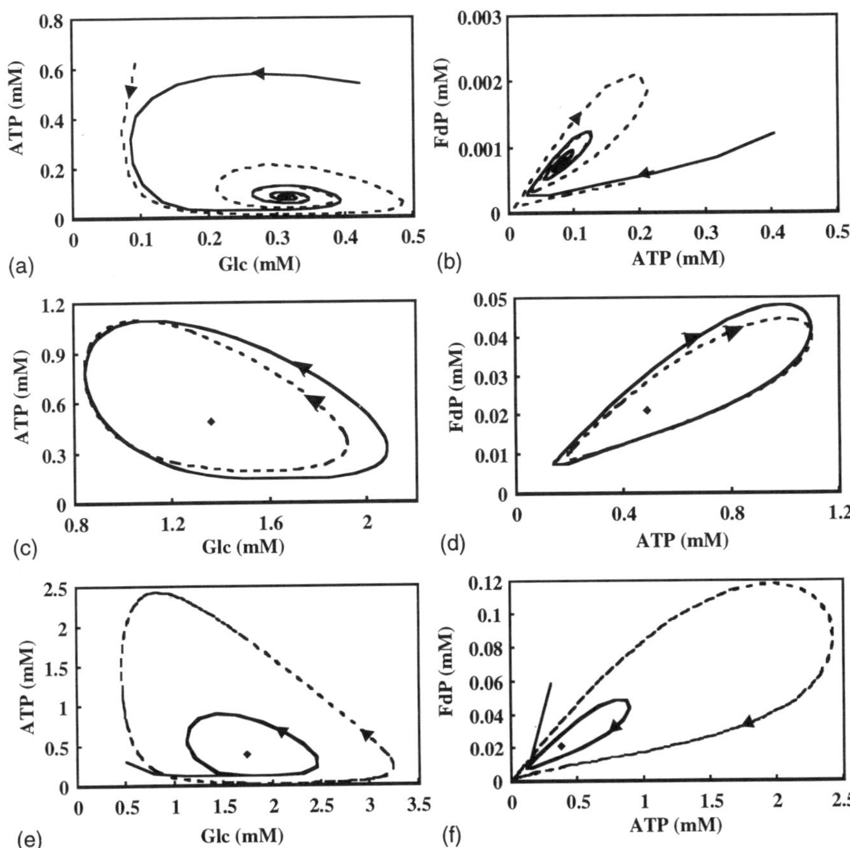

Figure 1.14 Phase space analysis for the parameter sets corresponding to the three points indicated with dots in Figure 1.13a. The plots represent the trajectories of the state variables for: (a,b) k_P = 0.3905 mM s⁻¹; k_1 = 0.3 mM⁻¹ s⁻¹; k_3 = 0.1 mM⁻² s⁻¹; V_{in} = 0.0075 mM s⁻¹; and (c,d) k_P = 2.033 mM s⁻¹; k_1 = 0.3 mM⁻¹ s⁻¹; k_3 = 0.1 mM⁻² s⁻¹; V_{in} = 0.2 mM s⁻¹; and (e,f) k_P = 2.5 mM s⁻¹; k_1 = 0.3 mM⁻¹ s⁻¹; k_3 = 0.1 mM⁻² s⁻¹; V_{in} = 0.2 mM s⁻¹. In (c,d) the diamonds indicate the reference concentration that corresponds to the unstable steady state solution. Dashed lines represent the behaviour of the kinetic model from which the linear thermodynamic model has been derived. The values of the constants were: C_A = P_t = 10 mM; K_M = 2 mM. (Reproduced from *Biophysical Journal*, 1991, **60**, 794–803, with permission of The Rockefeller University Press.)

by thermodynamic linearity but by the particular values of the kinetic parameters. Therefore, it is not the linearity of the flow–force relations but the kinetic structure, contained implicitly in the phenomenological coefficients (L) of the NET equations, which has the information about the dynamics of a given system.

APPENDIX 1A

Anaerobic glycolysis was modelled as described in Cortassa *et al.* (1990a). The reaction steps comprised by the model are:

- a constant input of substrate (V_{in});
- ATP-consuming processes of the 'upper' part of glycolysis, V_1 (hexokinase, phosphofructokinase lumped in the rate constant k_1);
- ATP-producing reactions of the 'lower' part of glycolysis, V_3 (phosphoglycerate kinase, pyruvate kinase, depicted by k_3);
- ATP consumption by ATP-consuming processes and by ATPases (active transport), V_9, and V_P, respectively;
- irreversible utilization of free-energy rich intermediates for synthesis, V_5;
- irreversible output of lactic acid, V_7.

The fluxes through each step are described by the following expressions:

$$V_1 = k_1[\text{ATP}][\text{Glc}] \tag{1.15}$$

$$V_3 = k_3[\text{ADP}][\text{I}]([\text{P}_t - 2[\text{I}] - [\text{ATP}]) \tag{1.16}$$

$$V_5 = k_5[\text{I}] \tag{1.17}$$

$$V_7 = k_7[\text{L}] \tag{1.18}$$

$$V_9 = k_9[\text{ATP}] \tag{1.19}$$

$$V_P = k_P \frac{[\text{ATP}]}{K_M + [\text{ATP}]} \tag{1.20}$$

where k_1, k_3, k_5, k_7 and k_9 are rate constants.

The reaction describing the ATP consumption by an H^+ pump, V_P, was assumed to follow Michaelis–Menten behaviour as a function of ATP (the parameter d_P corresponds to the product $V_M\tau(1 - dp)^6$ as described in Cortassa *et al.*, 1990a). The system is described by the following set of ODEs, and includes a conservation relationship for adenine nucleotides:

$$\frac{d[\text{Glc}]}{dt} = V_{in} - V_1 \tag{1.21}$$

$$\frac{d[\text{ATP}]}{dt} = -2V_1 + 4V_3 - V_9 - V_P \tag{1.22}$$

$$\frac{d[\text{I}]}{dt} = V_1 - V_3 - V_5 \tag{1.23}$$

$$\frac{d[L]}{dt} = V_3 - V_7 \tag{1.24}$$

$$C_A = [ATP] + [ADP] \tag{1.25}$$

The thermodynamic functions were evaluated through the instantaneous level of metabolites by the following equations:

$$J_a = -2V_1 + 4V_3 \tag{1.26}$$

$$\Delta G_a = \Delta G^{\circ}_a + RT\ln\frac{[ATP]}{(P_t - 2[I] - [ATP])(C_A - [ATP])} \tag{1.27}$$

$$J_{sl} = V_3 \tag{1.28}$$

$$\Delta G_s = \Delta G^{\circ}_s + RT\ln\frac{[L]^2}{[Glc]} \tag{1.29}$$

Accordingly, the thermodynamic efficiency, η, was defined as:

$$\eta = -\frac{\Delta G_a J_a}{\Delta G_s J_{sl}} \tag{1.30}$$

where ΔG_a and ΔG_s are the standard free-energy differences of ATP synthesis and catabolism (transformation of glucose, Glc, into lactic acid, L). J_a and J_{sl} are taken as the net production rate of ATP by glycolysis and the rate of lactate formation, respectively. (Notice the formal resemblance of equations 1.12 and 1.30.) The dissipation function, σ, for the stoichiometric glycolytic model is given by:

$$\sigma = (-\Delta G_s)J_{sl} + (-\Delta G_a)J_a \tag{1.31}$$

The system of differential equations 1.21–1.25 was integrated numerically using SCoP with an Adams integration method (Duke University, 1989). The stability analysis was performed with the AUTO package (Doedle, 1986).

To address the question about the advantage of oscillatory behaviour, the average thermodynamic efficiency was calculated in two different ways: either from the average of instantaneous fluxes and forces over a period (equation 1.32) or from the average of instantaneous dissipations (equation 1.33), the latter taking into account the instantaneous product of the flow and its conjugated force:

$$\eta = \frac{\int_0^P J_a dt \int_0^P \Delta G_a dt}{\int_0^P J_{sl} dt \int_0^P \Delta G_s dt} \tag{1.32}$$

$$\eta = \frac{\int_0^P J_a \Delta G_a \, dt}{\int_0^P J_{sl} \Delta G_s \, dt} \tag{1.33}$$

The calculations based on equation 1.32 gave results similar to those computed from the concentrations that the system would exhibit at the unstable steady states provided by the stability analysis (section 8.9).

APPENDIX 1B

Glycolysis is depicted as resulting from individual fluxes corresponding to different steps of the pathway related through conservation equations. To obtain the non-equilibrium thermodynamic description, first, kinetic rate equations (Cortassa *et al.*, 1990a, 1991) were written with linear dependences on the substrate concentrations. V_1 and V_P are given by equations 1.15 and 1.20, respectively. V_3 was slightly modified as follows:

$$V_3 = k_3 (C_A - [ATP])[FdP]P_t \tag{1.34}$$

To make the transition to the nonlinear thermodynamic model, the concentrations were then expressed into the corresponding chemical potentials (Rottemberg, 1973; Van der Meer *et al.*, 1980; Rothschild *et al.*, 1980), using:

$$\frac{[M]}{[M]_r} = \exp\left(\frac{\mu_M - \mu'_M}{RT}\right) \tag{1.35}$$

$[M]_r$ and μ_M being the concentration and chemical potential of metabolite M at the reference state. The linear non-equilibrium thermodynamic description was obtained by expanding each exponential function of a chemical potential as its Taylor series around an arbitrary reference state:

$$\exp\left(\frac{\mu_M - \mu'_M}{RT}\right) = 1 + \frac{\mu_M - \mu'_M}{RT} + \frac{1}{2}\frac{(\mu_M - \mu'_M)^2}{RT^2} + \frac{1}{6}\frac{(\mu_M - \mu'_M)^3}{RT^3} \tag{1.36}$$

and neglecting all but the first two terms. As in the MNET formalism (Westerhoff and Van Dam, 1987), the flux through each of the steps, J_i, is expressed as a function of the chemical potentials of the substrates of that particular step:

$$J_1 = k_1 [ATP]_r [Glc]_r \left(1 + ln\frac{[ATP]}{[ATP]_r} + ln\frac{[Glc]}{[Glc]_r}\right) \tag{1.37}$$

$$J_3 = k_3 P_t (C_A - [ATP]_r)[FdP]_r \left(1 + \ln\frac{(C_A - [ATP])}{(C_A - [ATP]_r)} + \ln\frac{[FdP]}{[FdP]_r} \right) \quad (1.38)$$

$$J_P = k_P \frac{[ATP]_r}{K_M + [ATP]_r} \left(1 + \frac{K_M}{K_M + [ATP]_r} \ln\frac{[ATP]}{[ATP]_r} \right) \quad (1.39)$$

To simulate the dynamic behaviour of this linear non-equilibrium thermodynamic system, the time dependence of the free variables was written as the sum of the linear flow–force relations (equations 1.37–1.39). Interpreting the chemical potentials as driving forces, this led to the following expressions for the rate of change of metabolite concentrations:

$$\frac{d[Glc]}{dt} = V_{in} - J_1 \quad (1.40)$$

$$\frac{d[ATP]}{dt} = -2J_1 + 4J_3 - J_P \quad (1.41)$$

$$\frac{d[FdP]}{dt} = V_1 - V_3 \quad (1.42)$$

The resulting system of differential equations (1.40—1.42) was integrated numerically using the SCoP package from the National Biomedical Simulation Resource (Duke University Medical Center, Durham). The Adams subroutine was used with an integration step of 0.01 s.

Dynamic organization in cellular systems   2

The relationship between events occurring at the molecular–supramolecular levels, such as gene expression, cellular energetics, metabolism and 'supra' properties such as cellular growth, proliferation and differentiation, is an unsolved problem in the whole biological literature. We conceive those 'supra' properties as emerging from the autonomous dynamics and spatio-temporal coordination of events occurring at the molecular–supramolecular level. Within this conceptual framework it may be postulated that cell growth, proliferation and differentiation involve a sequence of processes coordinated in space and time, i.e. should be performed at the 'right' time and space inside the cell or a tissue (Dean and Hinselwood, 1966). Thus, the autonomous dynamics of cellular processes and their spatio-temporal coupling is at the heart of cell function (Aon and Cortassa, 1993).

We conceive dynamics as relating cell structure and function. In fact, if cell function is considered from the point of view of the allometric relationship of spatio-temporal levels of organization, emerging from the transitions between those levels, and if organization of cellular structures is also the result of the appearance of new levels of organization, then dynamics is at the heart of understanding cell structure–function relationships. In other words, the transition from the primary molecular mechanism elicited by, say, an environmental signal to the generalized cellular response at a higher level of organization may be visualized as an instability in the dynamics of the processes involved at different levels of organization. The task is then to identify processes in which dynamics triggered by environmental signals could be transduced in a coherent cellular response emerging at higher levels of organization. We have previously suggested that the appearance of macroscopic coherence at the cellular level, revealed as new functional properties, would occur in processes at spatial dimensions of micrometres and relaxing in the order of minutes (Aon and Cortassa, 1993). Microtubules may reach lengths in the order of micrometres and depolymerize within minutes (Kirschner and Mitchinson, 1986; Erickson and O'Brien, 1992).

We have shown that the presence of assembled or unassembled microtubular protein in the physiological concentration range, either *in vitro* or intracellularly, modulates fluxes through chained reactions catalysed by enzymes related to carbon metabolism (Cortassa *et al.*, 1994a). Thus, a systemic property such as metabolic fluxes might depend on a general feature of cytoplasmic organization, i.e. the polymeric status of microtubular protein and its concentration. Additionally, the sort of spatial arrangement of F-actin (bundles or random filaments) exerts a specific effect upon the kinetic parameters of pyruvate kinase (Aon *et al.*, 1996a). The interest in the dynamic organization view of cellular function is stressed by the fact that structural self-similarity, i.e. fractality, appears to exist in biological macromolecular lattices and the groundplan of living cells, as recently proposed (Rabouille *et al.*, 1992; Aon and Cortassa, 1994).

2.1 THE CONCEPT OF DYNAMIC ORGANIZATION

A cell comprises an ensemble of levels of organization which result from the interaction between processes of different natures, i.e. chemical, electrical and mechanical, such as solute transport, enzyme activity, protein synthesis, electron transport, gene expression and organelle function (Aon and Cortassa, 1993). We define **dynamic organization** as the evolution of the spatio-temporal organization of biological processes between successive levels of organization. The evolution of a dynamic regime of a biological process into another is achieved through instabilities (bifurcations) which give rise to the emergence of collective behaviour of, for example, molecules, enzyme activity or gene expression, spatio-temporally coherent as in the example presented in Figure 1.8. In unravelling cellular function it is crucial to understand how the transition between levels of organization comprising processes at different spatio-temporal scales occurs. In turn, to understand this transition we need to know how these processes are coupled or interact at the same or different levels of organization (see below and Chapters 7 and 12).

Kinetically, biological systems are nonlinear because of their multiple interactions; for example, feedbacks (such as substrate inhibition), feed-forwards (product activation), cross-activation or cross-inhibition. In the case of chemical reactions they may be arranged in linear, branched or cyclic pathways, or combinations of these basic topologies. As discussed in Chapter 1, due to the intrinsic nonlinearities built into the network of chemical reactions and compartmentalization, biological systems are able to change their dynamic behaviour, i.e. to bifurcate toward new steady states or attractors. In thermodynamic terms, the multiple interactions exclude biological processes from the near-equilibrium domain where equations 1.6 hold. Thus, also, the stability properties of equations 1.8 and 1.9 may not be valid, and the system may achieve self-organized states.

When biological systems bifurcate, they evolve toward other attractors of their dynamic state space (Kauffman, 1989). Coupling between dynamic subsystems of cellular systems is achieved through common intermediates (Chapter 1). In such interacting networks, a variable in one dynamic subsystem, because of its quick relaxation toward perturbations, may act as a parameter of another dynamic subsystem that relaxes slowly (section 1.5.1). This leads the system to shift between attractors (or 'quasi-attractors') all the time following perturbations (Haken, 1991).

2.2 RELAXATION TIME: A QUANTITATIVE CHARACTERIZATION OF THE DYNAMIC BEHAVIOUR OF CELLULAR PROCESSES FUNCTIONING AT DIFFERENT TIME SCALES

Goodwin (1963) defined the relaxation time of a system as 'the time required for the variables to reach a steady state after a 'small' disturbance'. For a quantitative description of the relaxation time of cellular processes, a mathematical description of the system under study should be at hand. The size of the 'small' disturbance should fulfil the requirement that the perturbation allows a linearization of the system of equations in the neighbourhood of a steady state. The eigenvalues measure the exponential behaviour of the trajectories of dynamic systems in the neighbourhood of a steady state (Chapter 1); their reciprocals define the relaxation time and indicate the time needed by the system to return to a steady state after a perturbation (Chapter 1) (Goodwin, 1963; Heinrich *et al.*, 1977). The system contains as many eigenvalues as variables and if it is hierarchically structured in time, its eigenvalues are widely different. However, it should be emphasized that the eigenvalues characterize the system's dynamics only in the close neighbourhood of the steady state where the linear approximation is applicable (Heinrich *et al.*, 1977). The local behaviour is characterized by the eigenvalues of the Jacobian matrix. The eigenvalues measure the exponential behaviour of the trajectories of the dynamic system in the neighbourhood of a steady state. The logs of the eigenvalues are called the Lyapunov exponents, which measure the sensitive dependence on initial conditions (Chapter 1).

2.3 THE LIVING COMPLEXITY AND LEVELS OF ORGANIZATION

One of the most distinctive features of the complexity of living systems is their structural and functional organization at many simultaneous levels (Figure 2.1; Churchland and Sejnowski, 1988). Cell function is an emergent property from transitions between levels of organization arising at bifurcation points in the dynamics of biological processes (Aon and Cortassa, 1993). What could be considered as biological evidence of the structural and functional organization of living systems at many simultaneous levels? The main evidence comes from information 'collected' by different perturbation probes, depending on the level of organization in the biological system under study. When lower levels of organization are affected, the higher levels reflect those changes by altering their organization and qualitative dynamics, often with the appearance of new functional properties.

The characteristic spatial dimension, E_c, and relaxation time, T_r, of distinct levels of organization involved in microbial and plant cell growth, solute transport, energy transduction, neuron firing and enzyme activity range from 10^{-11} s to 10^4 s and from 10^{-10} m to 10^{-1} m. Each process identified by its E_c and T_r will be said to belong to a level of organization (Table 2.1).

Taking enzyme activity as an example familiar to biochemists, let us show how the concept of level of organization can be applied physically and conceptually. The steady operation conditions of enzyme-catalysed reactions involving a single substrate are displaced when they are perturbed at the steady state. After perturbation, the relaxation of such a chemical system to its previous steady state (if stable) will be characterized by relaxation times, T_r, which can be related to the rate constants of the several reaction steps involved:

$$T_r = \frac{[1+(S/K_s)(1+V_s/V_p)]}{(V_s/K_s + V_p/K_p)} \tag{2.1}$$

in which S is the steady state concentration of the substrate; K_s and K_p are the Michaelis constants of the substrate and the product, respectively; and V_s and V_p are the maximum velocities of the forward and backward reactions. If an enzyme molecule is perturbed by a temperature jump or a laser beam, the lateral chains of amino acids will relax more rapidly (i.e. low T_r) than a conformational change of monomers (or oligomers) and the enzyme's reaction will be even slower (Table 2.1 and Figure 2.2).

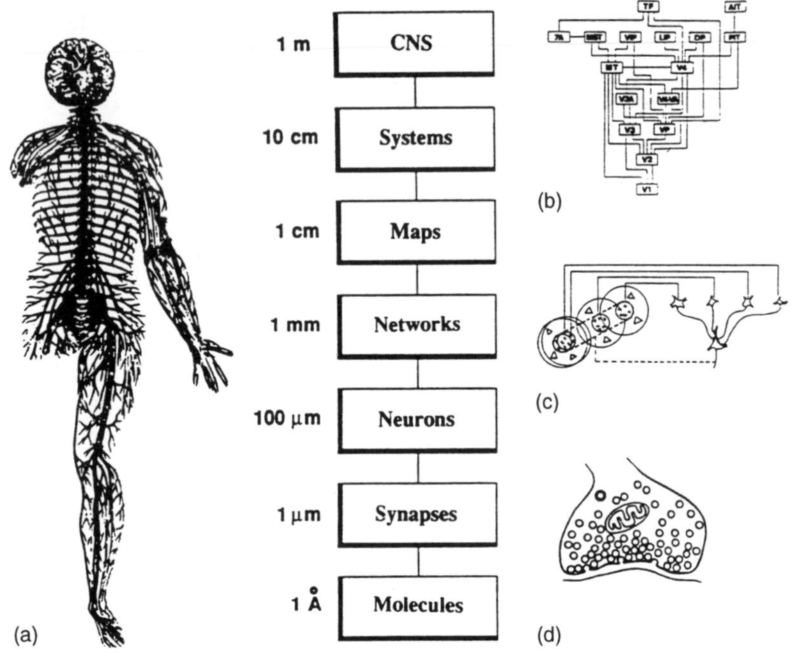

Figure 2.1 Structural levels of organization in the nervous system. The spatial scale at which different anatomical organizations can be identified varies over many orders of magnitude. (a) Drawing by Vesalius of the human brain, the spinal column and the peripheral nerves; (b) a processing hierarchy of visual areas in monkey visual vortex; (c) a small network model for the synthesis of oriented receptive fields of simple cells in visual cortex; (d) the structure of a chemical synapse. (Reprinted with permission from Churchland and Sejnowski, *Science*, **242**, 741–745, copyright 1988 American Association for the Advancement of Science.)

Temporally, therefore, a protein molecule with enzymatic activity may be described by distinct levels of organization according to the T_r following a perturbation. E_c describes the spatial length coordinate in angstroms (atomic, i.e. ionization equilibria and conformational changes of the E–S complex), nanometres (molecular, i.e. interaction between effectors and the E–S complex) and micrometres (supramolecular, i.e. substrates diffusion, cellular macroscopic spatio-temporal structures). Thus, globally, the biological process enzyme activity can be described in levels of organization that are characterized spatio-temporally. Although levels of organization can be conceptually detached, obviously they are not physically so. In fact, enzyme activity is the result of the interaction between all those levels of organization.

Table 2.1   Spatio-temporal dimensions of biological processes

Process	Log relaxation time (T_r) (s)	Log characteristic space (E_c) (m)
Enzyme activity		
Ionization equilibria	−10	−10
Enzyme–substrate local conformational motion	−10	−9
Enzyme–substrate/enzyme–enzyme lifetime	−3	−9
Enzyme catalytic turnover	−1	−7
Metabolite diffusion	1	−8
Spatial patterns	4	−2
Microbial growth metabolism		
Water bound to biopolymers	−11	−10
Enzyme–substrate local conformational motion	−10	−9
Diffusion small molecules (cell)	−6	−10
Equilibrium of catalytic groups with solvent	−7	−9
Metabolite transit times between enzymes	−5	−9
Covalent enzyme–substrate intermediate	−3	−9
Transient rates of NADH transfer in bienzyme complexes	−3	−9
Protein–DNA synthesis	−3	−8
Enzyme catalytic turnover	−3	−7
Association–dissociation between protein–nucleic acids	−2	−8
Cellular efflux of solutes	1	−6
Cation efflux down a gradient	1	−5.5
Enzyme concentration changes	2	−8
Osmotic swelling and turnover of glycolytic intermediates	2	−6
Cell generation time	3	−6
Ethanol production[a]	4.55	−1
Plant cell growth		
Pigment excitation	−13	−12
Stretch-potential activated channel	−10	−8
Quantum absorbance	−9	−11
Reaction centre capture	−7	−9
Electron transport	−6	−8
Protonation–deprotonation reactions	−4	−8.3
PSII–PSI turnover rates	−3	−8
Passive ion fluxes	−2	−7
Potential-gated ion channels	−2	−8
PSII–PSI proton flux	−1	−7
Cell elongation	1	−5
Δp–ΔpH building	1	−5.5
Membrane potential change	1	−5.3
Wall deformation	1	−8
Growth rate change	2	−5
Macromolecule polymerization	2	−7

Table 2.1 *continued*

mRNA synthesis	2	−8
Changes in A(T-D-M)P during microtubule gelation–contraction	3	−4
Turgor pressure–wall synthesis	3	−5
Polysaccharide degradation	3	−6
Anaerobic accumulation of end-products	4	−2
Photosynthesis–energy transduction		
Pigment excitation	−13	−12
Quantum absorbance	−9	−11
Reaction centre capture	−7	−9
Electron transport	−6	−8
Protonation–deprotonation reactions	−4	−8.3
PSII–PSI turnover rates	−3	−8
Passive ion fluxes	−2	−7
PSII–PSI proton flux	−1	−7
Δp–ΔpH building	1	−5.5
Membrane potential change	1	−5.3
Spatial patterns	2	−3
Solute transport		
Cellular efflux of solutes	1	−6
Δp–ΔpH building	1	−5.5
Lipid diffusion in membranes[b]	1	−5
Cation efflux down a gradient	1	−6
Osmotic swelling	2	−5.5
Cation efflux against a gradient	2	−6
Neurone firing		
Trans-gauche isomerization	−7	−9
Lipid exchange near proteins	−7	−8.5
Spike potential (firing)	−1	−5
Δp–ΔpH building	1	−5.5
Lipid diffusion in membranes[b]	1	−5

Space and time characteristic dimensions are expressed in seconds (s) and metres (m).
(a) The characteristic spatial dimension (in this case an extensive property) corresponds to a one-litre reactor and the characteristic relaxation is the time taken by a microorganism in a batch culture to reach the stationary phase of production.
(b) The relaxation time was calculated according to the expression: $t = x^2/2D$, in which D is the diffusion coefficient (10^{-8} cm^2 s^{-1}); x is the distance estimated as for a circumference of 5 μm radius and the relaxation is the time for a lipid molecule to diffuse through x.
See Aon and Cortassa (1993) for references. (Reprinted from Aon and Cortassa, 1993, by permission of Kluwer Academic Publishers.)

Most of the main functional properties of cells, such as energy transduction, solute transport, action potentials in neurons, macromolecule polymerization and cell growth and division, are placed in T_r ranges of 1 to 3 (seconds to several minutes on a logarithmic scale),

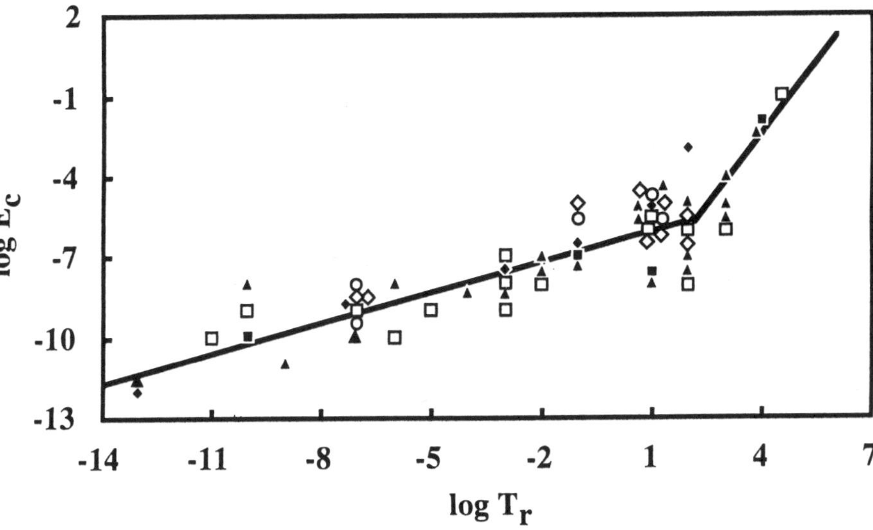

Figure 2.2 Allometric relationship between characteristic spatial dimensions, E_c, and relaxation times, T_r, of biological processes at different levels of organization. The logarithms of E_c and T_r of all biological processes presented in Table 2.1 were plotted and the points fitted by the logarithmic form of the allometric equation (2.7, 2.8, 2.14, 2.15): (log E_c = –0.4197 + 0.381 log T_r) and (log E_c = –0.052 + 1.8812 log T_r). Filled square, enzyme activity; open square, microbial growth–metabolism; filled triangle, plant cell growth; open diamond, solute transport; filled diamond, photosynthesis–energy transduction; open circle, neuron firing. (Reprinted from Aon and Cortassa, 1993, by permission of Kluwer Academic Publishers.)

and E_c of –7 to –6 (around micrometres). Molecular properties involved in those processes span T_rs of –6 to –10 (ms to ps) and E_cs of –8 to –10 (nanometres to angstroms). These are wide spatio-temporal spans; thus, for molecular properties to extend their range of action to higher spatio-temporal dimensions, some organizing principles implying their coherent spatio-temporal behaviour should be invoked. Without those organizing principles, it is impossible to understand how an apparently purposeful action, i.e. cellular function, could arise at higher spatio-temporal dimensions, i.e. levels of organization. We have proposed (Aon and Cortassa, 1993) that self-organization implying bifurcations in the spatio-temporal behaviour of living systems (Nicolis and Prigogine, 1977; Haken, 1978; Kauffman, 1989) may help to rationalize transitions between levels of organization (Chapter 6). After bifurcations arise, new macroscopic properties emerge. Hopefully, those new macroscopic properties belong to new functions when integrated to a 'wilful' machine such as a cell.

When analysing cell function (e.g. macromolecule secretion) we should be aware of the existence of and be able to distinguish between:

- the level of observation or information given by the particular method of perturbation;
- the phenomenon we search to explain;
- the level of explanation chosen for a particular phenomenon.

2.4 CHARACTERIZATION OF BIOLOGICAL PROCESSES AT DIFFERENT LEVELS OF ORGANIZATION

The space, E_c, and relaxation time, T_r, that characterize biological processes at different levels of organization contain dynamic information about those processes (Figure 2.2 and Table 2.1). The relaxation time, T_r, contains implicit information about the dynamics of the processes. In a linear system of the type:

$$\frac{dy_1}{dt} = a_{11}y_1 + a_{12}y_2; \quad \frac{dy_2}{dt} = a_{21}y_1 + a_{22}y_2 \tag{2.2}$$

T_r is related to the reciprocal of the eigenvalues describing how long it takes state variables (y_1, y_2) in equations 2.2 to relax to a steady state after a perturbation (Pratt, 1974; Jeffrey, 1990). Around the steady state, the trajectories followed by the state variables relaxing are defined by:

$$y_1 = C_1 e^{\lambda t}$$
$$y_2 = C_2 e^{\mu t} \tag{2.3}$$

where λ and μ are the eigenvalues of equations 2.2.

When several state variables are interdependent, the system dynamics will be ruled by the variable with the smallest eigenvalue and the fast relaxing variables will be 'entrained' by the slower ones.

T_r integrates dynamic information (rate constants and affinities) as well as gene expression since maximum velocities are proportional to the enzyme concentration (Chapter 11). E_c is related to the spatial dimension where the processes under consideration are taking place. E_c may be understood as the 'critical length' in dissipative structures generated through reaction–diffusion mechanisms (Nicolis and Prigogine, 1977). For instance, in the case of reactions ruled by reaction–diffusion mechanisms, the dynamics is implicit in the non-dimensional Thiele modulus, ρ (Cortassa *et al.*, 1990b):

$$\rho = \frac{V_M e^2}{K_M D} \tag{2.4}$$

where V_M is the enzyme maximal rate; e is the width of the spatial coordinate, e.g. a membrane, where the reaction is taking place; D is the diffusion coefficient; and K_M is the Michaelis constant of the enzyme or the substrate concentration giving half-maximal rate of enzyme. The Thiele modulus, ρ, integrates spatio-temporal information about reaction (rate constants and affinities) and the coupling of diffusion (diffusion coefficient) and spatial dimensions.

Overall, T_r and E_c are the result of the spatio-temporal integration of the inherent dynamics of cellular processes involved in biological phenomena. The importance of the dynamic information contained implicitly in E_c and T_r becomes apparent when one approaches the problem of evolution between successive levels of organization. When the dynamics of a process is perturbed at a certain level of organization, it may become unstable at bifurcation points. Under those conditions the system evolves toward a new level of organization exhibiting a new sort of macroscopic coherence; for example, (a)symmetry changes, sophisticated dynamics, or evolution toward a new steady state (Cortassa and Aon, 1994b).

The specific mechanisms determine the range of parameter values for which spatio-temporal changes in the organization of biological processes may arise. For the latter it becomes crucial to know how biological processes are coupled to each other (Chapter 12).

2.5 LEVELS OF PERTURBATION AND LEVELS OF EXPLANATION

2.5.1 LEVELS OF PERTURBATION

The perturbation methods used to study biological phenomena reflect the dynamics of the levels of organization at which they integrate into the living system under study (Table 2.1 and Figure 2.3). According to the relaxation time and length scale of the probe used to sense a biological process, information will be obtained from a certain level of organization of the cell (Table 2.2 and Figure 2.3). If an allometric function acceptably describes the relation between the spatio-temporal dimensions of biological processes, then a similar relationship should exist between the spatio-temporal dimensions sensed by the different probes. Figure 2.4 shows that when relaxation times and length scales sensed by different probes are represented in a double logarithmic plot, an allometric relationship with a double slope appears. This is a similar result to that described between the characteristic relaxation time, T_r, and space dimension, E_c, of biological processes of different natures spanning several levels of organization (Figure 2.2).

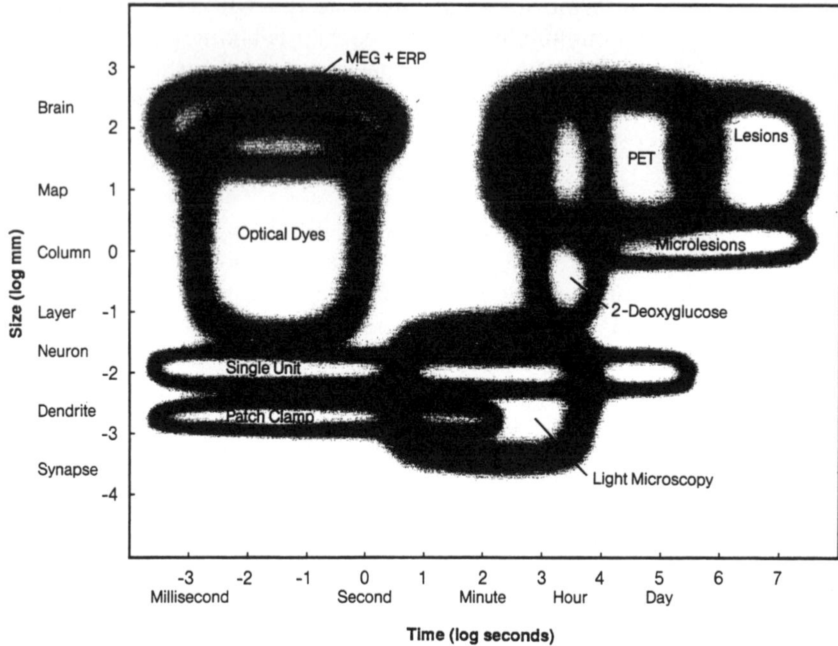

Figure 2.3 Spatio-temporal resolution of various experimental techniques for studying the function of the brain. The vertical axis represents the spatial extent of the technique, with the boundaries indicating the largest and smallest size of the regions from which the technique can provide useful information. For instance, single-unit recording can only supply information from a small region of space, typically 10–50 μm on a side. The horizontal axis represents the minimum and maximum time intervals over which information can be collected with the technique. Thus, action potentials from a single neuron can be recorded with millisecond accuracy over many hours. (Reprinted with permission from Churchland and Sejnowski, *Science*, **242**, 741–745, copyright 1988 American Association for the Advancement of Science.)

In trying to explain the appearance of macroscopic spatio-temporal patterns such as those observed in a photobiochemical system (Aon *et al.*, 1989b; Aon and Cortassa, 1993), the question is: what shall we look at? Table 2.1 shows that several spatio-temporal levels of organization may be distinguished in this particular system, ranging from pigment excitation to spatial patterns. It is known that if the water-splitting system is inhibited, macroscopic patterns do not arise (Aon *et al.*, 1989a). This is at the molecular level of organization. However, neither are patterns observed if a gradient of the electron acceptor is absent, even with an active water-splitting system. Therefore, the appearance of the macroscopic spatio-temporal patterns depends on events occurring at the

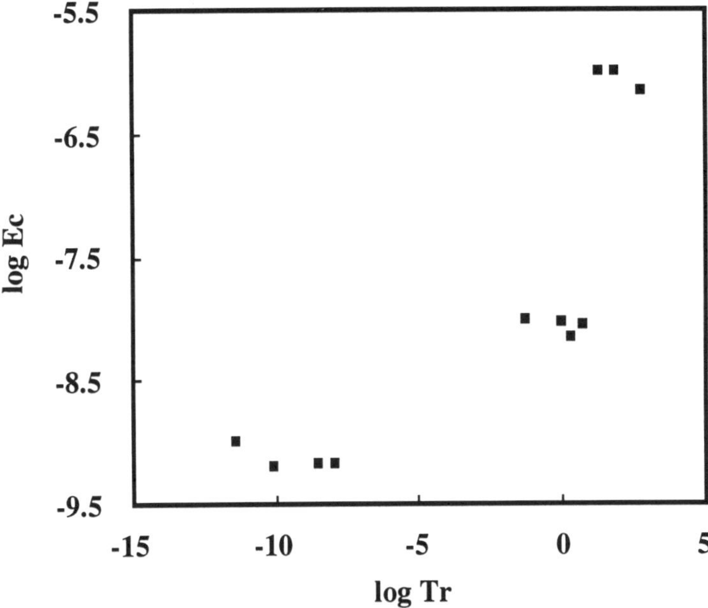

Figure 2.4 Relationship between characteristic spatial dimensions, E_c, and relaxation times, T_r, of perturbation probes sensing physiological processes at various levels of organization. The logarithms of E_c and T_r of the perturbation probes presented in Table 2.2 were plotted and the points fitted by the logarithmic form of the allometric equation: ($\log E_c = -8.084 + 0.1072 \log T_r$) and ($\log E_c = -8.116 + 0.9112 \log T_r$). Indicated spatial and temporal characteristic dimensions correspond to the mid value of the interval from which they provide information (see also Figure 2.3).

molecular level and a macroscopic cue, i.e. the gradient. Additionally, the intrinsic kinetics of the photobiochemical system is nonlinear, which gives rise to bistable dynamics that explains the appearance of the alternating bands of blue (oxidized dichloroindophenol, DCIP) and green (reduced DCIP) (Aon *et al.*, 1989a; Aon and Cortassa, 1993). A direct consequence of the arguments discussed above is that in order to explain a biological phenomenon no single level of organization has a 'priority' over the others. In other words, there is no fundamental level of observation to explain, for example, the appearance of the spatio-temporal patterns in the photobiochemical system; no single level of organization contains all the information needed to account for the observed coherent behaviour (Peacocke, 1983; Churchland and Sejnowski, 1988). This is probably also true for cellular systems since we are dealing with functions which emerge from the whole, i.e. either cell, organ or organism.

Table 2.2   Characteristic spatio-temporal dimensions of physiological processes probed by different perturbation methods in cellular systems

Method	Relaxation time (s) (Particle diameter (m)[a])	Probe and cells	Viscosity (with respect to water)	Reference
Time-resolved fluorescence microscopy	$2\text{–}4 \times 10^{-9}$	BCECF (MDCK, Swiss 3T3)[b]	1.1–1.2	Dix and Verkman, 1990
Picosecond polarization microfluorimetry	$0.3\text{–}25 \times 10^{-9}$ (0.65×10^{-9})	BCECF (3T3 fibroblasts)	1.2–1.4-fold higher	Fushimi and Verkman, 1991
Electron spin resonance (ESR)	0.09×10^{-9} (0.64×10^{-9})	PCAOL (Swiss 3T3, BHK, BALB c-MCA3T3)	2–3-fold higher	Mastro *et al.*, 1984; Mastro and Keith, 1984
Fluorescence recovery after photobleaching (FRAP)	2–8 $(8.2\text{–}9.6 \times 10^{-9})$	F-BSA-tubulin (sea urchin, embryonic)	8-fold higher	Salmon *et al.*, 1984a
FRAP	19	DTAF–tubulin (sea urchin embryos)	2–3-fold higher	Salmon *et al.*, 1984b
FRAP and video image processing[c]	$1.8 \times 10^{-3}\text{–}1.0$ (1×10^{-8})	Dextrans (Swiss 3T3)[d]	6.5-fold higher	Luby-Phelps *et al.*, 1986
Nuclear magnetic resonance spectroscopy (NMR)	$T_1 = 275 \times 10^{-3}$ $T_2 = 53 \times 10^{-3}$	Water (Artemia cyst)	~10-fold less (T_1) ~30-fold less (T_2)	Clegg, 1984
FRAP	1–2 F-IgG: (10.8×10^{-9}) F-G-actin: (7×10^{-9}) F-tubulin: (9.6×10^{-9})	Fluorescein-IgG, F-BSA, carboxy fluorescein (erythrocytes, human fibroblasts)	68-fold higher (F-IgG, F-BSA) 17-fold higher (carboxy-fluorescein)	Wojcieszyn *et al.*, 1981
Quasielastic neutron scattering (QNS)	4×10^{-12} (1×10^{-9})	Water (Artemia cyst)	4-fold higher	Clegg, 1984
Dielectric measurements (DM)	$10\text{–}25 \times 10^{-12}$ (1×10^{-9})	Water (Artemia cyst)	2–3 fold higher	Clegg, 1984
Magnetometry of twisted particles	$333\text{–}1000$[e] (0.7×10^{-6})	Maghemite (Fe_2O_3) particles (0.7 μm diameter)	1×10^{6}-fold higher	Valberg and Albertini, 1985

Table 2.2 *continued*

| Differential interference contrast light microscopy (DIC microscopy) | 22 (shortening)– 72 (elongation) (1×10^{-6}) | Microtubules | Erickson and O'Brien, 1992 |

(a) In parentheses, length scale of the probe; without parentheses, probe's relaxation time.
(b) Hydrodynamic radii of hydrated BCECF = 0.6–0.7 nm (Dix and Verkman, 1990).
(c) In Luby-Phelps *et al.* (1986) it was shown that the mobility of dextrans in the cytoplasm of Swiss 3T3 cells was inversely dependent on molecular size. The cytoplasmic diffusion coefficient decreased linearly as a function of R_g (estimated radius of gyration) up to R_g = 141 Å. In Valberg and Albertini (1985) it is shown that, as the shear rate goes down, the measured viscosity goes up and that as a general trend larger particles yield a larger relative viscosity.
(d) Average pore size of the cytoplasmic mesh = 20–40 nm Luby-Phelps *et al.* (1986)
(e) Estimated from the shear rate whose dimensions are s^{-1} Luby-Phelps *et al.* (1986).
Abbreviations: BCECF = 2,7-bis-(2-carboxyethyl)-5-(and-6-)carboxyfluorescein; DTAF = dichlorotriazinyl-aminofluorescein; FITC = fluorescein isothiocyanate; PCAOL = 2,2,5,5-tetramethyl-3-methanolpyrroline-N-oxyl; BSA = bovine serum albumin; F-BSA = fluorescein isothiocyanate-labelled BSA; F-IgG = fluorescein isothiocyanate-labelled IgG.

2.5.2 LEVELS OF EXPLANATION

According to our spatio-temporal characterization of a level of organization, the level of explanation of a biological phenomenon depends on the spatio-temporal dimensions of the processes involved in such a phenomenon. One appropriate question should be: does our perturbation method provide information about the level of organization corresponding to the phenomenon we want to explain?

To be more precise, the concepts above are exemplified by a concrete case. Let us say that we want to explain how the cell membrane invaginates to endocyte a particle, a process in which many levels of organization are implied. Let us suppose that we use a fluorescent molecular probe, that integrates to the membrane and changes its emission properties following the endocytotic process. Those changes in emission properties will reveal an average result of changes in membrane fluidity or the thermotropic mesophormism of lipids around the probe. Are those changes related to the endocytosis process? Definitely, yes. Do those changes represent an adequate level of explanation, supposing our question is how endocytosis happens? We conjecture not, since endocytosis is a supramolecular phenomenon and we are probing the membrane at the molecular level. This is true even for microenvironments, sensed by the probe within the membrane, that are dependent on the membrane's macroscopic properties. The membrane's curvature radii or the interfacial tension are properties of the whole ensemble of molecules as the trans-gauche isomerization of the hydrophobic chains of phospholipids is a molecular property (Maggio, 1985; Maggio, 1994, and references therein). Therefore, according to our question about how endocytosis happens, what we seek to understand is how the plane of the membrane bends, drastically changing its curvature radii to englobe a particle or a patch of molecules in the interior of the cell. All these phenomena that happen in the spatio-temporal 'window' of the endocytotic phenomenon are influenced by, for example, the molecular composition of the bilayer, ionic concentration or catalytic properties of the surface.

2.6 TRANSITIONS BETWEEN LEVELS OF ORGANIZATION AND CELL FUNCTION

The E_c of a process taking place at molecular, supramolecular and cellular levels is related to its respective T_r through the logarithmic form of an allometric law (Figure 2.2). Two curves with different slopes appear in a double logarithmic plot. The transition between both curves occurs at spatio-temporal dimensions of micrometres and minutes, respectively, and is referred to as the 'transition point' (Figure 2.2). Elsewhere we have suggested (Aon and Cortassa, 1993, 1994), and supported with experimental evidence (Cortassa *et al.*, 1994a), that polymerization–depolymerization of cytoskeleton components at the 'transition point' are the sort of subcellular processes that in the spatio-temporal 'window' of micrometres and minutes may induce the appearance of macroscopic coherence at the cellular level (Chapters 6 and 7). Assembly–disassembly of cytoskeleton components was shown to play the role of a coherent inducer of changes in systemic properties such as metabolic fluxes. The effect on metabolic fluxes shown in the presence of polymerized or non-polymerized microtubular protein in the physiological range (1–15 μM) was concentration-dependent.

The experimental evidence supporting the existence of transitions between levels of organization in living cells is that the inhibition of processes at lower levels inhibits the appearance of spatio-temporal macroscopic organization. It also reveals that those transitions imply discontinuities or bifurcations in the dynamic behaviour of biological processes which, depending on their position in the parametric space, may arise as self-organized macroscopic coherence (Chapter 1). The coherence may be temporal, spatial or spatio-temporal, depending on whether the analysis is of homogeneous or heterogeneous systems, respectively. The following series of examples suggests the existence of transitions between levels of organization, i.e. after affecting lower levels, the appearance of macroscopic order at higher levels is looked for.

- Spatial patterns observed in a photobiochemical system arose in an unstirred homogeneous suspension of thylakoids with an imposed linear concentration gradient of an electron acceptor. The banding pattern was shown to depend on the activity of the water-splitting system. When thylakoids were subjected to treatments known to selectively block or destroy the water-oxidation step, no spatial patterns were observed (Aon *et al.*, 1989b); see Chapter 5 for further discussion and modelling.

- During yeast glycolytic oscillations, inhibition with pyrazole of alcohol dehydrogenase II, the first enzyme of the ethanol oxidative pathway in yeast, provoked the disappearance of macroscopic oscillations in highly dense cell suspensions (Aon *et al.*, 1992). The ethanol oxidation step being a promoting factor of the oscillations at high cell densities, its inhibition quenched the oscillations.

- The third example concerns an ovomucin gel which forms a fibrillar macromolecular lattice or hydrogel with definite solvent properties and water-holding capacity (Rabouille *et al.*, 1990). The rheological properties of such a hydrogel are modified by perturbing hydrophobic or electrostatic interactions, probably reflecting changes in the geometry of the lattice (Rabouille *et al.*, 1992). Diverse macroscopic morphologies were shown by polymers upon slow dehydration at different salt concentrations. These morphologies were described by different fractal dimensions, D (Figures 2.5 and 2.6). Bound water and polymer interactions at the molecular level are likely to participate in microscopic selection of different macroscopic morphological features (Aon and Cortassa, 1994); see also Chapter 6.

These examples support the possibility of a link between the spatio-temporal scaling properties of biological processes, and the transitions between levels of organization in cellular systems that may, for example, display fractal geometry (Chapter 6).

2.7 AN ALLOMETRIC FUNCTION DESCRIBES TRAJECTORIES OF DYNAMIC SYSTEMS

The purpose of this section is to show how an allometric equation is derived from two-dimensional linear homogeneous systems of differential equations. The allometric equation is of considerable biological interest (Rosen, 1967; Sernetz *et al.*, 1985; Reiss, 1989; Pagel and Harvey, 1989) and describes the trajectories of systems in which two state variables varying simultaneously are each growing exponentially.

Let us consider the linear homogeneous dynamic system whose equations of motion are of the form:

Figure 2.5 Dendritic-like fractal patterns in an ovomucin gel/salt system. The ovomucin gel (6 mg protein ml⁻¹) was sonicated in the presence of 0.1 M NaCl. (a) Dendritic-like patterns observed in a surface equivalent to two thirds of the dried drop. (× 50). (b) Electron microscopy of dendritic-like patterns. (× 1200). Notice that a self-similar structure is apparent when (a) and (b) are compared: the same general shape is observed after a 24-fold magnification. (From Rabouille, Cortassa and Aon, 1992.)

$$\frac{dx_i}{dt} = \sum_{i=1}^{n} a_{ij} x_j \tag{2.5}$$

where a_{ij} are constants. For the more specific case of two state variables, equation 2.5 will be written as in equation 2.2.

In order to describe the trajectories of x and y we rewrite equation 2.3 for C_1 and C_2 (both positive) as follows:

$$ln\frac{x}{C_1} = \lambda t$$

$$ln\frac{y}{C_2} = \mu t \tag{2.6}$$

Combining 2.5 and 2.6 to eliminate t gives:

$$\frac{1}{\lambda} ln\frac{x}{C_1} = \frac{1}{\mu} ln\frac{y}{C_2} \tag{2.7}$$

Then the equation of the trajectories can be obtained:

(a)   (b)   (e)   (f)

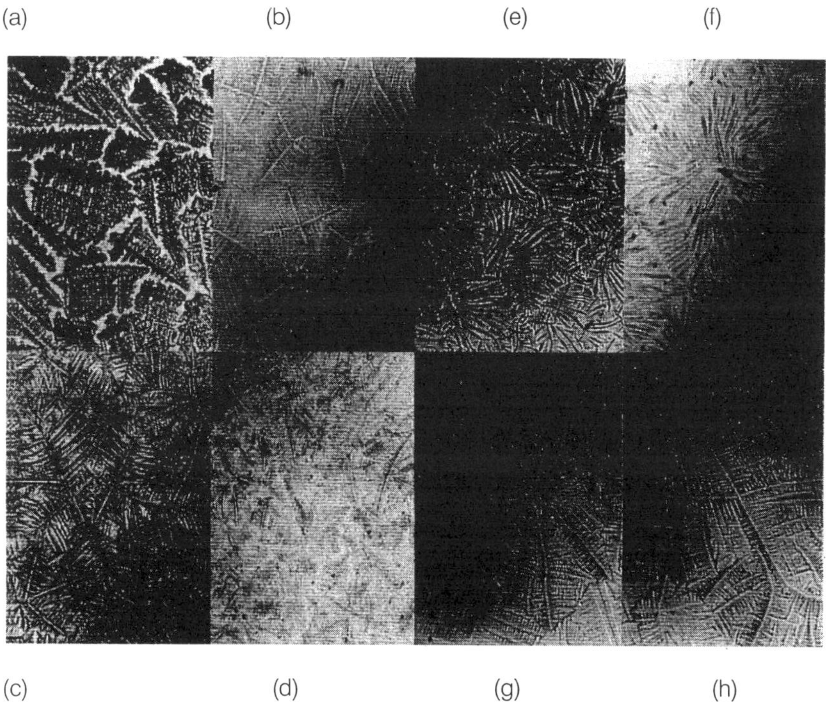

(c)   (d)   (g)   (h)

Figure 2.6 Changes in the fractal dimension, D, of dendritic-like macromolecular patterns obtained by modification of the salt and macromolecular nature of the lattice. An ovomucin gel (6 mg protein ml^{-1}) was sonicated in the presence of (a) 25 mM KI, or (b) 100 mM KCl, (c) NaBr, (d) NH$_4$Cl, and the patterns obtained as described in Figure 2.5. ($\times$ 30). Macromolecular fractal patterns obtained with (e) albumin, (f) fetuin, (g) ovalbumin and (h) starch, in the presence of 0.1 M NaCl. ($\times$ 10). (From Rabouille, Cortassa and Aon, 1992.)

$$y = \beta x^{\alpha} \tag{2.8}$$

with:

$$\beta = \frac{C_2}{C_1^{\mu/2}} \tag{2.9}$$

and:

$$\alpha = \frac{\mu}{\lambda} \tag{2.10}$$

which is the allometric equation (Rosen, 1970; Pratt, 1974). Needham found considerable evidence for the existence of allometric relations between concentrations of specific substances in growing systems and the total weight of the systems (quoted in Rosen, 1967).

The exponent of the allometric equation (2.8) is the ratio of eigenvalues (equation 2.10) which describe the stability of singular points and of trajectories in the neighbourhood of steady state solutions (Rosen, 1967; Thom, 1972; Guckenheimer and Holmes, 1983; Jeffrey, 1990). Several interpretations have been given to β (ratio of initial values: Pagel and Harvey, 1989); rate constants (Cortassa *et al.*, 1991); and α (ratio of rate constants: Pagel and Harvey, 1989; kinetic order of a reaction: Savageau, 1976). We will retain the meaning of the exponent α as the ratio of eigenvalues (μ, λ) of a Jacobian matrix (a_{ij}) arising from the description of linear homogeneous dynamic systems or linearized nonlinear systems around steady states, that describe their stability properties around singular points (Rosen, 1967; Thom, 1972; Lewis *et al.*, 1977; Kaufmann, 1989).

2.8 AN ALLOMETRIC INTERPRETATION OF THE RELATIONSHIP BETWEEN THE CHARACTERISTIC SPACE DIMENSION AND THE CHARACTERISTIC RELAXATION TIME OF BIOLOGICAL PROCESSES

There is an 'isomorphism' between the equations describing the trajectories followed by dynamic homogeneous systems of two state variables described by ODEs (equations 2.2, 2.3, 2.6–2.8) and equations 2.12–2.15 below, relating the characteristic space dimension with the characteristic relaxation time. In order to understand this isomorphism, let us assume the existence of an independent variable – a sort of scaling factor, L, that increases with the level of organization such that it provides an arbitrary quantitative scaling of a level of organization. Both the characteristic space dimension, E_c, and the characteristic relaxation time, T_r, are functions of the independent variable, L, as follows:

$$\frac{dE_c}{dL} = \gamma E_c ; \quad \frac{dT_r}{dL} = \rho T_r \tag{2.11}$$

Equations 2.11 constitute a linear system of ODEs accepting the general solution:

$$E_c = C_1 e^{\gamma L}$$
$$T_r = C_2 e^{\rho L} \tag{2.12}$$

in which γ and ρ are necessarily positive since both T_r and E_c are variables that grow exponentially with the level of organization, as shown in Figure 2.2 and Table 2.1. This property agrees with the fact that T_r and E_c are 'unstable' with respect to the independent variable L. In other words, as a process 'travels' to a higher level of organization, T_r and E_c change to larger values.

At this point, note the formal resemblance between the equations describing the dependence of T_r and E_c on the variable L (equations 2.12), and a trajectory of a state variable at an unstable steady state (characterized by an eigenvalue with a positive real part) as a function of time (equation 2.6).

By dividing equations 2.11 the scaling factor, L, is eliminated as follows:

$$\frac{dE_c}{dT_r} = \frac{\gamma E_c}{\rho T_r} \tag{2.13}$$

the integral form of which gives rise to an allometric relation:

$$E_c = \frac{C_2}{C_1^{\gamma/\rho}} T_r^{\gamma/\rho} \tag{2.14}$$

with C_1 and C_2 having a meaning similar to that described in equation 2.9. Equation 2.14 may be written in logarithmic form:

$$\frac{1}{\rho} ln \frac{T_r}{C_1} = \frac{1}{\gamma} ln \frac{E_c}{C_2} \tag{2.15}$$

Equations 2.14 and 2.15 are formally similar to equations 2.8 and 2.7, respectively, emphasizing the likeness between the equations that describe the trajectories of dynamic homogeneous systems of two state variables and those relating the characteristic space dimension to the characteristic relaxation time.

2.9 AN EXPERIMENTAL INTERPRETATION OF COHERENCE FURTHER FROM THE 'TRANSITION POINT'

An allometric relationship between the characteristic spatial dimension, E_c, and the characteristic relaxation time, T_r, implies that these quantities grow exponentially with the level of organization (Figure 2.2). The exponent of the allometric relationship is a ratio of values γ and ρ in equation 2.12 depicting the sensitivity of E_c and T_r to variations in the level of organization; γ and ρ correspond to the exponents of the equation describing how E_c and T_r vary with the level of organization (equations 2.12 and 2.14). The faster a given dimension (spatial or temporal) grows with the level of organization, the larger is the value of its exponent (γ or ρ, respectively). In Figure 2.2 at lower levels of organization, i.e. before the 'transition point', E_c grows roughly 1/3 with respect to T_r ($\alpha = 0.38$). This means that the spatial coordinates of processes, namely at molecular and supramolecular levels of organization, grow slower with respect to their relaxation times. Further from the transition point a drastic change in the slope is verified (α = 1.88), the meaning of which is clear: the spatial coordinates of processes at cellular, tissue and organismal levels grow much faster than the relaxation time. We have interpreted this change in the slope as the spatio-temporal 'window' (micrometres and minutes) at which macroscopic coherence emerges in cellular processes (Aon and Cortassa, 1993). This macroscopic coherence at the cellular level might be driven chiefly by processes exhibiting conspicuous increments of their spatial coordinates.

For instance, under conditions at which macroscopic coherence emerges, i.e. further from the transition point, E_c undergoes a larger change (as shown by the larger slope in Figure 2.2) than T_r upon a shift from subcellular to cellular or supracellular levels. As shown in one of our examples, single yeast cells oscillate synchronously in homogeneous suspensions, resulting in a macroscopic coherent oscillation at the cell population level with an oscillatory period of 1 to 2 minutes (Figure 2.7a) (Aon *et al.*, 1991, 1992). At higher cell concentrations a decay in the amplitude of the oscillations revealed a coherent supracellular synchronization phenomenon which was shown to be dependent upon cellular metabolism keeping yeast cells longer in phase among them (Figure 2.7b) (Aon *et al.*, 1992). We show how coupled enzymatic reactions (glycolysis) exhibiting autocatalysis at lower levels of organization, i.e. E_c of nanometres with T_r in the order of a few milliseconds, are involved in the appearance of a dynamically self-organized, oscillatory phenomenon at the cellular level. The spatio-temporal dimensions in enzymatic reactions are nanometres (E_c) and a few milliseconds (T_r), whereas at the cellular levels E_c corresponds to micrometres (the diameter of a yeast cell) and T_r corresponds to minutes (the oscillatory period). This coherent cellular behaviour becomes supracellularly organized as the cell concentration is increased (E_c of a centimetre, i.e. the side length size of the test tube containing the synchronized cell suspension, and T_r of minutes, i.e. the larger oscillation period of Figure 2.7a, i and ii). Therefore, at the higher (supracellular) levels of organization and further from the transition point, E_c shows a higher sensitivity (from μm to cm) than T_r (from s to min) to the change in level of organization given by the self-organized phenomenon.

In other words, further from the transition point the temporal relaxation of the dynamics of biological processes implied would be restricted in the long range to the region where spatial patterns appear. This restriction would provoke that 'essential' dynamic variables remain bounded or that attractors reduce to small ones. The latter is in agreement with synchronization properties: for the same processes occurring wider apart in a spatial sense, they will be able to settle their dynamics through 'clamping' of their variables to a bounded (probably

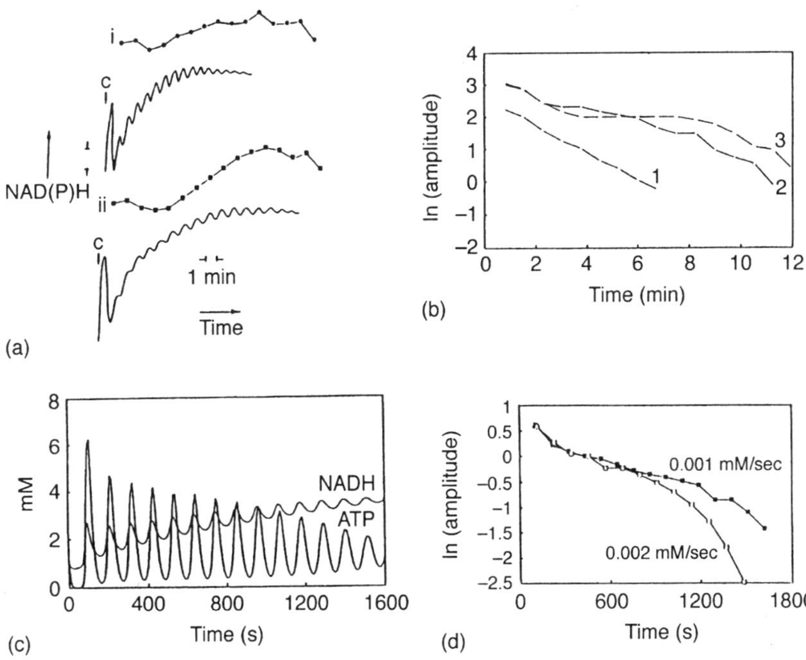

Figure 2.7 Transition from molelcular to supracellular levels of organization in a spatially homogeneous cellular system: cell-to-cell interaction during glycolytic oscillations in yeast cell suspensions. Yeast cells were grown, harvested and starved, and the kinetics of NAD(P)H in yeast cell suspensions at different cell densities was followed as described in Aon et al. (1991). (a) Oscillations were triggered by addition of 10 mM KCN after a glucose pulse (20 mM) at 10^7 cells ml^{-1} and 15 mM KCN at $3-5 \times 10^7$ cells ml^{-1}. Plots (i) and (ii) were obtained by subtracting the natural log of the amplitudes of curve 1 from curves 2 and 3 in (b). (b) The amplitude decay of oscillations was analysed as a function of time and cell concentration, assuming a single exponential decay ($A_{NADH} = A_o e^{-\gamma t}$, where γ is the damping factor, A_o is the initial amplitude of the oscillating fluorescence signal and t is time). (c) Numerical simulation of the experimentally observed oscillations. The ODEs system described in Appendix 12A (equations 12.10–12.19) was integrated with the following parameter values: V_{in} (mM s^{-1}) = 0.12; V'_{in} (mM s^{-1}) = 0.001; k_1 (mM^{-1} s^{-1}) = 0.025; k_3 (mM^{-3} s^{-1}) = 0.05; k_5 (s^{-1}) = k_7 (mM^{-1} s^{-1}) = 0.1; k_8 (mM^{-1} s^{-1}) = 6.25×10^{-4}; k_9 = 0.05 (s^{-1}); k_{11} (mM^{-1} s^{-1}) = 0.01; τ = 0.83; $\Delta\mu/\Delta p°$ = 0.01. The kinetic parameters of the proton pump were: K_M = 2 mM; V_M = 0.5 mM s^{-1}. The total nucleotide (C_A) and phosphate pool (P_t) were both 10 mM, while the pyridine nucleotide pool (C_N) was 4 mM. The initial values of the state variables (in mM) were the following: Glc = 0.157; ATP = 1; I = 0.0031; NADH = 1; Pyr = 0.394; EtOH = 1; AcCoA = 0.01. (d) The kinetics of amplitude decay of oscillations in the numerical simulations was performed as described in (b). The two curves correspond to different ethanol uptake rates (mM s^{-1}) by the cells. NAD(P)H scale (between arrows) is 15 μM. (Reprinted from Aon and Cortassa, 1993, by permission of Kluwer Academic Publishers.)

small) range of variation. This notion also supports homeostatic properties of the integrated dynamic system.

Several authors have observed that the remarkable stability properties of developing organisms resemble those exhibited by trajectories of general dynamic systems moving towards stable attractors (von Bertalanffy, 1987; Rosen, 1970; Ashby, quoted by Kauffman, 1989). A cell being a dissipative dynamic system, when forced, its dynamics may either become unstable or, if bounded (stable), the dynamic behaviour may approach that of an attractor. Thus, the existence of an attractor is related to the stability of the dynamic trajectories. Attractors are of obvious importance since they represent the asymptotic long-term behaviour of dynamic systems (Kauffman, 1989; Chapter 1). Global cellular responses such as whether to stop or to resume cell division, or to choose between divergent developmental paths, could be induced by changes in environmental parameters. This topic is further explored in Chapter 6 as applied to supramolecular networks supporting catalysis and in Chapters 9 and 10 for cell growth, division and development.

2.10 THE ALLOMETRIC POWER LAW AS IMPLIED IN GROWING SYSTEMS

Given two state variables N and M that in growing systems may represent N, a population of microbes, and M, a population of macromolecules synthesized in a constant proportion to the microbial biomass, the growth laws of N and M are proportional to their magnitude at any instant:

$$\frac{dN}{dt} = k_1 N$$
$$\frac{dM}{dt} = k_2 M$$

(2.16)

Equations 2.16 have solutions:

$$N(t) = N_0 e^{k_1 t}$$
$$M(t) = M_0 e^{k_2 t}$$

(2.17)

where N_0 and M_0 are the initial population.

Systems which obey equation 2.17 are said to grow exponentially. Now let N and M be two functions that satisfy equations 2.16. It is possible to eliminate time by dividing equations 2.16:

$$\frac{dN}{dM} = \frac{k_1 N}{k_2 M}$$

(2.18)

the integral form of which gives rise to an allometric relation:

$$N = \frac{M_0}{N_0^{\,k_2/k_1}} M^{k_2/k_1} = \alpha M^\beta \qquad (2.19)$$

Thus, as far as growing systems are concerned, whenever any two state variables of the system are growing exponentially they are necessarily related by an allometric law (Rosen, 1967) (equations 2.8 and 2.14).

Allometric growth laws have been analysed to explain morphogenesis of complex systems, where morphogenesis is conceived as a differential growth of some parts of an organism (Savageau, 1979). Specific cases were considered, such as oscillations in allometry or continuously changing allometry. Interestingly, Savageau analysed the case where sharp breaks in allometry occurring at critical times of organization, such as metamorphosis, can be described. Then groups of variables may exhibit an allometric relationship with a different slope with respect to that exhibited by the allometric relationship ruling another group of variables (Savageau, 1979). These sharp breaks could be reminiscent of the 'transition point' found in the double allometric relationship between the relaxation time and characteristic space of many biological processes (Figure 2.2; Aon and Cortassa, 1993).

Figure 2.8 shows that, in a semilogarithmic plot, the increase in biomass or macromolecules is linearly related to the doubling time of the cell population. This pattern of synthesis of the main macromolecules is referred to as **exponential**. An exponential rate of mass synthesis during the cell division cycle means that the absolute rate of synthesis is continuously increasing; the amount of mass synthesized per unit time is not constant throughout the division cycle (Cooper, 1991). According to equations 2.16 and 2.17, if two variables that are growing exponentially, such as biomass and a macromolecule, are plotted in a double logarithmic plot a linear (allometric) relationship appears according to equation 2.19. Thus, a defined relationship apparently exists between the whole (cell) and each of its constituents (macromolecules). Our finding of an allometric relationship between spatio-temporal dimensions of several (sub)cellular processes implies that biological systems scale their functioning in space and time in an exponential way (Figures 2.2 and 2.4). Furthermore, this finding suggests that the allometric relationship is not an *ad hoc* mathematical model but it may reflect a principle of cellular organization (section 6.7).

The allometric constants α and β are the ratio of initial values and the ratio of a pair of constants, respectively. The net effect of the formal transition from the rate equations 2.17 to 2.19 is to make the allometric

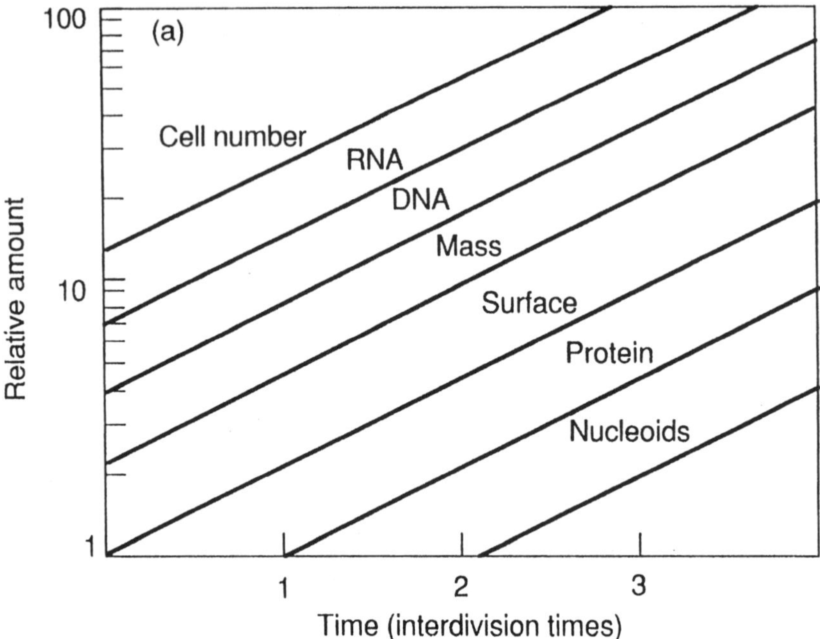

Figure 2.8 Exponential bacterial growth. When bacteria are grown at different growth rates by adding different carbon sources or different supplements of amino acids, vitamins or nucleosides, the macromolecular composition of the cells changes. The abscissa gives the doublings per hour, with 3 indicating rapidly growing cells (20 min doubling time) and 1 denoting slower cells (60 min doubling time). The differences in slope indicate that the ratio of the different components changes at each growth rate. (Reproduced from Cooper, 1991, *Bacterial Growth and Division*, by permission of Academic Press.)

formalism explicitly independent of time. However, implicitly equation 2.18 depends on time through the rate constants of the system, e.g. the growth rate.

2.10.1 THE LOGISTIC EQUATION AND THE LOGISTIC GROWTH OF POPULATIONS

Differential equation models of population dynamics, or of growth of populations, whether ordinary, delayed, partial or stochastic, imply a continuous overlap of generations (Murray, 1989). Many species have no overlap whatsoever between successive generations and so population growth is in discrete steps. These steps can be quite short for primitive organisms, in which case a continuous (in time) model may be a reasonable approximation.

Equation 2.20 depicts the differential form of the logistic equation (Edelstein-Kashet, 1988; Murray, 1989; Stewart and Golubitsky, 1992):

$$\frac{dN}{dt} = rN\left(1 - \frac{N}{K}\right) \tag{2.20}$$

which models the population growth continuously in time. Autocatalysis is given by the first term on the right-hand side and nonlinearity by the second term proportional to N^2.

The expression $N(t) = rN(1 - N/K)$ is an estimate for the next-generation population, in terms of the present population. In this model, the *per capita* birth rate is $r(1 - N/K)$, that is, dependent on N. The constant K is the carrying capacity of the environment, usually determined by the available sustaining resources (Murray, 1989; Stewart and Golubitsky, 1992). When the population is low, and there is no competition for space or food, each generation grows relative to the previous one by a factor of r. Then, at very low levels the population grows roughly exponentially: $N(t) = rN$, which represents exponential growth. But when the population becomes larger, competition cuts down the rate of growth by an amount rN^2, and this becomes more and more important the larger the population N becomes. The parameter r is a measure of the rate constant at which the size of the stable steady state population is reached; that is, it is a measure of the dynamics, i.e. the time scale of the response of the model to any change in the population.

For larger step lengths between successive generations, models must relate the population at time $t + 1$, denoted by N_{t+1}, in terms of the population N_t at time t. The analogous difference equation of 2.20 is:

$$N_{t+1} = rN_t\left(1 - \frac{N_t}{K}\right), \quad r > 0, \quad K > 0 \tag{2.21}$$

which models population growth in discrete time stages, i.e N_{t+1} is an estimate for the population in the next period in terms of the population at the present period. If $N_t > K$, then $N_{t+1} < 0$, which is an obvious drawback of this model. A more realistic model should be such that for a large N_t there should be a reduction in the growth rate but N_{t+1} should remain non-negative:

$$N_{t+1} = N_t \exp\left[r\left(1 - \frac{N_t}{K}\right)\right], \quad r > 0, \quad K > 0 \tag{2.22}$$

Since t increases by discrete steps there is, in a sense, an inherent delay in the population to register change. Because difference equations and delay differential equations share discrete time intervals for describing growth phenomena, the model depicted in equation 2.22 allows both types of equation to relate. Equation 2.22 is the logistic map that is studied on the unit interval and leads to very complicated dynamics, or eventually to chaos, depending on the parameter r (May, 1976; Hess and Markus, 1987).

2.11 A POWER LAW DESCRIBES THE TRANSITION FROM LOCAL TO GLOBAL CONNECTEDNESS IN PERCOLATION CLUSTERS

Percolation is a process of transport through a lattice where diffusion is restricted to certain directions or, in other words, where randomness is frozen (Feder, 1988). For a further discussion, see Chapter 6.

In percolation processes there is a characteristic percolation threshold, P_c, below which the spreading process is confined to a finite region of the lattice. The percolation probability, $P(p)$, is defined as the probability that a fluid injected at a site chosen at random will wet infinitely many pores (Plate 1). For low concentrations, p, of wet pores we find that $P(p)$ is negligible. As p increases the probability of belonging to the largest cluster increases drastically near the percolation threshold ($P_c = 0.593$, for a quadratic lattice) and then $P(p)$ increases almost linearly to 1 as $p \rightarrow 1$. The critical probability, P_c, is defined as the largest value of p for which $P(p) = 0$. Thus, by definition we have $P(p) = 0$ for $p < P_c$.

Near the critical probability, P_c, as p is increased, the probability of belonging to the largest cluster increases drastically. The percolation process undergoes a transition from a state of local connectedness to one where the connections extend indefinitely (Feder, 1988).

The percolation probability vanishes as a power law near P_c:

$$P(p) \sim (p - P_c)^{\beta}, \text{ for } p > P_c, \text{ and } p \longrightarrow P_c \qquad (2.23)$$

This is analogous to what happens at magnetic phase transitions, where the local order of dipoles increases in range as the temperature is lowered to the transition temperature, T_c, of the material. Below T_c, the magnetic moments are aligned, on the average, throughout the sample and we have a magnet (Feder, 1988).

A consequence of the percolation cluster nature of the cytoplasm of living cells is on the transition from microscopic (local) to macroscopic (global) behaviour as a sort of phase transition which may be described by a power law near the percolation threshold (Chapter 6).

2.12 THE POWER LAW: MORE THAN A MERE MATHEMATICAL ISOMORPHISM?

The power law describes the relationship between two exponentially growing populations (e.g. microorganisms, molecules), two state variables in a dynamic system, the phase transitions in the probability of fluids to wet pores in percolation clusters, phase transitions in changes of lipid mesomorphism or magnets, and the relationship between E_c and T_r of biological processes at different levels of organization. The power law function, $f(t) = bt^\alpha$, satisfies the homogeneity relation:

$$f(\lambda t) = \lambda^\alpha f(t) \qquad (2.24)$$

for all positive values of the scale factor λ. Functions that satisfy this relation are said to be **scaling**. Even though not all power laws represent fractals, the latter also show scaling given by the fractal dimension, D.

This mathematical isomorphism reveals a more profound characteristic of processes in which thousands or millions of particles involved (e.g. molecules, cells) are spatio-temporally coordinated. Also, the combination of exponentially growing and decaying processes, as in logistic growth, may give rise to chaos (see above and Chapters 1 and 3).

The subcellular processes taking place at different levels of organization show spatial as well as temporal scaling. At instabilities of their dynamics, subcellular processes bifurcate exhibiting transitions between levels of organization (Cortassa *et al.*, 1991; Aon and Cortassa, 1993).

The fact that widely different levels of organization (i.e. a microbial culture or a chemical reaction) may be described with similar phenomenological laws (i.e. the Monod–Michaelis–Menten or power law equations) stresses the existence of a functional self-similarity between biological processes operating at different levels of organization (Chapter 8). Independently of mechanistic details, the functional dependence existing between variables and parameters is isomorphic for either an enzymatic reaction or the complex chemical reaction to synthesize a microorganism, i.e. microbial growth. This is valid at two widely differing spatio-temporal coordinates: molecular and supracellular, respectively. This isomorphic functional transformation is the basis of the concept of functional self-similarity and may be postulated as an additional link between the spatio-temporal scaling exhibited by physiological processes and geometric self-similarity which is the main feature that characterizes fractal objects.

The analogy between the concepts of functional self-similarity and geometric self-similarity is an additional aspect of the link between fractal organization and spatio-temporal scaling of biological processes. This topic will be further treated in Chapter 6.

2.13 BIFURCATIONS AND LEVELS OF ORGANIZATION

The dynamic organization concept implies that, at the cellular level, coherence arises from instabilities in the dynamics of biological processes taking place at different levels of organization. Those instabilities coincide with bifurcations (defined in Chapter 1) in the spatio-temporal behaviour of a biological process (i.e. glycolysis, cell cycle, cell division, cell differentiation), giving rise to entirely new patterns of behaviour, such as the appearance of asymmetries or the emergence of self-organized spatio-temporal patterns.

Biological phenomena such as sporulation (cell differentiation) or cell division, e.g. in yeast, reveal 'massive' or multidimensional bifurcations at several levels of organization. Strictly speaking, from a dynamic point of view, a set of parameters is defined in a system with more than one control parameter (Abraham and Shaw, 1987). Thus, a bifurcation may be a massive change in the dynamics of cellular processes at several levels of organization following changes in several parameters. Figure 1.8 emphasizes the multidimensional character of functional regulation in cells, i.e. there is a concomitant balance of two parameters, C_0 and K_{pol} (Chapters 6 and 7).

Rhythms as a fundamental property of biological systems 3

Oscillations are ubiquitous temporal patterns observed in biological systems that may occur in different time scales; the periods range from seconds, as in the case of excitation and nerve pulse propagation, to a month or a year, as in the female ovulatory cycle or flowering cycle in higher plants (Lloyd and Rossi, 1993). The spatio-temporal dimensions at distinct levels of organization of biological phenomena, including oscillatory processes, were presented in Table 2.2.

From the viewpoint of kinetics, the conditions that should be met to get such a coherent behaviour is through the coupling of at least two variables in a scheme involving kinetic nonlinearities (such as autocatalysis: Chapter 1). Additionally, it has been proposed that the central oscillator of the circadian clock may be a controlled chaotic attractor (Lloyd and Lloyd, 1993). The latter is founded on the extreme sensitivity of chaotic systems to tiny perturbations – the so-called 'butterfly effect'. This property has been exploited both to stabilize regular dynamic behaviours and to direct chaotic trajectories rapidly to a desired state (Shinbrot *et al.*, 1993; Weiss *et al.*, 1994; section 3.4).

Kinetic models implying spatial homogeneity can account for long-term oscillations (i.e. longer than a few minutes) when diffusion is coupled to reaction by means of compartmentation, or delays are imposed on the system. In that case, longer oscillatory periods, in the order of hours or a day, may be achieved.

3.1 SOME DEFINITIONS

A **steady state**, which should not be confused with an equilibrium state, refers to a constancy in a certain property of a system variable (e.g. concentration, length) with time (Figure 1.1). Steady state is, in fact, a mathematical concept which may be practically detected in, for example, a system of ODEs, as an infinitesimal or very low value of the derivative of a state variable (section 1.1). For instance, in the ODEs of Lorenz (1963) describing the weather, the state variables are related to the temperature difference between ascending and descending currents, to the intensity of convective motion and to the distortion of the vertical temperature profile from linearity; or in the glycolytic model described in Chapter 1 the variables described by ODEs are the metabolite concentrations (equations 1.21–1.24).

Oscillations are a dynamic behaviour that describes a cyclic change in time of a system's state variable with a constant periodicity, i.e. waveform and period (Hess and Markus, 1987; Friesen *et al.*, 1993; Chapter 1). **Complex oscillations** (several periods) differ from **noise** or **chaos** (random variation in period and amplitude). Oscillations may be damped or sustained, according to whether the kind of attractor in the state space is a stable focus or a limit cycle (Figure 1.7a,f).

3.2 RHYTHMS

A **rhythm** is a variation of a property in time with a certain periodicity. Rhythms have very different time spans: seconds, as in action potentials and nerve impulse; minutes, as in glycolytic oscillations and cAMP pulses in *Dictyostelium discoideum*; hours, as in the cell cycle; 24 hours, as in circadian rhythms in *Euglena* and *Drosophila*; and weeks or months, as in the female cycle (ovulation) and in the number of blood cells in a patient with CGL (chronic granulocitic leukaemia) (Edmunds, 1984, 1988; Glass and McKey, 1988).

A very interesting and challenging piece of evidence is the dynamic behaviour of the product of *per* gene in *Drosophila* which is involved in the appearance of oscillations in different temporal scales such as seconds, in male courtship song frequency, or hours or days (circadian rhythm) in locomotor activity and the eclosion of adult individuals from pupae. A positive feedback on *per* RNA level by its own proteic product has been described (Hardin *et al.*, 1990). However, more kinetic data are needed to be able to design a model accounting for the observed behaviour as well as to identify the other 'gears' in the mechanism of this circadian clock.

A quantitative analysis of a biological oscillator suggests answers to the following questions (Friesen and Block, 1984; Friesen *et al.*, 1993):

- What are the biological variables involved in the dynamics of the oscillator?

- How do these variables interact?
- Can these interactions lead to oscillations?

The analysis of biological oscillators encourages the study of the complete system instead of dissecting its components, because oscillations exist by virtue of interaction between the elements.

3.2.1 CIRCADIAN RHYTHMS

Circadian rhythms (CRs) are one of the more conspicuous manifestations of periodicity in biological organization (Figure 3.1). A CR is an oscillation in biochemical, physiological or behavioural variables, or all of them, with the following basic features (Dunlap, 1990):

- In nature it has a period length of exactly 24 h (i.e. it is entrained by the diurnal cycle of warmth/light and cool/darkness), but the oscillation can continue under constant environmental conditions with a period that is close but usually not exactly equal to 24 h.
- The rhythm can be reset by a brief interruption of the constant regime.
- The period length of the oscillation is relatively independent of temperature within the physiological range.

As shown for *Euglena gracilis* Klebs (Figure 3.1), several activities comprising different levels of organization (Chapter 2) are entrained by circadian rhythmicity.

Circadian organization is widespread in plant and animal kingdoms in either unicellular or multicellular organisms and includes, for example, photosynthetic capacity, cell division, organelle shape and ultrastructure (chloroplasts) and enzymatic activities. Edmunds (1984) proposed several features to characterize circadian rhythms: a ≈ 24 h period, ubiquity, entrainability, persistence after the periodic stimulus has ceased, phase-shiftability, modulation of free-running period by light intensity, temperature-compensation and genetic basis. It has been hypothesized that adaptive features by organisms related to, say, solar periodicities have been developed as endogenous, autonomously oscillatory biological clocks (Edmunds, 1984).

Circadian systems contain at least three elements: an input pathway or set of input pathways, a circadian pacemaker and an output pathway or set of output pathways (Takahashi, 1993). Input and output pathways of the circadian clock within each organism appear to be specific, whereas the core mechanism of the pacemaker is apparently fundamentally

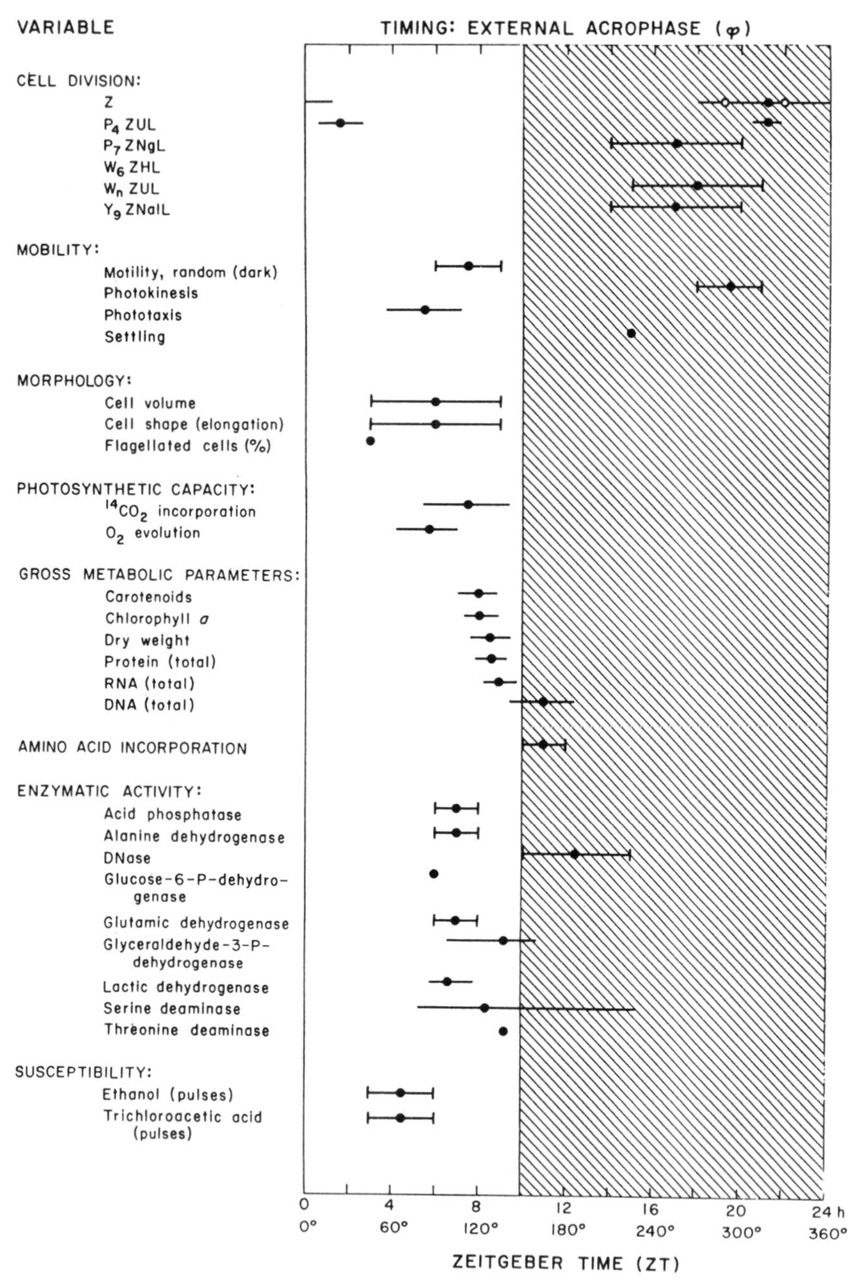
CIRCADIAN SYSTEM OF EUGLENA
VARIABLE
TIMING: EXTERNAL ACROPHASE (φ)
CELL DIVISION:
Z
P4 ZUL
P7 ZNgL
W6 ZHL
Wn ZUL
Y9 ZNalL
MOBILITY:
Motility, random (dark)
Photokinesis
Phototaxis
Settling
MORPHOLOGY:
Cell volume
Cell shape (elongation)
Flagellated cells (%)
PHOTOSYNTHETIC CAPACITY:
14CO2 incorporation
O2 evolution
GROSS METABOLIC PARAMETERS:
Carotenoids
Chlorophyll a
Dry weight
Protein (total)
RNA (total)
DNA (total)
AMINO ACID INCORPORATION
ENZYMATIC ACTIVITY:
Acid phosphatase
Alanine dehydrogenase
DNase
Glucose-6-P-dehydro-
genase
Glutamic dehydrogenase
Glyceraldehyde-3-P-
dehydrogenase
Lactic dehydrogenase
Serine deaminase
Threonine deaminase
SUSCEPTIBILITY:
Ethanol (pulses)
Trichloroacetic acid
(pulses)
0   4   8   12   16   20   24 h
0°   60° 120° 180° 240° 300° 360°
ZEITGEBER TIME (ZT)

similar in all organisms (Takahashi, 1993). Some circadian rhythms found in nature are analysed in the following paragraphs.

Unicellular organisms can express circadian rhythms (Lloyd *et al.*, 1982). In *Euglena*, circadian rhythms of NAD^+ appear to occur. These diurnal oscillations were attributed to NAD kinase and NADP phosphatase. With the use of pulses of specific inhibitors it could be determined which elements were involved in the circadian rhythmicity, depending upon the moment of the cycle in which the inhibitors were applied (Lloyd *et al.*, 1982). A satisfactory modelling of the state variables involved in the *Euglena* circadian rhythm is a prerequisite for the thorough understanding of the functioning of the clock but has not yet been attempted.

In *Drosophila melanogaster* a circadian rhythm of the genic product of the *per* locus related to locomotion activity has been described (Baylies *et al.*, 1987; Hardin *et al.*, 1990). Mutants lacking the product of the *per* gene are viable but either display arrhythmic locomotor activity or do not show rhythms in the eclosion for the emergence of adults from pupae and in the male courtship song. An interesting feature of the *per* gene that encodes for a mRNA of 4.5 kb is its involvement in a mechanism of periodical function spanning very different temporal scales. According to immunoelectron microscopic analysis, *Per* has been shown as a predominantly nuclear protein in the lateral neurons and the glial cells of the adult central nervous system of *Drosophila* (Ewer *et al.*, 1992; Liu *et al.*, 1992). The expression of *per* itself is circadian, and both *per* mRNA abundance and *Per* protein levels oscillate, both rhythms being interdependent (Takahashi, 1993). The *Per* protein rhythm appears to be regulated at two levels: transcriptional and post-transcriptional. The peak of the *per* RNA cycle precedes the peak of the *Per* protein cycle by about

Figure 3.1 (see facing page) Acrophase chart for *Euglena gracilis* showing the time relations among circadian rhythms. The *E. gracilis* system provides an excellent case for temporal differentiation: a large number of diverse behavioural, physiological and biochemical activities are partitioned along the 24 h time axis. This circadian time structure is illustrated by the acrophase chart, which provides a convenient way of indicating time relationships of periodicities both to a synchronizing light–dark (LD) cycle (or other Zeitgeber) and to each other during either entrainment or free-running conditions. The external acrophase relates the peak of the fitted curve to the onset of light in a synchronizing LD cycle and to subjective dawn in continuous darkness (DD) or dim continuous illumination (LL) (when light would have come on had the LD cycle been continued). LD are entraining and LL are for free-running conditions. Zeitgeber is environmental time cue. (Reproduced from Edmunds, 1988, *Cellular and Molecular Bases of Biological Clocks*, with permission of Springer-Verlag.)

6–8 h. It appears that *Per* acts as a negative regulator of its own transcription (Takahashi, 1993). In terms of dynamic organization, *per* may be contributing as a bifurcating parameter inducing different dynamics, as in the case of phosphodiesterase and adenylate cyclase activities for *Dictyostelium discoideum* (Goldbeter and Segel, 1980) (section 2.2).

A useful dictum that can be extracted from the observations about circadian rhythmicity is that an effector (e.g. the product of an oncogene, a growth factor) may influence the relative value at which some relevant physiological parameter reaches a bifurcation point in the biological system under consideration, inducing phase shifts or changes in period. Under those conditions, it would be worth investigating not only which are the 'gears' of the clock and what are the differences between transformed and normal cells but also what kind of dynamics could arise from the interactions and coupling between those 'gears'. Hypothetically, the process of oncogenic transformation may be viewed as a shift in the relative position of a bifurcation point, i.e. a change in the sensitivity towards regulators.

(a) Analysis of circadian oscillators at the organ, cellular and subcellular levels

The elucidation of the nature of circadian oscillators has been attempted at the level of organ, cellular or subcellular function. The cellular nature of circadian oscillators in multicellular organisms has only recently been established. Evidence available indicates that circadian rhythms can be expressed by primary cell cultures derived from the pineal gland of birds and reptiles, the hypothalamic suprachiasmatic nuclei of rodents and the retina of amphibians and birds (Takahashi, 1993, and references therein)

The high frequency ticks of the short-period biochemical oscillations have been considered by workers in the field either as the escapement for low-frequency circadian clocks or not closely related to circadian phenomena (Ehret *et al.*, 1973; Edmunds, 1988).

Feedback loops in energy supply and intermediary metabolism, as in the yeast glycolytic oscillator or the cAMP oscillations of *D. discoideum*, have been demonstrated as mechanisms able to generate periodicity in the ultradian domain (periods <10 min) (Lloyd *et al.*, 1982; Aon *et al.*, 1991; Lloyd and Stupfel, 1991). The ultradian level or metabolic domain (periods in the order of minutes) appears as dissociated from the circadian one. There is instead a halfway in the ultradian level, that is given by rhythmicities whose period, although still ultradian, is in the order of several hours. These rhythms comprise regulatory circuits involving transcription and translation and, therefore, extend into both the cell cycle (genetic) and circadian domains, i.e. periodic enzyme synthesis (Lloyd *et al.*, 1982; Edmunds, 1988).

(b) Modelling of circadian oscillators

One key question in this subject may be formulated as follows: how is the relatively long period of circadian rhythms generated? The answer to this question has been attempted at a dynamic and a genetic level. The former approach was inspired by the notion of frequency reduction mechanisms in coupled oscillators (Winfree, 1967; Pavlidis, 1969), while the latter approach was exemplified by the postulated existence of accurate 'clocks' at the cellular level (Ehret and Trucco, 1967; Edmunds, 1988).

Concerning frequency reduction, a population of weakly coupled oscillators with slightly different frequencies tends to synchronize to a common frequency that is not very different from that observed in individual, non-interacting oscillators (Winfree, 1967). Another approach (Pavlidis, 1969) stressed the strength of the coupling between individual oscillators as a possible frequency-reduction mechanism. When coupled in an inhibitory manner, at higher strength of the coupling, oscillators are mutually entrained to a new frequency much lower than the one at which they would have oscillated individually. A cluster of about 100 neurons of the basal retina of the marine mollusc *Bulla*, single-cell cultured, showed a circadian rhythm in membrane conductance. Isolated *Bulla* neurons are sufficient for the generation of circadian rhythms. Although intercellular communication is not necessary for the generation of the rhythm, it would play an important role in mutual coupling and synchronization of the oscillators' population to maintain a coherent output (Michel *et al.*, 1993, quoted by Takahashi, 1993). Synchronization in the dynamics of cells or macromolecular populations is involved in the appearance of macroscopic oscillations (Mandelkow and Mandelkow, 1992; Aon *et al.*, 1992) (Chapters 6 and 12; Figure 12.16).

The notion of cellular 'clocks' is based on evidence indicating that eukaryotic or prokaryotic organisms tend to divide at more or less constant intervals (Edmunds, 1988). The so-called **chronon model** is based on a circadian time-keeping postulate of transcriptional regulation. A genetic algorithm of linear sequential replication – especially linear sequential transcription, cistron by cistron, of a very long polycistronic unit termed the **chronon** – provides a cell with a remarkably simple interval-measuring and event-enumerating device (or 'clock') (Ehret *et al.*,

1973). According to the chronon model, within every cell there exist hundreds of replicons, called chronons, each of which comprises a polycistronic strand of nuclear (or organellar) DNA some 200 to 2000 cistrons in length (reviewed in Edmunds, 1988). The transcription rate would be limited by feedback loops in which a given cistron would not be transcribed before the synthesis of a specific precursor catalyst or enzyme, the synthesis of which, in turn, depends on the presence of an RNA transcript from the most recently read cistron. Thus, in the chronon model a precise sequence of chained events will act as a 'pacemaker' of the circadian clock.

According to the chronon model (Ehret and Trucco, 1967) the presence of multiple replicons in eukaryotic chromosomes has permitted in all eukaryotes the preservation of naturally selected primitive polycistronic sequences acting as escape mechanisms for the circadian period. In prokaryotes or eukaryotes growing exponentially, these 'clocks' are extremely temperature-dependent and unless the temperature is constant they are poor time-keepers.

Under the slow mode of growth, the **infradian mode,** in all eukaryotes natural selection has pruned the chronon to precisely the number of cistrons to provide the circadian period (~24 h). Ehret and Trucco (1967) suggested that the relative temperature independence of circadian outputs in the infradian mode of growth could be accounted for by a switchover mechanism. This 'switchover' in cellular regulatory mechanisms would occur because of mass limitations on net reaction rates taking place during the ultradian mode to diffusion limitations during the more nutritionally stringent conditions of infradian growth.

The **intron-delay hypothesis** (Gubb, 1986, 1993) emphasizes cell-cycle regulated gene expression at the transcriptional level rather than replication as in the chronon model described above. The hypothesis postulates that the length of the transcription unit is important in modulating gene expression in a cell-cycle dependent manner. This would affect the amount of mature transcript synthesized during rapid division cycles (Shermoen and O'Farrell, 1991; Gubb, 1993). This topic is extensively treated in Chapter 12, where it is applied to the yeast cell cycle.

(c) An energy-dependent circadian 'clock'?

Ehret *et al.* (1973) discarded the possibility that the high frequency oscillations described first in glycolysis were not related to circadian phenomena on the grounds that the timekeeping capacity of a clock (e.g. a mechanical one) is independent of its energy source. Taking into account that every action in the cell is energy-dependent, one could argue that in the end the metabolic ultradian domain will be more or less directly involved with circadian rhythmicity.

The role of cellular energy metabolism in the generation of circadian rhythms would be through the activity of metabolic pathways transducing information between 'clock' and 'hands' by means of feedback regulation and energy supply, depending on the metabolic state of the cell (Sel'kov, 1975; Edmunds, 1988). The key question now is: how are those rhythms emerging at different time domains coupled to each other? One approach to this problem has been to consider that a day-period rhythm is the result of coupling between multiple, short period, intracellular oscillators with a collective circadian output (Pavlidis, 1969).

3.2.2 ULTRADIAN RHYTHMS

Strictly speaking, **ultradian rhythms** comprise oscillatory phenomena exhibiting periods of less than 24 h while **infradian rhythms** are those with periods greater than 24 h. In the ultradian domain are biological processes whose periodic dynamics cover time scales ranging from milliseconds, seconds or minutes to the circadian-controlled events of about 24 h (Edmunds, 1988; Lloyd and Stupfel, 1991). To the lower extreme of the time scale belong, for example, high-frequency oscillations of ionic concentrations and metabolic intermediates, whereas to the other side of the ultradian domain pertains the cell division cycle of several organisms, apart from circadian periods (Lloyd and Stupfel, 1991). Therefore, the ultradian domain extends from ultrafast (milliseconds) to slower (20 h) periods, in that way covering a wide range of physiological processes.

(a) Rhythmic oscillations in neurons and their role in network oscillations

Over the last 15 years it has been recognized that the overall activity of the nervous system emerges from the interplay of synaptic activity and intrinsic electrophysiological properties of neurons (Sejnowski $et\ al.$, 1988; Llinás, 1990). Two possible mechanisms might determine the frequency of rhythmic oscillations in a population of neurons: the intrinsic oscillatory properties of the individual cells, and the time course of synaptic currents (Figure 3.2). In several types of nerve cell (e.g. cardiac sinoatrial node cells, inferior olivary neurons) their intrinsic rhythms are determined by the kinetic properties of membrane Ca^{2+} and K^+ currents (Traub $et\ al.$, 1989). Oscillations in olivary and thalamic cells are dependent on the activation of calcium and potassium conductances. The intrinsic electro-responsiveness of neurons, which makes them not functionally interchangeable (i.e. a thalamic neuron cannot replace, functionally, a neuron of another type, such as an inferior olivary cell), is responsible for the membrane potential oscillations – one of the most remarkable properties of the central nervous system.

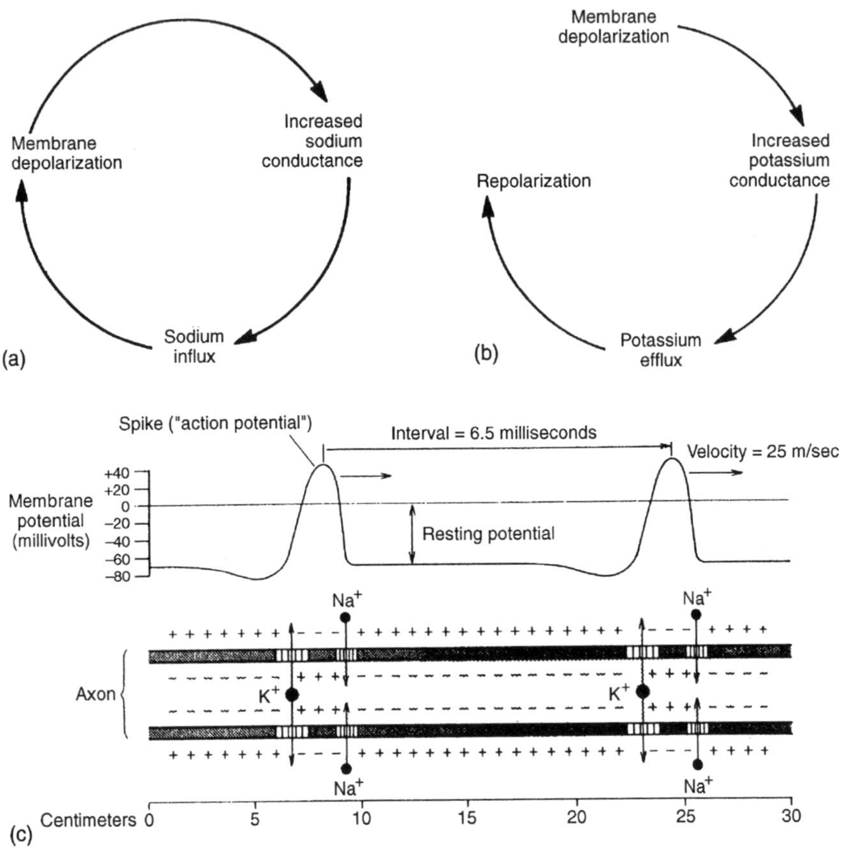

Figure 3.2 Action potentials and membrane excitability in neurons. (a) Positive feedback effect resulting from above-threshold depolarization of the membrane; (b) restoration of the membrane's resting potential; (c) propagation of a nerve impulse along the axon. The large change in potential is initiated by a small influx of sodium ions which opens voltage-sensitive sodium channels, changing the potential further. The membrane's resting potential is restored as the sodium channels are inactivated, and potassium channels open to permit an outflow of potassium ions. This sequence of events begins at the axon hillock and continues down the length of the axon. (Reproduced from Churchland, 1989, *Neurophilosophy*, with permission of MIT Press.)

When the cells are coupled together via electrotonic junctions (e.g. heart and inferior olive) or via inhibitory synapses, population oscillations emerge at the same or similar frequencies. Complex spatial patterns emerge in the initiation of population oscillations and the system is able to regulate the spatial coherence of population bursts by changing the mean excitability of the cells (Traub *et al.*, 1989). The amplitude and frequency of the emergent rhythmic activity at a neuronal network depend on intrinsic cellular properties and the connectivity and strength of both excitatory and inhibitory synapses (Traub *et al.*, 1989).

(b) Fast oscillations in starved slime moulds

The cAMP pulses released by *Dictyostelium discoideum* are a chemotactic signal in response to starvation (Gerisch and Hess, 1974; Devreotes, 1989) (Figure 3.3). Starved cells of this slime mould exhibit the capacity of responding to a pulse of cAMP by synthesis and relay of cAMP or of showing oscillatory behaviour in the synthesis and relay of the chemoattractant (Figure 3.4). The model of Goldbeter and Segel (1977, 1980) takes into account the following steps. The receptor of cAMP may exist in two alternative states, one of which is capable of activating the adenylate cyclase that will use ATP to produce cAMP, which in turn will be released into the extracellular medium. The integration of the ODEs system in Goldbeter's model showed that there exists a region in which cells, after a pulse, will not be excited (steady state) – a region in which cells are capable of relay, i.e. will respond to a pulse of extracellular cAMP by a synthesis of cAMP in an oscillatory way and will relax again to a steady state. In a third kind of dynamic behaviour the system will oscillate, synthesizing and periodically releasing pulses of cAMP. In Figure 3.4 the three sorts of behaviour are represented as a function of two of the parameters of the equations: phosphodiesterase and adenylate cyclase activities. For instance, when a cell of *D. discoideum* is starved, an increase in adenylate cyclase activity is observed and the cell effectively shifts from a non-excitable type of dynamics to excitability, relay and oscillations.

Another behaviour predicted by Goldbeter's model is arrhythmic or random-period oscillations or chaos. In the latter, strange attractors are obtained (section 3.4).

(c) Cytosolic calcium oscillations

For several systems comprising activating eggs or fully active cells, solitary or periodic pulses of calcium travelling from one pole of the cell

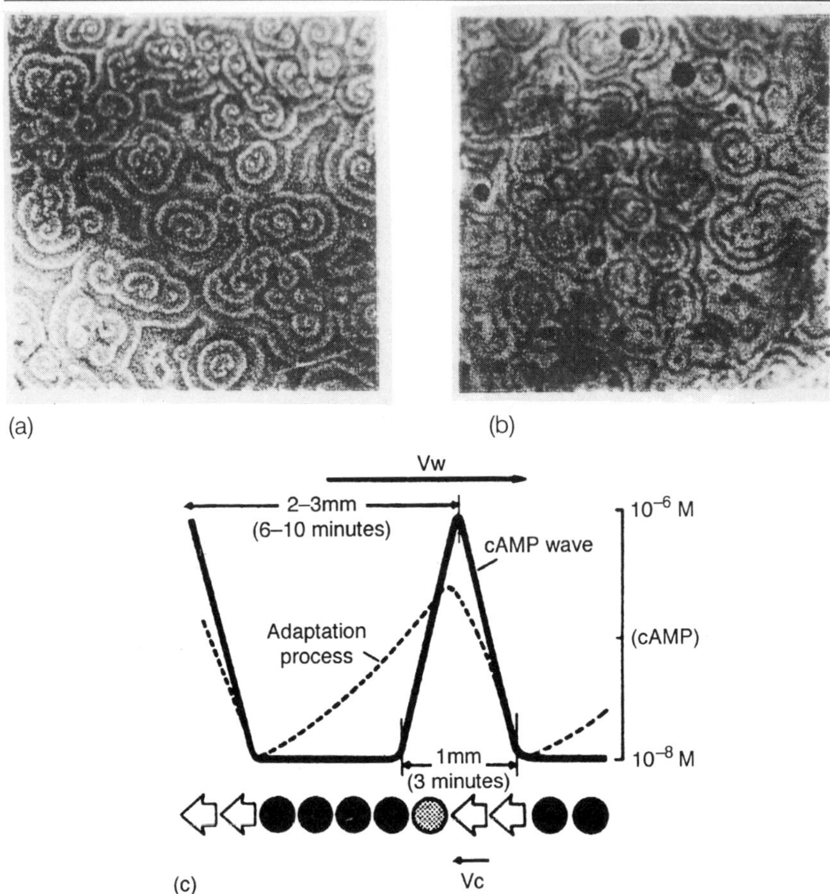

Figure 3.3 Chemotaxis and cAMP signalling mediate aggregation in *Dictyostelium discoideum*. (a) Dark-field photograph of aggregating cells at the 5 h stage of development; territories containing 1 million cells are actually 1–2 cm in diameter; (b) fluorographic image of cAMP waves within a monolayer of aggregating cells; (c) dynamics of signal relay and chemotaxis. The heavy line represents the cAMP concentration. Symbols represent a single radial line of cells: open arrows, cells moving towards the centre; shaded circles, randomly oriented cells; arrow vectors, the speed and direction of motion of the cAMP wave (Vw = 300 µm min[-1]) and the moving cells (Vc = 20 µm min[-1]). (Reproduced from Devreotes, 1989, by permission of Elsevier Science.)

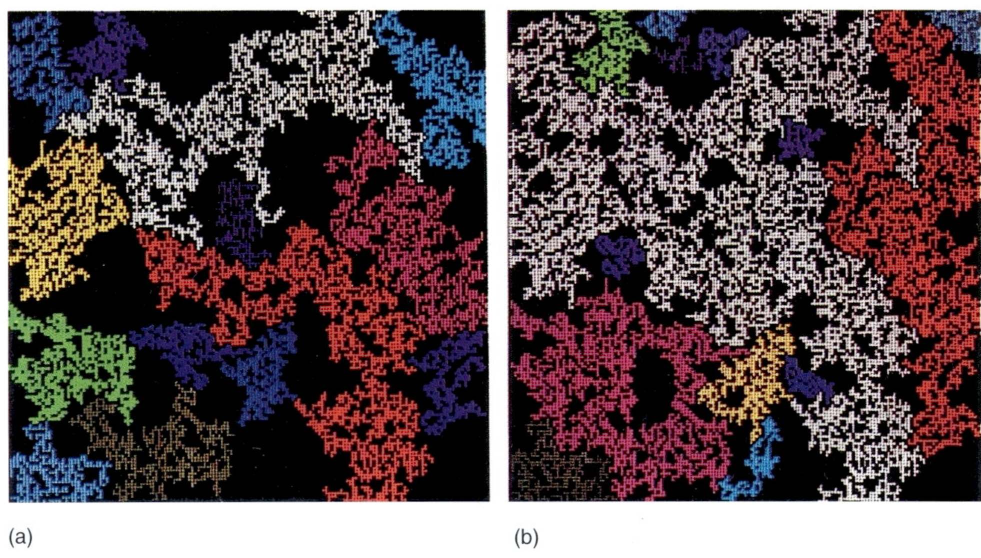

Plate 1 Percolation clusters. Unlike diffusion in liquids, randomness in percolation clusters is frozen into the medium since fluid spreading can only follow predetermined paths, i.e. fluid behaves like coffee in a percolator. A percolation cluster is a type of 'random fractal'. The plate shows a quadratic lattice which consists of pores in a matrix, and neighbouring pores connected by small capillary channels. A fluid injected into any given pore invades another pore that is directly connected to that pore to form a cluster. The effect of increasing the occupation probability P(p) on a 160 x 160 quadratic lattice is shown: (a) p = 0.58; (b) p = 0.6; (c) p = 0.62. In each of the three panels, the largest cluster appears white. Smaller clusters are shown in darker shades. Unoccupied sites are black. (Reproduced from Feder, 1988, *Fractals*, by permission of Plenum Publishing Corp.)

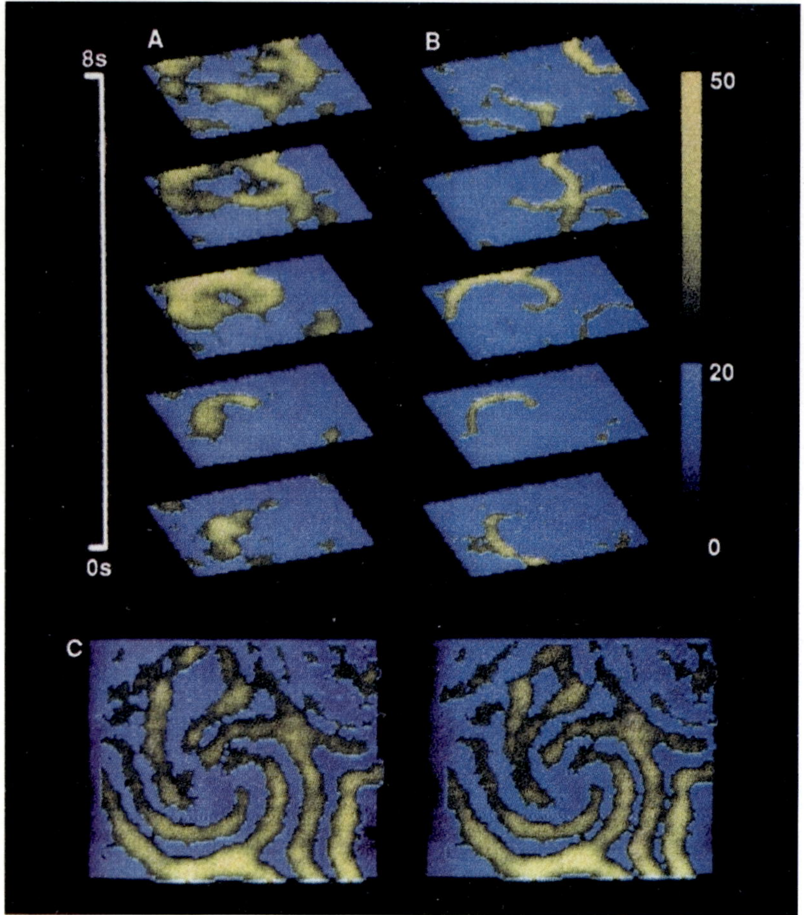

Plate 2 Intracellular spiral Ca^{2+} wave propagation and annihilation.
(A) Spatial patterns of Ca^{2+} release image at 2 s intervals at the
surface membrane of the oocyte. This series of optical slides (650 × 650
× 40 μm) was recorded 2 min after the initial application of acetylcholine
(ACh, 1 μM) in an oocyte expressing muscarinic acetylcholine receptors
(mAChRs). Resting concentrations of free Ca^{2+} are depicted on a blue
scale. Increases in [Ca^{2+}]$_i$ are depicted in green-yellow, on the same
scale. (B) Active wavefronts of Ca^{2+} release for the corresponding
images shown in (A) determined by sequential substration. (C) Stereo
view (cross-fusion) of the five images shown in (B) observed normal to
the x–y plane of the oocyte. This view follows the completion of one
period of spiral wave propagation (10 s). (Reproduced with permission
from Lechleiter *et al.*, *Science*, **252**, 123–126, copyright 1991 American
Association for the Advancement of Science.)

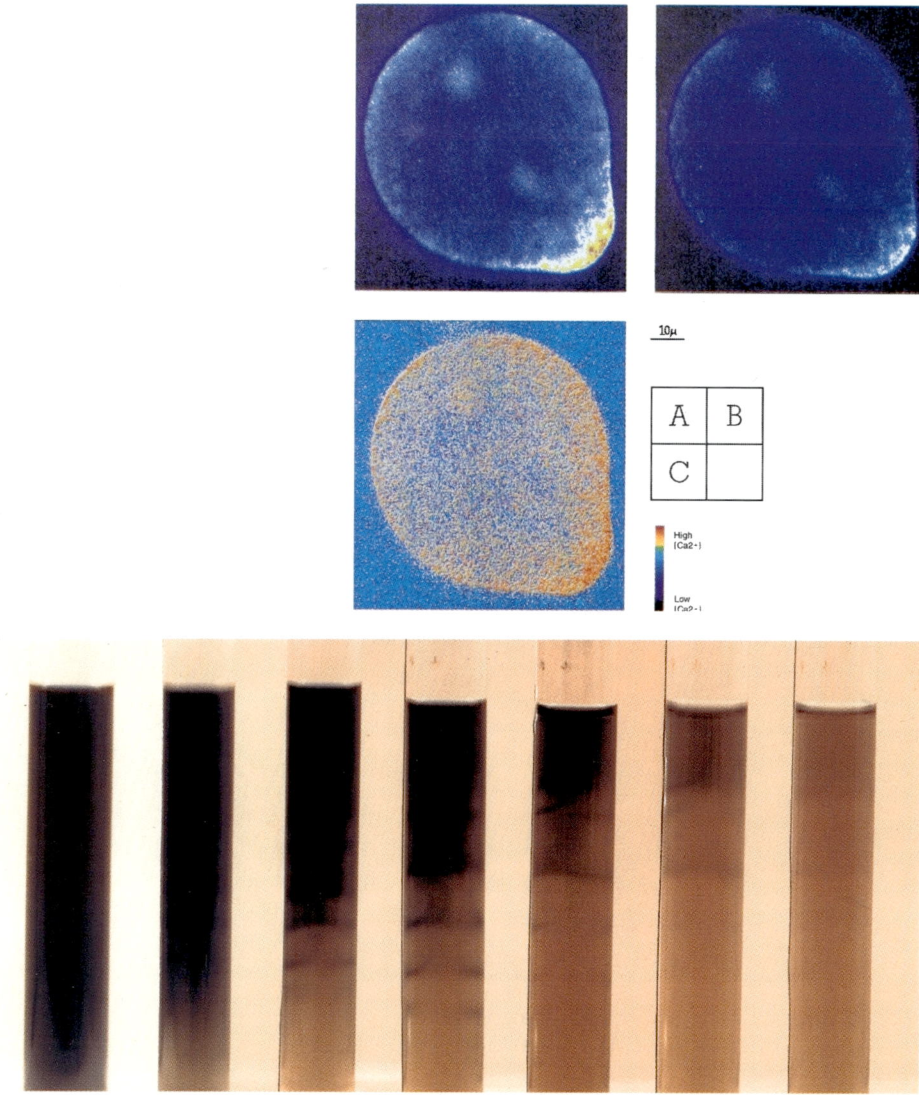

Plate 3 Cytoplasmic calcium gradients in polarized 18 h-old *Fucus* zygotes. A zygote was microinjected after polarization with an equimolar combination of 10 000 mol wt dextran-conjugated Calcium Green (A) and SNARF (B) and ratio A/B (C). Out of 22 cells, 19 exhibited a sharp gradient with elevated [Ca^{2+}] at the tip of the emerging rhizoid. This sharp gradient was also observed in rhizoids of 18 zygotes (18–36 h old) microinjected 3–5 h after fertilization. Images were taken 30 min after injection. The gradient did not usually extend further than 5 μm beyond the tip, and is approximately two orders of magnitude higher than the basal [Ca^{2+}]. (Reproduced from Berger and Brownlee, 1993, by permission of Cambridge University Press.)

Plate 4 Spatio-temporal evolution of the one-dimensional banded pattern of the electron acceptor DCIP, in a homogeneous suspension of thylakoids. The DCIP gradient has the following parameters: a = 0.5; b = 0.1. Photographs of the one-dimensional pattern were taken at intervals of 5 min; the first pattern (time zero) is on the left. (From Aon, Thomas and Hervagault, 1989.)

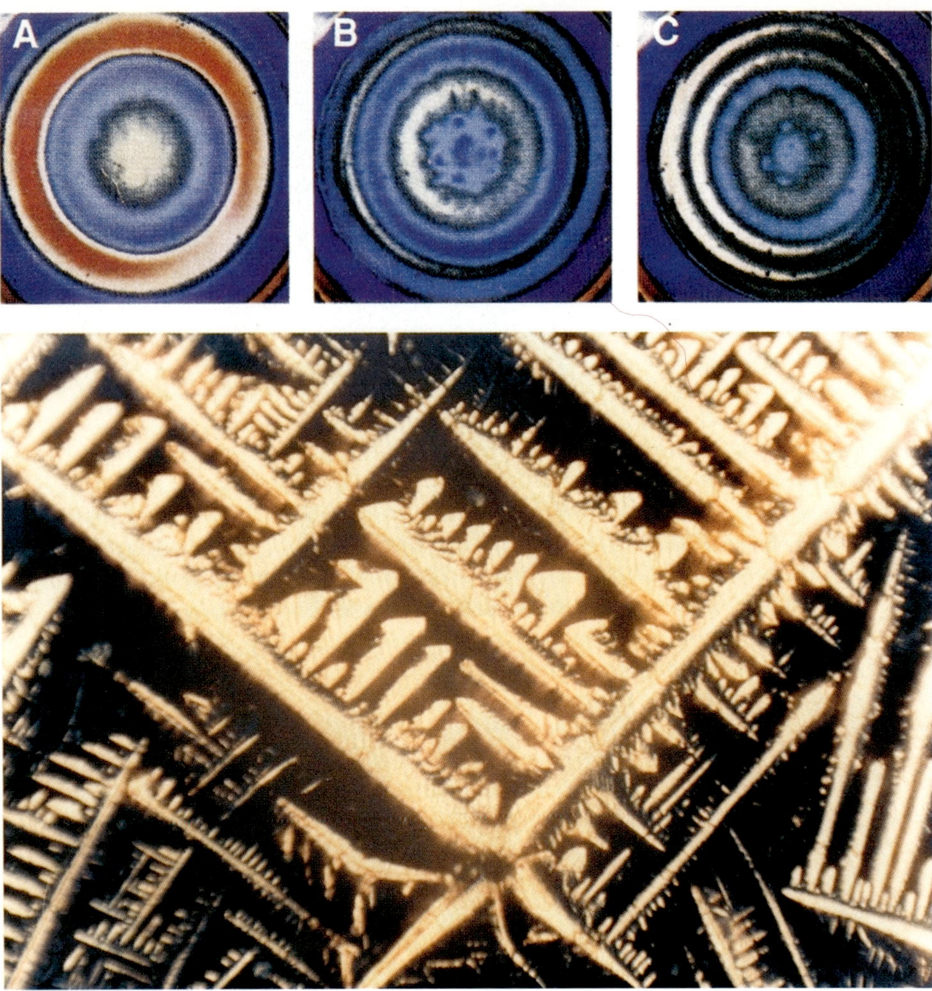

Plate 5 Spatial patterns of oscillating microtubules. A thin layer of tubulin solution (46 mg ml^{-1}) was subjected to a temperature jump and observed in transmission mode by a two-dimensional UV spectrophotometer at 390 nm. Images were taken at (A) 182, (B) 499 and (C) 949 s after the temperature shift. The maxima of turbidity (pseudocolour green to light blue) correspond to maxima in microtubule assembly; they are separated by regions of assembly minima (red to yellow). (Reprinted with permission from Mandelkow *et al.*, *Science*, 246, 1291–1293, copyright 1989 American Association for the Advancement of Science.)

Plate 6 Dendritic-like fractal patterns of an ovomucin gel/salt system in the presence of Coomasie Blue. An ovomucin gel (6 mg ml^{-1}), prepared as described in Rabouille *et al.* (1989), was resuspended and sonicated in the presence of 0.1 M NaCl and Coomasie Blue. After slow drying of a 50 μl aliquot of the ovomucin gel on a glass slide, the circle formed was observed without coverslip under a light microscope and photographed. The picture shows that the dendritic-like patterns are given by the salt precipitates on an underlying macromolecular lattice of (glyco)proteic nature, i.e. the blue background of the picture. A fractal dimension D of 1.785 ± 0.018 (r = 0.997) was determined for the glycoproteic lattice (Rabouille *et al.*, 1992). (Photo courtesy of C. Rabouille.)

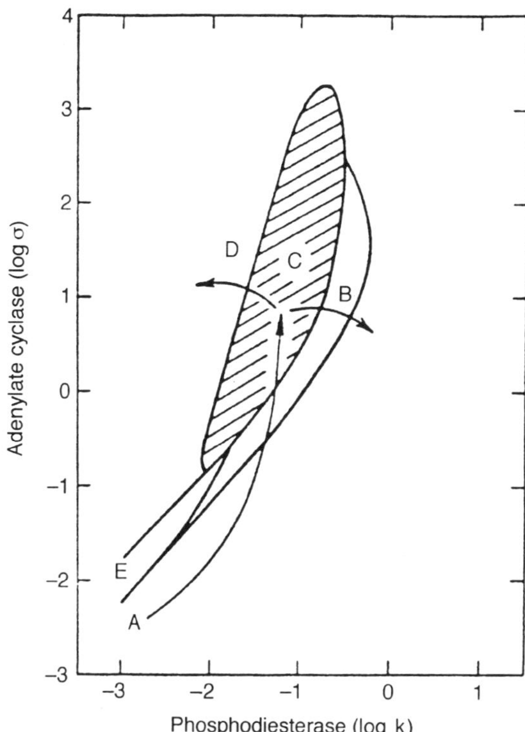

Figure 3.4 Dynamic behaviour domains of a model for the cAMP signalling system in *Dictyostelium discoideum*. Two parameters that play a central role in the model are plotted: the (normalized) maximum activity, σ, of adenylate cyclase and the activity, k, of phosphodiesterase. Several dynamic domains are delimited in the σ–k parameter plane: A = steady low level cAMP secretion; B–D = relay capability, autonomous oscillations and high steady secretion of cAMP, respectively; E = multiple steady states. During the course of each hour following starvation, the parameter state of the system is represented by a point that moves slowly along the developmental path (arrow). The path crosses successively the no-relay, relay and oscillatory domains; upon ceasing to oscillate, the system may either return to the excitability domain or settle to a high, steady cAMP secretion. (Reproduced from Goldbeter and Segel, 1980, *Differentiation*, by permission of Springer-Verlag.)

to the other, or from the cell periphery inward, have been observed (Jaffe, 1991). A steep rise in cytoplasmic Ca^{2+} concentration has been determined shortly after fertilization which is apparently correlated to the triggering of development, or after physical injury to the sarcolemma (Jaffe, 1980; De Loof *et al.*, 1992; Ishide *et al.*, 1992; Berger and Brownlee 1993). Ca^{2+} oscillations are often associated with the propagation of Ca^{2+} waves within the cytoplasm (Berridge *et al.*, 1988; Berridge, 1990, 1993). Several models have been proposed to explain Ca^{2+} oscillations and waves (see Atri *et al.*, 1993, for a critical assessment). As one of the most widespread cellular oscillatory phenomena (Dupont and Goldbeter, 1992), calcium oscillations remain attractive since periodic behaviour is thought to be a frequency-encoding information mechanism, whose resistance to degradation by noise and enhanced control precision have long ago been recognized (Rapp *et al.*, 1981). Cytosolic Ca^{2+} oscillations occur in a large number of cell types, either spontaneously or after stimulation by hormones or neurotransmitters.

Jaffe has proposed that those intracellular calcium pulses may take the form of cytosolic calcium waves which in turn are propagated by calcium-induced calcium-release (Plate 2) (Fluck *et al.*, 1991; Lechleiter *et al.*, 1991). The velocities of these waves are remarkably consistent, i.e. 10 μm s^{-1} in activating eggs and 25 μm s^{-1} in other cells (Jaffe, 1991; Fluck *et al.*, 1991). A basic equation for reaction–diffusion waves, i.e. the Luther equation (quoted by Jaffe, 1991), gives the velocity of such waves:

$$v = \alpha \sqrt{\frac{D}{t_r}} \qquad (3.1)$$

where v is wave velocity, D is the diffusion constant of the autocatalytic species (here that of free cytosolic Ca^{2+} ions, which is known to be ≈ 600 μm s^{-1}), t_r is the reaction time or time taken for Ca^{2+} to rise e-fold, and α is a factor that largely depends on the kinetics of the autocatalytic reaction (Jaffe, 1991).

The oscillation period of Ca^{2+} ranges from seconds to tens of minutes, depending on the cell type, whereas the velocity of Ca^{2+} waves varies from 10 μm s^{-1} on the surface of oocytes and 30 μm s^{-1} in hepatocytes, to 100 μm s^{-1} in the cytoplasm of cardiac cells (for review and references: Berridge *et al.*, 1988; Berridge, 1993; Dupont and Goldbeter, 1992). Perturbation methods with time resolution in the order of milliseconds and spatial resolution in the micron range are required to monitor rapid local changes in the Ca^{2+} concentration associated with nerve excitation and thus to overcome the blurring effects of diffusion (Figure 2.3) (Hernandez-Cruz *et al.*, 1990). Brief Ca^{2+} currents produce an inward wave of increased calcium that reaches the centre of the cell in ≈ 300 ms.

Subsequently, intracellular calcium equilibrates across the cell to resting values within 6 to 8 s (Hernandez-Cruz *et al.*, 1990).

Conventional microscopic studies of calcium dynamics employing the photoprotein aequorin or the ratiometric dye fura-2 have determined that the resting level of cytoplasmic free calcium in whole, unfertilized sea urchin eggs is approximately 100 nM, but directly after fertilization $[Ca^{2+}]_i$ rises to 1–5 μM. Confocal laser microscopy of sea urchin eggs after fertilization indicates a global elevation in free calcium throughout the whole egg rather than just in the cortex, though the relative rise in calcium tends to be slightly higher and more prolonged in the peripheral ooplasm than in the centre of the egg (Stricker *et al.*, 1992). Using selective microelectrodes, Grandin and Charbonneau (1991a,b), reported that cell division in *Xenopus* embryos was accompanied by periodic oscillations of the $[Ca^{2+}]_i$ level (70 nM amplitude around a basal $[Ca^{2+}]_i$ level of 400 nM) which occurred with a periodicity equal to that of the cell cycle (30 min). $[Ca^{2+}]_i$ oscillations did not take place in the absence of cell division in artificially activated eggs or when cell division was arrested (Figure 3.5). Eggs artificially activated by pricking in spite of not dividing, undergo almost all the metabolic events characterizing fertilized eggs, including surface contraction waves, cycling of maturation promoting factor activity and pH_i oscillations. Blocking of cell division with nocodazole induced arrest of $[Ca^{2+}]_i$ cycling (Grandin and Charbonneau, 1991a,b).

Differences in the resting levels of nuclear versus cytoplasmic free calcium, together with a rapid increase in nuclear free-calcium levels in several cell types, have been previously noted (Hernandez-Cruz *et al.*, 1990; Przywara *et al.*, 1991; Stricker *et al.*, 1992). Ca^{2+} signals in the nucleus were consistently larger and decayed more slowly than those in the cytosol after electrical stimulation of voltage-clamped sympathetic neurons isolated from the bullfrog, according to studies combining confocal laser-scanned microscopy and long-wavelength calcium indicators. These results suggest that the nonlinear kinetic mechanism underlying the observed amplification involved Ca^{2+}-induced release of Ca^{2+} (Hernandez-Cruz *et al.*, 1990). Apparently, there are diffusional barriers to the movement of Ca^{2+} ions from the cytoplasm through the nuclear envelope; however, Ca^{2+} stored in the endoplasmic reticulum–nuclear envelope lumen plays a role in macromolecular transport (Bustamante, 1994a, and references therein).

(d) The glycolytic oscillator

Oscillations in metabolic intermediates are a dynamic behaviour shown by glycolysis arising under certain physiological conditions. Yeast cells have provided an invaluable experimental tool for the elucidation of the

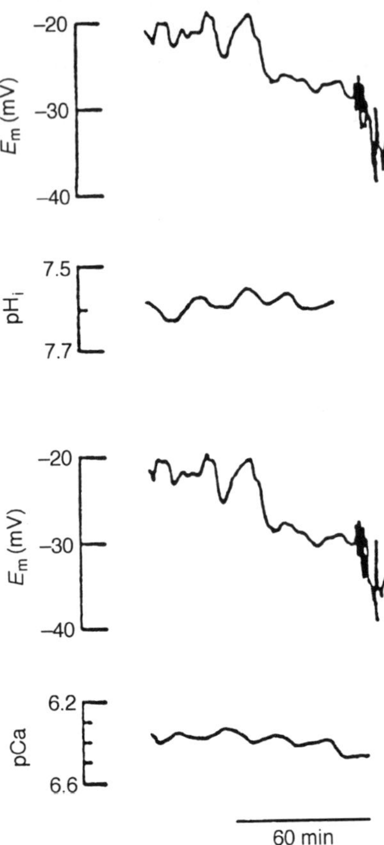

Figure 3.5 Temporal relationship between pH$_i$ oscillations and Ca^{2+} oscillations in *Xenopus* embryos; the oscillations were simultaneously measured in a single embryo. *Xenopus* embryos display pH$_i$ and Ca^{2+} oscillations with a period equal to that of the cell cycle. Both [Ca^{2+}]$_i$ and pH$_i$ oscillated around their basal levels, 0.31–0.50 μM (pCa 6.5–6.3) and pH 7.55–7.65, respectively, for four or five cell cycles, the amplitude of the oscillations being around 50–100 nM [Ca^{2+}]$_i$ and 0.04–0.06 pH unit. In this embryo, the duration of the cell cycle was 21–24 min. The peaks of the Ca^{2+} oscillations were found to occur 9–12 min after the acidic peaks of the pH$_i$ oscillations. (Reproduced from Grandin and Charbonneau, 1991, by permission of The Company of Biologists Ltd.)

underlying mechanism. The glycolytic oscillations observed in the yeast *Saccharomyces cerevisiae* have been characterized as an ultradian, i.e. short-period (minutes), metabolic oscillator (Edmunds, 1988; Lloyd and Stupfel, 1991).

The glycolytic flux and the concentration of glycolytic intermediates in intact yeast cells oscillate under specific conditions (Chance *et al.*, 1964; Hess and Boiteux, 1971; Shulman, 1988). Furthermore, yeast cells can show sustained oscillations when continuously supplemented with glucose (Hess and Boiteux, 1973), or in the transition from glucose to ethanol consumption phases during growth in batch cultures (Richard *et al.*, 1993). Since the prescient studies of NAD(P)H oscillations in intact yeast cells by Chance and collaborators, subsequent efforts have concentrated mainly on yeast extracts in order to explain the mechanisms involved in the observed glycolytic oscillations. The main conclusion was that the allosteric properties of phosphofructokinase (PFK) were responsible for the oscillations (Hess and Chance, 1978).

In cell-free extracts of yeast it was demonstrated that PFK was involved (Hess and Boiteux, 1971, 1973; Pye, 1973; Higgins *et al.*, 1973; Boiteux *et al.*, 1975; for reviews, see Goldbeter and Caplan, 1976, and Winfree, 1980). From these yeast studies, the long-held view of PFK as the primary oscillophor emerged (Hess and Boiteux, 1971; Hess, 1977; Hess and Chance, 1978). It was implicitely assumed that the sophisticated regulatory mechanisms of PFK were the necessary and sufficient condition to explain oscillations not only *in vitro* but also *in vivo* (Hess and Boiteux, 1971; Goldbeter and Caplan, 1976). This is already accepted in important textbooks (Alberts *et al.*, 1989).

The pioneering studies of Chance *et al.* (1964) also reported for the first time that inhibition of oxidative phosphorylation triggered yeast glycolytic oscillations, monitored by fluorescence changes in intracellular NAD(P)H. Further studies in whole yeast cells starved after growth on glucose suggest that *in vivo* the regulation of glycolysis dynamics, e.g. oscillations, may involve several potential sites of regulation (Aon *et al.*, 1991). One novel aspect appears to be the contribution of mitochondrial functions with several levels of regulation of glycolysis dynamics, at least in the ultradian domain (periods ranging between 1 and 2 min). Essentially, mitochondrial functions would affect ATP and NADH cytoplasmic levels, which in turn finely tune glycolysis dynamics (Figure 3.6). The regulation of adenine and pyrimidine nucleotides appears to be effected by:

- the mitochondrial proton motive force, through the regulation of ATP synthetic or hydrolytic fluxes;
- the adenine nucleotide translocator, presumably through the regulation of the rate of cytoplasmic–mitochondrial ATP/ADP exchange;
- mitochondrial dehydrogenases, by regulating the rate of NADH reoxidation (Aon *et al.*, 1991).

A further regulatory level in glycolysis when yeasts are operating at high ATP loads may be provided by the alcohol dehydrogenase (Aon *et al.*, 1991).

In the work of Aon *et al.* (1991) a quantitative model of glycolysis and its interactions with mitochondrial functions was allied to the experimental work. A more thorough analysis of the spatio-temporal regulation of glycolysis and oxidative phosphorylation *in vivo*, along with modelling studies, is presented in Chapter 7.

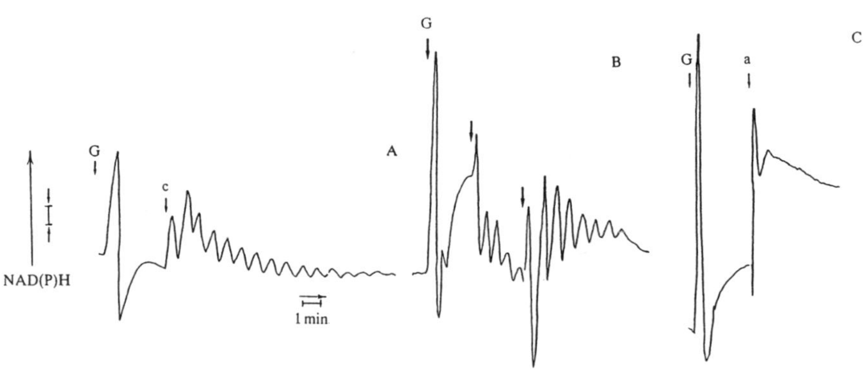

Figure 3.6 Effect of respiratory inhibitors on the appearance of glycolytic NAD(P)H oscillations in *Saccharomyces cerevisiae* cells. A suspension of yeast cells (10^7 cells ml^{-1}) grown, harvested and starved as described in Aon *et al.* (1991) were (A) pulsed with 20 mM glucose in the cuvette of the fluorimeter at 25°C with stirring, followed by 10 mM KCN, or (B) preincubated with 4 μM myxothiazole (1 min) and followed by successive glucose additions (arrows) of 2 mM each, or (C) addition of antimycin A (1 or 3 μM) after the glucose pulse. G = glucose; c = KCN; a = antimycin A. The remaining extracellular glucose 3 min after the 20 mM pulse was of 80% and an additional 10% was consumed 15 min after KCN addition. The scale of fluorescence corresponds to 1.5 μM NAD(P)H. (Reproduced from Aon *et al.*, 1991, by permission of The Company of Biologists Ltd.)

(e) A critical appraisal of the mechanisms underlying yeast glycolytic oscillations *in vivo*

As described above in section (d), in yeast extracts the allosteric properties of PFK were shown to be the source of the oscillations (Hess and Boiteux, 1968; Higgins *et al.*, 1973; Boiteux *et al.*, 1975) and the primary nonlinear property giving rise to oscillations (Higgins *et al.*, 1973; Boiteux *et al.*, 1975; Goldbeter and Caplan, 1976).

Almost at the same time that the PFK-based generation of glycolytic models was born, Sel'kov (1975) introduced a stoichiometry-based model of glycolysis. This approach was manifestly abandoned in favour of the 'PFK paradigm'. Sel'kov's model suggested the possibility that the stoichiometric nature of the glycolytic pathway could account for the oscillatory behaviour when one of the pathway products is used to prime the first reactions (Sel'kov, 1975; Cortassa *et al.*, 1990a). This stoichiometric model served as the basis of a proposal further developed to include the interactions with mitochondrial activity (Aon *et al.*, 1991) (Figure 3.7; Appendix 3). The nonlinear (autocatalytic) kinetic mechanism in this model is given by the stoichiometry of ATP production in anaerobic glycolysis, i.e. the number of ATP molecules produced in the second half of the pathway exceeds the number consumed in the first half (Cortassa *et al.*, 1990a, 1991; Aon *et al.*, 1991). Oscillations were obtained only in the presence of that autocatalytic feedback loop (Chapter 1).

The question arises as to whether the triggering of glycolytic oscillations *in vivo* can also be explained by allosteric effects on PFK activity similar to those shown in cell-free yeast extracts. Experimental evidence is now available to make the latter possibility doubtful:

- The control of PFK under Pasteur effect conditions could not be ascribed to changes in any one particular effector but rather to contributions from a variety of effectors (Table II in Reibstein *et al.*, 1986). Furthermore, fructose 2,6-bisphosphate (Fru 2,6-P_2) was the only effector that changed substantially between the anaerobic and aerobic conditions, being undetectable in the former situation. These observations suggest that, *in vivo*, the allosteric properties of the PFK could hardly be manifest, at least in aerobic–anaerobic transitions.

- It was additionally shown that the autocatalytic mechanism given by the stoichiometry of glycolysis, coupled to mitochondrial activity through the ATP and NADH cytoplasmic pools, may contribute as another regulatory level of glycolysis *in vivo*.

- In yeast extracts, the presence of complex oscillations apart from sinusoidal, i.e. spike-like or square-type, were experimentally

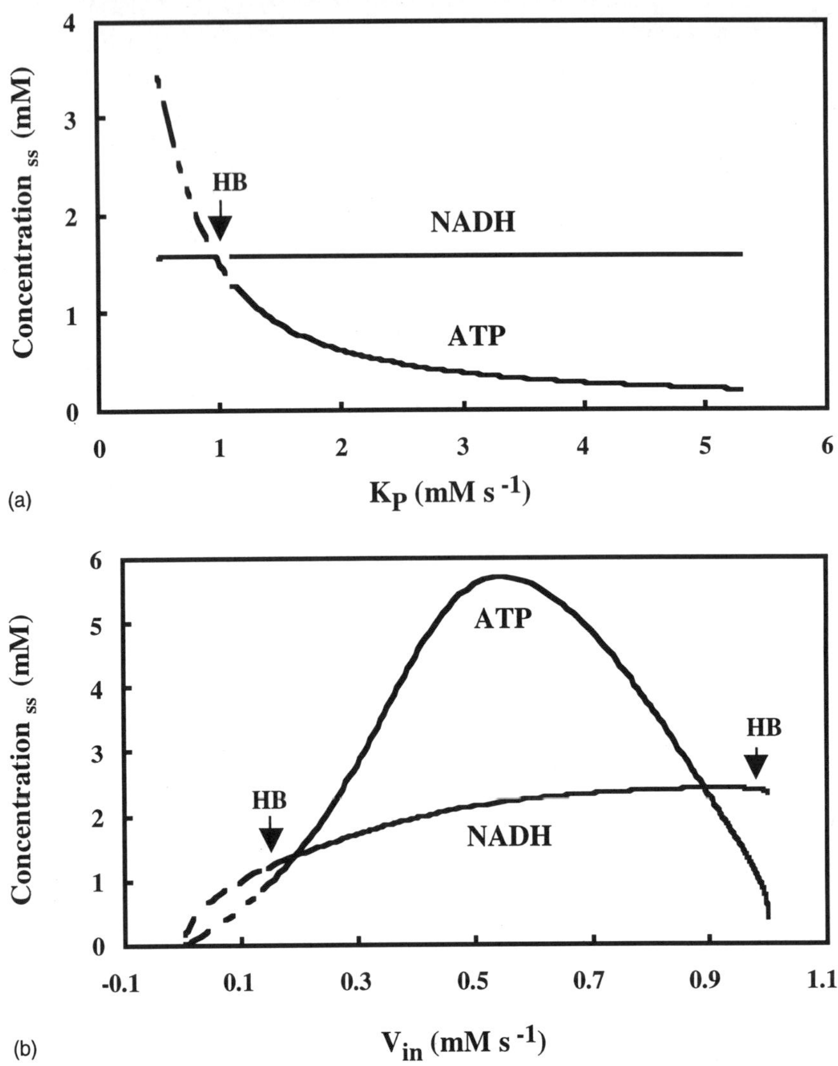

demonstrated (Hess and Boiteux, 1968, 1973). The stoichiometric model, without taking into account allosteric effects, was able to exhibit sinusoidal, square and spike-like oscillations. Furthermore, complex oscillations with period doubling in ATP were concomitant with the square-like oscillations in NADH (Figure 7.2).

- A great deal of experimental evidence supports the notion of several control points in glycolysis (Den Hollander *et al.*, 1986; Reibstein *et al.*, 1986).

- Existing experimental and theoretical evidence suggests that the control of the glycolytic flux in *Saccharomyces cerevisiae* is on the uptake and apparently not on PFK or other irreversible steps, e.g. pyruvate kinase (Heinisch, 1986; Schaaf *et al.*, 1989; Cortassa and Aon, 1993a, 1994a).

According to our results, the dynamics of glycolysis in starved yeast cells under semi-anaerobic conditions may attain oscillatory regions in the parametric space through a concerted action of, at least, the following relevant physiological parameters:

- the rates of substrate input and substrate phosphorylation;
- the load induced by the non-glycolytic ATP-consuming processes;
- the rate of cytoplasmic NADH reoxidation.

Stability analysis of the stoichiometric model showed that, at low substrate input rates, a decrease in the leak of phosphorylated intermediates, e.g. through glycerol formation, can induce sustained oscillations (Aon *et al.*, 1992; Cortassa and Aon, 1994b).

It is likely that the extreme conditions of starvation allowed fine regulatory mechanisms of glycolytic oscillations by mitochondrial activity to become evident. The question then arises whether this fine-tuning exists for non-starved and aerobically growing yeast cells. In fully aerobic cultures in the presence of excess glucose, glycolytic oscillations would probably be buffered by mitochondrial activity by keeping high cytoplasmic ATP and NAD levels, and this would explain why

Figure 3.7 (see facing page) Stability analysis of the glycolytic model. The stability analysis of the ODEs system (3.3–3.6; Appendix 3A) was performed numerically with AUTO. The stable (solid line) and unstable (dashed line) branches of the steady state solutions and the Hopf bifurcations were computed automatically. The bifurcation diagram was obtained with the following parameter values. (a) V_{in} (mM s^{-1}) = 0.25; k_1 (mM^{-1} s^{-1}) = 0.0949; k_3 (mM^{-3} s^{-1}) = k_5 (s^{-1}) = k_7 (mM^{-1} s^{-1}) = 0.1; k_9 = 0.05 (s^{-1}); τ = 0.83; $\Delta\mu_H°$ (V) = −0.3. The kinetic parameters of the proton pump were: K_M = K'_M = 2 mM; V_M = V'_M = 0.5 mM s^{-1}. The total nucleotide (C_A) and phosphate pool (P_t) were both 10 mM, while the nicotinamide nucleotide pool (C_N) was 5 mM. The steady state (ss) ATP and NADH values represented on the y-axis were obtained as a function of the bifurcation parameter K_p (= τ V_M [1 − $\Delta\mu_H/\Delta\mu_H°$]^5) (mM s^{-1}), i.e. the ATP load introduced into the system by non-glycolytic ATP-consuming processes. (b) K_p (mM s^{-1}) = 0.751. The rate constants, kinetic parameters of the proton pump, total nucleotide (C_A) and phosphate pool (P_t) were as in (a). The steady state (ss) ATP and NADH values represented on the y-axis were obtained as a function of the bifurcation parameter V_{in} (mM s^{-1}), i.e. the substrate input rate. Arrows point to the values of the bifurcation parameters for which a Hopf bifurcation (HB) appears. (Reproduced from Aon *et al.*, 1991, by permission of The Company of Biologists Ltd.)

oscillations were not found in aerobic, non-substrate limited yeast cell suspensions (Shulman, 1988).

Evidence has been presented in support of fructose 2,6-bisphosphate (a powerful activator of yeast 6-phosphofructo-1-kinase) not being an essential component of the oscillatory mechanism of glycolysis in extracts and cells of *S. cerevisiae* (Yuan *et al.*, 1990). It was indeed observed that Fru 2,6-P_2 altered the dynamic properties of 6-phosphofructo-1-kinase, but the endogenous minute amount of the activator and the phase relationship of the oscillations compared with other metabolites led the authors to conclude that the activator had practically no effects on PFK1 (Yuan *et al.*, 1990).

3.3 THE CELL DIVISION CYCLE

Eukaryotic cells duplicate their DNA during a period called S (**DNA synthetic period**); after a gap (**G_2**), nuclei divide in another period called **mitosis** (M), followed by cytokinesis. Between mitosis and the next DNA synthetic phase, there is a period called **G_1**, the length of which is variable. The time taken for one complete cycle is called **generation time**; for most growing mammalian cell lines in tissue culture the interval between divisions is 10–30 h (Pardee *et al.*, 1978); it is 10–30 min in certain prokaryotes (e.g. *Benekea natriegens, Escherichia coli*) and 2–3 h in the yeast *Saccharomyces cerevisiae* or the ciliate *Tetrahymena pyriformis* (Edmunds, 1988). The high degree of heterogeneity in the mean generation time, τ_d (Chapter 11) is not only confined to different cell types but may arise within a single cell population of the same species from environmental shifts (e.g. nutritional, temperature, pH, etc.) as well as from karyotypic variability (Edmunds, 1988). The changes in generation time appear to be due to variation in the length of G_1, with the duration of S (6–8 h) + G_2 (2–6 h) + M (1 h) being relatively constant in mammalian (Pardee *et al.*, 1978) or yeast cells (Beck and von Meyenburg, 1968). The numbers in parentheses correspond to mammalian cells.

At present the cell cycle is viewed according to two models: the dependent and the independent pathways (Hartwell *et al.*, 1974; Mitchison, 1989; Murray and Kirschner, 1989; Hartwell, 1991). According to the dependent pathway model any periodic event in growth is thought to be related to or controlled by the main periodic events of the DNA-division cycle. In view of the independent pathways model, control is by a timer or oscillator rather than by dependent sequences.

The view of a sequence of dependent ordered events of the cell cycle in yeast emerged from studies with temperature-sensitive cell division cycle (cdc) mutants of *S. cerevisiae*. Those studies revealed that the execution of late events in the cell cycle depends on the prior completion of early events (Hartwell, 1991, and references therein). Temperature-sensitive cdc mutants of *S. cerevisiae* were defined because of the property to arrest division at a unique stage of the cell cycle regardless of their stage at the time when they were shifted from permissive to the restrictive temperature (Hartwell, 1991) (Chapter 11). The behaviour of cdc mutants showed that the execution of late events in the cell cycle depended on the prior completion of early events (Hartwell *et al.*, 1974). The studies concerning the functions executed by the heat-sensitive cdc gene products suggested that the dependent order of the cell cycle events was a result of an underlying order of gene product function (Hartwell, 1991).

In terms of their primary defects, the cdc mutants of *S. cerevisiae* fall into three classes (Harold, 1991). The first class comprises genes that specify structural materials or the enzymes required for their synthesis: tubulin, actin, chitin, DNA, 10 nm filaments. The second class consists of genes that code for enzymes in central metabolic pathways, such as pyruvate kinase and phosphoglucose isomerase. The third class contains genes whose products govern the rate of reaction at the branch or regulatory points of chemical pathways, e.g. adenylate cyclase, protein kinase, calcium-binding proteins (Harold, 1991; Aon *et al.*, 1995) (Table 11.3). For several of these cdc mutants (cdc28, cdc35, cdc19, cdc17, cdc21) we have provided experimental evidence which shows that carbon and energy uncoupling, together with activation of the fermentative pathway, were associated with the arrest of cell proliferation, when the mutants were blocked at different stages of their progression through the mitotic cycle (Aon *et al.*, 1995).

The cell division of the early *Xenopus* embryo presents a striking contrast to the view of a cascade of dependent events. The main evidence for the independent view of control of the cell cycle is that periodic events persist after a block to the DNA division cycle (Mitchison, 1989). Mitosis does not depend on DNA replication, as it does in many other cells, because inhibition of DNA synthesis does not prevent nuclear division (Kimelman *et al.*, 1987). Moreover, inhibition of mitosis does not prevent successive rounds of DNA replication as it does in other cell types (Kimelman *et al.*, 1987). The activated but enucleated *Xenopus* egg exhibits contractions with the same periodicity as the divisions of a nucleated egg, suggesting the presence of a cytoplasmic clock that controls cell division (Hara *et al.*, 1980).

A single biochemical mechanism underlying the cell cycle in all eukaryotic organisms has been proposed as a unified view: between clocks and dominoes, i.e. independent and dependent pathways (Murray and Kirschner, 1989; Hartwell, 1991). According to genetic and molecular biology studies performed in the last 15 years or so, the cell cycle of yeast and *Xenopus* is thought to be driven by a 'clock' whose biochemical basis appears to be the periodic synthesis and degradation of cyclin coupled with activation and inactivation of the *CDC2pombe/CDC28cerevisiae* 34-kd serine/threonine protein kinase (p34^{cdc2}) by dephosphorylation and phosphorylation mechanisms, respectively. Both models are reconciled on the basis that the cdc2/cdc28 kinase oscillator runs the cell cycle but, in addition to this, somatic cells and eukaryotic microorganisms have checkpoint controls that feed forward to the next event to ensure that it does not occur if the previous event has not been completed. Checkpoints are thought of as signal transduction pathways that generate an inhibitory signal in response to delayed upstream events and target this signal to the next downstream event (Hartwell, 1991). The evidence for checkpoints came from observations around the dependence of mitosis upon prior DNA replication (or upon the repair of DNA damage). For instance, inhibition of DNA replication with hydroxyurea induces cells to arrest in early S-phase and they do not undergo mitosis. This dependence of mitosis on prior completion of DNA replication is believed to be due to the action of specific gene products (Hartwell and Kastan, 1994). Inactivation of these genes by mutation relieves the cell of this dependence, and such mutants will enter mitosis with incompletely replicated DNA. Eliminating the RAD9 gene of *S. cerevisiae* allows cells to enter mitosis without first completing DNA replication.

At least two checkpoints detect DNA damage: one at the G_1/S transition and one at the G_2/M transition (Hartwell and Kastan, 1994). Loss of the RAD9 gene checkpoint decreases mitotic fidelity 10–20-fold in an unperturbed cell and has a much greater effect if the cell is experiencing DNA damage or defects in DNA replication (Hartwell, 1991). If checkpoints exist to ensure mitotic fidelity, how did the early embryos of *Drosophila* and *Xenopus* come to dispense with these controls? A possible explanation is that early embryos require rapid and synchronous division. In that case, checkpoints would act antagonistically to these needs because they delay division to permit repair (Hartwell, 1991).

3.4 CHAOS

3.4.1 THE BORDERLINE OF CHAOS

Often, under the action of strong forces, macroscopic systems show very complex and turbulent behaviour – for example, waterfalls or storms. Before reaching a highly turbulent condition, many systems pass through a series of intermediate states in which the time dependence of their state variables is already erratic. Such systems can be successfully described by mathematical models involving a finite, and even small, number of dynamic variables (Guckenheimer and Holmes, 1983; Procaccia, 1988). Those dynamic systems can be described by ODEs (Chapter 1).

The sensitive dependence on initial conditions which is the primary characteristic of strange attractors requires at least one eigenvalue being greater than zero (Rietman, 1989). The path followed by the solution-trajectories x(t) in parameter space by different dynamic systems at the borderline of chaos are identical irrespective of the nature of dynamic variables (Procaccia, 1988) (Chapter 1).

3.4.2 CHAOS IN DYNAMIC SYSTEMS

A dynamic system with n independent variables is described by n coupled differential equations of the type of equation 1.11 (Chapter 1 and Appendix 1). Two conditions must be fulfilled for chaos to occur in such equations:

- There must be at least three independent variables.
- The right-hand side of the equations must contain at least one nonlinear term (Olsen and Degn, 1985).

Two main features characterizing chaos must be fulfilled:

- For defined parameter values, almost all initial conditions give rise to aperiodic dynamics.
- Arbitrarily close initial conditions display independent temporal evolution as time proceeds.

The latter is called **sensitive dependence on initial conditions** (Glass and Mackey, 1988). The '**butterfly effect**' is so named because even a butterfly flapping its wings could disturb the initial conditions enough for a chaotic (e.g. weather) system to cause a completely different weather pattern to evolve over time (Weiss *et al.*, 1994).

For a diagnosis of chaos one may directly inspect a time series, i.e. the graph of a variable as a function of time (Chapter 1, Figure 1.3). Either amplitudes or periods may be scrutinized. If there are alternating high and low amplitudes (or different periods) or some repetitive pattern of amplitudes of different heights, then it is advisable to change some parameter slightly and see if the periodicity is doubled or cut in half. If this is the case, changing the parameter in the direction of period doubling is likely to lead to chaos (Olsen and Degn, 1985). However, this is not always the case: it is possible for a system exhibiting period doubling not to evolve to chaos (Olsen and Degn, 1985). In some cases, the transition to chaos may take place abruptly due to a crisis or through intermittency (Olsen and Degn, 1985; Markus *et al.*, 1985; Hess and Markus, 1987).

3.4.3 'STRONG' AND 'WEAK' CHAOS

Sensitivity to initial conditions characterizes **'strong' chaos** and destroys predictability (Procaccia, 1988). This means that on iteration there is a scheme of the sort:

$$x_{n+1} = f(x_n) = 2x_n \tag{3.2}$$

The distance between the orbits, $x_n = |x_n - x'_n|$, grows exponentially as $2^n x_0$ until the two orbits diverge widely.

'Weak' chaos has to do with the abundance of possible different dynamic asymptotic states such as periodic orbits, which are allowed even after all transients have died out (Procaccia, 1988). The exponential growth in the number of orbits of longer period, also called **period doubling**, is the hallmark of weak chaos and is a necessary condition and prerequisite for strong chaos. The abundance of asymptotic states is only a necessary condition for strong chaos; it can happen that one of the available orbits is stable, in which case chaotic motion is not observed. Chaotic motion can be perceived if all the available orbits are unstable (Procaccia, 1988).

3.4.4 CHAOS IN BIOLOGICAL SYSTEMS

Chaos is ubiquitous in biology, at several levels of organization (for reviews see Olsen and Degn, 1985; Hess and Markus, 1987; Montero and Morán, 1992). Chemical systems may exhibit chaotic behaviour if they contain certain types of feedback such as enzymatic systems, e.g. peroxidase and glycolysis (Markus *et al.*, 1984, 1985), excitable cells, cellular signal transmission, heart beat, epidemics and brain activity (section 3.4.2) (Figure 3.8).

Neurological disorders and cardiac arrhythmias may be viewed as originating from chaotic dynamics of physiological parameters such as electrical activity (Mackey and Glass, 1977; Olsen and Degn, 1985; Glass and Mackey, 1988; Montero and Morán, 1992). A recent hypothesis, based on chaos theory and the associated geometrical concept of fractals, proposes that physiological ageing is associated with a generalized age-related loss of complex variability in multiple processes including cardiovascular control, pulsatile hormone release and electroencephalographic potentials (Lipsitz and Goldberger, 1992).

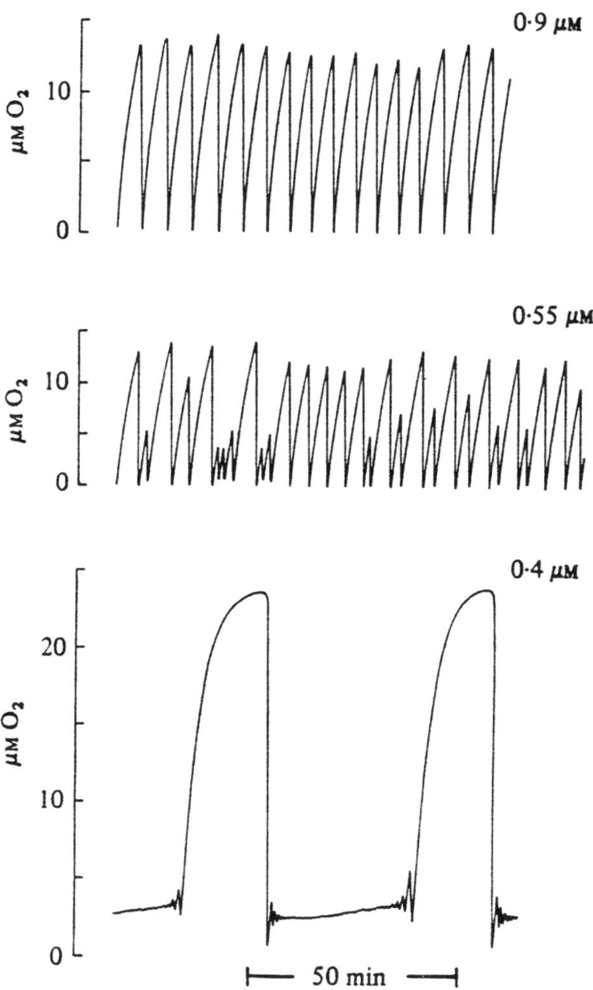

Figure 3.8 Periodic and chaotic oscillations of the concentration of O_2 in the peroxidase–oxidase reaction in open systems at three different enzyme concentrations (0.9 μM, 0.55 μM, 0.4 μM). (Reproduced from Olsen, 1983, by permission of Elsevier Science.)

While chaos is intriguing, it is not clear what role it plays in real chemical processes of living systems. Some recent work on control of chaotic dynamics has thrown up useful hypotheses about its putative role (Ott *et al.*, 1990; Peng *et al.*, 1991; Shinbrot *et al.*, 1993; Petrov *et al.*, 1993). Let us explore this topic.

One of the fundamental aspects of chaos is that many different possible motions are simultaneously present in the system. A chaotic system traces a strange attractor in the phase space in which an infinite number of unstable limit cycles are embedded, each characterized by a distinct number of oscillations per period (Peng *et al.*, 1991; for a review, see Shinbrot *et al.*, 1993). Orbits are unstable in the sense that the smallest deviation from the periodic orbit, e.g. due to noise, grows exponentially rapidly in time, and the system orbit quickly moves away from the original periodic orbit into another one. In addition to periodic orbits, it is common for continuous time dynamic systems to have unstable steady states embedded in chaotic motions (Shinbrot *et al.*, 1993). Ott, Grebogi and Yorke have demonstrated that unstable periodic orbits can be stabilized by introducing small, controlled perturbations to a system constraint (Ott *et al.*, 1990). This approach focuses on 'chaos control' rather than 'chaos elimination' (Weiss *et al.*, 1994). If a system is not chaotic, small parameter perturbations will merely change the orbit slightly. Only after large parameter alterations might the system exhibiting periodic solutions (limit cycles) move towards another attractor. On the contrary, selection of stable periodic outputs from a chaotic attractor requires only small time-dependent perturbations of a system parameter (Ott *et al.*, 1990; Lloyd and Lloyd, 1993). Techniques based on these concepts have been developed and applied to the control of chaotic behaviour of ventricular tachycardia induced by the cardiac glycoside ouabain in an intact cardiac muscle preparation (Weiss *et al.*, 1994).

A single system configuration can be accommodated by multiple-use systems that are employed for different purposes at different times. In agreement with the latter concept, and on the basis of the mechanism to control chaos, it has been proposed that rhythmic outputs from a controlled chaotic attractor may account for the generation of biological rhythms (Lloyd and Lloyd, 1993). This putative mechanism may explain the intrinsic variability of complex biological systems, e.g. variations from 16 h to 30 h observed in the expression of the genic product of the *per* locus of *Drosophila* when mutated, without need to invoke noisy limit cycle operation (Lloyd and Lloyd, 1993). From these and other observations, it has been suggested that some mutational events may perturb the regulatory devices normally employed to stabilize a periodic orbit within the chaotic attractor.

3.4.5 DYNAMIC ORGANIZATION, (CONTROLLED) CHAOS AND FRACTALS

Dynamically, chaos differentiates with respect to periodic phenomena because of its sensitive dependence on initial conditions, and because the dynamic regime is characterized by 'strange' attractors in phase space. In open systems, periodic as well as chaotic dynamics subsist by the continuous supply of matter and energy, in this way belonging to the widely encompassing range of dissipative, spatio-temporal structures. The fact that strange attractors are embedded in unstable limit cycles that, when deliberately controlled via feedback, may be stabilized around certain orbits opens the possibility that controlled chaos participates in the self-regulation of living systems (Peng *et al.*, 1991). Chemically unstable systems such as the Belousov–Zhabotinskii reaction can be kept oscillating regularly by applying a sequence of small adjustments to the concentration of some of the chemicals involved in the reaction (Petrov *et al.*, 1993) (Chapter 5).

Often the link between chaos and fractals has been established due to the fractal geometrical structure of variable trajectories for strange attractors. It is now clear that nonlinear systems can exhibit strange attractors that may be quantitatively described by fractal geometry (Glass and Mackey, 1988). On every scale examined, strange attractors resemble fractals with a fractal dimension, D, of ≈ 1.3 (Mandelbrot, 1982, 1993; Shinbrot *et al.*, 1993). Biologically, it could be relevant to consider that the fractal geometry of the airway tree of the lung and vascular trees is an adequate design to sustain flows of fluids (e.g. air, blood) that can be locally laminar, whereas globally those flows exhibit chaotic, turbulent, dynamics with improved and rapid mixing of air and blood (Mandelbrot, 1982, 1993; Sernetz *et al.*, 1985; Glass and Mackey, 1988; Weibel, 1991).

The existence of a borderline between points, cycles and chaos stresses the fact that periodic as well as erratic behaviour in the time dependence of variables is another kind of possible dynamic regime shown by dynamic systems.

3.4.6 SYMMETRY AND CHAOS

Both symmetry and chaos – pattern and disorder – can coexist naturally within the same mathematical framework. Chaotic patterns produced by iteration of difference equations look complicated, yet they are prescribed by a short computer program and a few numbers (Stewart and Golubitsky, 1992). Complex spatio-temporal patterns may ensue when complex dynamics coexist in systems that are spatially coupled (Winston *et al.*, 1991; Cross and Hohenberg, 1994). Those systems are not reducible to a model with a small number of degrees of freedom such as the logistic equation or the Lorenz model. Systems that are said to be large display spatio-temporal chaos (Cross and Hohenberg, 1994). Diffusive coupling of chemical waves of the B-Z oscillatory reaction gives rise to the spontaneous appearance of spiral waves which entrain themselves spatio-temporally in long time scales (Winston *et al.*, 1991).

In the standard Rayleigh–Bénard convective system, for some parameter values an ordered state of straight or weakly curved rolls breaks down to spatio-temporal chaos. This chaotic state consists of elementary spiral structures which appear and disappear in an irregular fashion (Cross and Hohengberg, 1994). These results may be significant for population dynamics since most environments are spatially subdivided, or patchy. Under those conditions, the relationship of the local to global (or metapopulation) dynamics of populations becomes a timely topic. A remarkable range of dynamic behaviours is shown by mathematical models for host–parasitoid interactions where, in each generation, specified fractions of the host and parasitoid subpopulations in each patch move to adjacent patches (Hassell *et al.*, 1991). These models with limited diffusive dispersal may exhibit complex patterns of spiral waves or spatially chaotic variation, or static 'crystal lattice' patterns, or they may become extinct (Chapter 6). This range of behaviour is obtained when the local dynamics is deterministically unstable, and with a constant host reproductive rate and no density dependence in the movement patterns (Hassell *et al.*, 1991).

APPENDIX 3

A model taking into account the interactions of glycolysis and mitochondria was developed. The model is described by a set of four nonlinear ODEs (3.3–3.6) and three conservation equations (3.7–3.9) (for further details, see Aon *et al.*, 1991):

$$\frac{d[Glc]}{dt} = V_{in} - V_1 \tag{3.3}$$

$$\frac{d[ATP]}{dt} = -2V_1 + 4V_3 - V_9 - V_{P1} - V_{P2} \tag{3.4}$$

$$\frac{d[I]}{dt} = V_1 - V_3 - V_5 \tag{3.5}$$

$$\frac{d[NADH]}{dt} = 2V_3 - 2V_7 \tag{3.6}$$

$$C_A = [ATP] + [ADP]; \quad C_N = [NAD] + [NADH];$$
$$P_t = P_i + 2[I] + [ATP] \tag{3.7–3.9}$$

with the following fluxes:

$$V_1 = k_1[\text{ATP}][\text{Glc}] \tag{3.10}$$

$$V_3 = k_3[\text{ADP}][\text{I}][\text{P}_i][\text{NAD}] \tag{3.11}$$

$$V_5 = k_5[\text{I}] \tag{3.12}$$

$$V_7 = k_7[\text{NADH}][\text{NADH}] \tag{3.13}$$

$$V_{P1} = K_P \frac{[\text{ATP}]}{K_M + [\text{ATP}]}; \quad K_P = \tau V_M \left(1 - \frac{\Delta\mu_H}{\Delta\mu_{H^0}}\right)^5 \tag{3.14}$$

$$V_{P2} = K_P' \frac{[\text{ATP}]}{K_M' + [\text{ATP}]}; \quad K_P' = \tau V_M' \left(1 - \frac{\Delta\mu_H}{\Delta\mu_{H^0}}\right)^5 \tag{3.15}$$

Symmetry in dynamic biological organization

4

This chapter deals with the question of how millions of molecules can be spatio-temporally organized in morphogenetic fields such that form arises in developing organisms at macroscopic scales extending from micrometres to millimetres – the characteristic lengths over which biological form is generated. At present, the predominant view is that of molecular biologists: the genetic programme metaphor and the implicit belief that the ultimate explanation of morphogenesis resides in the molecular realm, i.e. that all relevant information is encoded in the DNA. Thus, a basic implication of the genetic programme metaphor is that genes, via their products, determine macroscopic biological form. One should not forget, on the one hand, that genes only encode proteins that participate either as catalysts or effectors of metabolic fluxes or as structural elements such as cytoskeleton or membrane components (Chapter 11). On the other hand, the spatio-temporal scaling of molecular events is very different from that corresponding to morphogenesis in embryonic fields. Thus, the activity of thousands of molecules in developing systems is somehow subjected to spatio-temporal coordination.

We propose an alternative approach to understanding spatio-temporal coordination which is rooted in the concept of dynamic organization developed in Chapter 2. The starting point of this alternative view is to recognize the most prominent feature of biological systems: its functional organization at many simultaneous levels (Table 2.1). Some aspects of this approach have conspicuous predecessors (Needham, 1934; Turing, 1952; Goodwin, 1963; Rosen, 1970; Nicolis and Prigogine, 1977; Kauffman, 1989).

According to the concept of dynamic organization, new functional properties emerge in living systems by transitions that occur between levels of organization. They are self-organized since those transitions occur at instabilities in the dynamics of biological processes encompassing spatio-temporal coherence of hundreds of thousands of molecules or supramolecular structures. At this point we introduce the notion of symmetry. When transitions between levels of organization occur, spatio-temporal coherence manifests itself as a change in symmetry which may be spatial, temporal or both. Therefore, changes in symmetry are a clear manifestation of the appearance of new functional properties in living systems by transitions between levels of organization.

Let us now introduce the concept of symmetry and specify its kinetic and thermodynamic roots, and finally rejoin it to the concept of dynamic organization.

4.1 SYMMETRY

Everyone has an intuitive perception of the concept of symmetry that appears repeatedly in everyday life. We are surrounded by symmetrical objects and we are ourselves bilaterally symmetrical. The Greek word *symmetros* means 'regular', 'well proportioned', 'harmonious' (quoted in Eigen and Winkler, 1981). Mathematically, symmetry may be defined as a transformation of an object that leaves the object invariant, apparently unchanged (Stewart and Golubitsky, 1992).

The simplest form of symmetry is based on mirror images on either side of a straight line (Figure 4.1a). This is the 'right–left' or **bilateral symmetry** that occurs frequently among higher organisms (Eigen and Winkler, 1981). The two hands in a human body are mirror images; they seem identical, yet they are not superimposable. The only difference between them is that one can be described by tx, ty, tz transformed coordinates, whereas the other by –x, –y, –z coordinates. Thus, they are spatially inverted forms (Garay, 1987). Another important way of creating symmetrical forms is the progressive spatial repetition of a basic pattern. This process can consist of a translation or rotation through an angle equal to 360 degrees divided by some integer (Eigen and Winkler, 1981). All spatial symmetries can be derived from a combination of these two basic operations (Figure 4.1).

Because of the symmetry property, an arbitrary axis may be defined with respect to which a rotation or a translation (transformations) may be performed leaving the object (or pattern, or motif) without an apparent change. The resulting object (pattern, motif) is superimposable with respect to the original one before the transformation. If the object is superimposable in any spatial position or orientation, it is homogeneous,

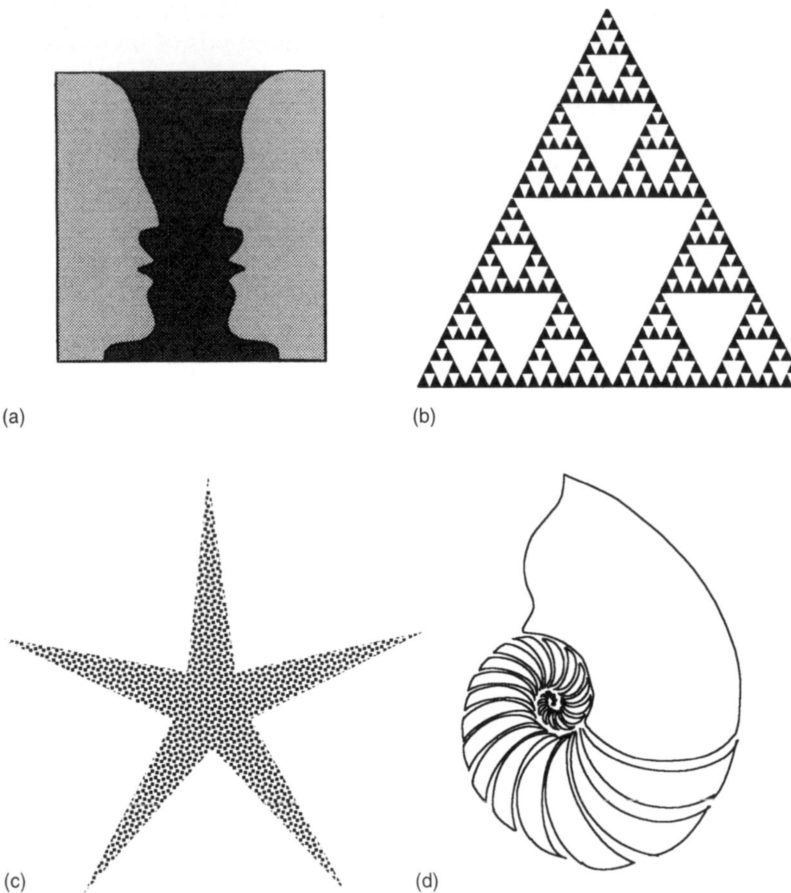

Figure 4.1 Symmetrical patterns. (a) Bilateral symmetry: a vertical axis defines the transformation (rotation) that converts one side of the symmetrical object into the other. (b) Sierpinski gasket: an identical object may be obtained by 120° rotation and additionally exhibits self-similarity. (c) Starfish: an identical object may be obtained by 72° rotation. (d) *Nautilus* cross-section, representing a discrete logarithmic spiral symmetry.

completely symmetrical and isotropic. If an axis can be defined and two identical but not superimposable patterns appear, the object is said to present bilateral symmetry. In fact, two axes are needed to define a transformation that will leave a bilaterally symmetrical object unchanged. If the pattern is able to be superimposed but only after rotating it in a fixed angle around an axis, the motif is said to be **radially symmetrical** (Figure 4.1b,c).

Symmetry breaking occurs when changes take place in the number of axes with respect to which transformations may be performed. Also, changes in symmetry implying a spatial reorganization may occur, e.g. the passage of particles distributed at random in the surface of a sphere to particles forming patches. In other words, the average distribution is still symmetrical but the local distribution is not, because the original symmetrical pattern was not stable when a parameter of the system (e.g. the size of the field) increased over a certain value.

4.2 SYMMETRY AND BIOLOGICAL ORGANIZATION

A primary feature of the structural polarity of vertebrates is the fact that members of this group exhibit bilateral symmetry. Vertebrates share this common organizational feature with a number of invertebrate groups (annelid worms, arthropod phylum which includes crustaceans, arachnids, insects). In strong contrast is the radial symmetry of coelenterates and echinoderms, whose body parts radiate out from a central axis like the spokes of a wheel. The degree of activity of animals appears to be correlated with their type of symmetry. The radiate echinoderms and coelenterates are in general sluggish types, slow-moving or fixed to the bottom or, if free-floating, mainly drifters with the current rather than active swimmers. Vertebrates, arthropods and marine annelids are, on the other hand, generally active animals.

Although the vertebrate body is essentially a bilaterally symmetrical structure, there are many exceptions to this general statement. Organs which primitively lay in the midline may be displaced: the heart may be off centre; the abdominal part of the gut – stomach and intestine – is usually twisted and the intestine may be convoluted in a complicated asymmetrical fashion. Again, in paired structures those of the two sides may differ markedly; for example, in birds only one (the left) of the two ovaries is functional in the adult. A still greater asymmetry is seen in the flounders, where the whole shape of the body is affected by the substitution of the two sides for the normal top and bottom of the animal.

Asymmetry is also present at the molecular level in the form of chirality or handedness (Pasteur, 1884; Mason, 1984). The metabolism of living cells clearly prefers to make use of compounds that are characterized stereochemically by one direction of rotation, i.e. handedness or chirality. However, we will not stress this sort of asymmetry since we are mainly dealing with symmetry-breaking phenomena which depend upon the collective dynamic properties of molecules or macromolecules.

4.3 TEMPORAL SYMMETRY AND BIOTHERMOKINETICS

The concept of symmetry is easily associated with space. Nonetheless, temporal (a)symmetry may also be identified in our familiar experience. Any periodic phenomenon that happens repeatedly is exemplifying temporal symmetry, such as oscillations and biological rhythms (Chapter 3). Any moment within the cycle will be reproduced, or an equivalent state of the system will be recovered after a lapse, equal to the period of the phenomenon.

Asymmetry defines directions either in space or in time. In that respect, a thermodynamic equilibrium state is, within the perspective of temporal symmetry, equivalent to a homogeneous space for spatial symmetry. Within the equilibrium state, any temporal transformation (shift in forward or backward direction) will leave time unchanged. An irreversible process introduces an arrow in time (Prigogine, 1967) which defines temporal asymmetry. The state of any system undergoing an irreversible process is different if we look at it with a certain lapse of time, backward or forward, with respect to a specified instant.

Symmetry, either temporal or spatial, is related to thermodynamics. Such a relationship may be understood in terms of identical mutual interactions between different variables in a system, e.g. of two metabolites. Translated into thermodynamic terms, identical mutual interactions mean symmetry in the matrix of phenomenological coefficients associated with the description of its dynamic behaviour (Chapter 1). We have already discussed Prigogine's demonstration that in the linear domain of flows and forces, which implies a symmetrical matrix of phenomenological coefficients, it is not possible to get any coherent behaviour (i.e. self-organization) whether temporal, like oscillations, or spatial symmetry-breaking, such as the appearance of any ordered pattern (Chapter 1).

4.4 CURIE'S PRINCIPLE OF SYMMETRY

Curie (1894) made a fundamental contribution to our understanding of symmetry. He presented two statements of a general principle (quoted by Stewart and Golubitsky, 1992):

> If certain causes produce certain effects, then the symmetries of the causes reappear in the effects produced.
> If certain effects reveal a certain asymmetry then this asymmetry will be reflected in the causes that give rise to them.

This principle may be conceptually appreciated in a process of transport of charged species coupled to a chemical reaction. In this system two forces may be identified: the gradient of electrical potential, $\Delta\Psi$, and the affinity of the chemical reaction, $A = -\Delta G$.

In the near equilibrium domain we may write:

$$J_d \text{ (transport flux)} = L_{11}F(-\Delta\Psi) + L_{12}A \qquad (4.1)$$

$$V \text{ (reaction rate)} = L_{21}F(-\Delta\Psi) + L_{22}A \qquad (4.2)$$

According to the Onsager reciprocity relationship in the near equilibrium domain:

$$L_{12} = L_{21} \qquad (4.3)$$

Additionally, when the gradient of electrical potential is null, i.e. there is a symmetrical distribution of charged species i, the transport flow will be zero irrespective of the affinity of the chemical reaction and then:

$$L_{12} = L_{21} = 0 \qquad (4.4)$$

which means that if there is no anisotropy, then neither is there coupling between reaction and transport. This is an example where we may recognize Curie's principle: the effects (i.e. transport flow) cannot be asymmetrical ($\neq 0$) if there is no asymmetry in the force (i.e. electrical potential gradient) driving the phenomenon.

However, Curie's principle does not hold for every situation as stated. Sometimes, asymmetrical effects are observed upon symmetrical causes (forces). This is due to the fact that symmetry should be considered together with the stability problem (Chapter 1). A symmetrical state may well be unstable and a set of asymmetrical stable states (e.g. two states of opposite asymmetry) will have an equal probability of occurring, though in a given system just one of those states with equal probability will be observed at a time. Let us consider as an example a ruler standing horizontally on which two equal compression forces are applied at its extremes. At the begining the ruler will stay horizontal (symmetrical); when the force passes over a threshold, the ruler will buckle towards, say, the bottom. It is equally probable that the ruler will buckle towards the top but only one of the positions will be observed at a time.

Symmetry-breaking and the 'violation' of Curie's principle emerges when a symmetrical state is no longer stable. At this point, the relationship between thermodynamics and symmetry becomes self-evident. The thermodynamic conditions for stability described in Chapter 1 are no longer fulfilled when there are coupled processes exhibiting kinetic nonlinearities under far from equilibrium conditions. Under those conditions it has been demonstrated that Onsager reciprocal relations do not hold, so that the matrix of phenomenological coefficients is not symmetrical (Prigogine, 1967; Chapter 1). Thus, for symmetry-breaking to occur, the matrix of phenomenological coefficients should also be asymmetrical.

4.5 ASYMMETRY AND TRANSPORT: ANISOTROPY AND VECTORIALITY

Anisotropy is a closely related concept to that of asymmetry. Unevenness in the intrinsic properties of a system, when different directions in space are considered, is known as anisotropy. It is associated to **vectoriality**, which implies a passage between scalar (homogeneous, isotropic) to vectorial objects or processes (exhibiting spatial directionality). In the case of reaction coupled to transport of matter, the locally-crossed phenomenological coefficients in a membrane (L_{12} and L_{21} in equations 4.1 and 4.2) have vectorial character since they relate a scalar process (the chemical reaction) to a vectorial one (the flux) (Figure 4.2) (Katchalsky and Spangler, 1968; reviewed in Caplan and Essig, 1983).

Mitchell (1968) stated that macroscopic transport occurred spontaneously when the catalysts or substrate and group diffusion were arranged anisotropically relative to structures such as membranes. He coined the concept of **intrinsic anisotropy** of a group transfer system to define the anisotropy with respect to the accessibility of group-donor and group-acceptor molecules, and **extrinsic anisotropy** as a result of an unequal distribution of the group-acceptor species.

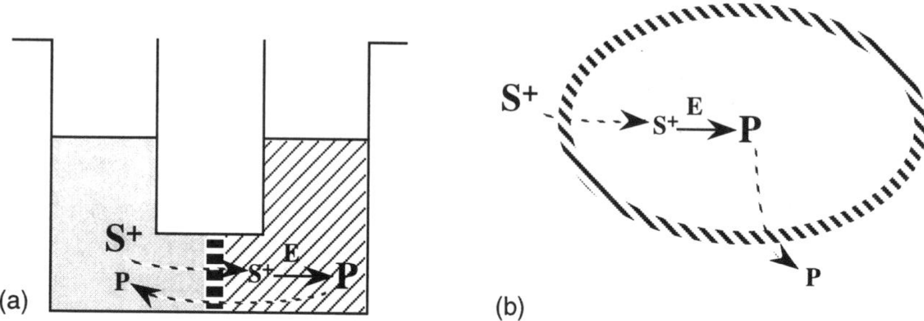

Figure 4.2 Vectoriality in transport systems. (a) Reactor showing coupled scalar (chemical reaction) and vectorial (transport) processes, and representing a metaphor of (b) a cell bathing in, for example, a culture medium from which substrates are transported into the cell and are chemically processed inside. The size of S^+ and P symbols indicates the relative concentration of the charged substrate(s) and product(s), respectively, in each compartment. The catalyst of the reaction, E, is present only in the right-hand compartment of (a). Transport and reaction processes are indicated with dashed and solid arrows, respectively.

The near-equilibrium formulation of thermodynamics in membrane transport processes, coupled to chemical reactions, retains the vectorial nature of mass flow through the membrane and the scalar character of chemical reactions. The set of global phenomenological coefficients thus obtained satisfy the Curie principle. It was demonstrated in a symmetrical membrane that if the proper choice of global flows and forces is made, the cross-phenomenological coefficients relating mass flow to reaction flow will always be zero and therefore coupling is not possible (Caplan, 1973). This provides a method for examining global symmetry experimentally. Coupling in membrane transport processes comes about as a direct consequence of (a) the membrane asymmetry and (b) the stationarity condition, which imposes a mutual linear dependence on certain of the flows.

An analogue of Curie's principle is provided by linear systems in which the phenomenological coefficients are a function of the equilibrium properties of the membrane components as exemplified by equations 4.1 and 4.2. If such equations hold locally, i.e. at each point within the membrane, the local cross-coefficients must have vectorial character since they relate a scalar (chemical reaction) to a vectorial (flux) process. The latter would imply anisotropy, i.e. intensive properties are different according to the spatial direction in which they are studied. Conversely, if a medium is isotropic and the processes occurring are linear, local vectorial coefficients cannot exist (Caplan, 1973).

4.6 REACTION–DIFFUSION AND SYMMETRY-BREAKING

Turing's pioneer work (1952) demonstrated that an autocatalytic reaction occurring in an initially uniform or isotropic field, when coupled to the transport of matter through diffusion, could give rise to symmetry-breaking and the appearance of a spatially organized state such as a gradient. He coined the word **morphogen** to name the chemicals participating in the reaction and diffusing through a field. Since Turing's (1952) postulation of a reaction–diffusion theory for pattern formation, many hypotheses have been proposed to explain from a theoretical point of view how stable, spatial, non-uniform concentration gradients of substances can arise from an initially uniform distribution of them, when subjected to an external perturbation.

The system proposed by Turing fits all thermodynamic and kinetic constraints for symmetry-breaking and self-organized structures to occur. The models assumed an evenly distributed infinite source of a substance, which is equivalent to considering the system to be open to the flux of matter (Meinhardt, 1986) (Figure 4.3). There should be two processes occurring simultaneously (coupled), namely reaction and diffusion. Another condition concerns the nonlinear kinetics of at least one of the processes which is satisfied by the autocatalytic nature of the reaction involved.

Reaction–diffusion models provide putative mechanisms that explain the appearance of spatial heterogeneity that may throw light on morphogenesis itself. This is related to developmental biology and organization of the cytoplasm and may also account for uneven distributions of subcellular structures and organelles (Chapters 6 and 10).

4.7 THE CONCEPT OF MORPHOGEN

The concept of morphogen refers to a chemical able to generate a spatial heterogeneity in an initially homogeneous field (Slack, 1987). However, the very idea of an initial homogeneous field is biologically counter-intuitive. In living organisms, heterogeneity is the rule and the orientational cues for initiation of the establishment of structures are ubiquitous. The initial homogeneity assumed by Turing raises biologists' suspicions since it introduces an element of chance in biological organization, with the concept of random perturbation giving rise to an uncertainty component in the spatial organization. This is in contrast to the well established fact that development and cytoplasmic organization are highly conserved processes. In that respect, it should be taken into account that the environment where the egg develops (equivalent to boundary conditions for pattern formation) is highly conserved too. Therefore, it is not unexpected that the product of development with both highly conserved genetic background and environment is also highly reproducible.

The elusive nature of morphogens (Tsonis, 1987; Slack, 1987) with the features attributed by Turing is perhaps one additional factor that has contributed to the reluctance of developmental biologists to adopt a kinetic preconception (Harrison, 1987). In this sense, retinoic acid in chick limb bud and DIF in *Dictyostelium discoideum* were shown to induce the formation of digits 2, 3 and 4 and stack differentiation, respectively, in a concentration-dependent manner (Thaller and Eichele, 1987). These two chemicals are candidate morphogens.

In our opinion, Turing has been interpreted too literally. A morphogenetic field may well be a pattern of fluxes through main metabolic pathways within a tissue (Chapters 10 and 11). Alternative morphogenetic fields would be provided by gradients of adhesiveness as proposed by Oster *et al.* (1983). At the cellular level, the cell's position in a tissue could be sensed by cell cytoarchitecture, since the genome itself

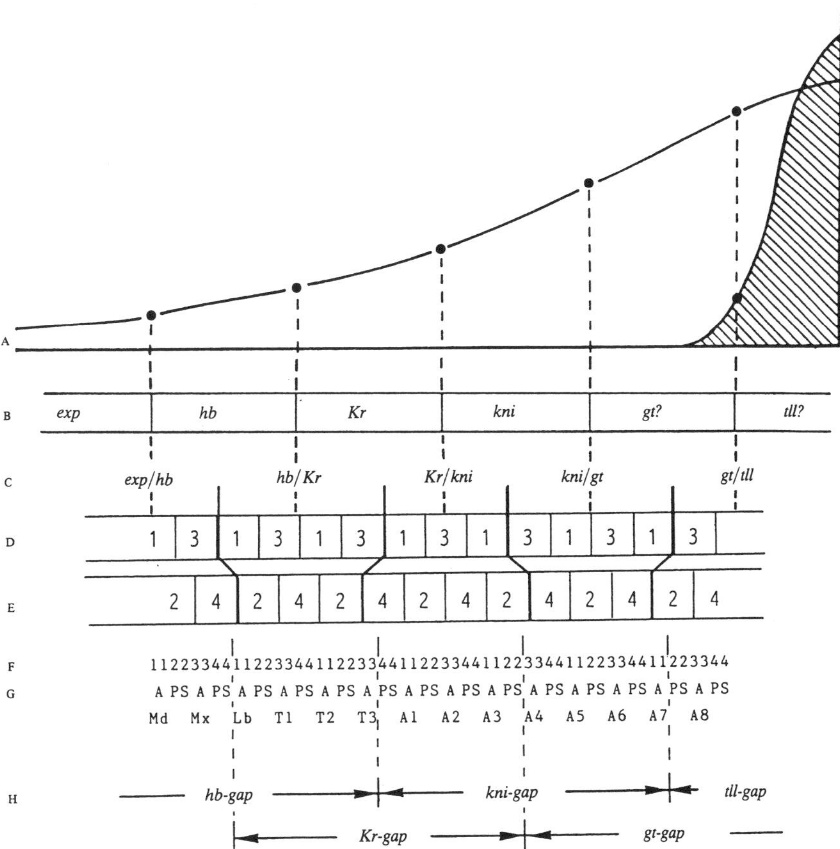

Figure 4.3 A model for segmentation in *Drosophila*. According to the model proposed by Meinhardt (1986), the primary anterior–posterior gradient (A) is assumed to activate four position-dependent cardinal genes (B) (*hb*, *Kr*, *kni* and possibly *gt*). Two further genes are activated at both margins (*exp* and *tll*) (C). The borders between these cardinal regions (*hb/Kr*...) organize a portion of 3–5 segments (enframed) of the two binary sequences ...131... (D) and ...242... (E). Both sequences are out of register. A seven-fold repetition of four cell states (...1234...) represents the double-segment pattern. It results from the ...131... and ...242... sequence either by merging or by induction (F); a tentative assignment of the 1, 2, 3 and 4 cell states and pair-rule loci. Each 1234 sequence induces two SAP sequences, the pattern of individual segments (G), the predicted gap sizes if one of the cardinal genes is lost. The actual gaps can be somewhat smaller due to an elongation of the 1234 sequences from both margins (H). (Reproduced from Meinhardt, 1986, by permission of The Company of Biologists Ltd.)

appears to be responsive to major controlling signals such as cell shape and surface contact from the intracellular macromolecular networks (Penman *et al.*, 1981; Shinohara *et al.*, 1989). This issue is developed further in Chapters 10 and 11.

4.8 LEVELS OF ANALYSIS OF DEVELOPMENTAL PROCESSES

A problem persists in the whole literature which may be stated as follows. What is the appropriate level of organization at which to search for an explanation of the appearance of cells with different fates from a common undifferentiated ancestor?

One of the most distinctive features of the complexity of living systems is their structural and functional organization at many simultaneous levels. Biological evidence of the existence of levels of organization was presented in Chapter 2. In developmental processes, several levels of organization should coexist. Explanations of biological phenomena may be taken from one of those levels, e.g. molecular or cellular. As a result, each has evolved as a discipline on its own with little exchange with the others.

Development may be studied from the angle of what kind of phenomena give rise to organization. It may be through self-organization (dissipative structures) or through the achievement of a certain spatial segregation (for example, heterogeneity in the distribution of organelles, cytoskeleton components or subcellular particles) attained because of particular environmental conditions (such as chemical gradients, gravity or electric fields) that compel a particular type of organization. In any case, order is achieved by a response of the dynamics of the processes involved.

Development may be broadly classified as mosaic or regulative (Davidson, 1990) though most cases found in nature lie somewhere in between these two extremes. **Mosaic development** refers to a complete independence of each part of a developing embryo from its neighbouring parts. Thus, if a part in an embryo is translocated (transplanted) to a different region, it will evolve to render the same structure that would have arisen from that part when placed in its right position. According to mosaic models of embryonic patterning, a rigid and static localization of distinct determinants is invoked. It had been believed that worm development exhibited rigid cell lineages, with almost no inductive interactions between cells of different cell lineages during development, but this notion had to be dismissed after the work of Schierenberg (1987) with *Caenorhabditis elegans*.

Regulative development corresponds to that in which any missing part in a developing embryo will be recovered since its neighbouring regions will reset their developmental fates to render the missing structure. This type of developmental strategy requires strong cell-to-cell interactions throughout development. Regulative models postulate morphogen gradients or inductive cascades of cell-to-cell interactions.

From the perspective of the dynamics of morphogenesis, it is the stability of biological organization that is achieved that determines whether development will be mosaic or regulative, or if the specification of cellular fate is autonomous or conditional, respectively (Davidson, 1990). In this context, the problem of stability refers to how large is the parametric space in which a certain stable state is achieved. Figure 4.4 shows a dynamic interpretation of biological development in terms of a system displaying bistability. The upper branch of stable steady states stands for the differentiated state; the lower branch represents the pluripotent form. When the parametric space compatible with a certain organization is narrowed (e.g. the upper branch of stationary states in Figure 4.4b) upon a relatively small perturbation, the state of the system will be lost and it will undergo a transition to another state, allowing regulative development and reshaping. In case the parametric space is wide enough, it will be difficult for the system to find its way to reach another branch of states, and organization will be stable. In that case, development will be referred to as mosaic (Figure 4.4a).

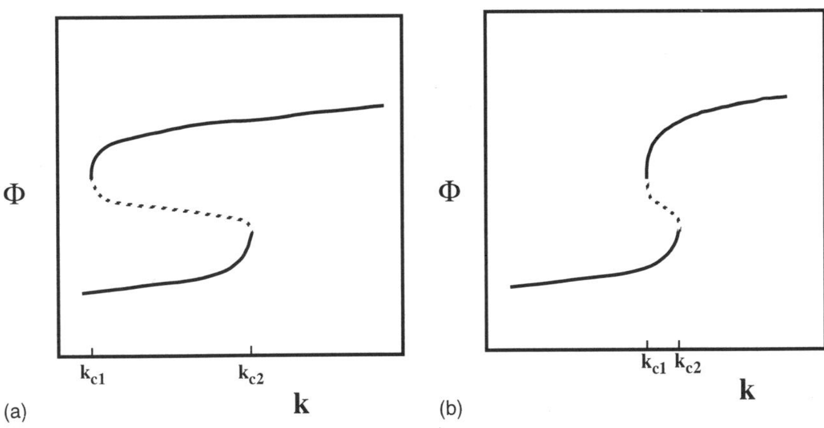

Figure 4.4 Regulative and mosaic development interpreted as stability of developing systems showing bistable behaviour. Two bistable curves are displayed with different widths in the parametric range (k) which separates both branches of stable states (continuous lines) through a region of unstable states (dashed lines). In (a) a wide parametric region (k_{c1}–k_{c2}) must be spanned by the system before changing its state at critical values of the parameter under study (k_c); whereas (b) shows that a narrow span suffices to get the system out of a particular state (see text). This behaviour is described for two-coupled dynamic systems in Figures 1.8 and 6.11.

With respect to the question about formation of the main axis of symmetry in developing systems, experimental data indicated that it is organized either during oogenesis (at least with respect to one axis) as occurs in all animals, or immediately after fertilization, as occurs in the case of brown algae and fucoids (Kropf, 1992; Goodner and Quatrano, 1993). In both cases, there is an instability leading to a rupture of symmetry, ending in the establishment of an initial polarity. At this stage, in the case of animal cells the mother provides a heterogeneous environment (boundary conditions) which will determine that the egg organizes spatially asymmetrically, undergoing what Prigogine called an **assisted bifurcation** (Prigogine and Stengers, 1983). The organization will certainly occur, primed by environmental conditions to render a reproducible pattern and to allow a frog's egg to yield another frog. If symmetry-breaking occurred through random fluctuations driving to self-organization (e.g. through a reaction–diffusion mechanism), the appearance of patterns in developing systems not driven by a predefined environment would not have been reproducible as found in nature.

Taking into account that development is highly conservative, it is doubtful that a random fluctuation is a spatial cue for the establishment of polarity. It has been shown for some eggs that fluctuations arising from external stimuli (the site of sperm entry or the direction of illumination) could be the trigger for symmetry-breaking events that provoke polarized growth. With these few exceptions we do not know the nature of the polarizing signal (section 4.14).

The next level is to identify developmental processes of particular systems that may mediate in the triggering of instabilities of a certain state to allow the emergence of a new type of organization. This level of analysis, together with that presented in previous paragraphs, searches for processes in which characteristic spatio-temporal dimensions lie further from the 'transition point' in the allometric curve relating different levels of organization (Chapter 2, Figure 2.2). The processes that have to be identified should occur in spatial dimensions of micrometres which relax within a few minutes in order to give rise to macroscopic order (Aon and Cortassa, 1993). At this level of analysis, unequal distribution of ion pumps and channels within the plasma membrane is a clear manifestation of cell polarization. Developmental events in fucoid eggs suggest that relevant processes for cell polarization are electrophoresis of channels driven by Ca^{2+} currents and stabilized through interaction with cytoskeleton components (Kropf, 1992). Alternatively, transcellular currents may result from local regulation of transporters such that they are active at only one end of, say, a zygote, instead of asymmetrical distribution (Kropf, 1992).

A thorough survey of the key cellular events leading to the establishment of polarity in developing systems of different animal species has been performed (Davidson, 1990). The main questions to be answered at this level of analysis concern whether the cleavage pattern should be invariant; which axes exist at the moment of fertilization; the way in which the secondary axis is laid down and the autonomous *versus* conditional specification of cell lineage; and what the processes are through which the differentiated fates of the progeny of cleavage-stage blastomeres are first established – what is called specification (Davidson, 1990). The latter concepts are related to mosaic and regulative development.

Another level of study is the molecular dissection of the elements intervening in a particular mechanism and the characterization of the properties of a molecular entity that plays a certain role in a differentiation process. This is, in fact, the level that has been most extensively worked out. Since an integrative picture provided by the other two levels of description is missing, for the time being it is merely a collection of empirical information without a theoretical framework to interpret the data. Moreover, it is not always easy to distinguish between a chemical that participates in processes involved in symmetry-breaking and self-organization and other compounds differentially expressed as a consequence of the appearance of polarity in a developing embryo.

4.9 SYMMETRY AND DEVELOPMENT

This section deals with the emergence of molecular coherence in such a way that developing organisms acquire a determined form in the spatial scale of micrometres to millimetres. Development of the whole organism spans a time scale of days or weeks but each process of regional specification may take only a few hours (Slack, 1987). More precisely stated, the problem is to understand how molecules (proteins), macromolecules and supramolecular structures are organized dynamically and structurally over characteristic dimensions so that, for example, specific shapes emerge. Diffusion has been proposed as a mechanism to link local-range (regional specification) to global developmental processes (e.g. organs or whole organism) (Crick, 1970).

Classical embryological experiments suggest that specification of cell fates in many systems is controlled by the asymmetrical partitioning of cytoplasmic components present in the egg. The principle of symmetry, first stated by Curie (1894), was:

> Lorsque certains effets révèlent une certaine dissymétrie, cette dissymétrie doit se retrouver dans les causes qui lui ont donné naissance.

This was paraphrased almost 90 years later by Kirschner (1982) for developing systems:

> The asymmetry found in the adult organism reflects asymmetry that either exists within the egg or develops in the early embryo.

Let us take a concrete developing system to visualize the problem posed in the paragraphs above. The amphibian oocyte is actually asymmetrical. Thus, the question is raised about the way in which this single cell acquired its spatial organization during oogenesis. Indeed, heterogeneity is the rule in living organisms. Some elements that help to answer the previous question come from studies in a variety of systems which suggest that cytoskeletal structures participate in the asymmetrical distribution of cytoplasmic elements together with 'developmental potential'. These studies are germane to our proposal that polymerization–depolymerization of cytoskeleton components may be involved in the establishment of a self-emergent (a)symmetry axis (Chapters 2 and 5). Cytoskeleton components may self-organize in the spatio-temporal 'window' of micrometres and minutes for which transitions from microscopic to macroscopic organization at the cellular level happen (Chapters 6 and 7).

The following sections demonstrate the dynamics of the distribution of cytoskeleton components in the macroscopic asymmetrical organization of *Caenorhabditis elegans* and *Xenopus laevis* embryos, as well as in the lower unicellular eukaryote *Saccharomyces cerevisiae*.

4.9.1 ESTABLISHMENT AND MAINTENANCE OF ASYMMETRY IN *CAENORHABDITIS ELEGANS* ZYGOTES

During the first cell cycle following fertilization, the embryo of *C. elegans* undergoes a series of events: contraction of the cell membrane, pronucleus migration, completion of meiotic division of the egg pronucleus and migration of the asters that determine an asymmetrical positioning of the mitotic spindle (for a review, see Goldstein *et al.*, 1993). All asymmetry manifestations described in Figure 4.5 could be inhibited by disruption of microfilaments by continuous treatment of the zygote with cytochalasin D (CD), an inhibitor of microfilament assembly (Strome and Wood, 1983; Hill and Strome, 1988).

Continuous treatment of the one-cell embryo of *C. elegans* with CD, beginning either before or during the early stages of pseudocleavage, inhibits all of the manifestations of embryonic asymmetry (Hill and Strome, 1988). These data strongly suggest microfilament-mediated events during the first cell cycle. Figure 4.5 shows the main events occurring after fertilization in the one-cell embryo of *C. elegans* (Hill and

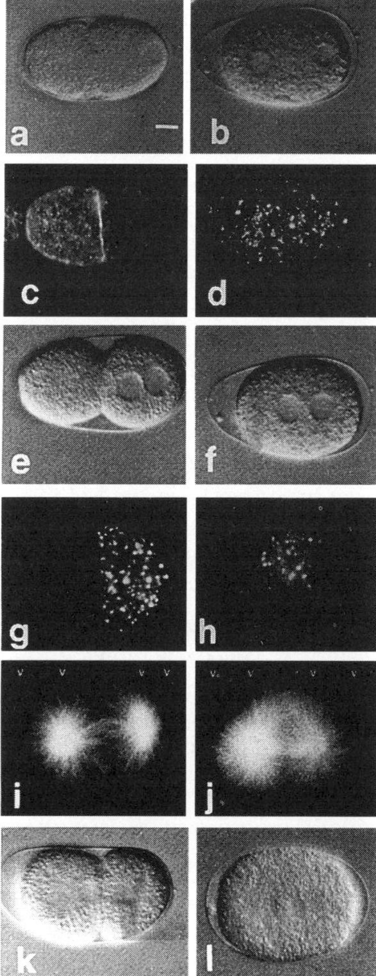

Figure 4.5 Role of microfilaments in the establishment and maintenance of asymmetry in *Caenorhabditis elegans* zygotes. Comparison of events that occur in (left) norrnal embryos and (right) embryos treated continuously with 2 μg ml^{-1} cytochalasin D (CD) after meiosis. (a,b) CD treatment inhibits pseudocleavage; (c,d) CD treatment disrupts the normal array of microfilaments; (e,f) pronuclear meeting occurs in the posterior hemisphere of untreated embryos and in the centre of CD-treated embryos; (g,h) CD treatment inhibits the posterior segregation of germline-specific P granules, which instead coalesce in the centre of the zygote; (i,j) in CD-treated zygotes the mitotic spindle does not become asymmetrically positioned (arrowheads indicate positions of centre of spindle asters and the cell periphery); (k,l) CD treatment inhibits cytokinesis. Bar = 10 μm. (Reproduced from Hill and Strome, 1988, by permission of Academic Press.)

Strome, 1988). Approximately 1 h after fertilization, the meiotic divisions of the egg pronucleus are completed at the anterior end of the zygote. This is followed by contractions of the anterior cell membrane, known as early pseudocleavage (Hill and Strome, 1988) (Figure 4.5a). This contraction leads to a constriction perpendicular to the anterior–posterior axis that resembles a true cleavage furrow, i.e. late pseudocleavage (Hill and Strome, 1988) (Figure 4.5e). During pseudocleavage, the egg pronucleus begins to migrate toward the posterior. The sperm pronucleus slowly moves away from the posterior periphery, and the two pronuclei meet in the posterior hemisphere (Figure 4.5e) and then migrate to the centre of the embryo. Concomitant with pseudocleavage and pronuclear migration, germline-specific P granules, visualized by indirect immunofluorescence, become localized in the posterior cortex of the zygote (Figure 4.5g). The mitotic spindle, initially positioned symmetrically along the anterior–posterior axis, becomes asymmetrical in both position and structure as it grows. By anaphase the spindle is positioned slightly closer to the posterior end of the cell and the posterior aster appears flattened relative to the anterior one (Figure 4.5i). Cytokinesis occurs at a position determined by the position of the mitotic spindle and generates a small posterior cell (P1) and a larger anterior cell (Figure 4.5k).

The finding that cell size, cell cycle times and positioning of the P granules were correlated in *C. elegans* embryos suggests that a single microfilament-mediated event establishes embryonic asymmetry late in the first cell cycle (Hill and Strome, 1988; Goldstein *et al.*, 1993).

4.9.2 INVOLVEMENT OF CYTOSKELETON IN THE CHANGE OF SYMMETRY BY SUBCORTICAL ROTATION OF CYTOPLASM RELATIVE TO THE SURFACE OF *XENOPUS* EGGS

The *Xenopus* egg is a huge cell with a distinct animal–vegetal polarity that develops into dorsal–ventral polarity following fertilization. The amphibian egg undergoes a 30° rotation of its subcortical contents relative to its surface during the first cell cycle (Ancel and Vintemberger, 1948). Movement over an approximate distance of 350 μm (equivalent to 30°) takes place during the second half of the first cell cycle, a 50 min period (at 18°C) (Vincent *et al.*, 1986, 1987). The rotational movements in the vegetal hemisphere are directly implicated in the specification of the embryonic dorso-ventral axis. Cytoskeletal proteins change their distribution following oogenesis and fertilization (Figure 4.6). Subcortical rotation is dependent on microtubules throughout its entire course since it is prevented when *Xenopus* eggs are injected with microtubule-depolymerizing drugs – colchicine, vinblastine or nocodazole (Vincent *et al.*, 1987). These results show that rotation is highly sensitive to inhibitors of microtubule polymerization. On the other hand, grey crescent formation is remarkably insensitive to CD, suggesting the independence of subcortical rotation on microfilament assembly (Manes *et al.*, 1978; Vincent *et al.*, 1987).

4.9.3 YEAST CELL POLARITY

Cell polarity is evidenced in yeast in two vital processes: bud emergence during cell cycle, and cell elongation during 'mating'. In the latter, the direction of elongation follows an external cue given by the position of the mating partner. The cues acting on the establishment of polarity during budding are still obscure. There are two types of budding patterns according to the ploidy of yeast: axial or bipolar (Chant, 1994).

Cytoskeleton components (namely, actin and microtubules) become polarized along the mother–bud axis. Also, proteins known to interact with actin, such as class V myosin or calmodulin, are believed to play a role in the polarized secretion of vesicles (Chant, 1994). Other proteins have been identified as being differentially distributed in the bud tip or at the neck between mother cell and bud. In the latter group there are proteins binding GTP or GDP and changing their subcellular distribution according to the bound nucleotide (Chant, 1994). BUD1 and CDC42 are representative proteins that participate in GTP–GDP exchange cycles, providing a clear link between cellular energetics and the mechanisms intervening in polarized growth in yeast.

4.10 FLUCTUATING ASYMMETRY AND DEVELOPMENT

Development is characterized by stepwise evolution toward a stable, predictable and reproducible final state. Nevertheless, instabilities are verified in developing embryos. Fluctuating asymmetry is a measure of such developmental instabilities and arises when deviations from bilateral symmetry occur. In a population of organisms, this fluctuating asymmetry is normally distributed such that the mean is zero (Graham *et al.*, 1993). These authors model fluctuating asymmetry with a system of four ODEs for two chemicals, an activating and an inhibitory one, ruled by autocatalytic expressions (Turing, 1952). The dynamic behaviour of this model is chaotic, i.e. it exhibits sensitivity to the initial conditions. Thus, if the initial concentration of the activator is different in the right and left sides of the developing organism, the model evolves with deviations from the right–left symmetry, accounting for fluctuating

	Oogenesis				Maturation	Fertilization	First cell cycle	

(See Reference 1 for details of morphology)

MORPHOLOGY

Stage I oocyte	*Stage II-III oocyte*	*Stage III-IV oocyte*	*Stage V-VI oocyte*	*Mature egg*	*First half*	*Second half*	*First cleavage*
50-300µm dia. Mitochondrial cloud on presumptive vegetal side of nucleus.	300-600µm dia. Yolk, then pigment start to appear. Mitochondrial cloud disperses. Germ plasm moves to cortex.	600-1000µm dia. Pigment accumulates preferentially in animal cortex. Yolk in radial islands.	1000-1300µm dia. Unpigmented band forms at equator. Yolk in radial islands in animal half only. Oocyte arrested at meiotic prophase I.	Germinal Vesicle (GV) breaks down. Major cytoplasmic reorganisation. Egg arrested at meiotic metaphase II.	Actin based contraction of animal cortex. Pronuclear migration and fusion Cytoplasm fluid (72).	Cytoplasm has microtubule based viscosity (72). Cortex rotates 30 degrees relative to cytoplasm.	Cytoplasm has microfilament based rigidity (72).

TUBULIN

Stage I oocyte	*Stage II-III oocyte*	*Stage III-IV oocyte*	*Stage V-VI oocyte*	*Mature egg*	*First half*	*Second half*	*First cleavage*
Tubulin (αβ) found concentrated in perinuclear masses and mitochondrial cloud in some (8,9) but not all (7) studies. Microtubules (L) in mitochondrial cloud (17). Tubulin in cortex (9,8).	Tubulin in remnants of mitochondrial cloud and in radial yolk-free cytoplasmic strands (8,7,12).	Tubulin in symmetrical radial array (8,7,12). Tubulin not tyrosinable despite constant level of tubulin tyrosine ligase activity (83).	Level of polymer high (20) but asters not inducible (22,68,84,84). Microtubules detected around GV (10,63), in cortex (11) and rarely in cytoplasm (26). Radial tubulin array in animal hemisphere (8,12). Tubulin in GV (8). α-Tubulin tyrosinable (83) and able to assemble *in vitro* (26).	Level of polymer high (20,21). Asters inducible (24,25,68,84,85) Microtubules detected in spindle (10,86) and cortex (11,87,19). Less stable than in St. VI oocyte (20). α-Tubulin tyrosinable (83).	Tubulin mostly unpolymerized (21). Sperm aster grows in animal hemisphere. Active X-MAP enhances microtubule elongation (28).	Level of polymeric tubulin high (21). Microtubules aligned in parallel array beneath vegetal cortex (38).	Microtubules depolymerize (21) and vegetal array thins out (38). X-MAP inactivated at mitosis (28).

Figure 4.6 Changes in distribution of tubulin in the *Xenopus* egg: deployment and function of tubulin during changes in cellular organization accompanying oogenesis, oocyte maturation, and following fertilization. For references (in parentheses) see original work of Elinson and Houliston (1990). (From Elinson and Houliston, 1990, by permission of W.B. Saunders Co.)

asymmetry (Graham *et al.*, 1993). A developing organism will amplify initial asymmetries to render a body with many different cell types organized in various tissues which in turn form organs fulfilling diverse functions.

4.10.1 DEVELOPMENTAL HOMEOSTASIS AND PROTEIN HETEROZYGOSITY

Fluctuating asymmetry, developmental stability and homeostasis are concepts related to heterozygosity (Mitton, 1993). In animal populations of various species a relationship between heterozygosity and developmental homeostasis has been found: the higher the heterozygosity, the lower the phenotypic variability and fluctuating asymmetry, which results in higher developmental stability (Mitton, 1993).

A systematic search of enzyme polymorphism influencing physiological phenotypes accounting for the above correlation was undertaken (Mitton, 1993). Physiological efficiency increases with protein heterozygosity, as could be estimated through the oxygen uptake rate as a means to evaluate the basal metabolic cost. Resting metabolic rates were lower in heterozygous individuals than in homozygous ones. In some species it was even found that heterozygosity correlated with increased growth rates or reproduction efficiency. These observations may be understood through the following rationale: protein heterozygosity decreases the maintenance metabolic cost and therefore allows more energy to be apportioned to sustain growth and development. Developmental instability leading to fluctuating asymmetry arises when embryos face stressful environmental conditions that reduced the fitness of an organism. The higher developmental stability exhibited by heterozygous individuals may be understood in terms of the differences in maintenance energy between homo- and heterozygous organisms becoming important under environmental stress and resulting in lower fluctuating asymmetry associated with higher heterozygosity (Mitton, 1993).

4.11 FRACTALS, SYMMETRY AND DEVELOPMENT

In terms of the organization of cellular cytoplasm, the fractal concept has provided new insights from the interpretation of the organization of the ground of living cells as percolation clusters (Aon and Cortassa, 1994; Chapter 6). A percolation lattice may be 'wet' in a number of sites. When the number of 'wet' sites reaches the percolation threshold, a percolation cluster appears that exhibits the property of global connectedness, i.e. a liquid injected at an arbitrary site in the percolation cluster may wet a region far away from the injection point (Feder, 1988). Because of this property, it is not far-fetched to conceive that the products or effectors of the dynamics in an embryonic domain can reach other embryonic regions if the latter belong to a percolation cluster (Chapter 6). In this respect, one may imagine that the growth in size of a developing field is equivalent to the growth of the lattice in number of sites. The number of the lattice's wet sites may eventually fall below the percolation threshold and the system may lose its property of global connectedness, giving rise to regionalization. Likewise, candidate morphogens such as retinoic acid and the fibroblast growth factor, with the capacity to act as mesoderm-inducing factors in the early amphibian embryo, may show a genuine threshold response (Slack, 1987, and references therein). It has been postulated that threshold responses occur in percolation clusters as a consequence of the property of percolation threshold exhibited by those 'random fractals' (Feder, 1988; Kopelman, 1988; Aon and Cortassa, 1994) (Chapter 6).

4.12 PATTERN FORMATION

In the field of developmental biology, pattern formation refers to the orderly spatial establishment of body structures throughout embryonic life. It is widely accepted that the body's pattern, at least in its major axis, depends on the establishment of a monotonic gradient (Lawrence, 1988) of a substance (morphogen). The essence of the modern theory states that the scalar value of the gradient at each point in the field determines the local pattern element (head, abdomen, tail) that will be formed (Lewis *et al.*, 1977; Saunders and Kubal, 1989). Otherwise stated, the concentration profile of the morphogen becomes translated into a body pattern along an axis, e.g. antero-posterior (Lawrence, 1988). Thus, a theoretical prediction is that the height and inclination of the gradient landscape become expressed in the order scale and polarity of the elements of the body. In a photobiochemical system it has been shown that the number of bands given by an electron acceptor in the presence of a monotonous gradient depends upon the values of the gradient parameters related to the size and shape of the landscape (Chapter 5).

In the newly laid eggs of the fruit fly, anterior and posterior determinants exist that interact during cleavage and, gradually, a gradient is established between them. The two main axes of the *Drosophila* embryo are the antero-posterior (AP) and the dorso-ventral (DV). The AP axis comprises three systems: anterior, posterior and terminal, whereas the DV axis is determined by one system only. The process of determination of the embryonic axis depends on about 30 genes (Nüsslein-Volhard, 1991). All four systems share the following biochemical mechanisms (Nüsslein-Volhard, 1991):

- In the freshly laid egg, each system shows a localized gene expression (mRNA accumulation).
- This spatial heterogeneity of mRNAs results in an asymmetrical distribution of proteins (gene products) that function as transcription factors.
- This transcription factor distributes in a concentration gradient that defines limits of expression of one or more zygotic genes.

The products of genes *bicoid*, *gene Y* and *dorsal*, with the exception of *nanos*, are considered morphogens in the classical sense: they are distributed in gradients and determine positions along the axes in a concentration-dependent manner (Nüsslein-Volhard, 1991). These maternal morphogens presumably act as transcription factors that control the spatial domains of transcription of zygotic pattern genes. The combined action of the four systems defines the expression of zygotic target genes in at least seven distinct regions along the AP axis and at least three in the DV axis (Nüsslein-Volhard, 1991).

4.12.1 MOLECULAR GRADIENTS

During early *Drosophila* development, the single nucleus of the zygote undergoes a series of 13 largely synchronous divisions before cellularization. These divisions define 14 stages, numbered according to the nuclear division which completes each stage. The protein products of the *Drosophila* homeobox gene caudal (*cad*) accumulate in a concentration gradient spanning the AP axis of the developing embryo. During stage 7, most of the nuclei begin to migrate towards the periphery and this process is complete by the end of stage 9. In the period corresponding to stages 7–9, Cad protein accumulates dramatically, especially in the posterior half of the body, forming a smooth concentration gradient along the AP axis (Figure 4.7a,b) tightly localized at nuclei. During stages 10–13 the concentration gradient persists and spans the length of the embryo (Figure 4.7b–d) (Macdonald and Struhl, 1986).

Beginning in early stage 14, the pattern of Cad protein distribution changes dramatically, its expression persisting or being enhanced in the posterior half of the embryo (Figure 4.7e). This results in a transient bipartite staining pattern in which the anterior 40% of the embryo appears devoid of Cad protein, with relatively high levels of staining remaining in the posterior 50%. The Cad protein is progressively lost in

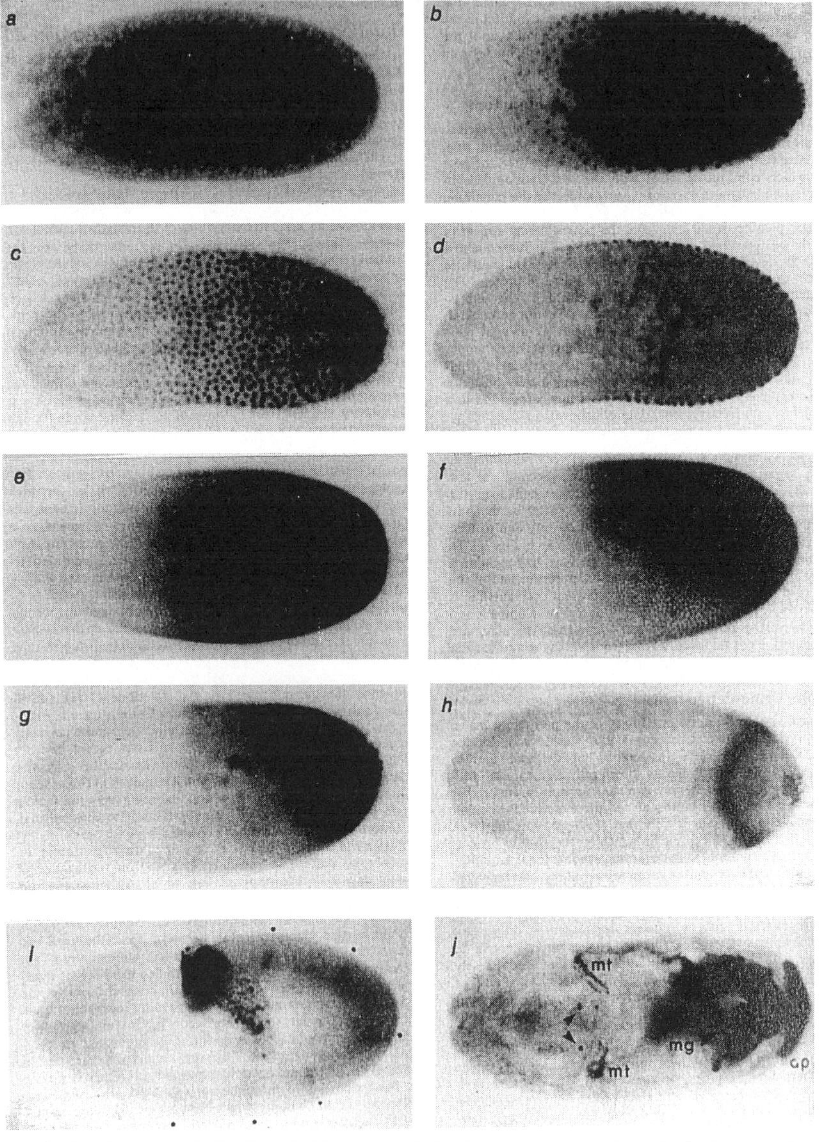
a
b
c
d
e
f
g
h
i
j
mt
mt
mg
cp

the posterior half of the embryo (Figure 4.7f,g) until finally it generates a diffuse posterior ring which progressively sharpens into a discrete posterior ring three to four cells wide (Figure 4.7h).

Mutations in the *cad* gene that reduce or eliminate the gradient cause abnormal zygotic expression of at least one segmentation gene (*fushi tarazu*) and alter the global body pattern (Macdonald and Struhl, 1986).

The bicoid (Bcd) protein has been shown to be distributed in a concentration gradient in the sincytial blastoderm. The gradient of Bcd appears to be generated from translation at a localized source of mRNA and generalized degradation of the protein. In other words, the gradient is probably formed by diffusion away from the local source and dispersed decay. The shape of the gradient could be varied through the genic dose

Figure 4.7 (see facing page) A molecular gradient in early *Drosophila* embryos and its role in specifying the body pattern: immunohistochemical localization of Cad protein during embryogenesis. (a) Stage 8 embryo. All of the nuclei are still located within the central yolky region of the egg. Staining of the nuclei with the Cad antibody is most intense in the nuclei. Graded distribution of the Cad protein is already apparent, as displayed by the gradual reduction of staining towards the anterior pole. (b) Stage 10. Nuclear migration to the periphery is complete and the pole cells have budded off. The Cad protein gradient now extends along the entire body axis, peaking at the posterior pole. This pole-to-pole concentration gradient is maintained until the beginning of stage 14. (c,d) Stage 12. The Cad protein is confined predominantly to nuclei and is apparent both in peripheral and yolk nuclei. (e) Early stage 14. During this stage the distribution of Cad protein undergoes a dramatic change, becoming progressively more polarized until it is restricted to a single posterior ring 3–4 cells wide at the onset of gastrulation. (f) Early-mid stage 14. Progressive loss of Cad staining continues, extending more quickly along the ventral surface. Cad staining now appears predominantly in a diffuse posterior ring which extends dorsally to about 55% embryonic length. (g) Mid stage 14. Cad staining continues to decline from the anterior and dorsal directions, though intense staining persists in the posterior ring and in the pole cells. (h) Late stage 14. By the onset of gastrulation, Cad protein is found in a tightly restricted posterior ring ($\approx$3–4 cells wide around the entire embryo) and in the pole cell nuclei. (i) Germ band extension. (j) 15 h embryo. The Cad staining can be seen in cells giving rise to portions of the anal pads (ap), the posterior midgut (mg) and the malpighian tubules (mt) which wander in and out of the plane of focus. All embryos are oriented with anterior to the left and the plane of focus through the middle of the embryo (a,d,i,j) or at the periphery (b,c,e,f,g,h). Embryos before gastrulation are staged according to the nuclear division cycle. (Reprinted with permission from *Nature*, **324**, 537–545, copyright 1986 Macmillan Magazines Ltd.)

and this affects the distribution of the banding pattern of Eve protein in gastrulating embryos, suggesting that the scalar position in the bicoid gradient determines the position in the *Drosophila* embryo along the AP axis (Driever and Nüsslein-Volhard, 1988a,b). Consensus exists at present that the concentration profile of the bicoid morphogen becomes faithfully translated into the embryo pattern along the antero-posterior axis (Lawrence, 1988). Driever and Nüsslein-Volhard (1988) state that a striking property of Bcd activity is its 'long-range' effect on neighbouring regions. It was claimed to behave as an organizer, determining polarity and pattern with long-range influence (Nüsslein-Volhard, 1991). In *bcd* embryos not only are the structures normally formed at the deleted site of mRNA localization, but also the 'anlagen' of the entire anterior half is missing. This property can best be explained by invoking a gradient mechanism in which different concentrations of the *bcd* gene product determine the position of a series of different structures along the anterior part of the embryo (Driever and Nüsslein-Volhard, 1988a).

The interpretation of experimental data mentioned in the preceding paragraph belongs to one of the two main hypotheses proposed to clarify pattern formation during development of living organisms: the 'positional information' and the 'isomorphic pre-pattern' concepts. The former hypothesis has been worked out at length by Wolpert (1969) and proposes that a monotonic gradient of a chemical along the major axis, between the edges of a morphogenetic field, would inform the cells in the field about their position with respect to the edges. The genome of cells would interpret that information according to a series of thresholds that would result in a variety of possible fates. However, the question about the mechanism giving rise to the gradient is not addressed. In this respect, Crick (1970) proposed that a substance diffusing between a source and a sink could generate a monotonic gradient in the spatio-temporal order of magnitude required for the establishment of positional information in the developing organism. The main condition that has to be fullfilled is for the diffusion coefficient to be larger than 10^{-8} cm^2 s^{-1} (Chapter 5). Again this mechanism does not address the way in which the source became localized.

The positional information hypothesis proposed by Wolpert (1969) has also gained strength in the experimental approach, especially after the demonstration of graded distributions of products of expression of homeotic genes (*caudal, bicoid*) (Macdonald and Struhl, 1986; Mlodzik and Gehring, 1987; Driever and Nüsslein-Volhard, 1988b) and the gradient of retinoic acid in the chick limb (Thaller and Eichele, 1987, 1990).

The isomorphic pre-pattern hypothesis (Nagorcka, 1989) claims that for each element of the body pattern a pre-pattern is needed that will be interpreted according to a single threshold to specify one or at most two cell types. This hypothesis emphasizes the mechanisms through which the pre-pattern is generated. As already discussed, the kinetic equations needed to generate the pre-pattern are nonlinear and the distribution profile of the chemical may not be intuitively recognized as in the case of a gradient generated by diffusion. This feature is probably one of the reasons why this approach has been less worked out experimentally.

There are already kinetic models attempting to explain the appearance of belt-like structures (a pre-pattern) that could, in turn, trigger the expression of genes in a predictable way (Meinhardt, 1982, 1988). A model, built to simulate the metameric pattern during fruit fly development, requires that four or even five chemicals interact to produce striped structures (Figures 4.8 and 4.3). It has been claimed that many different kinetic models may give a gradient. Among the variety of kinetic models proposed up to now, there are not many different ways of pattern generation involving realistic situations from a biological point of view, i.e. not involving trimolecular steps (as in the Brusselator) or disappearance of molecules that are not yet present, as in Turing's model (1952). The stability of model solutions and the size-regulation capabilities of the model are also criteria not easily met by many possible pattern generation equations.

Recently, complex symmetrical patterns of chemotactic bacteria growing on soft agar plates on succinate have been demonstrated experimentally (Budrene and Berg, 1995). The regular, radially symmetrical patterns were formed as a chemotactic response to aspartate, which is formed from succinate. Bacterial growth will amplify the spatial inhomogeneity of aspartate concentrations. Since cells are mobile they are able to form aggregates that detach from the growing swarm ring giving rise to either a pseudo-rectangular or hexagonal lattices. From these results one may suggest that a specialized genetic programme is not required since processes like cell proliferation, secretion of attractant substances and cell motility can generate complex patterns such as those found in developing systems (Budrene and Berg, 1995).

Despite the importance of the theoretical concepts exposed in preceding paragraphs, the interaction between theoretical and experimental approaches has been scarce, perhaps due to the difficulties in identifying the morphogens (section 4.7). On the other hand, the belief that the developmental programme is entirely coded in the genome has stimulated the accumulation of more information about genes and their expression products, expecting with such an (inductive) approach the emergence of new principles (Holliday, 1988).

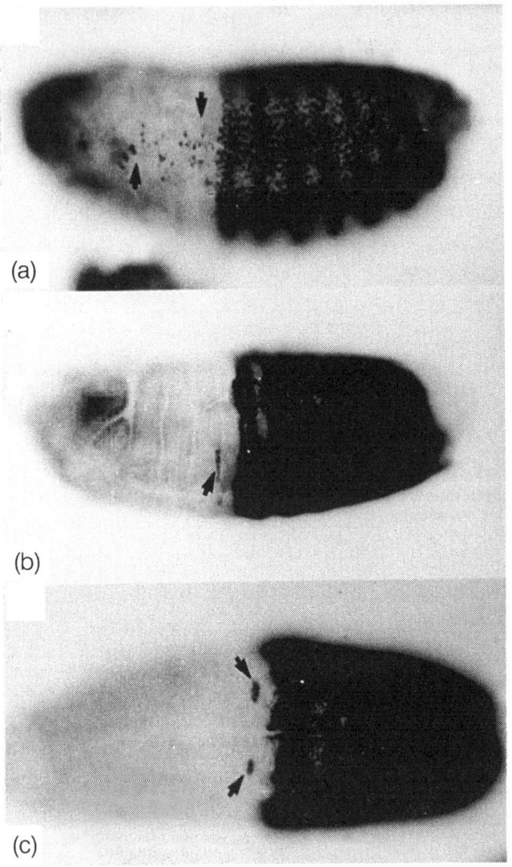

Figure 4.8 Striped wild type pattern expression of the protein products of the abdominal-A (*abd-A*) gene, member of the bithorax complex of *Drosophila*. Abd-A protein products wild type expression were revealed with a specific polyclonal antibody. The normal expression domain of *abd-A* extends from parasegments 7 to 13 (heavily stained parasegments to the right of the embryo). The effect of mutations on *abd-A* expression are shown: (a) infra-abdominal (*iab*) mutation *iab-2^k*, homozygous embryo, stage 11; arrows point to scattered cells anterior to the normal *abd-A* domain showing presence of the Abd-A antigen; (b) *iab-3^{Uab4}* infra-abdominal mutation; embryo of similar age as in (a); arrow points to cells of parasegment 6 with *abd-A* product; (c) *iab-3^{Uab4}* embryo at the extended germ band period expressing Abd-A protein in parasegment 6 (arrows). (Reproduced from Macías, Casanova and Morata, 1990, by permission of The Company of Biologists Ltd.)

4.13 SOME CASE STUDIES OF SYMMETRY IN DEVELOPING SYSTEMS

4.13.1 FUCOID ZYGOTES AND *PELVETIA* EMBRYOS

There are some simple developing systems where the studies attempting to characterize the sort of processes participating in the establishment of organization have been more successful. A careful survey of the events taking place at selection, formation and fixation of a developmental axis in fucoid zygotes and the subsequent expression of polarity has been reported (Kropf, 1992; Goodner and Quatrano, 1993). It has been known for a long time that when spherical zygotes of the brown alga *Fucus* are subjected to a gradient of light, the plane of the first cell division is always perpendicular to the light axis. The resulting cell plate divides the zygote into two unequal cells: the smaller rhizoid cell emerges from the shaded portion of the gradient, whereas the larger thallus cell is on the lit side (Goodner and Quatrano, 1993). The polar axis imposed on the zygote by an external gradient establishes the polarity and division plane of the two-celled embryo and the developmental axis for the whole organism. The basic processes involved in the generation of cell asymmetry are polarized secretion of vesicles, transcellular currents of Ca^{2+} ions and the involvement of cytoskeleton components such as actin microfilaments. Apparently, the first sign of asymmetry that can be detected in a zygote grown in unilateral light is an electrical current that flows through the zygote during the formation of the axis (Goodner and Quatrano, 1993). This transcellular current is carried in part by Ca^{2+} with local accumulation at the rhizoid tip giving rise to a cytoplasmic Ca^{2+} gradient (Plate 3).

Once the main events have been characterized, the involvement of molecules is more easily integrated into a working scheme to provide a plausible explanation of the establishment of cellular polarity in fucoid zygotes. Two main problems may still be distinguished. The first one is related to the mechanism through which a morphogen induces a change in a cell's commitment. The second one concerns how a molecule produces, say, changes in shape, since the spatio-temporal characterization of the molecular level is far smaller and faster than the supracellular levels of organization (Chapter 2). It is at higher spatio-temporal levels of organization (micrometres to millimetres and minutes to several hours) that morphological modifications will become macroscopically visible. In *Fucus* zygotes, microfilaments of actin are involved in the stabilization of the initial asymmetries generated by the light gradient (Kropf, 1992; Goodner and Quatrano, 1993). The first stable asymmetry observed in the zygote is given by the redistribution of actin to the site of rhizoid formation, which is temporally correlated with the fixation process and spatially related to the site of polar elongation. Disruption of the actin cytoskeleton with cytochalasins prevents axis fixation (Goodner and Quatrano, 1993). Changes in symmetry of the cytoskeleton entraining changes in shape are a clear manifestation of the macroscopic spatio-temporal reorganization at the cellular level (Figures 4.5 and 4.6) (see also Chapter 6).

In *Pelvetia*, once the developmental axis is established the resulting embryo exhibits a mosaic behaviour. Upon ablation of the rhizoid or the thallus cells, the remaining cell divides repeatedly but does not regenerate the missing structure (Kropf *et al.*, 1993). Digestion of cell wall produced a rhizoid cell that was round and without polarity. This observation indicated that the polarity is maintained through interactions between membrane components and the cell wall: once the latter is destroyed, polarity is lost. The loss of asymmetry is also reflected in the organization of microtubules which are arranged in a characteristic elongated morphology in rhizoid and become delocalized and expanded over basal regions of the nuclear envelope upon cell wall digestion (Kropf *et al.*, 1993). These results indicated that the establishment and fixation of symmetry axes are dynamic processes occurring at a specific stage during embryo development: later, either the polarizing signals are absent or the brown algae cells are no longer responsive to their action. The continuance of polarity depends on the integrity of the elements whose dynamics gave rise to the symmetry-breaking.

4.13.2 FROG OOCYTES: ORGANIZATION OF THE EGG AND EMBRYO

Another well known sign of asymmetry, in fertilized eggs of *Xenopus*, is the germ plasm – a collection of small basophilic masses positioned just below the surface of the vegetal hemisphere. These masses aggregate during the first few cleavages to form a small number (usually around four) of larger ones. Each one of them is inherited by one of the vegetal blastomeres of the dividing embryo. These blastomeres give rise to the progenitors of the germ line, the primordial germ cells later in development. When germ plasm is examined by transmission electron microscopy, it is found to consist of mitochondria and electron-dense granulofibrillar masses known as germinal granules (Wylie *et al.*, 1986).

Frog oocytes are very large single cells whose cytoplasmic movements and asymmetries are fundamental to the correct development of the subsequent embryo. The amphibian egg has spherical symmetry, whilst the gastrula is bilateral. To impose bilateral symmetry on a sphere, two polarities (vectors or axes) are required. These axes should join opposite points on the surface going through the centre of the sphere and being possibly, but not necessarily, perpendicular to each other. Two axes of symmetry are necessary to specify the spatial location of every point on the surface of an egg. In the frog, one axis (the animal–vegetal axis) is established during oogenesis; the other (the dorsal–ventral axis) is established after fertilization. One of the axes of symmetry is generated in the maternal environment, and the other from the sperm entry site (section 4.1).

Experiments performed with eggs and early embryos of *Xenopus* have shown that large-scale cytoplasmic movements take place during oocyte differentiation, oocyte maturation and the first cell cycle after the egg's fertilization. During the first cell cycle following fertilization (about 75 min in *Xenopus*), a complex series of cytoplasmic movements take place (see below). Following these movements, the egg becomes bilaterally symmetrical (Gerhart *et al.*, 1983; Wylie *et al.*, 1986). Asymmetries produced by these cytoplasmic movements are involved in the earliest developmental decisions (Wylie *et al.*, 1986). There are several reports of asymmetries in the distribution of both RNA and protein species in the egg. The heterogeneous distribution of proteins along the animal–vegetal axis has been identified by transversely slicing large numbers of eggs and analysing posteriorly soluble proteins by SDS-PAGE (Wylie *et al.*, 1986). The developmental significance of these asymmetrically distributed macromolecules is unknown. All these asymmetries in the egg must be initiated during the period of oocyte differentiation.

The unfertilized frog egg is organized with radial symmetry about an animal–vegetal axis. Bilateral symmetry is introduced during the second half of the first cell cycle by a cytoplasmic reorganization originally termed the 'rotation of symmetrization' (Ancel and Vintemberger, 1948; Houliston and Elinson, 1991). The process of establishment of bilateral symmetry comprises a rotation of the egg cortex with respect to the vegetal yolk mass by 30°, and results in the formation of a 'grey crescent'. The grey crescent marks the future site of the dorsal lip of the blastopore, the 'organizer' of the amphibian embryo, and the rotation is necessary for the localization of 'dorsal information' to this region (Figure 10.4) (Elinson and Kao, 1989).

The effects of the reorganization of the egg's contents along the new axis are seen to be manifold as the intracellular organization of the egg becomes the pancellular organization of the blastula and gastrula. In these stages, the 5000–20 000 cells cleft from the egg materials gain motility, adhesive properties, gene expression and developmental autonomy in a patterned way, such that on the dorsal side all these aspects of cell biology are expressed earlier and more strongly than on the ventral side (Gerhart *et al.*, 1983; Gerhart and Keller, 1986).

(a) The G_2-like period and the effect of cold, pressure and ultraviolet irradiation

The development of dorsal organs in the embryo is affected when the egg is subjected to brief exposures to low temperature, high pressure or UV irradiation, before the time of 0.72 of its first cell cycle but not later (Scharf and Gerhart, 1983) (Figure 4.9). The treated egg cleaves apparently normally, but gastrulates symmetrically instead of starting the lip on the prospective dorsal side, and then never neurulates. It forms a 'ventral embryo' or 'belly piece', lacking central nervous system, notochord, somites and other organs of the dorsal side (Gerhart *et al.*, 1983; Gerhart and Keller, 1986). This ventral development is found at highest doses of UV irradiation, thus being an extreme effect, while at intermediate doses dorsal structures are completely deleted in a dose-dependent way in anterior to posterior order. As the strength of the perturbation is increased, the embryos are, first, headless, then headless and trunkless, and finally headless, trunkless and tailless.

Cold and pressure are known to depolymerize microtubules, and pressure has been reported to depolymerize actin filaments (Gerhart *et al.*, 1983). The cortical/cytoplasmic rotation producing the grey crescent can be prevented with colchicine and other microtubule destabilizers (Manes *et al.*, 1978; Vincent *et al.*, 1987; Elinson and Rowling, 1988) as well as with UV light (Manes and Elinson, 1980; Vincent *et al.*, 1987).

The transient array of parallel microtubules (MTs) believed to be part of a rotation motor which sustains the 'rotation of symmetrization' described above, i.e. a 30° rotation of the entire thin, outer cortex relative to the deeper cytoplasm, has been shown to be sensitive to UV and colchicine treatments (Elinson and Rowling, 1988). The parallel MTs appear at the beginning of rotation and disappear at its completion. They are oriented parallel to the direction of rotation and are located in the vegetal hemisphere, where the force is probably generated (Elinson and Rowling, 1988). The parallel MTs are found 1–3 mm deep from the plasma membrane.

Ultraviolet irradiation of *Rana pipiens* or *Xenopus laevis* eggs during the cell cycle following fertilization, at 0.35–0.4 (fixed at 0.71–0.74) or 0.25–0.3 (fixed at 0.7), respectively, prevented the appearance of the cortical array of MTs and cortical rotation (Elinson and Rowling, 1988) which correlated with dorso-anterior indexes (DAIs) of 0.07 and 0.06,

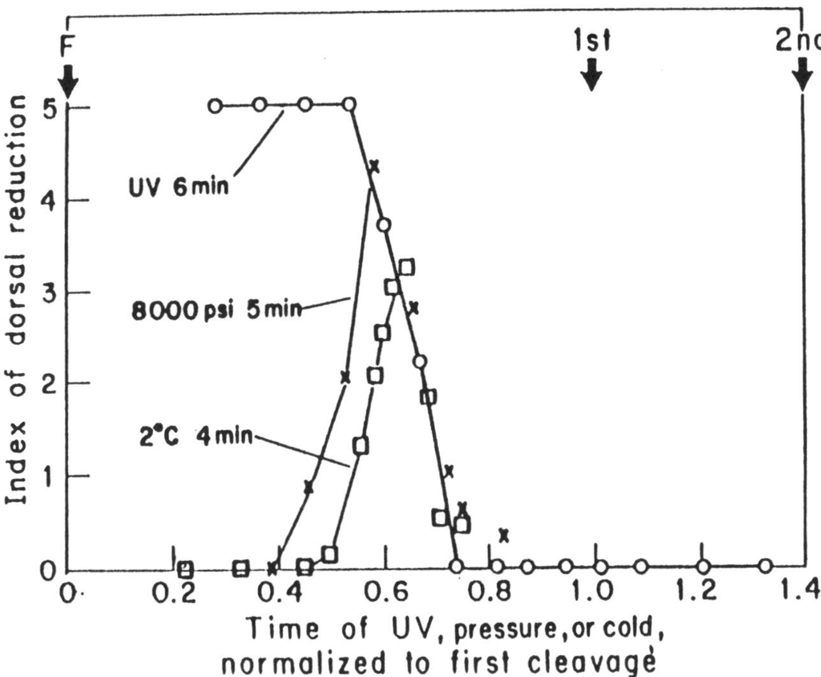

Figure 4.9 Time course of sensitivity of the egg to treatments interfering with subsequent dorsal development in *Xenopus*. Eggs were exposed to UV radiation on the vegetal hemisphere, or to cold or pressure, for an interval of 4–6 min at the times indicated. The index of dorsal reduction is a numerical score indicating the severity of the effect of the treatment on dorsal development. A score of 5 means no dorsal structures were formed, the embryo being a radially symmetry 'belly piece' possessing a short gut, red blood cells and a ciliated epidermis. A score of 0 means a normal tadpole. By the time 0.72, the egg becomes resistant to all the treatments. (Reproduced from Gerhart *et al.*, 1983, with permission.)

respectively (Elinson and Rowling, 1988). The most extreme embryo is a radially symmetrical, ventralized embryo, also called the 'ground state', scored as a grade 0 on the DAI scale where a normal embryo is a DAI grade 5 (Elinson and Kao, 1988) (Figure 4.9).

MT polymerization begins at 0.3 normalized time in fertilized eggs and at 0.5 normalized time in electrically activated eggs. The array of parallel MTs forms over the entire 3 mm^2 vegetal surface in less than 10 min. It appears that at 0.5 normalized time of the first cell cycle, conditions become favourable for MT polymerization, and MTs are initiated from many centres (Elinson and Rowling, 1988). Recently, a direct link between the growing sperm aster and the cortical array of parallel MTs was seen to exist (Houliston and Elinson, 1991). Unusual non-spindle MTs were found in the unfertilized egg in which they appeared to be evenly distributed throughout the cytoplasm, except in the immediate vicinity of the meiotic spindle (Houliston and Elinson, 1991). Furthermore, a rapid loss of MTs is associated with the transition to anaphase (Houliston and Elinson, 1991) which is opposite to the change observed in other cell cycles (Kirschner and Mitchinson, 1986; Verde *et al.*, 1990).

4.13.3 THE FRUIT FLY CASE

In the development of *Drosophila melanogaster*, a thorough dissection of most genetic elements participating in pattern formation has been successfully achieved, especially those related to the establishment of the main axis (antero-posterior and dorso-ventral) and with the metameric body organization (section 4.12.1).

A series of reports has described the appearance of graded distributions of proteins coded by *bcd* and *cac*, two maternal polarity genes, and 'zebra' patterns (striped) for the proteins coded by pair ruled genes, *h* and *runt*, and posteriorly *ftz*, *eve*, *en* (Ingham, 1988; Nüsslein-Volhard, 1991) (Figure 4.10). With the exception of the work in the Bcd protein, these patterns are not discussed in terms of the existing developmental hypothesis. Driever and Nüsslein-Volhard (1988b) proposed that the exponential gradient of Bcd protein would arise from a localized source, diffusion through the syncitial cytoplasm and generalized degradation (section 4.12). This represents a dynamically self-organized phenomenon; the maintenance of the spatial inhomogeneity requires a continuous supply of free energy, e.g. ATP hydrolysis spent in protein synthesis. The *Drosophila* embryo, as all living systems, is open because there is a net entry of nutrients and an exit of metabolic waste products. Concerning pattern itself, the rates of protein synthesis and degradation, as well as its diffusion, are coupled processes with different affinities. However, in this case the authors exclude the possibility of nonlinear processes, such as autocatalytic synthesis of the protein (Driever and Nüsslein-Volhard, 1988b).

According to a model proposed by Meinhardt (1988), the correct activation of genes at a particular position requires two steps: the initial trigger from the pattern superior in the hierarchy and the stabilization of the pattern. The stabilization of the pattern may be achieved by local autocatalysis and long-range inhibition with a saturation of the autocatalysis or by mutual activation of cell states which exclude each other locally (Figure 4.3). Both mechanisms are able to generate stripes, albeit a tuning of the parameters demands that the initiation of stripe formation requires a strong trigger (Meinhardt, 1988). However, other explanations which are based on reliable experimental evidence are also possible. In fact, it has been shown that microtubules are involved in the localization of *bcd* RNA to the anterior pole of *Drosophila* oocytes (Pokrywka and Stephenson, 1991). When egg chambers are treated with microtubule destabilizing drugs like colchicine, nocodazole and tubulozole-C, *bcd* RNA becomes unlocalized in nurse cells and the oocyte. In contrast, the microfilament inhibitor cytochalasin D did not have any effect on *bcd* message localization (Pokrywka and Stephenson, 1991). Furthermore, the sensitivity of Bcd localization to taxol, which stabilizes and promotes the growth of existing microtubules, suggests that correct regulation of microtubule dynamics is crucial to the proper localization of *bcd* RNA (Pokrywka and Stephenson, 1991). The localization of P-granules in *C. elegans* was also shown to rely on cytoskeletal elements, i.e. microfilaments (section 4.9.1) (Hill and Strome, 1988).

At this point we loop to our proposal that polymerization–depolymerization of cytoskeleton components may be involved in the passage from microscopic to macroscopic functional coherence at the cellular level, achieved through self-organization of cytoskeleton components (Chapters 6 and 7). We have also postulated that in fractal percolation cytostructures, at their threshold level of occupancy, there are transitions between local and global connectedness (section 4.11). According to this postulation, below the percolation threshold mRNAs would stay localized, i.e. the 'embryonic field' would behave as loosely coupled or uncoupled dynamic subsystems. Above the threshold level of occupancy of the lattice, the internal egg processes' dynamics within appropiate boundaries will distribute anisotropically molecular components (Figure 4.7). That is to say, only the dynamics of the processes, and not molecules themselves (i.e. fluxes or movements of molecular, supramolecular or subcellular structures), are able to create spatial anisotropy in a long-range scale such as that needed for developmental polarity to appear (section 4.14). One should not forget that genes only encode proteins that participate either as catalysts or effectors of metabolic fluxes or as structural elements such as cytoskeleton or membrane components (Chapter 7).

Self-organization arises in systems with built-in nonlinear mechanisms such as autocatalysis far distant from thermodynamic equilibrium (Chapters 1 and 5). Thus, there should be a nonlinear process probably participating in the localization of *bcd* mRNA at the anterior pole of the fruit fly egg. The search for this nonlinearity will probably lead to an

understanding of the way in which the gradient is laid down. For the time being, there is no evidence that explains a non-homogeneous distribution (polarization) of *bcd* or *osk* mRNA, or the cues of polarity that allow them to become localized at the anterior and posterior ends of the embryo, respectively. It is known that *bcd* mRNA is already expressed in nurse cells and is subsequently injected into the cytoplasm of the oocyte through a cytoplasmic bridge located at the anterior pole. However, this fact does not explain the anterior localization since *osk* mRNA, also injected in this way, becomes concentrated in the posterior pole. The egg's cytoplasm is already polarized by maternal influences, which make any symmetry-breaking phenomena deterministic, giving rise to a specific pattern.

We would emphasize that the spatio-temporal level of organization in which the molecules are trapped lags far below those able to generate order at tissue or even cellular levels (Chapter 2). It appears that there are two methods of information storage: the highly conservative DNA one (genetic information), and an epigenetic one that, unlike DNA, would be

Figure 4.10 (see facing page) Kinetics of molecular patterns in *Drosophila*: examples of the sequence of molecular patterns that generate periodicity along the antero-posterior axis of the early *Drosophila* embryo. Each row, (a)–(f) shows four adjacent sections of the same embryo hybridized with ^{35}S-labelled probes for transcripts of *bcd*, *Kr*, *h* and *wg*, respectively. (a) *bcd* RNA is easily detectable at the anterior pole (left) but transcripts of the three zygotically expressed genes cannot yet be detected. (b) About 10 min later, *Kr* transcripts can be seen to have accumulated in the central third of the embryo between about 30% and 60% EL. (c) After a further 10 min the levels of *Kr* transcript have increased; by this stage, low level *h* expression is just detectable in most of the peripherally located nuclei. (d) After the final nuclear division before cellularization *bcd* RNA is beginning to dissipate, reflecting the end of its functional period. The domains of *Kr* and the other gap genes *hb* and *kni* (not shown) are now well established and their combined activities begin to organize the expression of *h*. Thus, peaks of *h* transcript appear along the antero-posterior axis in a somewhat graded fashion, a major peak at this stage coinciding with *Kr* domain. (e) During the next 10–15 min, *bcd* RNA disappears completely. Two new domains of *Kr* expression appear (arrowed), one at the anterior and one at the posterior pole of the embryo, while the central domain narrows somewhat. The modulated pattern of *h* is fully resolved into one of seven regularly spaced stripes between 70% and 20% EL and, in addition, an antero-dorsal patch of expression. At this stage, expression of the segment polarity gene *wg* is first detectable in three distinct domains: at the anterior pole; antero-dorsally, overlapping the *h* patch; and in a stripe, posterior to the last *h* stripe. (f) *Kr* and *h* expression persists with the onset of gastrulation; note additional (eighth) stripe of *h* expression (arrows). By this stage, *h* activity has organized *ftz* expression (not shown) which, together with *eve*, regulates expression of *wg* to be transcribed in 14 stripes, reflecting the establishment of the parasegment primodia. (Reprinted with permission from *Nature*, **335**, 25–24, copyright 1988 Macmillan Magazines Ltd.)

highly dependent on environmental conditions and the system's previous history. At this level, the information would be stored in epigenetic circuits, i.e. the dynamic behaviour of metabolic networks in the cell (Chapters 5, 10 and 11). The fact that the environment is also conservative suggests that both levels of information, genetic and epigenetic, interact to give a normal development.

4.14 LONG-RANGE ORGANIZATIONAL CUES

A basic implication of the genetic programme metaphor is that genes via their products determine macroscopic biological form. The genetic programme concept is actually based on short-range forces which essentially characterize molecular interactions. Genes are responsible for short-range interactions in the order of angstroms or at most nanometres (i.e. Van der Waals, electrostatic, hydrophobic, coulombic forces), and embryonic fields are in the order of millimetres (Goodwin, 1986). Moreover, molecular events happen in the temporal range of nanoseconds to milliseconds, whereas development of the whole organism occurs over a time scale of days or weeks and each process of regional specification may only take a few hours (Slack, 1987). Seemingly, the spatio-temporal events taking place at the molecular level cannot account for organization spanning scales of 10 to 100 microns in length.

Several spatio-temporal scales must be spanned before molecular, supramolecular or even supracellular levels become compatible for an explanation of embryonic development. Patterns of long-range cytoplasmic forces are not programmed in the genes; on the contrary, cytoplasmic spatio-temporal organization as a component of morphogenesis may be described in thermokinetic terms. Genes produce proteins (catalysts or structural proteins) that by changing parameters or boundary conditions determine the system dynamics that may give rise to order in living organisms. According to Goodwin, the only way to understand morphogenesis is within the context of an appropiate field theory that correctly and quantitatively answers the following question (Goodwin, 1986):

> what consequences of (known or unknown) natural laws are responsible for the characteristic spatial order that is observed in organisms? Within the context of an appropiate field theory that correctly and quantitatively answers this question, it will be possible to understand how particular gene products, affecting certain parameters or boundary values, can affect wavelengths amplitudes and physical shapes.

What are the long-range organizational cues that allow discontinuous transition between levels of organization? These organizational cues belong to the supramolecular–cellular levels of organization and their properties are:

- ability to influence the spatio-temporal genetic–epigenetic expression far away from their origin;
- ability to be translated along large distances, allowing a coherent spatial expression;
- that they are ubiquitous.

These long-range forces sharing the properties described above constitute the driving forces that coordinate the spatio-temporal expression of macromolecules and supramolecular structures, functionally organized.

4.14.1 WHAT IS THE NATURE OF LONG-RANGE FORCES?

In morphogenetic fields, combined chemical, electrical and mechanical forces are expected to occur and interact. Electromagnetic fields occur in cells when charged particles (e.g. molecules) or aggregates or larger structures (e.g. vesicles) are subjected to movement. Membrane-bound ATPases may transduce energy from dynamic electrical fields through electroconformational coupling (Tsong and Astumian, 1988). The energy of an electric field can be transduced to perform chemical work, such as active transport or ATP synthesis. Thus, electrical currents are good candidates to be involved as long-range organizational cues since they provide the possibility that electrochemomechanical stimuli are transmitted over great distances, provoking rapid communication between remote cells or embryo regions (Robinson, 1985).

Cellular electrical phenomena comprise ionic concentration/activities, ionic/voltage gradients, transcellular or nuclear electrical (ionic) currents, gradients of charged macromolecules which may arise as the result of self-electrophoresis and ionic compartmentation (Jaffe, 1977; De Loof, 1986; Gow, 1989; Aon and Cortassa, 1989; Vanden Broeck *et al.*, 1992; Bustamante, 1994). Transcellular ionic fluxes are, at least in part, a consequence of the ability of cells to segregate ion pumps and channels in different domains of the plasma membrane (Wessels, 1986; Raven, 1987; Gow, 1989) (Chapter 12). Developing cells coming from both the plant and animal kingdoms exhibit transcellular ionic currents associated with polarized growth (Nuccitelli, 1984; Wessels, 1986) which also applies to microorganisms (Gow, 1989). The plasmalemma H^+ pump could account, at least in part, for the endogenous electrical fields associated with the natural H^+ currents (Jaffe, 1979; Aon and Cortassa, 1989) and transmembrane electrical potential. It is known that electrophoretic mechanisms modulate the stabilization and dynamics of membrane topography, e.g. polarized growth (Brown and Montezinos, 1976; Giddings *et al.*, 1980; Poo, 1981; Lin-liu *et al.*, 1984; Reiss *et al.*, 1984; Chapman and Staehelin, 1985; Herth, 1985; Schnepf *et al.*, 1985). Weak endogenous electrical fields, of the order of 1 mV per cell diameter, could redistribute membrane proteins (Jaffe, 1977; Nuccitelli, 1984). The electrical field across the plasmalemma of plant cells could well account for the short range-order of cellulose synthase complexes in the plane of the membrane or electrical coupling of root cells (Cortes, 1992). Voltage drops ranging from −80 mV to −200 mV have been measured across the plasmalemma of higher plant cells. These voltage drops would induce electrical field strengths in the range of 80–200 kV cm^{-1} (for an average membrane thickness of 10 nm) (Neumann and Katchalsky, 1972). Electrical field strengths of 20–100 kV cm^{-1} have been shown to induce realignment of particles with a diameter of 20–30 angstroms (Takashima and Schwam, 1985). Thus, cytoplasmic electrical fields raised by the voltage gradients generated by transcellular ionic currents could also be involved in the vectorial extrusion of extracellular cell wall material and membrane components (Kropf, 1986) by exocytosis and fusion with the plasmalemma of secreted Golgi vesicles (Roland, 1973; Chrispeels, 1976; Giddings *et al.*, 1980).

Gene expression at the nuclear level is influenced by cytoplasmic stimuli of intra- and extracellular origin (Chapter 9). Recent patch-clamp detection of ion channel activity at the nuclear envelope (for a review, see Bustamante, 1994a) puts into focus the potential regulatory properties by nuclear ion signalling. Nuclear ion channels have a large conductance (up to 1000 pS) which gives them a potential modulatory role in nuclear structure and function (Bustamante, 1994b). Apparently, the nuclei behave like a barrier to the propagation of intracellular-free Ca^{2+} waves and other ions or cAMP-dependent protein kinase. Transport of Ca^{2+} across the nuclear envelope is thought to occur through pumps (for a review, see Bustamante, 1994a).

Several lines of evidence support the notion of electrically induced anisotropy in developing systems. The site of appearance of the rhyzoid in *Pelvetia fastigiata* and the tip of grown pollen tubes of *Lillium longiflorum* were anticipated by electrical currents. These currents may be the result of a polarization of pumps in membranes, triggered in turn by self-electrophoretic mechanisms (Jaffe, 1977, 1979; Poo, 1981). Oster *et al.* (1983) proposed mechanical forces giving rise to spatial inhomogeneities which could play a role in haptotaxis, chemotaxis or contact guidance.

4.15 SYMMETRY AND MORPHOGENESIS: A SYNTHESIS

Following the paradigm introduced by Turing in the 1950s, diffusion has been proposed as a mechanism to link local to global-range coherence in developmental processes (Crick, 1970). Turing's pioneering work demonstrated that reaction–diffusion mechanisms could give rise to symmetry-breaking and the appearance of a spatially organized state. In our opinion, the interpretation of Turing's proposal has been too literal. Morphogenetic fields may well be a pattern of fluxes through main metabolic pathways (Chapters 9–11), ionic gradients or gradients of adhesiveness (Oster *et al.*, 1983) instead of chemicals. We believe it is necessary to move a step further to enrich Turing's legacy. That enrichment should allow us to explore alternative intellectual paths to pose the problem of form and function in a biologically meaningful way.

One of the most distinctive features of the complexity of living systems is their structural and functional organization at many simultaneous levels (Churchland and Sejnowski, 1988; Aon and Cortassa, 1993). According to the concept of dynamic organization, the intrinsic dynamics of the processes involved at different levels of organization give rise to biological function and form (Chapters 2 and 11). It is the stability of biological organization achieved that determines whether development will be mosaic or regulative or, in biological terms, if the specification of cellular fate is autonomous or conditional, respectively (Davidson, 1990).

Thermokinetically, the principles of self-organization could account for transitions between levels of organization so that a certain spatial segregation (e.g. heterogeneity in the distribution of organelles or subcellular particles in an egg) may be attained (Aon and Cortassa, 1994). Self-organization happens in thermodynamically open systems far removed from equilibrium. Built-in kinetic nonlinearities in biological systems are essential to amplify initial fluctuations and dynamically generate new macroscopic patterns. We have proposed that transitions between levels of organization occur at bifurcation or limit points in the dynamic behaviour of growing or developing systems (Aon and Cortassa, 1993, 1994). Polymerization–depolymerization of cytoskeleton components take place in the spatio-temporal 'window' in which the passage from microscopic to macroscopic organization at the cellular level occurs (Aon and Cortassa, 1993) (Figures 2.2 and 4.6). For transitions to happen, the dynamics of processes must become unstable. Those transitions may be spatially visualized through symmetry changes. Symmetry-breaking and the 'violation' of Curie's principle emerge when a symmetrical state is no longer stable. At this point, the relationship between thermodynamics and symmetry becomes self-evident. The thermodynamic conditions for stability described in Chapter 1 are no longer fulfilled when there are coupled processes exhibiting kinetic nonlinearities in a system operating under far from thermodynamic equilibrium conditions.

Cellular organization essentially occurs in heterogeneous media. An instability in the dynamics may lead to a rupture of symmetry, ending in the establishment of polarity. In the case of animal cells, the mother provides a heterogeneous environment (boundary conditions) which will determine that the egg organizes spatially in an asymmetrical way. Heterogeneity may arise in an asymmetrical landscape that in turn introduces orientational cues. Gradients are basic asymmetrical landscapes of great biological significance that represent from a thermodynamic point of view 'dissipative' structures (Nicolis and Prigogine, 1977) or 'kinetically maintained' structures (Harrison, 1982). These structures are 'dissipative' or 'kinetically maintained' since a continuous supply of free energy by chemical reactions and transport processes is required in order to sustain the spatial inhomogeneity (Chapter 5). The presence of gradients has been demonstrated for the bicoid protein in *Drosophila* embryos (Driever and Nüsslein-Volhard, 1988a).

A monotonic gradient of a chemical between the edges of a morphogenetic field would inform the cell of its position in the field ('positional information') with respect to the edges (Wolpert, 1969). The genome of the cell would interpret that information according to a series of thresholds that would result in a variety of possible fates. In other words, the concentration profile of the morphogen becomes translated into a body pattern along an axis, e.g. anteroposterior (Chapter 5) (Lewis *et al.*, 1977; Lawrence, 1988; Saunders and Kubal, 1989; Aon and Cortassa, 1993). Nevertheless, a main problem may still be distinguished. It concerns the mechanism(s) through which a morphogen induces a change in the cell's commitment. At the cellular level, a cell's position in a tissue could be sensed by cell cytoarchitecture, since the genome itself appears to be responsive to major controlling signals such as cell shape and surface contact from the intracellular macromolecular networks (Penman *et al.*, 1981; Shinohara *et al.*, 1989). Chapters 6 and 10 develop this issue further.

Dynamic organization in biologically oriented artificial systems

5

The aim of this chapter is to present some useful examples of self-organized structures in artificial systems that help to visualize concepts needed to interpret dynamic organization in living systems. The term 'artificial' refers to analogues of biological systems; for example, systems composed of isolated enzymes (either soluble or immobilized), or isolated biological membranes or organelles, or inorganic systems exhibiting similar properties to biological ones. We attempt to present some models representing the artificial system and the biological analogues with the aim of emphasizing in each case the main conditions that should be fulfilled in order to obtain dynamic organization.

5.1 CONDITIONS FOR A SYSTEM TO EXHIBIT SELF-ORGANIZATION

Autonomous processes able to lead to spatio-temporal self-organization have to operate under conditions that are far from thermodynamic equilibrium. Although some reaction steps in cellular metabolism seem to meet near-equilibrium conditions, actually the overall state of metabolic pathways is far from equilibrium. As pointed out in Chapter 1, organization cannot arise either at equilibrium or in the near-equilibrium domain where the relations between fluxes and their driving forces are linear and symmetrical.

The constraints for self-organized structures to appear are:

- that the systems in which they arise should be open to fluxes of energy and matter;
- that there should be several coupled processes through some intermediary of any nature (chemical, electrical, mechanical);

- that at least some processes should exhibit a kinetic nonlinearity in the dependence of the rate on reactant concentrations.

From the point of view of kinetics, the same constraints may be expressed in terms of the properties of systems of ODEs (Chapter 1).

One main characteristic of biological systems is the presence of several interrelated or coupled processes. It is known that a single equation may describe bistability only for processes where reaction and diffusion are coupled, e.g. enzymatic ones (Kernevez *et al.*, 1979; Kernevez, 1980; Cortassa *et al.*, 1990b). The latter is true even for zero-dimensional systems, i.e. those that do not include partial differential equations (Kernevez, 1980; Aon and Cortassa, 1993). For complex temporal patterns such as oscillations and chaos, at least two or three ODEs, respectively, are necessary to describe them. In addition, only nonlinear kinetic equations, such as those including feedback or feed-forward, can give rise to organized patterns. A nonlinear behaviour would imply that the change in a given flux is not proportional to the change in a variable or a parameter influencing the behaviour of that flux (Chapters 1 and 3).

In the following sections different sorts of self-organized behaviour will be emphasized rather than the systems in which these behaviours were observed. Of those systems we will simply call attention to aspects fulfilling conditions for self-organized behaviour.

Chapter 1 introduced deterministic systems of differential equations (ODEs or PDEs) that describe the temporal evolution of variables, called state variables, as a function of their parameters. According to parameter values, the behaviour of state variables may change at bifurcation points for which the stability properties change (for example, from stable to unstable steady states). Dynamic bifurcation theory is a main tool for detection and analysis of self-organization since it allows qualitative characterization of the dynamic behaviour of any system in terms of the type and stability of steady states exhibited by that system as a parameter is varied (Chapter 1). Different qualitative behaviours (e.g. bistability, oscillations, chaos) may be detected, which is of great help since biological systems (e.g. metabolic networks) may exhibit those various dynamic behaviours apart from asymptotic steady states.

An ensemble view of steady states and the localization of bifurcation points is especially relevant in the case of multiple stationary states, i.e. bistability. This is the topic of the next section.

5.2 BISTABILITY

Bistability is essentially characterized by two branches of stable asymptotic steady states continued by another branch of unstable states (Figure 5.1) (Kernevez *et al.*, 1979; Kernevez, 1980; Hervagault and Thomas, 1987; Aon *et al.*, 1989b; Aon and Cortassa, 1993). Transitions between those branches exist at limit (i.e. bifurcation) points where sudden transitions between both branches of steady states occur (Figure 5.1; see also Figure 1.8). Thus, bistability is a kinetic behaviour characterized by the existence of two stable steady states for the same set of control parameters, i.e. under the same experimental conditions (Neumann, 1973; Kernevez *et al.*, 1979; Kernevez, 1980; Hervagault and Thomas, 1987; Aon *et al.*, 1989b). The occurrence of one or the other steady state depends on the initial state, i.e. on the system's 'history' before attaining that particular parametric condition. Bistability is the simplest of the seven elementary catastrophes described by Thom (1972), i.e. the cusp catastrophe (see also Saunders and Kubal, 1989).

Bistability has been observed in chemical reactions and membrane electrical potential in neurons (Aihara and Matsumoto, 1983). One interesting topic that will be discussed is the appearance of discrete states arising out of continuous gradients in bistable processes (Lewis *et al.*, 1977; Aon *et al.*, 1989a,b; Aon and Cortassa, 1993).

Bistability can only arise in an open system where coupled processes occur under far from equilibrium conditions and with one of the processes exhibiting nonlinear behaviour with respect to one of its parameters. The latter results in a cubic dependence of the system's behaviour on a parameter, i.e. a third-degree polynomial (Thom, 1972; Hervagault and Thomas, 1987).

Bistability has been described in the catalytic activity of a soluble enzyme in a continuously stirred tank reactor (CSTR). The enzyme conformational changes were involved in the appearance of multiple steady states (Aon *et al.*, 1989a). Conformational changes in proteins have been shown to play a role, either regulatory or catalytic, in enzyme catalysis (Koshland, 1987). When such conformational changes occur slowly with respect to the catalytic rate, the resulting kinetics will be non-hyperbolic as it occurs when the desorption of the product from the active site is slow (Ricard *et al.*, 1974). This is one of the mechanisms that give rise to hysteretic behaviour which is different from bistability. Bistability was shown to occur when conformational transitions in an enzyme led to different final conformational states according to the enzyme's history, which is reflected in its activity.

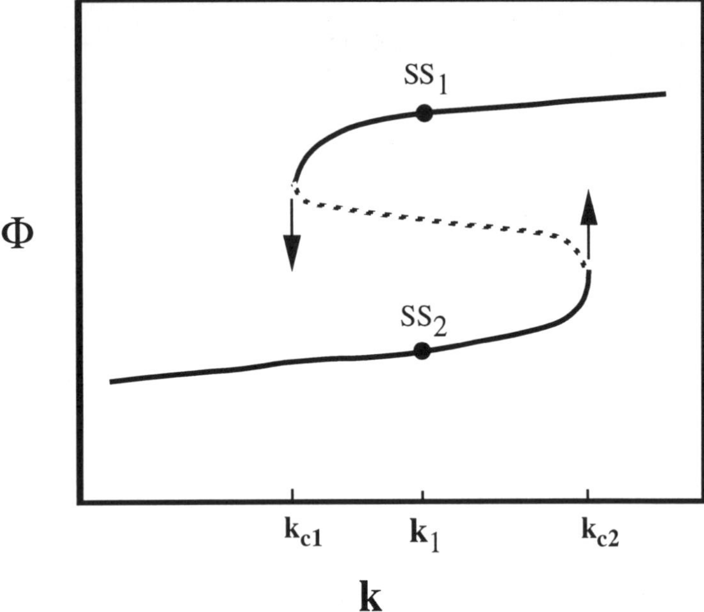

Figure 5.1 Bistability curve of a state variable, Φ, as a function of a parameter, k. The sort of behaviour represented here is typical of dynamic systems exhibiting multiple stationary states. Bistability has been utilized to explain memory devices or history-dependent changes in several systems of different natures (e.g. enzymatic, neuronal, microbial), the main reason for this being that for the same value of parameter k_1 two steady states of Φ are possible (ss_1 and ss_2) depending on the previous history of the system, i.e. for increasing or decreasing values of k. At critical values of the parameter k, the dynamics of the system jumps down or up the branch of steady states, depending on the value of the parameter, k_{c1} or k_{c2}, respectively, and the previous history of the system.

5.2.1 BISTABILITY OF GLUCOSE 6-PHOSPHATE DEHYDROGENASE ACTIVITY IN A CONTINUOUSLY STIRRED TANK REACTOR (CSTR)

In an open system, an enzyme exhibiting Michaelis–Menten kinetics, when subjected to a series of successive pH changes, showed a bistable dynamic behaviour (Aon *et al.*, 1989a). The glucose 6-phosphate dehydrogenase (G6PDH) activity, measured as stationary values of NADPH produced in an open reactor, resulted in bistable behaviour when the enzyme's solutions were treated at different pH values before injection into the reactor, the pH of which was constant at 7.5 (Figure 5.2). The stock solution of the enzyme was subjected to successive steps of acidification (forward direction) followed by alkalinization (reverse direction).

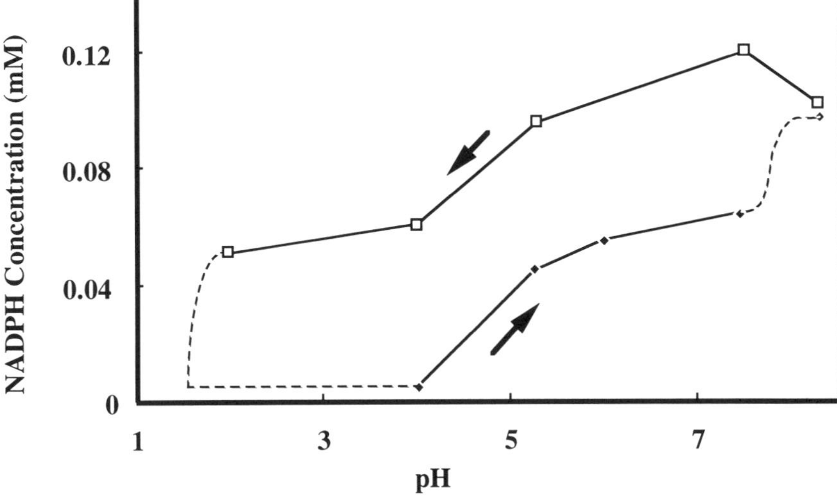

Figure 5.2 Stationary values of NADPH obtained as a result of glucose-6-phosphate dehydrogenase (G6PDH) activity (pH-history dependent) in a continuously stirred tank reactor. The reaction catalysed by G6PDH was analysed in a reactor operated under open conditions. The A_{340} of the reactor outlet flow was registered continuously. Steady state NADPH concentration was considered to be attained when the variation in OD_{340} was less than 0.02 h^{-1}. The time necessary for achieving the stationary values of NADPH was 2.0 ± 0.5 h – except for pH 8.3, either forward or reverse, which required 1.0 ± 0.15 h. The pH of the enzyme solution that was fed to the reactor was varied in the forward (open squares) or the reverse (filled squares) direction up to the indicated pH values before injection into the reactor. Dashed lines show the hypothetical pathway followed by G6PDH activity between both branches of steady states. (From Aon *et al.*, 1989, by permission of The Biochemical Society and Portland Press.)

The activity of G6PDH was maximal at pH 7.5 and decreased progressively as the pH of the enzyme solution was gradually lowered to 2.0. Subsequent alkalinization did not produce a recovery of the enzyme activity until the pH of the enzyme solution reached a value of 8.3 (Figure 5.2). Thus, two steady states of activity for each pH value in the open CSTR were observed upon successive pH variations of the G6PDH solution.

5.2.2 pH-INDUCED CONFORMATIONAL HYSTERESIS OF G6PDH

Intrinsic protein fluorescence results indicated that conformational changes of the enzyme at the time of entrance into the reactor were different at each pH and depended on the direction of pH change of the enzyme's solution (Figure 5.3). When the enzyme entered the reactor it apparently retained the previous pH record in its conformation.

Figure 5.4 shows that the enzyme activity was correlated with conformational changes as measured by protein intrinsic fluorescence after pH-dependent changes. It appears that between pH 8.3 and 5.3, the latter corresponding to the isoelectric point of G6PDH, there is no correlation between enzyme activity in the CSTR and quantum yield. However, after the pH value corresponding to the isoelectric point of G6PDH (forward direction), the enzyme activity and its conformation showed a strong interdependence. Another point of inflection occurs at pH 4.0 (reverse direction) after which activity and conformation are also tightly coupled. It is remarkable that after pH 5.3 (forward direction), the quantum yield always decreases.

A complex ensemble of states of a protein includes a series of potential energy minima, each separated by an energetic barrier, only some of which are enzymatically active. That this could be the case for the present study is shown in Figure 5.4, where different quantum yields (Figure 5.3),

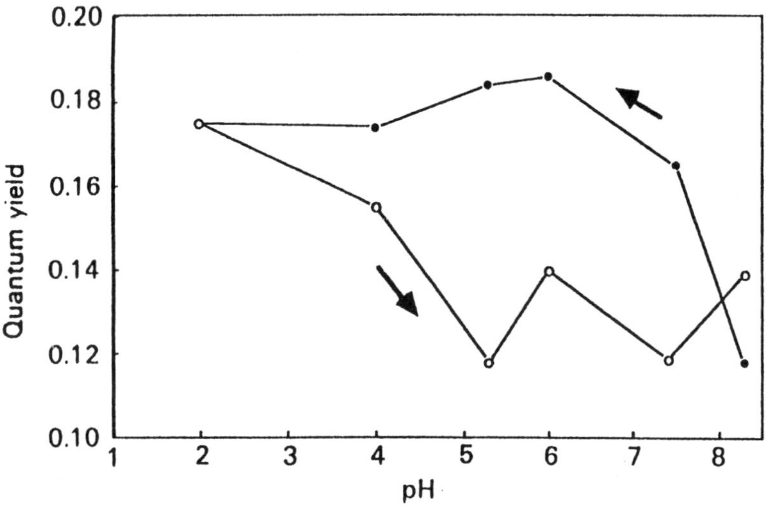

Figure 5.3 pH-history dependent changes of G6PDH quantum yield. The pH of the G6PDH stock solution was successively adjusted in the forward (filled circles) or reverse (open circles) direction up to the indicated value. A sample of the stock solution at each step was diluted with 9 vol. of 0.1 M Hepes containing 0.1 M $MgCl_2$, pH 7.5. The quantum yield was determined 60 min after dilution by comparing the integrated emission spectrum corrected for inner-filter effects with the corresponding corrected spectrum of the tryptophan. (Reproduced from Aon et al., 1989, by permission of The Biochemical Society and Portland Press.)

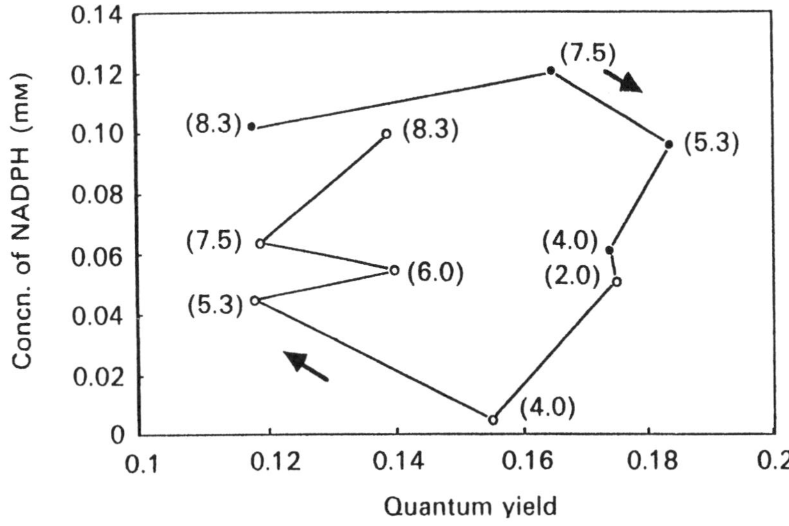

Figure 5.4 Kinetic structural relationships of G6PDH (pH-history dependent changes). NADPH concentrations at steady state (Figure 5.2) are plotted against quantum yield of the enzyme (Figure 5.3) at each pH value, indicated in parentheses, in the forward (filled circles) or reverse (open circles) directions. (Reproduced from Aon *et al.*, 1989, by permission of The Biochemical Society and Portland Press.)

depending on the pH history of the enzyme, corresponded to different activities (Figure 5.2). The described results reflect that different conformations may have the same activity; conversely, the activity level may suffer abrupt changes when relatively small conformational changes occur around tryptophan moieties (Figure 5.4). Therefore, the fluorescence of tryptophan residues appears to arise from different domains of the protein, either associated or not with the enzyme active site.

We suggest that both conformational changes and alterations in intermolecular associations are the nonlinear mechanism that gives rise to the dynamic bistable behaviour in the CSTR experiment. That a conformational change in an enzyme could have dynamic consequences has been demonstrated in an open futile cycle composed of two enzymes (Ricard and Soulié, 1982). The pH-induced process of G6PDH folding–unfolding (intramolecular), and association–dissociation near the isoelectric point (intermolecular), involving covalent (e.g. disulphide) bonds, would be the information-transducing mechanism of the previous record of the enzyme's pH. When this stored information is transduced, a bistable dynamic behaviour appears in an open reactor (Figure 5.2).

5.3 STATIONARY SPATIAL PATTERNS: REACTION–DIFFUSION MODELS

Several hypotheses have proposed rates of chemical change (reaction) coupled to transport of matter (diffusion) as a plausible mechanism to explain how biological organization arises. Chapter 4 discussed some aspects of the pioneering work of Turing (1952) demonstrating that in a system composed of two chemicals, one of them exhibiting autocatalytic kinetics and diffusing in a field, could give rise to spatial differentiation. In Turing's model, autocatalysis or reaction activation by its product or inhibition by its substrate under far from thermodynamic equilibrium conditions is a necessary condition for pattern formation (Turing, 1952; Nicolis and Prigogine, 1977; Meinhardt, 1982). The same was true for all models based on this pioneering work (Kernevez *et al.*, 1979; Murray, 1989; Cortassa *et al.*, 1990b).

The following section deals with the widely known oscillator of the Belousov–Zhabotinskii (B–Z) reaction, capable of exhibiting spatial organization (Tyson, 1976; Winfree, 1984). The behaviour of the B–Z reaction could be simulated by reaction–diffusion models based on Turing's formalism (Tyson, 1976).

5.3.1 TRAVELLING WAVES IN THE BELOUSOV–ZHABOTINSKII (B–Z) REACTION

Belousov (1959, quoted by Winfree, 1984) created the first chemical oscillator equivalent to the citric acid cycle with inorganic chemicals. He replaced the enzymes by the metals these enzymes carry as prosthetic groups and used cerium ions as catalysts and inorganic bromate instead of the coenzyme NAD.

The overall chemical reaction with malonic acid as substrate may be written as follows:

$$2BrO_3^- + 3CH_2(COOH)_2 + 2H^+ \Rightarrow 2BrCH(COOH)_2 \\ + 3CO_2 + 4H_2O \tag{5.1}$$

The reaction is catalysed by an oxidation–reduction couple such as Ce^{3+}/Ce^{4+} or Fe^{2+}/Fe^{3+} (either in ferroin or phenanthroline forms).

The basic phenomenon underlying the oscillatory behaviour in the B–Z reaction is that of excitability. Unlike bistability, in which two steady states are stable for a given set of parametric conditions, in the case of excitability there is just one stable steady state that may evolve upon a perturbation to an excited state which is not stable and the system will relax to the stable state again. The autocatalytic formation of a reduced brominated compound, bromous acid, is apparently the kinetic nonlinear mechanism giving rise to the self-organized behaviour in this system. Oscillations (chemical waves in two- and three-dimensional spaces) have been described in this reaction (Agladze and Krinsky, 1982; Winfree and Strogatz, 1984). It is not our aim to repeat here the detailed mathematical analysis trying to dissect the processes involved in the appearance of the spatio-temporal patterns (Figure 5.5). Nevertheless, we point out that the system is farther from thermodynamic equilibrium because of the large excess of malonic acid and bromate concentrations. The coupled processes are the reduction of bromate and the oxidation of malonate together with the redox transition in the catalyst (Ce^{3+}/Ce^{4+} or Fe^{2+}/Fe^{3+}).

Figure 5.5 is intended to show the striking resemblance that exists between the spatial patterns displayed by the pulsatile signalling of starving *Dictyostelium discoideum* (Figure 5.5d–f) and the waves of oxidation–reduction in the pure inorganic system of the B–Z reaction (Figure 5.5a–c (see also Chapter 3). These examples strengthen the

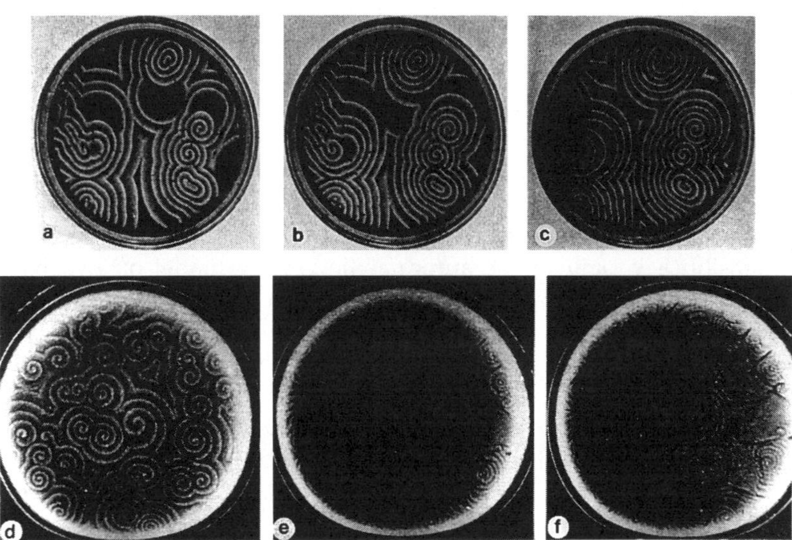

Figure 5.5 The striking similarity of the spatial patterns shown by the rings of chemical activity in the Belousov–Zhabotinskii reaction (a–c) and waves of cells of a layer of the slime mould amoebae (d–f). The petri dish in (d)–(f) appears to be divided into 'territories' and the waves of cells move inwards toward a central aggregation point. (Reproduced from (a–c) Winfree, 1974, with permission from A.T. Winfree; and from (d–f) Newell and Ross, 1982, by permission of the Society for General Microbiology.)

approach of artificial, dissected systems as a way to understanding biological, complex phenomena. The cAMP waves in *D. discoideum* and of redox activities in the B–Z reaction share the existence of autocatalytic mechanisms which give rise to the spontaneous appearance of symmetry-breaking, i.e. spatio-temporal patterns.

Several models have been proposed to unravel the oscillatory mechanism of the B–Z reaction (Tyson, 1976). The Brusselator is perhaps the most worked out (Nicolis and Prigogine, 1977). It is able to reproduce the outstanding features of the B–Z reaction, such as excitability and the spatial travelling waves in one-, two- and three-dimensional systems. The model assumes a constant supply of a reactant (A) that is converted to a compound (X), the latter participating in a trimolecular reaction step (i.e. a step involving the simultaneous participation of three different molecules) to produce more of itself (X). The latter step is the one conferring the nonlinear properties to the whole system and the source of instability leading to symmetry-breaking.

In earlier work, in order to simulate the travelling waves, the model required the diffusion coefficients of the chemicals participating in the reaction to differ by at least five (Nicolis and Prigogine, 1977) or 10 times. This model assumption is difficult to justify since all reactants involved in the putative mechanism have approximately similar physico-chemical properties. Another way to justify the difference in the values of diffusion coefficients has been to assume that these coefficients depend on the concentration (Malchow, 1988; Kagan *et al.*, 1989). Recent theoretical and experimental evidence shows that in one space dimension a variant of the autocatalytic Sel'kov model of glycolysis is able to show steady spatial patterns even with equal diffusion coefficients upon finite-amplitude perturbations (Vastano *et al.*, 1987; Pearson, 1993; see also Winfree, 1991, and Borckmans *et al.*, 1987). Patterns readily occur in two dimensions, ranging from regular hexagons to irregular, chaotic, spatio-temporal patterns, if participating reactants display unequal diffusion coefficients. If diffusion coefficients are equal there should be small asymmetries in the initial conditions amplified by the reaction dynamics for spatial patterns to appear (Pearson, 1993). The fact that nonlinear planar waves in two dimensions cannot be destabilized by diffusion when all diffusion coefficients are equal (Pearson, 1993) was suggested as an explanation for the latter observation. Nevertheless, Pearson stated that with ratios of diffusion coefficients as small as 2, symmetry-breaking could occur.

(a) Biological systems displaying spatio-temporal patterns similar to the B–Z reaction

As previously stressed, the self-organized properties exhibited by the B–Z reaction serve to demonstrate the main features that biological systems should possess to display coherent behaviour. Among rhythmic properties are the action potential waves through neuron axons (Aihara and Matsumoto, 1983), the peristaltic motion in intestines, the aggregation steps in the life cycle of the slime mould (*Dictyostelium discoideum*) (Goldbeter and Segel, 1980; Devreotes, 1989), to mention only a few relevant examples (Figure 5.5). The propagation of the action potential is an electrochemical process in which the far from equilibrium conditions are met by large ionic concentration gradients across membranes, mainly of Na^+ and K^+. The underlying coupled processes are the gating of the Na^+ and K^+ channels and the transport of both ions across the plasma membrane. The nonlinearity involved in the excitability phenomena is given by the non-ohmic dependence of the ionic current through the neuronal plasma membrane with the electrochemical potential difference across it (Aihara and Matsumoto, 1983). In a bioelectromechanochemical model representing meristematic cells growth, the membrane potential difference across the plasmalemma is represented by a formalism of modified Hodgkin–Huxley equations (Adelman and Fitzhugh, 1975; Aihara *et al.*, 1984) including the contribution of the K^+ and Na^+ passive diffusion and electrogenic H^+ pumps (Aon and Cortassa, 1989).

The slime mould *D. discoideum* lives as a unicellular amoeba feeding on bacteria. When starved, this organism displays an aggregation process leading to the formation of a fruiting body composed of stalk and spore cells. Among the first steps before the aggregation, the individual cells migrate toward an aggregation centre following a chemotactic signal, the chemical nature of which has been demonstrated to be cAMP (Gerisch and Hess, 1974) (Figure 5.5). The shape of the cAMP waves resembles those observed when the B–Z reaction is carried out in a petri dish (Figure 5.5). Several models have been able to simulate the chemical waves of *Dictyostelium* aggregation after starvation. The model that perhaps takes into account most of the existing experimental evidence is that based on the cAMP receptor desensitization after binding of its ligand (Goldbeter and Martiel, 1987). The model assumes that the receptor may exist in two alternative conformations, one of which is able to activate adenylate kinase to synthesize cAMP. The nonlinearity involved is given by a step in which two receptor molecules bound to cAMP interact with the catalytic unit of the adenylate cyclase, giving rise to a fully active enzyme. This step is trimolecular and such an example of nonlinear kinetics is able to account for the excitability property shown by the process. This model may reproduce the successive transitions through a cellular stage characterized by relay oscillations and by sustained oscillations later (Goldbeter and Segel, 1980; Goldbeter and Martiel, 1987) (Figure 3.4).

The next section looks at two experimental systems exhibiting transient patterns whose emergence is not triggered by an underlying oscillatory phenomenon. It will be shown that the patterns result from the dynamics of autocatalytic reactions taking place under definite boundary conditions.

5.3.2 pH-INDUCED ASYMMETRIC PATTERNS IN AN IMMOBILIZED BIENZYMATIC SYSTEM

Transient spatial patterns have been reported in a model system in which two enzymatic activities are immobilized in a membrane (Cortassa *et al.*, 1990b). The immobilization of enzymatic systems gives rise to diffusional limitations (mass transfer constraints) on the supply of substrates or removal of products in the immediate environment of the enzymes. In consequence, local metabolite concentration profiles may be established across the membrane thickness which will in turn determine the kinetic behaviour of the enzymes. In addition, immobilized enzymes constitute a well defined system from the point of view of composition, thereby allowing knowledge and control of all variables. Thus, modelling of these systems with reaction–diffusion equations is straightforward (Kernevez *et al.*, 1979; Kernevez, 1980; Hervagault and Thomas, 1987; Cortassa *et al.*, 1990b).

A reaction–diffusion model of two autocatalytic reactions catalysed by immobilized enzymes predicted a break of symmetry in the distribution of compounds (substrates and products) participating in the reactions (Cortassa *et al.*, 1990b). The enzymes chosen to demonstrate the patterns predicted by the model were urease and glutaminase. Both enzymes exhibit bell-shaped pH dependence curves with different pH optima. Since these enzymes catalyse the production (glutaminase) or consumption (urease) of protons, one of the sides of the bell-shaped pH dependence curve is autocatalytic (Appendix 5A). For the enzyme catalysing the production of an acid compound, the alkaline side of the curve is autocatalytic since as reaction proceeds in a poorly buffered medium the pH will decrease and the reaction rate will consequently increase (Figure 5.6). The opposite is true for urease that catalyses the reaction whose product is a basic compound. The reaction rate catalysed by glutaminase or urease exhibits maxima at pH 5.1 or 7.0, respectively. The alkaline branch of the glutaminase pH dependence curve is autocatalytic due to the balance of acid–base properties between substrates and products. Particularly, acidification of a reaction mixture at an initial pH > 5.1 occurs as the reaction evolves, driving the pH closer to its optimum and therefore increasing the reaction rate:

$$\gamma\text{-Methylglutamic acid} \xrightarrow{\;glutaminase\;} NH_3 + \text{glutaric acid monomethyl ester} \qquad (5.2)$$

$$\text{Urea} \xrightarrow{\;urease\;} NH3 + CO2 \qquad (5.3)$$

When diffusion is coupled to the autocatalytic reaction by means of immobilization, different patterns of the internal pH profile of the membrane appear. Modelling studies predict, and experiments confirm, that evolution of the pH profile of the membrane towards a nearly symmetrical steady state, through a transient asymmetrical pattern of pH, is triggered by slight asymmetrical pH perturbations at the boundaries. To allow the emergence of a transient spatial pH pattern, the system must fulfil the condition that the optimal pH of the enzyme, E_1, catalysing the acid production should be lower than the optimal pH of the enzyme, E_2, catalysing the step in which alkaline compounds are produced, i.e. the autocatalytic sides of their pH dependence curves should cross each other. The point where the two pH dependence curves cross each other is highly unstable. At pH 6.0, the system shows an unstable steady state, i.e. any perturbation drives the system towards a more alkaline or more acidic pH due to the autocatalytic behaviour exhibited by both enzymes with respect to pH. The bienzymatic system changes from a mode in which glutaminase activity prevails over urease to a mode in which urease activity predominates (Figure 5.6). This pH is extremely important because at this value the sense of the amplification driven by autocatalysis towards acidification or alkalinization will change abruptly. Thus, the reaction system behaves like a ball at the apex of a convex surface: the slighest disturbance will be amplified and the ball will roll downhill left or right, depending on the direction of the perturbation (Figure 1.5). In consequence, the behaviour of the system will strongly depend on the initial and boundary conditions of the assay.

(a) Visualization of pH profiles inside the membrane: combined modelling and experimental studies

According to numerical simulations, the model that represents the dynamic behaviour of the system predicts three possible situations of the

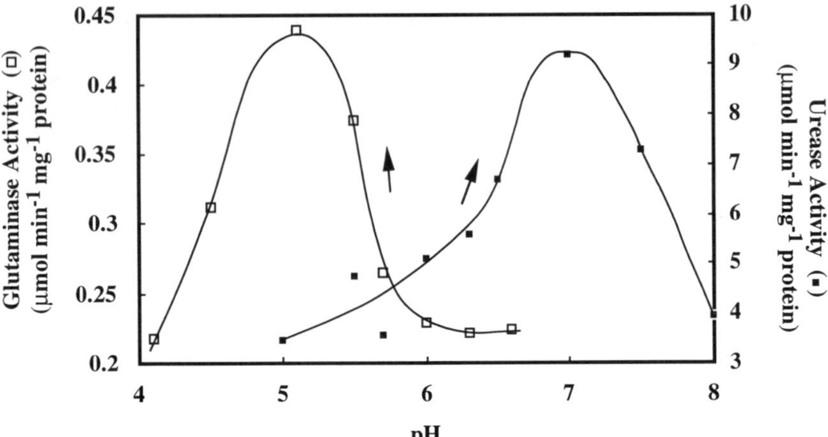

Figure 5.6 Specific pH-dependent activities of immobilized urease (curve on right) and glutaminase (curve on left); expressed as μmol of acid (urease) or alkali (glutaminase) consumed min⁻¹ mg⁻¹ protein. Arrows indicate the direction of evolution of the system when it crosses through the unstable steady state. In all cases each point is the average of four independent measurements and the standard error is less than 5% of the value. (Redrawn from Cortassa *et al.*, 1990, by permission of The Biochemical Society and Portland Press.)

evolution of the pH profile across the membrane, when the system is placed at the unstable steady state. If a small disturbance takes place at the boundaries, it will be translated into a local pH perturbation inside the membrane. The numerical resolution of the system predicts that it will be amplified and propagated across the membrane thickness. Figure 5.7 shows the internal pH profile in the membrane obtained by numerical integration of equation 5.10 (Appendix 5A) for three different boundary conditions. The temporal evolution of pH inside the membrane passes through a transient pH profile which will be referred to as the asymmetrical pattern. The final state is an almost symmetrical distribution of pH.

Figure 5.7 (see facing page) Numerical simulation of pH distribution profile across width of membrane containing immobilized urease and glutaminase. The initial conditions of the simulation were: (a) a gradient between pH 6.3 and 5.7 (boundary conditions) for σ = 450; the temporal coordinate corresponds to a value of 0.35 θ and each curve represents the pH distribution across the membrane at 0.035 θ; (b) a gradient between pH 6.3 and 5.7 for σ = 250; the temporal coordinate corresponds to a value of 0.2 θ and each curve represents the pH distribution across the membrane at 0.02 θ; (c) a gradient between pH 6.4 and 5.9 for σ = 250; the temporal coordinate corresponds to a value of 0.2 θ and each curve represents the pH distribution across the membrane at 0.02 θ. In each case the space coordinate represents the thickness of the membrane (1 cm). (From Cortassa *et al.*, 1990b, by permission of The Biochemical Society and Portland Press.)

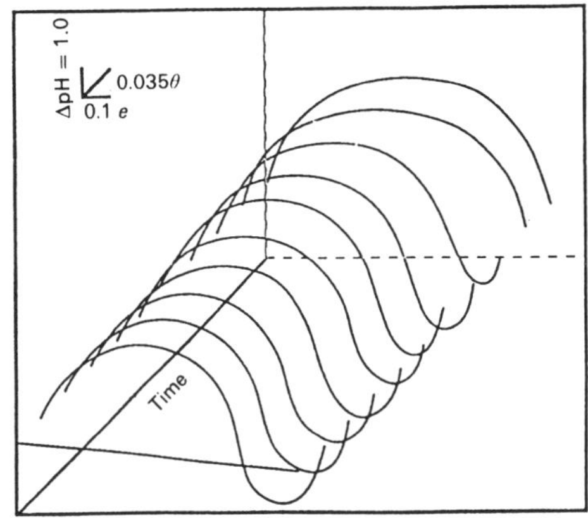

ΔpH = 1.0
0.035θ
0.1 e
Time
(a)

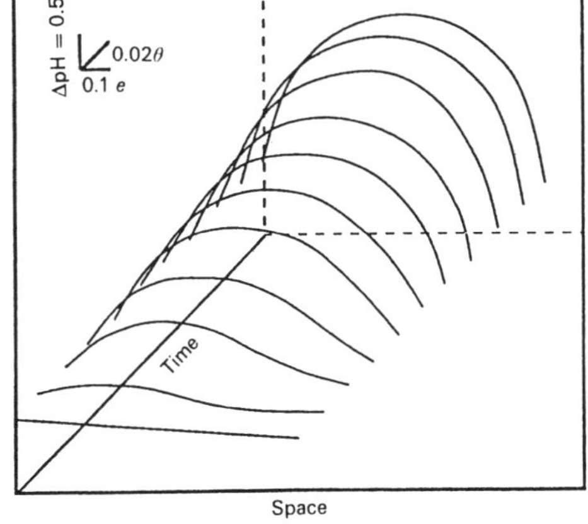

ΔpH = 1.0
0.02θ
0.1 e
pH
Time
(b)
ΔpH = 0.5
0.02θ
0.1 e
Time
(c)
Space

The asymmetrical pattern implies a difference between acid and alkaline regions of up 4 to 6 pH units, depending on the value of the Thiele modulus, σ, that contains diffusion, geometrical and kinetic information (Appendix 5A). The larger the value of σ, the greater is the difference between extremes of pH and the steeper the gradient (Figure 5.7).

Figures 5.8 and 5.9 show the experimental results obtained with parameter values corresponding to those in Figure 5.7. The phenomenon is triggered when the membrane equilibrated at pH 6.0 is placed in the reactor. The experiments shown in Figures 5.8 and 5.9 correspond to conditions in which either different perturbations were applied at each side of the membrane or for different concentrations of the immobilized enzymes.

An alkaline, nearly symmetrical and more rapid pattern is attained at relatively higher enzyme concentrations for the same small perturbation applied at the boundaries. In fact, the membrane with $1.8 \times 10^{-4}\,M\,s^{-1}$ arrived at a final, nearly symmetrical steady state after 10 h (Figures 5.8b and 5.9) whereas the membrane containing $3.3 \times 10^{-4}\,M\,s^{-1}$ of activity almost exhibits the final symmetrical pattern at 8 h (Figure 5.8a). However, the first spatial asymmetrical pattern takes more time to appear and the pH gradient is less steep (Figure 5.8b). The latter agrees rather well with theoretical predictions, since for $\sigma = 250$ the steady state is attained after 0.14θ, while for $\sigma = 450$ an equivalent of 0.28θ should pass before achievement of the final pattern (Figure 5.7).

5.3.3 A SPATIAL BANDING PATTERN OF PHOTOSYNTHETIC ACTIVITY

A spatial banding pattern from a continuous gradient of an electron acceptor was shown to be generated by the photosynthetic activity of thylakoid membranes. A photobiochemical system consisting of a homogeneous suspension of thylakoid membranes, with an imposed gradient of an electron acceptor (dichloroindophenol: DCIP or DCPIP), when illuminated was able to show a self-organized redox change in symmetry, i.e. passage from a monotonic gradient to a banded pattern (Aon *et al.*, 1989b). The temporal evolution of the pattern is shown in Plate 4.

The induction time for pattern formation was dependent on the ratio of gradient parameters, a/b; a is the concentration difference in electron acceptor between the top and the botton of the test tube, and b is the lower concentration limit of the electron acceptor gradient). For a/b < 3, the induction of patterning was strongly time-dependent with a constant distance between the bands. The self-organized redox transition was not observed when the electron acceptor was homogeneously distributed

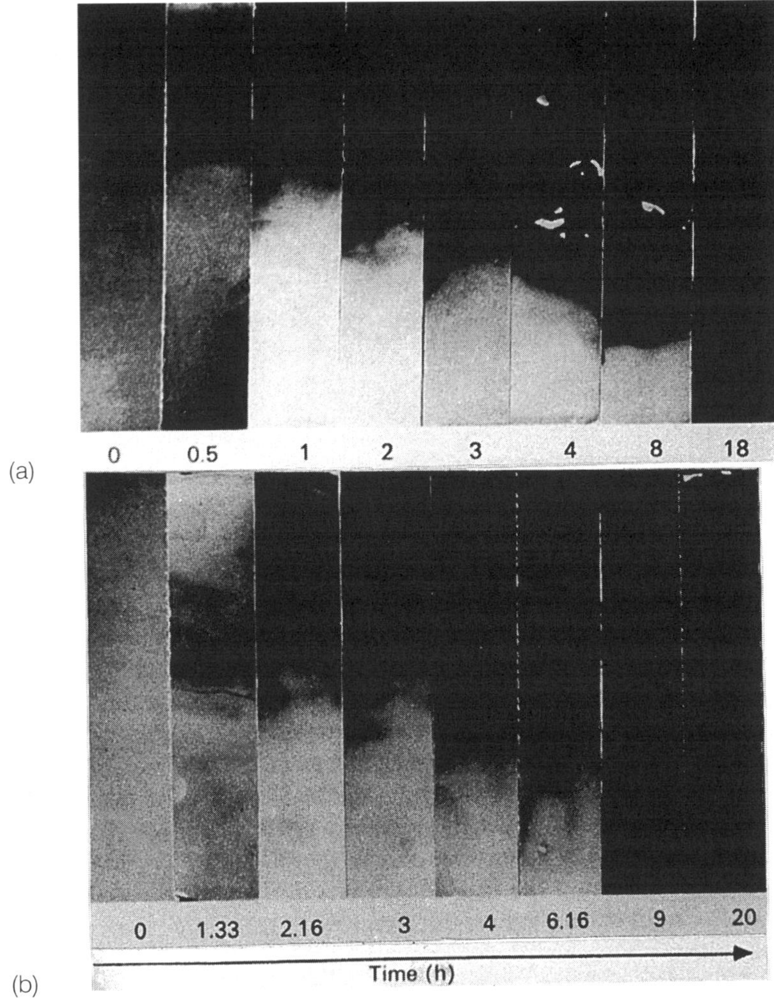

Figure 5.8 Photographs of the dynamics of pH-travelling waves in an immobilized bienzymatic membrane upon an antisymmetrical perturbation of pH. (a) The experiment was started by placing the membrane in the reactor and perturbing the pH in each compartment: 0.3 units of pH were applied to each compartment but in opposite directions, i.e. one compartment at pH 5.7 (pH = 6.0 − 0.3) and the other at pH 6.3 (pH = 6.0 + 0.3). The equivalent immobilized activity for each enzyme was 3.3×10^{-4} M s^{-1} in the membrane volume and a σ value of 430. (b) The equivalent immobilized activity for each enzyme was 1.8×10^{-4} M s^{-1} in the membrane volume and a σ value of 230. The progagation of the alkaline front is observed through the change in bromothymol blue colour. The distance between the two compartments was 1 cm. (From Cortassa *et al.*, 1990b, by permission of The Biochemical Society and Portland Press.)

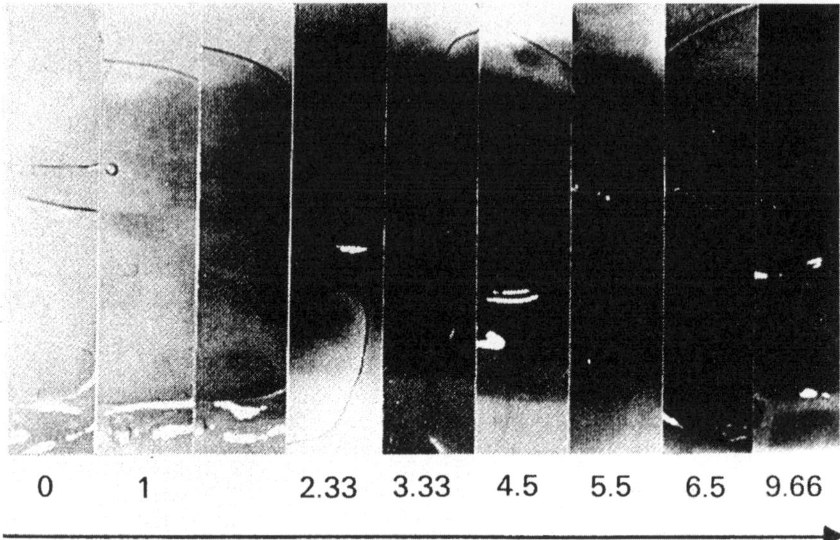

Figure 5.9 Dynamics of pH-travelling waves in an immobilized bienzymatic membrane upon an asymmetrical perturbation of pH. The reaction was started as described in Figure 5.8 by perturbing asymmetrically the pH (6.3 and 5.7) in both compartments. Enzymes were immobilized with an activity equivalent of 1.8×10^{-4} M s^{-1} in the membrane volume and $\sigma = 230$. The propagation of the alkaline front is observed through the change in bromothymol blue colour. The distance between the two compartments was 1 cm. (From Cortassa *et al.*, 1990b, by permission of The Biochemical Society and Portland Press.)

throughout the reaction mixture. Additionally, the number of bands observed varied when parameters a and b were changed, as well as when the concentration of active thylakoid membranes was modified. Since pattern formation was obtained only for certain relationships among parameters a and b, or thylakoid concentration, the phenomenon depends upon how far from thermodynamic equilibrium are the initial conditions imposed on the system.

In this photobiochemical system, the thermodynamic conditions for a system to achieve self-organization are easily identified. The system is open to an energy flux (light) and may be considered to be operating under pseudo-open conditions with respect to the flux of DCIP, due to the high initial concentrations of the acceptor. There are also coupled processes occurring with different affinities, namely the absorption of light intensity by the coloured DCIP in its oxidized form, its diffusion and the reaction that reduces it. Moreover, the flux (the rate of electron acceptor reduction) is not linearly related to its concentration: it looks like an inhibition by substrate excess with the rate of the electron acceptor reduction showing a sigmoidal dependence with light intensity (Breton *et al.*, 1985; Aon and Cortassa, 1993).

Under the experimental conditions described in Aon *et al.* (1989b), no oscillatory behaviour was detected at least for the reduction of the exogeneous electron acceptor by the photosynthetic activity. Furthermore, the obtained spatial periodicity is highly dependent on the water-splitting system. When thylakoids were subjected to treatments known to block or destroy selectively the water oxidation step (such as heat treatment or methylamine) no spatial patterns were observed. No influence on thylakoid activity, measured as the initial rate of electron acceptor reduction, could be detected under conditions of non-cyclic photophosphorylation with methyl viologen as the electron acceptor of photosystem I in the absence or presence of an uncoupler. It appears, then, that the electron transport from water through photosystem II to DCIP was one of the reaction steps involved in the establishment of the observed pattern.

For the same experimental design but in gelled medium, spatial patterning was observed. Although the same number of bands was obtained, the time of pattern appearance was two-fold longer than in the non-gelled medium. Therefore, the results obtained show that diffusion plays a key role in the observed phenomenon although at present convection cannot be completely excluded.

(a) Macroscopic organization and levels of organization in the photobiochemical system

Different levels of organization are involved in the spatio-temporal pattern formation observed in the photobiochemical system. Photosynthetic energy transduction comprises processes whose spatio-temporal dimensions range from 10^{-13} s to 10^{2} s and 10^{-12} m to 10^{-3} m: pigment excitation, electron transport, ion fluxes and building of a proton motive force, among others (Table 2.1). The water-splitting system of photosystem II involves processes occurring in some angstroms in space and 10^{-9} to 10^{-13} s in time. Meanwhile, the bands in the tube are in a spatial dimension of the order of millimetres or centimetres and their appearance evolves in the order of minutes, i.e. further from the 'transition point' in the plot relating the characteristic spatio-temporal dimensions of levels of organization (Table 2.1 and Figure 2.2). Since it has been demonstrated that spatial periodicity at the macroscopic level of the self-organized redox transition depends on the presence of a gradient of the acceptor (DCIP) (Aon *et al.*, 1989b), the microscopic level of photosynthetic activity (water-splitting system activity), even if necessary, is not sufficient to account for the appearance of structures that occur spatially and temporally in a higher level of organization. Thermal movement will tend to homogenize any arising macroscopic order; evidence for the latter is given by the fact that in the absence of an imposed gradient of the electron acceptor, the organized redox transition is not observed. It appears that the gradient acts as a macroscopic cue or a long-range anisotropic spatial restriction which couples an external stimulus (light) to the intrinsic autocatalytic reduction of the electron acceptor by photosynthetic activity, giving rise to the macroscopic self-organized change in symmetry.

Figure 5.10a shows the dependence on light intensity of thylakoid photosynthetic activity measured through the reduction of DCIP, while

Figure 5.10 (see facing page) Transition from molecular to equivalent dimensions of supracellular levels of organization in a photobiochemical system in the presence of a gradient. Experimental demonstration of autocatalysis in the photosynthetic reduction of dichloroindophenol (DCPIP). (a) Thylakoids were incubated in a thermostated reactor at 25°C and the photosynthetic activity was measured spectrophotometrically at 600 nm in continuous reduction by DCPIP. Activities (in mM DCPIP consumed per hour) correspond to initial rates. The experimental points were fitted according to equation 5.9 (Appendix 5B) ($r = 0.97$) by nonlinear regression analysis (Levenberg–Marquardt algorithm) with the following parameter values: $K_I = (30\ \mathrm{W\ m^{-2}})^3$; $K_A = 0.206\ \mathrm{mM}$; $V_m = 0.966\ \mathrm{mM\ h^{-1}}$. (b) DCPIP solutions at different concentrations containing heat-inactivated thylakoids (0.1 mg chlorophyll $\mathrm{ml^{-1}}$) were placed in a test tube (1.2 cm diameter) and irradiated perpendicularly ($I_0 = 30\ \mathrm{W\ m^{-2}}$). The light intensity in the absence of DCPIP ($I_0 = 16\ \mathrm{W\ m^{-2}}$) corresponds to the light scattering by a suspension of inactivated thylakoids equivalent to 0.1 mg chlorophyll $\mathrm{ml^{-1}}$. The experimental points were fitted by nonlinear regression analysis ($r = 0.97$) with equation 5.10 (Appendix 5B), with the following parameter values: $I_0 = 16\ \mathrm{W\ m^{-2}}$; $\alpha_1 = 0.309$; $\alpha_2 = 0.691$; $\epsilon_1 = 237.8\ \mathrm{mM^{-1}\ cm^{-1}}$; $\epsilon_2 = 1.977\ \mathrm{mM^{-1}\ cm^{-1}}$; e = 1.2 cm. (c) Thylakoids (0.1 mg chlorophyll $\mathrm{ml^{-1}}$) were incubated in a thermostated reactor of 1 cm diameter at 25°C and 30 W $\mathrm{m^{-2}}$ of light intensity. Activities (in mM DCPIP consumed per hour: filled circles) were measured as described above. Panel (c) also shows the simulations of the sensitivity to different light intensities of the autocatalytic reduction of the electron acceptor (equation 5.9; Appendix 5B). (From Aon and Cortassa, 1993, reprinted by permission of Kluwer Academic Publishers.)

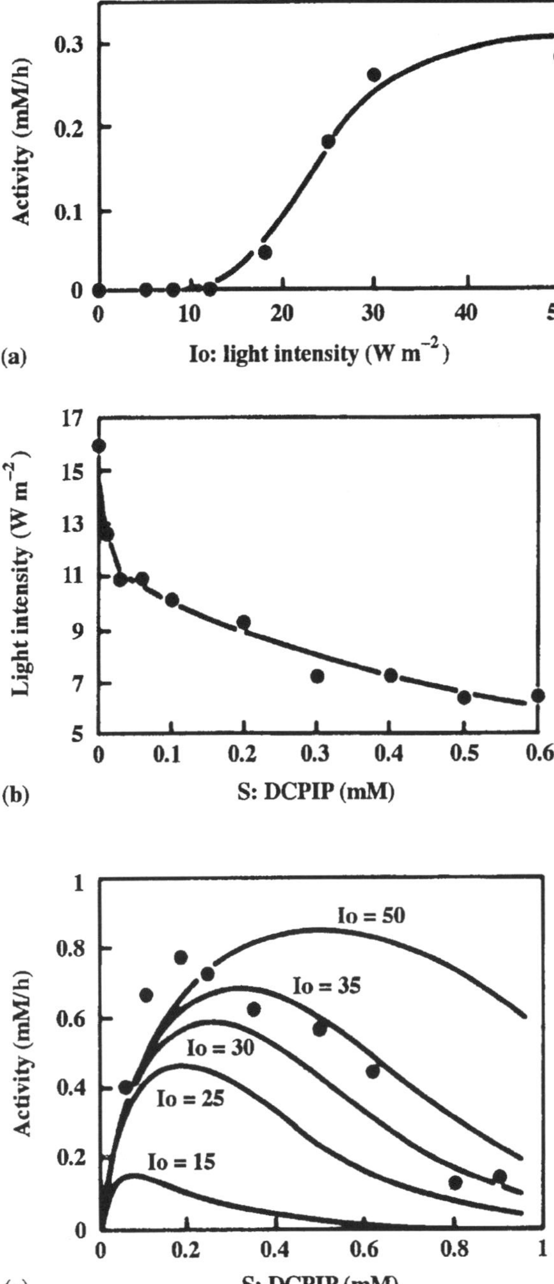
Activity (mM/h)
0.3
0.2
0.1
0
0   10   20   30   40   50
Io: light intensity (W m⁻²)
(a)
17
15
13
11
9
7
5
Light intensity (W m⁻²)
0   0.1   0.2   0.3   0.4   0.5   0.6
S: DCPIP (mM)
(b)
1
0.8
0.6
0.4
0.2
0
Activity (mM/h)
Io = 50
Io = 35
Io = 30
Io = 25
Io = 15
0   0.2   0.4   0.6   0.8   1
S: DCPIP (mM)
(c)

Figure 5.10b shows the monotonous exponential decrease in light intensity with the increase in DCIP concentration. When coupled, these two basic phenomena underlie the nonlinear autocatalytic reduction of the acceptor (Figure 5.10c).

A model to simulate the dynamics of the photosynthetic electron acceptor reduction was set up on the basis of experimental kinetic data. The model equations and their derivation are shown in Appendix 5B. From the numerical resolution of those equations a bistability phenomenon in the electron acceptor concentration appeared. The bistable behaviour shows that the photobiochemical system can switch to the lower or higher steady state branches of the acceptor concentration, as previously shown for the banded pattern in the test tube (Plate 4), depending on the rate of supply (Figure 5.11a) and gradient steepness of the acceptor (Figure 5.11b).

The results presented in this and the previous section suggest that energy transduction phenomena in non-homogeneous media (for example, in the presence of a gradient), involving autocatalysis and several levels of organization, may exhibit self-organized structures emerging at levels of organization further from the 'transition point' (Figure 2.2).

5.4 SPATIO-TEMPORAL PATTERNS DESCRIBED IN BIOLOGICALLY ORIENTED ARTIFICIAL SYSTEMS AS HEURISTIC TOOLS FOR INTERPRETING SIMILAR PHENOMENA OBSERVED IN LIVING SYSTEMS

5.4.1 SWITCHING BISTABLE MECHANISMS OF FLUX METABOLIC REDIRECTION

Several lines of experimental evidence suggest that, intracellularly, enzymes are organized in complexes whose formation as well as dissolution may be involved in metabolic regulation. Similarly, supramolecular structures of macromolecules inside the cell, such as those implicated with the cytoskeleton (e.g. microtubules), may be subjected to similar regulation (Mitchison and Kirschner, 1984a,b; Cortassa *et al.*, 1994a). Bistability is essentially a reversible dynamic regime whose existence may provide a metabolic system with a buffering effect and an increase in sensitivity by switching between two regimes, i.e. both branches of steady state fluxes. We have found conditions in which an irreversible switching between different branches of steady states for microtubular protein and pyruvate kinase activity were observed (for further discussion, see section 6.3.1d). Since the dynamics of a microtubular protein assembly–disassembly system displays

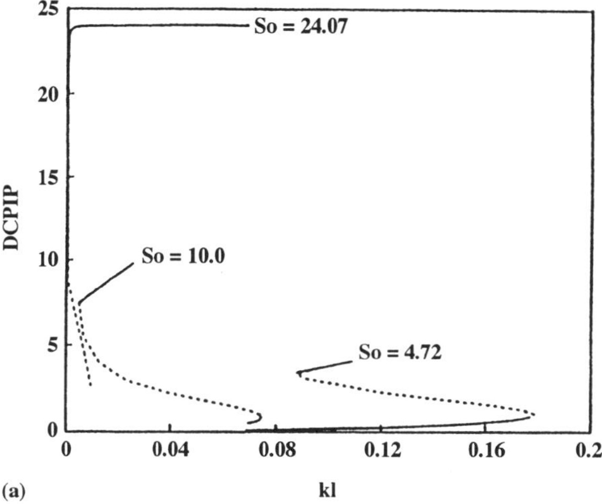

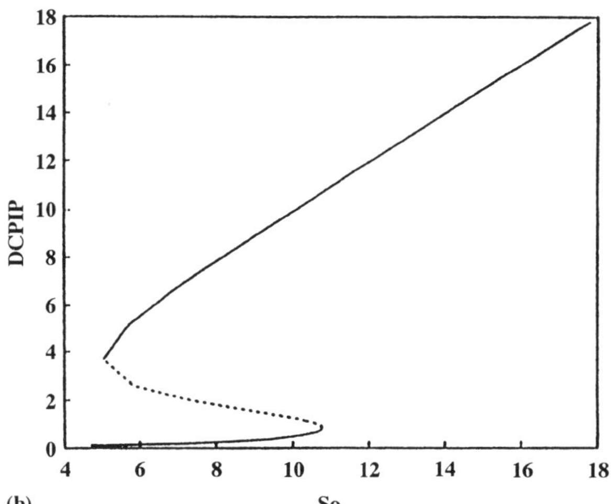

Figure 5.11 A model for the appearance of banded patterns in the photobiochemical system. The stability analysis and bifurcation properties of the autocatalytic photobiochemical system (equation 5.11) were performed numerically with AUTO. The stable (solid line) and unstable (dashed line) branches of the steady state solutions of equation 5.11 and the limit points (where the stability of the stationary solutions changes) were computed automatically. The bifurcation diagram was based on the parameter values obtained from the fitting of experimental data described for Figure 5.10. (From Aon and Cortassa, 1993, reprinted by permission of Kluwer Academic Publishers.)

bistability when isolated, and that of the isolated enzyme was monostable, we conclude that irreversible transitions only happen when both subsystems are coupled. Under conditions of dynamic coupling (Chapter 7), the dynamics of cytoskeletal protein entrains the dynamics of pyruvate kinase (Figures 1.8 and 6.10). Dynamic regulation results from bistable changes of intermolecular association–dissociation equilibria, i.e. a macromolecular bistable network, which alternate between two steady states implying, for example, high or low fluxes, along the same metabolic pathway (Chapter 6).

5.4.2 TRAVELLING pH WAVES DRIVEN BY AUTOCATALYTIC ENZYMATIC MECHANISMS

Qualitative agreement between theoretical predictions of a reaction–diffusion model and experimental results obtained in an immobilized bienzymatic system were observed in the phenomenon of pattern formation by travelling pH waves. For the same parameter values used in the experiments, such as σ and pH at the boundaries, the qualitative behaviour of the pH profile evolution inside the membrane is described by the model (Figures 5.7, 5.8 and 5.9; Appendix 5A).

The model used to represent the dynamic behaviour of an immobilized bienzymatic system cannot be considered a Turing model (Chapter 4). In the latter, it is assumed that the initial homogeneous state is stable and that the asymmetrical pattern arises because this homogeneous state becomes unstable when perturbations are triggered by the change of a control parameter. In the model of the bienzymatic system, the initial state is unstable, in order to be able to attain the asymmetrical pattern transiently. Additionally, in a Turing model the characteristic length of the pattern depends only on the reaction and diffusion features of the system, whereas in our case it is the geometrical characteristics of the system (namely, the membrane width) that determine the size and landscape of the pattern. The reaction and diffusion features of the system will determine the amplitude and temporal evolution of the phenomenon. It is worth noting that the appearance of coherent behaviour in the evolution of the pH profile across the membrane (i.e. the coordinated change of the pH towards the steady nearly symmetrical pattern, with a passage through a transient asymmetrical pattern) constitutes a transition from microscopic to macroscopic order. This latter transition emerges when the nonlinear properties of the implicated reactions are coupled to diffusion under well defined boundary conditions.

It has been suggested that the coupling of the nonlinear photosynthetic reduction of an electron acceptor (Breton *et al.*, 1985) to diffusion in the presence of a gradient was the underlying mechanism of the symmetry change of the electron acceptor in a photobiochemical system (Aon *et al.*, 1989b).

The pH wave did not appear to arise as a result of hydrodynamic instability because the reactions were performed in a solid matrix where enzymes were immobilized, and in which hydrodynamic currents were not possible. Although this example corresponds to a rather artificial system, the fact that hydrodynamics can be eliminated as a factor in the appearance of the patterns strengthens its biological significance. Due to the highly structured nature of biological media, convection and in general hydrodynamic phenomena are likely to be precluded. On the contrary, it is possible that reactions occur coupled to diffusion, thereby giving rise to the possibility of Turing instabilities. We do not mean to imply that these results support the notion that protons constitute a Turing's morphogen in a living cell but rather to suggest that any substance, participating in autocatalytic reactions, could eventually lead to transient asymmetrical patterns which might in turn condition the fate of a cell in a morphogenetic field.

The phenomenon described of a transient rupture of pH symmetry across a membrane may be considered as a spatial anisotropy-generating mechanism. It will be self-generated due to the nonlinear nature of the biological reactions occurring inside the membrane, and dynamic since it evolves in time. Several dynamic consequences may be envisaged from a 'dynamically self-generated anisotropy'. It could well be a mechanism for conditioning the activities of biological processes coexisting in space and thus a functional distribution in that field (Davies, 1973). Earlier work has shown that two compartmentalized enzymes in a selective membrane may produce a vectorial behaviour, namely, a phenomenon of active transport (Broun *et al.*, 1972, quoted in Cortassa *et al.*, 1990b). A dynamic anisotropy would have the same dynamic consequences, albeit in this case by compartmentalization of relevant biochemical activities. If a dynamically self-generated anisotropy of pH, or any other physico-chemical property, exists in the cytoplasm of a polarized cell, it would delimit several regions, or 'compartments', in which specific enzymic activities could be expressed, according to their activity profiles with respect to that physico-chemical property.

5.4.3 GRADIENTS TOGETHER WITH AUTOCATALYTIC MECHANISMS AS A PREREQUISITE FOR LONG-RANGE SPATIALLY BANDED PATTERNS

According to a reaction–diffusion mechanism, the interaction of diffusion transport and nonlinear chemical kinetics can lead to instabilities of a concentration field with the result that initially homogeneous domains develop into regional non-uniformities (Turing, 1952; Kernevez *et al.*, 1979; Murray, 1989). The autocatalytic nature of a photobiochemical reaction resulting from the coupling of two intrinsically linear processes, the reduction of DCIP by thylakoids and the light absorption by the same electron acceptor (Figure 5.10), have been demonstrated in a spatially compartmentalized system (Breton *et al.*, 1985). In those studies, the existence of multiple steady states of a single reaction in coupled open reactors, with mutual diffusion through an inert membrane, was shown by experimental and numerical methods. The autocatalytic nature of the photochemical reduction of DCIP was also found in the homogeneous and non-compartmentalized photobiochemical system. That the spatial patterns appeared only in the presence of the concentration gradient of the electron acceptor, that no oscillatory behaviour in the reduction of DCIP was detected with the experimental conditions used in this work and that diffusion is probably involved all suggest that the explanation of the observed spatial periodicity lies in the capability of the reaction mixture to exhibit multiple stationary states, i.e. bistability. The plausibility of the latter has been demonstrated in a zero-dimensional model of the photobiochemical system (Figure 5.11) (Aon and Cortassa, 1993). It has also been postulated that the coupling of autocatalysis with diffusion might account for periodic precipitation processes (Liesegang rings) in the absence or presence of reactant gradients (Kai *et al.*, 1982).

Experimental evidence has revealed the existence of gradients of substances which could function as spatial cues specifying information in early development. In the examples presented, a homogeneous biochemical system far from equilibrium with an imposed monotonous gradient of a substance can, when subjected to an external stimulus (light), generate a break in space symmetry which can act to a certain extent as a mechanism of asymmetrical distribution of morphogenetic substances, i.e. macromolecules. Without pretending to establish a direct analogy with embryology, the relative simplicity of the photobiochemical system can be used to make a profitable analysis to identify the type of nonlinearities participating in the appearance of banding patterns in developing embryos from both thermodynamic and kinetic points of view (see also Chapter 4 about symmetry in developing systems). This strategy may help us to visualize and conceptualize some features about the dynamics of pattern formation in developing systems. For instance, the ability of spatial patterns to give orientational cues in a morphogenetic field has been proposed especially in the form of a gradient profile (for a review, see Slack, 1987). A gradient could in turn provide the cells with information about their position in the tissue (Wolpert, 1969). Crick (1970) has calculated how a chemical substance, diffusing through a developing tissue from a localized source towards a sink located at the other end of the tissue, could produce a gradient. The measured diffusion coefficient of H^+ is well above 10^{-8} cm^2 s^{-1}, the threshold considered necessary for the establishment of a gradient across a distance of 1 mm in a time lapse as short as a few hours (Crick, 1970). These ideas are now beginning to be accepted by experimental embryologists (Lawrence, 1988; Oliver *et al.*, 1988). Several experimental demonstrations of the existence of gradients have been reported (Thaller and Eichele, 1987; Driever and Nüsslein-Volhard, 1988a,b; Oliver *et al.*, 1988).

Now, let us recall the subject of macroscopic order appearing in systems where molecules diffuse by thermal forces. The DCIP pattern appears to emerge through a reaction–diffusion mechanism, in which the autocatalytic (nonlinear) dependence of the DCIP reduction with light intensity coupled to diffusion could lead to the formation of bands of electron acceptor alternately with bands where it has been completely reduced. The gradient is a necessary intermediary as a long-range organizational cue to allow an inherent nonlinear mechanism to 'interpret' an external stimulus fluctuation through an asymmetry change. This asymmetry change may be a bifurcation point leading to oscillations immediately preceding the pattern formation or directly leading to spatial patterns. The reactions taking place at the spatio-temporal microscopic level suddenly become coherently self-organized at the macroscopic level. A kind of discontinuity between the microscopic (scalar, brownian) to the macroscopic (ordered, vectorial) organization occurs, i.e. a transition between levels of organization (Aon and Cortassa, 1993). By analogy in developing systems we should aim to identify the long-range cues that organize the activity of individual molecules in a macroscopic pattern (Goodwin, 1986). The development of *Drosophila melanogaster* exhibits several interesting spatially self-organized patterns, like the banding pattern in the protein products of many homeotic genes that participate in the definition of segments during the embryonic development of the fruit fly (*hairy, eve, fushi tarazu, sevenless, antenapedia, bitorax*) (Figure 4.10).

The banding pattern in the photobiochemical system and the establishment of the metameric pattern in *Drosophila* embryos both require an initial gradient, i.e. the electron acceptor gradient and the gradients of bicoid and nanos proteins in *Drosophila* (Driever and Nüsslein-Volhard, 1988a,b) (Chapter 4). It has been demonstrated that the slope of the gradient of the bicoid protein at early stages of the embryo determines the position and distribution of the stripes of the products of several homeotic segmentation genes (*hairy, eve, fushi tarazu*) in the blastodermal stage (Driever and Nüsslein-Volhard, 1988a,b, 1989).

5.4.4 GRAVITATIONALLY INDUCED SELF-ORGANIZED DISSIPATIVE MICROTUBULAR STRUCTURES

Living cells are open systems operating at some distance from thermodynamic equilibrium (Chapter 1). Polymerization of cytoskeleton components such as actin and tubulin occurs with energy expenditure (ATP or GTP, respectively: Chapter 6) and gives rise to self-organized spatio-temporal structures (Mandelkow *et al.*, 1989). The striking patterns formed by microtubules during mitosis of eukaryotic cells are a manifestation of their dynamically self-organized nature (Alberts *et al.*, 1989).

Spontaneous macroscopic space ordering of microtubular structures has been described (Mandelkow *et al.*, 1989; Tabony and Job, 1992). A simple biochemical system, consisting of a solution of purified tubulin containing GTP or a GTP-regenerating system, has been shown to assemble spontaneously into microtubules. After assembly in optical cells, microtubular structures show stationary periodic horizontal stripes of about 0.5 to 1 mm separation (Figure 5.12) (Tabony, 1994). In each band, the microtubules are oriented at either an acute or an obtuse angle with adjacent stripes showing opposite orientation (Tabony, 1994).

Apparently, the dissipatively self-organized striped pattern of microtubular structures arises from coupled reordering along with a chemical wave. It has been proposed that the chemical wave would comprise different concentrations of microtubules and free tubulin

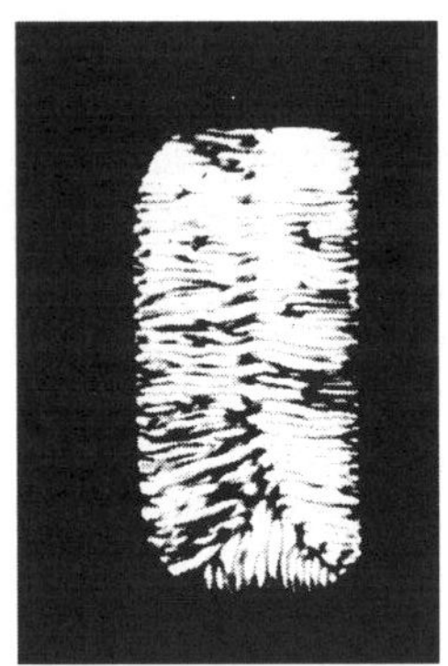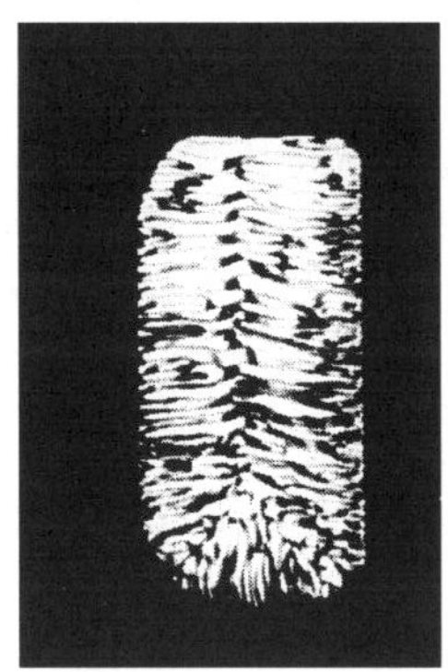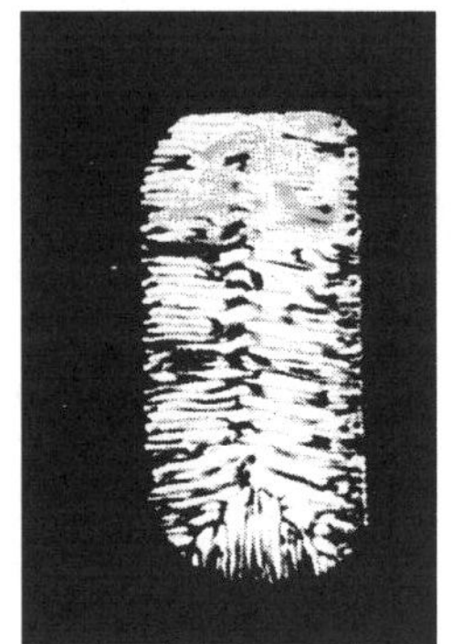

Figure 5.12 Gravitational symmetry-breaking in microtubular dissipative structures; microtubular solutions as viewed through crossed polars. The samples were prepared by warming tubulin in the presence of GTP from 4°C to 36°C in vertical cells (1 cm × 4 cm × 0.1 cm). The patterns take about 5 h to form. All samples were prepared in a room thermostated at 36°C, and no thermal convection occurred. (From Tabony and Job, 1992, with permission.)

(Tabony, 1994). The mathematical model described in Chapter 6 shows that a kinetic representation of microtubular assembly–disassembly has intrinsic nonlinear mechanisms such as the autocatalytic process of tubulin assembly. These mechanisms are responsible for nonlinear behaviour such as bistability, with either reversible or irreversible transitions (section 5.4.1).

The examples of self-organization in artificial systems presented in this chapter provide just a model of the way in which complex phenomena in real biological systems may be simplified to a certain extent in explaining the appearance of dynamic organization. Integrated with other conceptual tools, this will help to build a theoretical framework for the understanding of complex phenomena such as growth and development (Chapters 2 and 11).

APPENDIX 5A IMMOBILIZED BIENZYMATIC MODEL

The model considers an artificial protein membrane in which two enzymes, E_1 and E_2, are homogeneously immobilized (Cortassa *et al.*, 1990b). The product of the reaction catalysed by E_1 is an acid, while that produced by E_2 is a base. For the present experimental system, E_1 and E_2 will be glutaminase and urease, respectively. The reactions involved may be schematically represented as follows:

$$S_i \xrightarrow{\ E_i\ } P_i \quad i=1,2 \tag{5.4}$$

The model assumes that both enzymes obey Michaelis–Menten kinetics and that the substrates are present in excess to assure zero-order kinetics for both reactions. Moreover, the conditions of the reaction were assumed to allow substrate concentrations to remain constant inside the membrane, i.e. substrate consumption was assumed to be negligible compared with initial concentrations.

The specific activity ratio of immobilized urease/glutaminase is approximately 25, measured as the ratio of maximal velocities (V_M) at their optimal pH ($pH_{(opp)}$), in order to obtain the same V_M at the respective pH and to fulfil the condition $\sigma_1 = \sigma_2$ (see below).

The pH dependence of the enzymatic activity may be treated using the classical assumption that enzymes behave as dibasic acids, exhibiting two dissociation equilibria. Of the three forms in which each enzyme may exist, only the monoprotonated species shows activity. The pH dependence may be expressed by the following equation (Segel, 1975):

$$\alpha_i = \frac{V^i_{M(pH)}}{V^i_{M(opp)}} = \frac{K^i_{es1}10^{-pH}}{10^{-2pH}+K^i_{es1}10^{-pH}+K^i_{es1}K^i_{es2}} \quad i=1,2 \tag{5.5}$$

where K^i_{es1} and K^i_{es2} are the ionization constants of enzyme i, and $V^i_{M(opp)}$ and $V^i_{M(pH)}$ are the maximal rates at the optimum pH or at a given pH, respectively. This expression is valid only if   the relation between the dissociation constants of the enzyme $K^i_{es1}/K^i_{es2} > 10^2$ holds (Segel, 1975).

The optimum pH for E_i activity will occur at:

$$pH^i_{(opp)} = \frac{pK^i_{es1} + pK^i_{es2}}{2} \quad i = 1,2 \tag{5.6}$$

The pH optima for the activities of E_1 and E_2 should be different, with that of E_1 lower than that of E_2 (Figure 5.6) in order to obtain an asymmetric pattern (see below). Another condition required by the model is that the immobilized enzymes are homogeneously distributed and that only substrates and products can diffuse in the membrane.

The instantaneous pH is assumed to be described by the following expression:

$$pH = pH^* + \beta (P_2 - P_1) \tag{5.7}$$

with pH^*, the pH of the unstable steady state, given by:

$$\frac{pK^1_{es2} + pK^2_{es1}}{2} \tag{5.8}$$

Although equation 5.7 may look rather *ad hoc*, it may be obtained by expanding the function:

$$pH = pH^* + \beta \, \log\frac{(P_2 + 1)}{(P_1 + 1)} \tag{5.9}$$

as a Taylor series and neglecting the quadratic and higher order terms, an assumption which is valid for P_1, $P_2 \ll 1$. Moreover, the numerical values of pH may be verified experimentally for low values of P_2 and P_1, i.e. close to pH^*. From the experiments with the reaction products of glutaminase and urease the order of magnitude of β was estimated as 1 M^{-1}.

Under these conditions, the following kinetic equation allows a description of the pH dynamics:

$$\frac{\partial A}{\partial t} - \frac{\partial^2 A}{\partial x^2} - \sigma F(A) = 0$$

boundary conditions: $A(0,t) = A_0$ and $A(1,t) = A_1$

initial conditions: $A(x,0) = g(x)$

$$\tag{5.10}$$

where $A = P_2 - P_1$ and $F(A) = \alpha_2 - \alpha_1$; and:

$$\sigma_1 = \frac{V_M^1 e^2}{D_{H^+}} \quad ; \quad \sigma_2 = \frac{V_M^2 e^2}{D_{H^+}}$$

$$\sigma_1 = \sigma_2 \quad ; \quad \sigma = \beta\sigma_1 = \beta\sigma_2 \tag{5.11}$$

In the model we have chosen $\sigma_1 = \sigma_2$ in order to reduce the number of parameters to be investigated. In the equation describing σ_1 and σ_2, D_{H^+} was chosen instead of D_{P1} and D_{P2}, since the dissociation equilibria are assumed to be established faster than the diffusion of the products.

Equation 5.10 has been made non-dimensional by introduction of the factors e^2/D_{H^+} and e, the membrane thickness, as characteristic time and space units, respectively. This characteristic time is proportional to the time constant of the system, equal to $\theta = e^2/D_{H^+}$. Thus, the parameter σ contains kinetic and spatial information about the system.

APPENDIX 5B THE PHOTOBIOCHEMICAL SYSTEM

The equation describing the experimentally measured rate of DCIP reduction by the thylakoid photosynthetic activity as a function of DCIP concentration was obtained by combining measurements of light intensity and rates of electron acceptor consumption as previously described (Aon *et al.*, 1989b; Aon and Cortassa, 1993):

$$v(S,I) = V_M \frac{SI^3}{K_A I^3 + K_i^3 S + SI^3} \tag{5.12}$$

where the light intensity, I, may be considered as an additional substrate that interacts with the coloured electron acceptor, represented by S. K_A and K_i stand for the K_M of the electron acceptor and the light intensity, respectively.

The decay of the light intensity as a function of DCIP concentration is described by a double exponential law:

$$I = I_0 \left[\alpha_1 \frac{1 - e^{(-\varepsilon_1 eS)}}{\varepsilon_1 eS} + \alpha_2 \frac{1 - e^{(-\varepsilon_2 eS)}}{\varepsilon_2 eS} \right] \tag{5.13}$$

Equation 5.13 describes the mean light intensity as a function of the acceptor concentration and was obtained by integrating across the width of the test tube, a double exponential form of the Lambert–Beer law.

Equations 5.12 and 5.13 contain several parameters obtained from the fitting of the experimentally measured decay in light intensity (I) with DCIP (S) concentration (Figure 5.10b), namely α_1, ε_1, α_2 and ε_2. V_M, K_A and K_i were determined from the measured activities of DCIP photoreduction by thylakoid membranes as a function of DCIP concentration (Figure 5.10a). Other parameters such as I_0 and e depend on the experimental set-up. Equations 5.12 and 5.13 describe the experimental data shown in Figure 5.10c.

The following ODE is proposed to describe the dynamic evolution of the oxidized electron acceptor, S:

$$\frac{dS}{dt} = -v(S, I) + k_1(S_0 - S) \tag{5.14}$$

Equation 5.14 is a zero-dimensional spatial representation of the electron acceptor kinetics in the test tube. Figure 5.11a shows the bifurcation properties of equation 5.14 as a function of k_1. The kinetic constant, k_1, may represent the rate constant at which the electron acceptor is transported from, spatially, neighbouring regions, e.g. diffusion coefficient or other mass transport parameter, giving rise to different photosynthetic activities according to the position in the gradient (Figure 5.11a,b).

PART TWO
Living Cells under the Perspective of Dynamic Organization

Supramolecular structure and enzyme catalysis: the groundplan of living cells

6

Extracellular stimuli (e.g. light, hormones, pH, oxygen) trigger a coherent answer at the cellular level that leads the cell to different developmental paths, to stop division or to redirect metabolic fluxes. Chapter 2 argued that, at the cellular level, the passage from microscopic to macroscopic coherence could be given by instability of the dynamics of processes like polymerization–depolymerization of cytoskeleton components. One main piece of evidence that led us to suggest the autonomous dynamics of the cytoskeleton as an organizer of cellular activities was at the spatio-temporal 'window' of occurrence of polymerization–depolymerization processes that corresponded to the domain likely to give rise to macroscopic coherence at the cellular level, i.e. microtubules extend over several micrometres in a lapse of minutes (Kirschner and Mitchinson, 1986; Erickson and O'Brien, 1992). Instabilities in the autonomous dynamics of biological processes in a far from thermodynamic equilibrium domain give rise to self-organized behaviour. This behaviour could synchronize collective or systemic properties in cells such as metabolic fluxes (Cortassa *et al.*, 1994a; Cortassa and Aon, 1994b). Redirection of metabolic fluxes as a mechanism of carbon and energy uncoupling has been shown to be associated to cell division and differentiation (sporulation) of *Saccharomyces cerevisiae* (Aon *et al.*, 1995; Mónaco *et al.*, 1995; Aon and Cortassa, unpublished).

The view of the cell as a dynamic structural–functional unit, as is proposed in this book, was anticipated in the conception of Porter and Palade in the mid 1950s (Porter and Palade, 1957). These authors proposed that the reticulum acted as a channel to distribute metabolites to all parts of the cell. Additionally, because of the regions with special permeability, the membrane limiting the reticulum was considered similar to the plasma membrane from a structural and permeability point of view. This opens the possibility that the membrane potential maintained across the membrane by selective membrane permeability may transmit waves of depolarization of the sarcolemma and rapidly spread along the limiting membrane to all parts of the sarcomere (Porter and Palade, 1957).

This concept has since been further worked out, either experimentally or theoretically, by Skulachev and collaborators (Amchenkova *et al.*, 1988; Severina *et al.*, 1988; Skulachev, 1990). Closely related issues have been developed by Frolich (1975) whose ideas essentially deal with conformational changes in proteins entrained by dipole oscillations coupled to electromagnetic fields. Coherence will arise when excitation frequencies of oscillating electromagnetic fields harmonize with those of the proteins. Under those conditions, 'travelling polarization waves', with the consequent conformational changes in, for example, enzymes and foreseen reactivity, may be envisaged. In conjuction with Frolich's proposal, it has been suggested that the association–dissociation of enzymes from the actin network may be driven by electromagnetic fields (Clegg, 1983).

This chapter focuses on the coupled dynamics of enzyme-catalysed metabolic fluxes and supramolecular structures. The general approach will be that developed in Chapter 2. The structured view of the cytoplasm will be followed; and the conditions under which enzyme catalysis proceeds in heterogeneous media containing high concentrations of polymers will be analysed. Finally, possible paths are proposed, followed by the dynamics of intracellular processes so that macroscopic coherence might emerge, allocating new functional properties to the supramolecular catalytic networks of the cellular cytoplasm.

6.1 SUPRAMOLECULAR STRUCTURE AND ENZYME CATALYSIS

Several lines of evidence support a structured view of the cytoplasm and nucleus, giving rise to the question of the consequences for the dynamics of processes occurring in such a highly crowded but organized environment (Figure 6.1). At least one third of cellular proteins take part in the macromolecular networks and laminae of the cytoplasm and nucleus (Penman *et al.*, 1981). Most of the architectural components of the cell may be readily visualized when cultured cells are exposed to a sufficiently lipophilic non-ionic detergent, e.g. Triton X-100, in a suitable buffer (Penman *et al.*, 1981). Under these conditions, most lipids are solubilized, together with about two thirds of the cellular proteins (i.e. the 'soluble' fraction). When this separation is made, polyribosomes are retained on the cytostructure. The mRNA remains bound to the cytostructure, presumably via the proteins of the messenger ribonucleoprotein (mRNP) (Penman *et al.*, 1981).

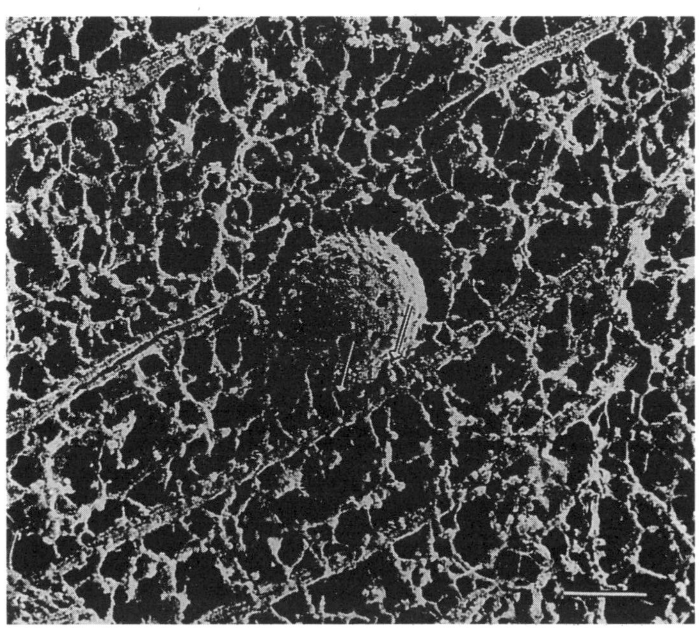

Figure 6.1 Molecular architecture of the cytoskeleton: high magnification view of membrane organelles and axoplasm in crayfish peripheral nerve axons permeabilized with saponin. Arrows indicate short cross-bridges between membrane organelles and microtubules. A fractal dimension, D, of 1.9072 (r = 0.999) was calculated for the cytoskeleton shown in this figure or a similar one (Figure 2 from Hirokawa, 1991: D = 1.9145, r = 0.999) (Aon, unpublished data). This D is somewhat higher than those obtained from photographs of immunofluorescent-stained microtubular lattices (Rabouille *et al.*, 1992). Apparently, the highly dense meshwork contributed by MAPs is responsible for the increase in the fractal dimension. D = 1.89 is characteristic of percolation clusters (Rabouille *et al.*, 1992; Aon and Cortassa, 1994). (From Hirokawa, 1991, with permission.)

Experimental evidence increasingly suggests that metabolism is strongly associated with cellular scaffolds (e.g. Clegg, 1984a,b, 1992; Masters, 1984; Ingber, 1993; Cortassa *et al.*, 1994a) and that these insoluble matrices and their dynamics in turn deeply influence the dynamics of chemical reactions (Ingber, 1993; Aon and Cortassa, 1994; Cortassa *et al.*, 1994; Pienta and Hoover, 1994) – for instance, the replication and transcription of DNA and the processing and translation of RNA (Penman *et al.*, 1981; Van Driel *et al.*, 1991; Suprenant, 1993); replication of viruses, endocytosis and secretion are intimately associated with profound reorganization of nuclear and cytoplasmic homo- and hetero-associations of macromolecular lattices (Penman *et al.*, 1981; Fulton, 1982; Hirokawa, 1991). Moreover, these nuclear functions appear to be highly localized within the nucleus rather than diffusely distributed (Van Driel *et al.*, 1991). The structural components of the nucleus are known to have a central role in the specific topological organization of the DNA (Getzenberg *et al.*, 1991). Direct evidence exists for specific three-dimensional organization of the DNA within the nucleus. Presumably, a specific spatial organization of the genome is required since regulatory sequences, e.g. enhancers, are located at a very large distance from a target gene (de Jong *et al.*, 1990). The nuclear matrix is the dynamic structural subcomponent of the nucleus that directs the three-dimensional organization of DNA into loop domains of approximately 60 kilobases (Getzenberg *et al.*, 1991).

Experimental results dealing with the pulse label of cells and chasing in the presence of an inhibitor of protein synthesis indicate that assembly of the cytostructure occurs primarily at the site of protein synthesis localization (Penman *et al.*, 1981). Several observations suggest that assembly of some cytoskeletal structures is difficult to explain solely by post-translational assembly. A model in which these proteins undergo co-translational assembly accommodates several experimental observations (Fulton and L'Ecuyer, 1993). Cytoskeletal proteins first associate with the cytoskeleton during translation, as nascent peptides. Moreover, the mRNAs for some cytoskeletal proteins are localized, suggesting that synthesis of these proteins may occur at sites appropriate for function or assembly. Many cytoskeletal mRNAs for actin, tubulin and intermediate filaments, among others, are also localized (Fulton, 1993) (Figure 6.1). The consequences of such results are far-reaching since the configuration of cell cytoarchitecture may influence gene expression. Cell cytoarchitecture, in turn under genomic control, must in some way sense the cell's position in a larger tissue pattern (section 10.3).

Arguments against the significance of intracellular organization for enzyme catalysis have grown around the concept that diffusion is rapid enough (time scale of milliseconds) and enzyme turnover slow (Srere,

1987; Hess and Mikhailov, 1994). The assumption that simple diffusion fills the cellular compartment space with components ready to react favours the supporters' view that the living cell's inside resembles more a stirred tank reactor than an unstirred one (Hess and Mikhailov, 1994; for a critical view of diffusion theory see Wheatley and Malone, 1993). These arguments are in favour of a homogeneous description of intracellular chemical reaction dynamics. It has been argued, notably by Srere (1987), that if Brownian movement of metabolites inside cells is strong, some energy input would be necessary to maintain a higher concentration of a product-substrate in the vicinity of an enzyme for it to display an appreciable reaction rate. Another argument that should be taken into account is that the hundreds of metabolites concentrated to the level of the K_Ms of the enzymes that transform them would require so much water that the cell would resemble a crystal with completely ordered water-solvating intermediary metabolites (Atkinson, 1969, quoted by Welch, 1977; Mendes *et al.*, 1992; but see Cornish-Bowden, 1991). Moreover, the 'homogeneous' description disregards the complex and dynamic scaffolds of the cell along with the fact that the groundplan of the cytoarchitecture may imply something about the dynamics of chemical reactions (Figure 6.1) (Clegg, 1984a; Luby-Phelps *et al.*, 1988). In section 6.4.1 we will try to highlight the links between geometry and dynamics.

We hypothesize that polymerization–depolymerization of cytoskeleton components may be involved in the emergence of coherence between different cellular activities. The question now is: can a change in the polymeric status of cytoskeleton components be reflected by a coherent change in catalysis? An answer to this question is provided by the known fact that cellular conditions favour both self-association of enzymes and the binding of enzymes to structural actin and tubulin polymers; the dynamic equilibrium displacement toward either bound or unbound forms of enzymes or different oligomeric states of enzymes with higher or lower activity will be essentially some of the cellular mechanisms driving the coherence in catalysis (Cortassa and Aon, 1994b; Cortassa *et al.*, 1994a). Moreover, the characterization of the cytoplasm as a fractal of the sort of percolation clusters would add new feasible behavioural possibilities regarding the dynamics of enzymatic catalysis in heterogeneous fractal media.

6.1.1 CYTOPLASMIC STRUCTURE, WATER CONTENT AND ENZYME CATALYSIS

The rate of metabolism in *Artemia* cysts does not depend on cell water content over a wide range. Results described by Clegg (1984a) showed that, well below full hydration, metabolic rates and cell water content are strictly dependent. Strict dependence on cell water content was in the range of 0.65 to 0.8 g H_2O g^{-1} dry weight (Clegg, 1984a). At more than 0.8 and up to 1.4 g g^{-1}, metabolic rate was nearly independent of cell water content, while below 0.65 g g^{-1} only metabolism of small molecules was evident. Although metabolic independence from water content was apparent for a wide hydration range, these observations strongly suggest that catalysis and metabolism occur in the interphase between polymers and 'structured' water (Clegg, 1984). The latter is a debated subject (section 6.2.2).

It has been demonstrated that the surface (phospholipids bilayers and mica sheets) clearly influences water layers over distances of as much as 50 angstroms (ca. 17 molecular layers of water). The calculations performed by Clegg showed that taking into account the effective distance over which water 'feels' the surface (i.e. 50 angstroms), 20–40% of the total water of the aqueous cytoplasm would be so influenced (Clegg, 1984).

Mammalian cells can be reversibly dehydrated with a consequent loss in volume of 65% without significant change in metabolism with respect to controls. A total water content of 0.6 g g^{-1} is still present in these cells which have lost 65% of their volume (Clegg, 1984). Taken together, these data show that:

- a great deal of metabolism can proceed in a cell containing around 40% of its total water content;
- this remnant water probably exists as an interphase between cytoplasmic supramolecular structures;
- both observations suggest that most of the catalytic activity may be occurring in that interphase.

6.1.2 ENZYME CATALYSIS IN WATER-STRUCTURED MICROENVIRONMENTS AND MECHANO-CATALYTIC COUPLING

Catalysis is stabilization of the transition state of a reaction. Destabilization of the ground state enzyme/substrate (ES) complex relative to the transition state is also necessary for enzymatic catalysis which prevents accumulation of the enzyme in the enzyme/substrate complex (Figure 6.2) (Herschlag, 1988).

Enzymatic catalysis will be increased by increasing the energy of the ES complex as long as the transition state energy is not also increased; this condition corresponds to an increase in K_M while the value of the ratio k_{cat}/K_M remains constant (Herschlag, 1988). The increased efficiency from increasing K_M can be explained as an advantage in minimizing the

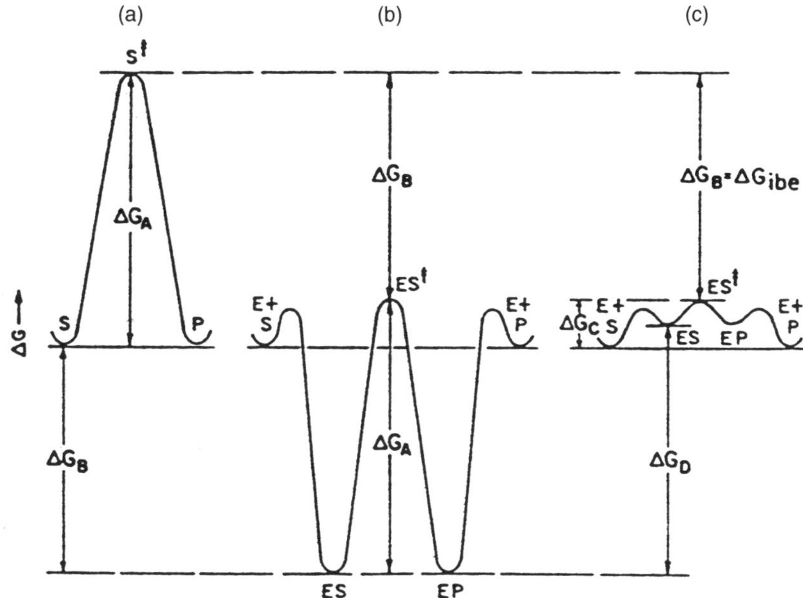

Figure 6.2 Comparison of the free-energy profiles for: (a) the non-enzymatic reaction of substrate S; (b) the enzymatic reaction of S in which the ground state and transition complexes are stabilized to the same extent; and (c) the enzymatic reaction of S in which the transition state complex is stabilized to a greater extent than the ground state complex. Catalysis occurs in the enzymatic reaction (c). All three reaction profiles are for the same standard state, which is a concentration of S that is less than K_M for the enzyme of (c). (Reproduced from Herschlag, 1988, by permission of Academic Press.)

amount of enzyme tied up as ES, thus maximizing the amount of enzyme available for catalysis of the reaction: $E + S \leftrightarrow E + P$. The increase in the observed rate from increasing K_M at constant k_{cat}/K_M is large when the substrate concentration is near or above K_M, because enzyme that is complexed with substrate when K_M is low becomes available for catalysis. When the substrate concentration is well below K_M a further increase in K_M at constant k_{cat}/K_M will yield no advantage because essentially all of the enzyme is free and already catalysing the reaction with the apparent second-order rate constant k_{cat}/K_M (Herschlag, 1988).

Rate enhancement in enzyme catalysis is accompanied by large changes in the volume of activation, probably from changes in the volume of hydration of residues at the surface of enzymes (Clegg, 1984b, and references therein). The hydration of protein groups that occurs when groups are transferred from a non-polar phase (the 'interior' of the protein) to an aqueous environment are accompanied by large changes in free energy, enthalpy, entropy and volume, reflecting water reorganization around both polar and non-polar groups (Greaney and Somero, 1979). In model transfer studies this occurs with a decrease in system volume due to water organization around the exposed groups. The energetics of these transfer processes are markedly influenced by salt composition, concentration of the aqueous phase and likely viscosity in the microenvironment of the enzyme, all of them intervening in rate enhancement.

For the transfer of peptide groups, changes in enthalpy and entropy, as well as the free energy, are all negative when transfer occurs from water to concentrated salt solutions, the favourable enthalpy change being the driving force. Conversely, when the transfer of non-polar groups is between water and salt solutions, the free energy, enthalpy and entropy of transfer are all positive. These transfers appear to occur with a high degree of enthalpy–entropy compensation, such that large changes in transfer enthalpy are associated with relatively small changes in transfer free energy (Greaney and Somero, 1979). One of the basic tenets of this model is that the transfer of protein groups at the water–protein interface during catalytic conformational changes contributes significantly to the energy (ΔG) and volume (ΔV) changes of catalysis (Greaney and Somero, 1979).

It has been suggested that slow enzymatic reactions are necessary for control processes. Herschlag (1988) showed that conformational changes can slow down enzymatic reactions by decreasing k_{cat}/K_M and k_{cat}. Another possible role of a conformational change in metabolic control is to increase the value of K_M in order to change the range of substrate concentration that gives the greatest change in rate (Herschlag, 1988). Enzymatic perfection is said to occur when an enzyme has a turnover number equal to the diffusion velocity of the substrate (Albery and Knowles, 1976). However, not all enzymes were selected for speed; for example, many futile cycles of phosphorylation–dephosphorylation are apparently set up to maintain a steady state of phosphorylation: the higher the turnover rate of those processes, the greater is the waste of ATP energy (Koshland, 1987).

In living cells, enzymes confront the problem of discriminating between catalysis of the reaction of two potential substrates in the presence of both compounds. This is the heart of the problem of biological specificity. It can be shown that the discrimination at any given concentrations of competing substrates is determined by the ratio of the values of k_{cat}/K_M for the two substrates, S_1 and S_2. This ratio is determined by the difference in free energy of the respective transition states, ES_1 and ES_2.

A protein may act as a free energy linkage device for coupling of chemical processes to the properties of the liquid medium (Lumry and Biltonen, 1969; Welch, 1977). **Intrinsic binding energy** (IBE) is the energy from binding that is used to stabilize the transition state. This energy is not fully expressed in the ES complex, but rather in the transition state complex. A catalytic advantage of a conformational change will only occur if the IBE realized from that conformational change is greater than the energy needed to drive the conformational change. In contrast, other types of conformational changes decrease catalysis by providing the energy to drive a conformational change instead of stabilizing the transition state, and these conformational changes do not enhance the specificity (Herschlag, 1988) (Figure 6.2). Hexokinase is the classic example of an enzyme that undergoes a conformational change. Crystal structures of similar isozymes of hexokinase revealed 'closed' and 'open' forms of the enzyme with and without bound glucose, respectively. The bound glucose molecule is virtually buried beyond reach of solvent in the closed structure and it is unlikely that it could dissociate without a conformational change of the enzyme. The increase in ATP hydrolysis catalysed by hexokinase with bound lyxose or xylose is consistent with a sugar-induced conformational change that activates the enzyme. The conformational changes of the kinase that surround the substrate on all sides may provide greater IBE than would otherwise be available in order to enhance catalysis and specificity. It has been suggested that the exclusion of water from the surroundings of the substrates can lower the dielectric constant of the active site in kinase, dehydrogenase and decarboxylase reactions (Herschlag, 1988).

6.1.3 MACROMOLECULAR CROWDING EFFECTS

Enzyme microenvironments in cells will contain a large variety of macromolecules, largely proteins which, taken together, will occupy a substantial fraction of the total cytoplasmic volume; this is called volume occupied or crowded (Minton and Wilf, 1981). Following changes in cell volume during osmotic regulation, molecular crowding and confinement directly affect the form and function of intracellular macromolecules (Minton, 1983; reviewed in Garner and Burg, 1994). From data obtained in red blood cells, it has been suggested that cells may use macromolecular crowding as a means of sensing their volume; what appears to be cell volume regulation is really a regulation of cytoplasmic macromolecular crowding (Parker, 1993; Haussinger *et al.*, 1994).

Apparently, protein concentration is what the cells seek to maintain, in order to approach an optimal degree of cytoplasmic non-ideality (Parker, 1993). The response of cells to hormones such as insulin and glucagon entrains a coherent metabolic response: swollen cells make protein and glycogen while shrunken cells catabolize these substances (Parker, 1993; Haussinger *et al.*, 1994). Concerning the latter results, a different explanation has been attempted by Haussinger *et al.* (1994) (section 6.2.1).

Crowding may have substantial effects upon the structure and catalytic activity of an enzyme which is itself present at very low concentrations. One possible mechanism proposes that crowding, through alteration of the average degree of self-association of the enzyme, will also change its average catalytic activity. This mechanism was shown to work for glyceraldehyde-3-phosphate dehydrogenase (GAPD) in the presence of ribonuclease A, β-lactoglobulin, bovine serum albumin and poly(ethylene glycol) (PEG, M_r 20 000) at additive concentrations of up to 30 g dl^{-1} (Minton and Wilf, 1981). The presence of high amounts of proteins promoted the formation of GAPD tetramers that were more active than the monomer, due to space-filling properties – also called 'molecular impenetrability' by Minton and Wilf (1981) – of the added species. The addition of high concentrations of unrelated globular proteins did not affect the activity of either monomer or tetramer but did affect the specific activity of the enzyme upon its state of subunit self-association, regulated in turn by the extent of macromolecular crowding (Minton and Wilf, 1981).

Aggregation and binding reactions that lower the volume occupied by the reactants are favoured in confined environments (Minton, 1992; Garner and Burg, 1994). **Confinement** refers to molecules that are not in solution but are organized in a network like a cytoskeleton or microtrabecular lattice. Qualitatively, confinement behaves like crowding: as the concentration of the matrix protein increases, the apparent activity of the test molecule becomes higher. Thus, aggregation and binding reactions are favoured in 'confined' environments. Unlike crowding, confinement favours the formation of long linear aggregates (such as actin filaments) rather than the compact globular species induced by crowding. *In vitro* simulation of molecular crowding with poly(ethylene glycol) (PEG) or a glycoprotein, ovalbumin, induced parallel bundling of actin filaments. An array of actin in bundles is likely to occur in the cytoplasm since the abundance of proteins should be able to have a similar effect to that observed *in vitro* (Suzuki *et al.*, 1989).

The kinetic properties of PFK seem to be regulated by the interaction of the enzyme with filamentous actin (Roberts and Somero, 1987). Apparently, actin stabilizes the active, tetrameric form of PFK provoking the observed kinetic activation. PFK activity would then reflect the equilibrium determined both by the self-association of PFK subunits and the association of PFK with actin. Studies of the effects of PEG on PFK subunit assembly have shown that stabilizing the PFK tetramer causes a decrease in the $K_{0.5}$ for fructose-6-phosphate (Roberts and Somero, 1987). Similar studies were performed with PFK and purified tubulin (Lehotzky *et al.*, 1993). Apparently, tubulin could bind to the monomeric and dimeric forms (but not the tetrameric form) of PFK, suggesting a perturbing effect of tubulin on the equilibrium of PFK oligomers (Lehotzky *et al.*, 1993). The authors interpreted their results assuming that the active tetramer of PFK dissociates into inactive species and in that way the overall specific activity of the enzyme decreases. When PFK concentrations were lowered from 15 μM to 3 μM (monomer) in the presence of 10 μM tubulin or microtubules, its relative activity decreased to 50% at the lowest concentration of the enzyme independently of the polymeric status of tubulin (Lehotzky *et al.*, 1993).

A very different picture was obtained with the enzymatic couple PK/LDH in the presence of microtubular protein (MTP: tubulin + MAPs) from rat brain (Cortassa *et al.*, 1994a). It was shown that the activation effect was largely due to the presence of MAPs (Figure 7.8b). With PK as the limiting enzyme of the flux, we could reproduce our experimental data by mathematical modelling. The models considered that MAPs in the polymerized or non-polymerized MTP lattice could displace the equilibrium of PK's oligomers to the pentameric (more active) forms of the enzyme; thus, the overall activity of the enzymatic couple was increased (Cortassa *et al.*, 1994a). When a third enzyme, PFK, was added to the couple PK/LDH with PFK as the limiting enzyme, the activation effects disappeared in the presence of non-polymerized MTP (Figure 7.8c). However, in the presence of polymerized MTP, PFK was activated. A possible explanation of these data may be that the presence of the polymeric lattice of microtubules removes the free tubulin from the medium and favours the oligomerization of PFK either because of molecular crowding or through specific effectors, MAPs or polymerized tubulin.

6.1.4 ENZYME INTERACTIONS WITH SUBCELLULAR STRUCTURES AND ORGANELLES

Intimate association of enzymes and ultrastructure as well as their dynamics have been considered of utmost importance in metabolic regulation (Welch, 1977; Clegg, 1984b; Masters, 1984; Srere, 1987; Srivastava and Bernhard, 1987). Several workers before were interested in the elucidation of enzyme binding properties with actin, its effect on enzyme kinetic properties and the interaction between enzymes in supramolecular aggregates and its influence upon transition times in consecutive catalytic steps (Clegg, 1984; Masters, 1984; Clarke *et al.*, 1985; Luther and Lee, 1986). Several studies have been reported on the interaction of glycolytic enzymes with tubulin or microtubules (Knull and Walsh, 1992; Ovádi and Orosz, 1992).

The concept of metabolic compartmentation or channelling contributed to answering a series of still-debated queries (see Welch, 1977, for a review; Cornish-Bowden, 1991; Mendes *et al.*, 1992). The physiological importance of the metabolic channelling concept relies on the following features (Welch, 1977; Ovadi *et al.*, 1989; Cheung *et al.*, 1989; Sumegyi *et al.*, 1991):

- Enzymatic active sites inside multi-enzyme complexes would be in close proximity, which implies segregation of competing metabolic pathways.
- High local concentration of metabolites at the active sites of their respective enzymes.
- Reduction of transit times.

These features would allow the maintenance of low concentrations of metabolites inside the cell which, in principle, avoids overwhelming of the solvent capacity of the intracellular milieu (Atkinson, 1969, quoted by Welch, 1977; Mendes *et al.*, 1992, but see Cornish-Bowden, 1991).

Figure 6.3 shows an actin-filament based enzyme cluster and supposed metabolic channelling exerted by glycolytic enzymes involving aldolase (ALD), glyceraldehyde 3-phosphate dehydrogenase (GAPDH), triosephosphate isomerase (TPI) and phosphoglycerokinase (PGK) (Clarke *et al.*, 1985). These functional clusters may also complex by piggy-backing (Bronstein and Knull, 1981; Clegg, 1984a; Clarke *et al.*, 1985). In fact, it has been noted that when purified enzymes such as TPI, phosphoglycerate mutase (PGM) and glucose-6-phosphate isomerase (PGI) were assayed individually, they did not bind to F-actin-tropomyosin. However, in a complex mixture containing all other glycolytic enzymes, significant binding of TPI, PGI and PGM was observed. This suggests that these enzymes were interacting indirectly by binding or 'piggy-backing' to other enzyme molecules already directly bound to the filament (Clarke *et al.*, 1985).

Dynamic association–dissociation of 'ambiquitous' hexokinase to mitochondria is able to regulate transients induced by, for example, glucose pulses in tumour cells, although not the steady state control of the glycolytic flux (Cortassa and Aon, 1994b). One important result of

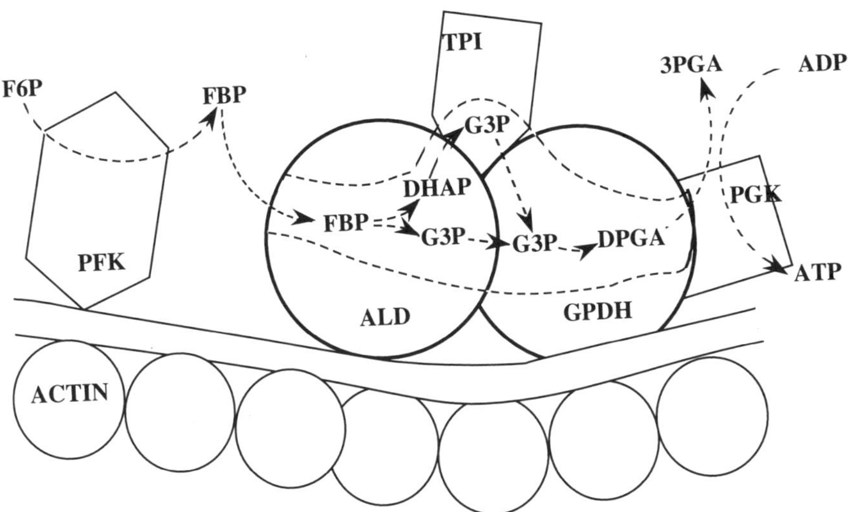

Figure 6.3 Proposed organization of glycolytic enzymes via the cytoskeleton. Glycolytic enzymes involving aldolase, glyceraldehyde 3-phosphate dehydrogenase, triosephosphate isomerase (TPI) and phosphoglycerate kinase may organize as functional clusters and also complex by piggy-backing, i.e. when enzymes are assayed in a complex mixture, significant binding of TPI, glucose phosphate isomerase and phosphoglycerate mutase is observed. This suggests that these enzymes were interacting indirectly by binding or piggy-backing to other enzyme molecules already directly bound to the filament. This organization has a potential for metabolic channelling and direct delivery of ATP to, say, the actin–myosin interface. The binding of a myosin head to an actin-containing filament is known to cause considerable changes in the filament structure; this mechanical signal might well induce multienzyme clusters. (From Clarke *et al.*, 1985, Glycolytic enzyme organization via the cytoskeleton and its role in metabolic regulations, 24, in *Regulation of Carbohydrate Metabolism*, Vol. 2, ed. R. Beitner, with permission from CRC Press, Boca Raton, Florida.)

reciprocal regulation between glycolysis and mitochondrial functions through hexokinase ambiquity is that it appears as a biochemical device more able to regulate transients than to control steady state behaviour. (See Chapter 7 for a more extensive discussion and illustration of this point.)

Changes in the kinetic properties of glycolytic enzymes have been reported upon binding to actin filaments (Masters, 1984; Luther and Lee, 1986; Keleti *et al.*, 1989) or cell membranes (Uyeda, 1992). Karadesh and Uyeda (1977) have shown changes in allosteric properties of PFK bound to erythrocyte membranes (Figure 6.4). Luther and Lee (1986), studying the role of phosphorylation in the interaction of rabbit muscle phosphofructokinase with F-actin, showed that the phosphorylated forms of PFK, either *in vivo* or *in vitro*, exhibit a distinct sigmoidal relationship of activity versus Fru-6-P concentration with respect to the dephosphorylated form, the relationship for the latter form being shifted to the left (Figure 6.5). The apparent Michaelis constant, $K_{0.5}$, for the dephosphorylated form was $74 \pm 7 \mu M$ (n = 1.6 ± 0.1) which was significantly lower than that of $120 \pm 20 \mu M$ (n = 2.3 ± 0.1) determined for the phosphorylated forms. Hence, dephosphorylated PFK has a higher apparent affinity for Fru-6-P and the interaction between enzyme and substrate is apparently less cooperative. It has been shown that PFK is able to interact with microtubules and tubulin: the PFK activity was inhibited by increasing amounts of tubulin and microtubules (Lehotzky *et al.*, 1993). Recent work has shown that the microcompartmentation of aldolase and GAPDH in four different cell types is associated with the actin cytoskeleton and that both enzymes exist in a soluble as well as in a structure-bound form (Minaschek *et al.*, 1992).

On the basis of the experimental evidence presented it may be conceived that the assembly–disassembly of cytoskeleton components could act as a macroscopic coherent inducer of changes in systemic properties such as metabolic fluxes. The supraregulatory effects of microtubular protein assembly–disassembly on the fluxes through coupled enzymatic reactions of carbon metabolism both *in vitro* and in a permeabilized yeast cell system have now been established (Cortassa *et al.*, 1994a; see also Chapter 7). A concentration-dependent stimulatory effect in the presence of polymerized or non-polymerized microtubular protein on metabolic fluxes sustained by enzymes of carbon metabolism has been shown (Chapter 7). The explored microtubular protein concentration range (1 to 15 μM) corresponded to the physiological range. In addition, there was evidence that PK, which was able to interact *in vitro* with the microtubular lattice, was sensitive to nocodazole in permeabilized yeast cells (Figure 7.12). Furthermore, a non-interacting system such as HK/G6PDH was insensitive to nocodazole. It was further shown that, intracellularly, the effect of nocodazole correlated with its expected effect on the microtubular lattice (Jacobs *et al.*, 1988; Solomon, 1991; Cortassa *et al.*, 1994a). Chapter 7 gives a more extensive discussion of this point.

6.2 THE GROUNDPLAN OF LIVING CELLS

There is compelling rheological evidence which shows that cytoplasmic polymers behave as a reversible non-covalent gel network which is constantly being formed and disassembled (Luby-Phelps *et al.*, 1986, 1988; Forgacs and Newman, 1994). As a protein crystal, the intracellular milieu favours self-associations and hetero-associations of enzymes or proteins,

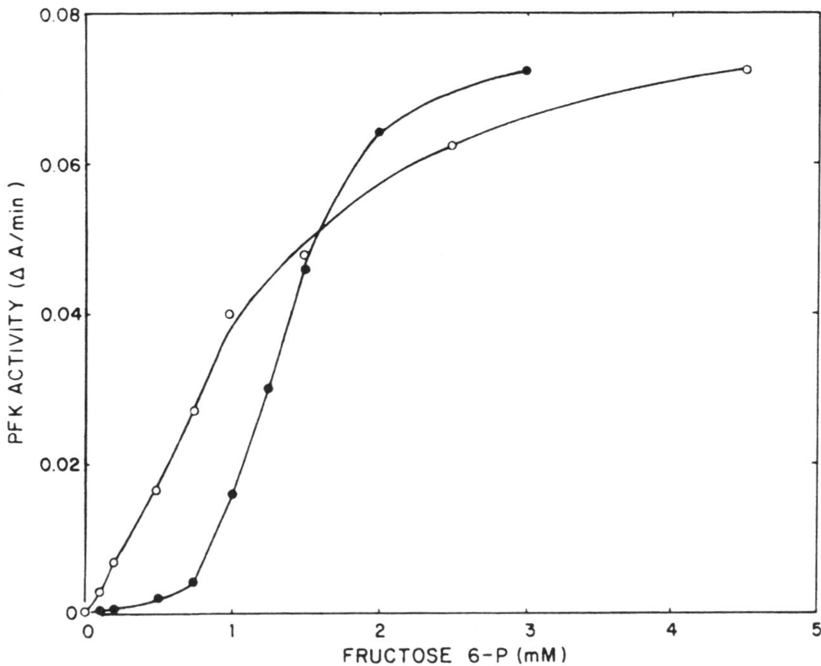

Figure 6.4 Changes in allosteric properties of PFK bound to erythrocyte membranes. Erythrocyte ghosts containing bound PFK were prepared by incubating approximately 100 U (1.2 mg) of PFK with the saline washed ghosts (4.7 mg of protein) in 5 mM potassium phosphate, pH 7.5, for 1 h at 0°C. Membranes were centrifuged, washed three times with 5 ml each of 5 mM potassium phosphate, pH 7.5, and suspended in 0.6 ml of the same buffer. Under these conditions, the amount of PFK bound was usually 10 U. Open circles, membrane-bound human erythrocyte PFK; filled circles, free PFK. (From Karadsheh and Uyeda, 1977, by permission of The American Society for Biochemistry and Molecular Biology.)

e.g. between microtubules and intermediate filaments, or even with subcellular organelles (Fulton, 1982; Hirokawa, 1991) (Figure 6.1). This complex cytoarchitecture has profound consequences for cell function. For instance, in *Escherichia coli*, protein comprises 55% of the dry weight and about 36% of the dry weight of the cell is dedicated to protein synthesis (Neidhardt, 1987; Goodsel, 1991).

The aqueous domain of the cytoplasm that occupies the space between the cytoskeletal network contains macromolecules and small organic and inorganic solutes (Fushimi and Verkman, 1991). The physical state, or more specifically the rheological characteristics of the

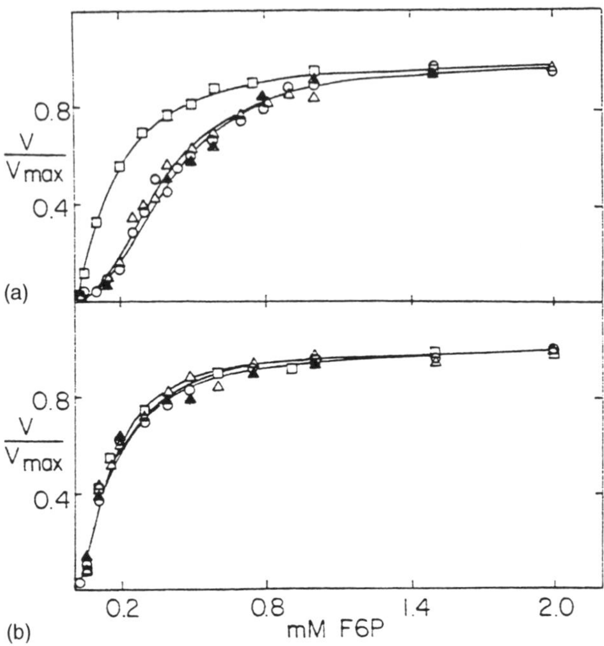

Figure 6.5 Role of phosphorylation in the interaction of rabbit muscle phosphofructokinase with F-actin. Initial velocity as a function of F6P concentration in (a) absence and (b) presence of F-actin. PFK activity was monitored by a coupled enzyme assay. The coupling of the muscle enzymes aldolase, glycerol-3-phosphate dehydrogenase and triosephosphate isomerase was controlled for their interaction with F-actin. In (b), reaction mixtures contained 0.1 ml of F-actin (2 mg ml⁻¹) and 0.1 ml of a PFK solution which had been diluted to give the required PFK concentration. After mixing, the samples were incubated at 23°C for 10 min. No further binding was observed after 10 min of incubation. After reaching equilibrium, 50 μl of the mixture was removed, diluted and assayed for PFK activity. The reaction mixtures were then centrifuged for 90 min at 110 000 g. After centrifugation, 50 μl of the supernatant was carefully removed, diluted and assayed for PFK activity as described. This activity represents the amount of free enzyme that was not forming a complex with actin. Knowing the respective amounts of total and free PFK, it is possible to calculate the amount of PFK bound to actin. Open circles, *in vivo* phosphorylated PFK; open triangles, *in vitro* phosphorylated PFK; filled triangles, *in vitro* phosphorylated PFK which was previously dephosphorylated; open squares, dephosphorylated PFK. (From Luther and Lee, 1986, by permission of The American Society for Biochemistry and Molecular Biology.)

aqueous domain of the cytoplasm, influences a number of intracellular dynamic processes including solute transport, diffusion-limited enzyme kinetics and cell motility. It is widely accepted by biologists that cytoplasm cannot be considered as a simple newtonian fluid (Clegg, 1984a,b, 1992; Luby-Phelps *et al.*, 1986). A more realistic picture of the rheological nature of the cytoplasm approaches that of a reversible, non-covalent gel network composed, in turn, of complex supramolecular networks, e.g. of actin filaments, microtubules, intermediate filaments and associated proteins (Luby-Phelps *et al.*, 1986, 1988; Rabouille *et al.*, 1990) (Figure 6.6).

Cytoplasm cannot be considered as a simple newtonian fluid (as could be suggested by its aqueous nature) because of the following properties (Luby-Phelps *et al.*, 1988):

- The concentration of protein in the cytoplasm of the average mammalian cells is 15–20% (g/100 ml).
- Cytoplasm has a jelly-like consistency.
- The measured dynamic viscosity of cytoplasm is inversely dependent on shear rate (as for a non-newtonian fluid).
- Brownian motion of intracellular particles suspended within cytoplasm is highly restricted.
- Cytoplasm contracts in the presence of higher than 10^{-6} M Ca^{2+} and mM concentrations of ATP.

A fractal view of the cytoplasm of living cells follows the 'structured' view but introduces new feasible behavioural possibilities with which we will be dealing later.

6.2.1 SPATIAL ORGANIZATION OF THE CYTOPLASM OF LIVING CELLS: THE LINKS BETWEEN RHEOLOGY AND GEOMETRY

The spatial organization of supramolecular structures composed by large macromolecules such as ovomucin depends on their biophysical (e.g. rheological) and morphological (e.g. geometrical) properties. There are intimate relationships between rheology and geometry in the spatial organization of a medium. Certainly, rheological data supply information about the viscoelastic properties of a fluid, allowing inferences to be made about the geometrical properties of molecules composing the fluid (Figure 6.7). When rheological changes are brought about by modification of the molecular structure by, say, enzymatic treatments, then biochemical, rheological and geometrical data can be combined to interpret the

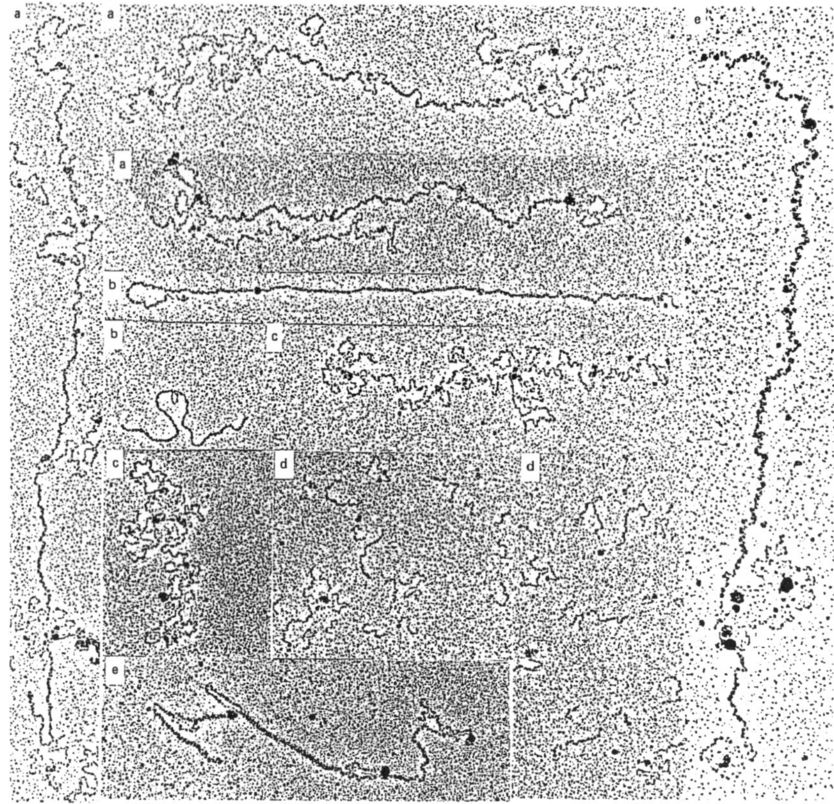

Figure 6.6 Supramolecular organization of ovomucin. Purified ovomucin was subjected to the macromolecular replica method: (a) ovomucin molecule with double-and single-stranded regions; (b) double-stranded ovomucin molecule; (c) single-stranded molecule; (d) sonicated ovomucin; (e) mucin purified from ovomucin gel without sonication. (Reproduced from Rabouille *et al.*, 1990, by permission of The Biochemical Society and Portland Press.)

biophysical properties of the organization of macromolecules (Rabouille, 1988; Rabouille *et al.*, 1989, 1990).

After treatment of the ovomucin gel with agents such as urea, GuHCl or SDS, which dissociate hydrophobic interactions or treatments that break electrostatic interactions, two kinds of rheological effects were noted: loss of the non-newtonian behaviour and a decrease in viscosity (Figure 6.7). Furthermore, a drastic change to a newtonian behaviour was shown by the ovomucin gel when sonicated (Rabouille *et al.*, 1989). Further analysis by electron microscopy showed that the ovomucin

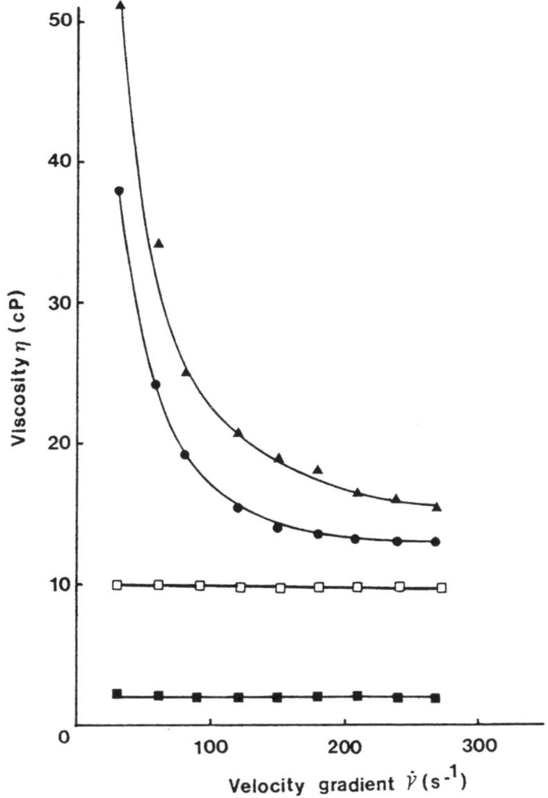

Figure 6.7 Rheological properties of an ovomucin gel: dependence of ovomucin gel viscosity (η) on velocity gradient (γ) after the ovomucin gel was subjected to different treatments. The ovomucin gel is non-newtonian with a viscosity between 52 and 17 cP when the velocity gradient varied from 0 to 270 s⁻¹. A dramatic change in rheological properties was observed after sonication or chymotrypsin treatment, which corresponded to the loss of non-newtonian properties (viscosity of 2.6 cP, independent of the velocity gradient). Filled triangle, untreated ovomucin gel; filled circle, neuraminidase-treated ovomucin gel; filled square, chymotrypsin-treated or sonicated ovomucin gel; open square, ovomucin gel treated with SDS, GuHCl or urea. (Reproduced from Rabouille *et al.*, 1989, by permission of Academic Press.)

molecules had greatly decreased their length and molecular weight during sonication (Rabouille *et al.*, 1990). The loss of the non-newtonian behaviour in favour of a newtonian type corresponds to a decrease in molecular asymmetry and deformation susceptibility and an increase in compactness (Figures 6.6 and 6.7).

An important suggestion emerging from these studies is that a spatial supramolecular organization through specific interactions between groups is needed for the ovomucin molecules to form a gel and not solely aggregation (Rabouille *et al.*, 1990). It has been postulated that the cytoskeleton could be tensionally integrated so as to provide a mechanism to distribute mechanical stresses to regulatory molecules and metabolic enzymes (Ingber, 1993). This concept may be relevant for, say, mechanically induced changes in cytostructure, either cytoplasmic or nuclear (Pienta *et al.*, 1989; Ingber, 1993). In the particular case of membrane proteins with biological activity, e.g. ionic channels, a sensitivity to mechanical tension has been demonstrated (Edwards and Pickard, 1987; Falke *et al.*, 1988; Morris and Sigurdson, 1989). The mechanical tension modifies the viscoelastic properties of plasmalemma that may be transduced into changes in lipid asymmetrical distribution and curvature radii, which is sensed in turn by membrane proteins (Maggio, 1985, 1994). Another example is provided by the effect of osmolarity on capping of plasmalemma ionic channels that appears to be mediated by metabolic energy, namely ATP (Zimmermann, 1978).

Reorganization of macromolecular assemblies by providing different spatial organization or orientational cues could in turn induce differential gene expression (Cook, 1989) or activation of metabolic fluxes (Cortassa *et al.*, 1994a; Aon and Cortassa, 1994; Haussinger *et al.*, 1994b). Hypotonic cell swelling or isotonic swelling in the presence of glutamine or insulin provoked a microtubule-stabilizing effect even in the presence of the microtubule inhibitor colchicine (Haussinger *et al.*, 1994a). It has been argued that, under similar conditions of cell shrinkage and swelling induced by hormones, cells seek to maintain the cytoplasmic macromolecular crowding through protein concentration (Parker, 1993). The kinetics and equilibria of biochemical reactions are strongly influenced by high concentrations of macromolecules that do have an apparent structural role (Parker, 1993; Cortassa *et al.*, 1994a). It has been conjectured that osmotically-induced shrinkage or swelling is coordinated by something other than cell volume itself (Parker, 1993). The microtubule-stabilizing effect exerted by cell swelling described by Haussinger *et al.* (1994a) was accompanied by increased levels of tubulin mRNA, which suggested a higher level of synthesis of tubulin monomer. These results support the view that cytoskeleton reorganization exerts global effects on the biochemistry of living cells. Changes in the organization of supramolecular structures (e.g. cytoskeleton components) may influence biochemical activities (e.g. enzymes) through different mechanisms. Those mechanisms may include conformational changes in proteins, with alterations in the degree of association between oligomeric proteins or binding properties to macromolecules. Chapter 7 includes an extensive discussion of this subject.

6.2.2 THE CONCEPT OF LEVEL OF ORGANIZATION AND THE STATUS OF INTRACELLULAR WATER

Chapter 2 presented evidence for the spatio-temporal scaling shown by distinct levels of organization in physiological processes such as cellular metabolism, energetics and electrophysiology as well as cell growth (Table 2.1 and Figure 2.2). Now that concept will be applied in discussing whether intracellular water has the properties of liquid water or rather behaves as frozen water. The rationale of levels of organization and their detection by probes will serve as a tool for arguing about the status of intracellular water. This is a controversial subject with views ranging between extremes: that almost all cell water is the same as pure water, or that none of the water in cells exhibits the same properties as pure water (Clegg, 1984a, and references therein). The rheological characteristics of the aqueous domains of cell cytoplasm have been studied by several biophysical methods. A large range of variation of cytoplasmic viscosity has been reported. Those values varied from 2 cP to >100 cP (Fushimi and Verkman, 1991) and have been obtained by several methods (Table 2.2).

One of the views supports the idea that a significant fraction of cell water is bound or organized by macromolecules. Rheological properties of water organized around macromolecules, such as density, viscosity or solvent properties, differ largely from those of normal water (Clegg, 1984a). The presence of intracellular organized water casts serious doubts on the utilization of thermodynamic information obtained *in vitro*. This outlook has been challenged by measurements obtained with picosecond polarization microfluorimetry methods of the probe BCECF. According to such data obtained by Fushimi and Verkman (1991), the cytoplasm is only 1.2–1.4 times as viscous as water, the activation energy for fluid-phase cytoplasmic viscosity being similar to that of water ($\sim$ 4 kcal mol^{-1}). Excluding significant compartmentation of the fluorophore, diffusional restrictions as well as significant binding to cytoplasmic structures, Fushimi and Verkman (1991) conclude that the fluid phases of cytoplasm and nucleoplasm are only 1.2–1.4 times as viscous as free water and are not influenced by major alterations in the cytoskeleton. They further suggested that the majority of cell water is not organized and has physical properties similar to those of free water. However, these results could well be explained by exclusion of the fluorophore probe from organized water. Observations by Cortassa and Aon (unpublished) with BCECF in the presence of microtubular protein (tubulin + MAPs), either polymerized or non-polymerized, strongly suggested that there is a significant 'squeezing' of the probe from the neighbourhood of the macromolecular lattice into the aqueous solution.

Cytoplasmic viscosities of 10–13 cP or 8–10 cP at 23°C or 37°C, respectively, have been measured in Swiss 3T3 fibroblasts (Dix and Verkman, 1990). These values were similar to those reported with fluorescence polarization methods (6–13 cP at 37°C) (Burns, 1969; Lindmo and Steen, 1977: quoted by Dix and Verkman, 1990), fluorescence recovery after photobleaching (4 cP at 37°C) (Luby-Phelps et al., 1986) or rotation of small molecules in cytoplasm measured by electron spin resonance (3–6 cP) (Mastro et al., 1984).

According to the data shown in Table 2.2 and Figure 2.4, water will be viewed as more or less mobile depending on the spatio-temporal window captured by the probe and the probe's thermodynamic properties determining where it will preferably partition. Probes give a picture of water relaxation toward perturbation which is characteristic of the level of organization from which they capture information (Table 2.2). The wide range of intracellular viscosities measured may be due to the fact that all 'sorts' of water may coexist at the cellular level. However, some timely distinctions should be made at this point. It may happen that water itself is the probe, which is the case for measurement techniques of quasi-elastic neutron scattering (QENS) and proton magnetic resonance (^{1}H NMR). Measurements of water mobility via QENS and ^{1}H NMR revealed that the diffusive motion of water was considerably lowered (three-fold and six-fold, respectively) with respect to pure water in the cytoplasm of *Artemia* cysts (Clegg, 1984a, and references therein). Those data might result from the average diffusive motion of water at all states of organization since water itself is the probe. As discussed above, depending on the probe its partition will be crucial. If the probe is 'squeezed' to an aqueous environment, the properties of water in that environment will be sampled irrespective of the degree of macromolecular organization. Let us pursue this point further.

Cytoplasmic probing of viscosity has been performed with fluorophores and steady-state anisotropy measurements (Fushimi and Verkman, 1991). The rotational correlation time τ_{ic} describes the rotational motion of the unbound fluorophore according to the Stokes–Einstein equation:

$$\tau_{ic} = \eta \frac{V_\eta}{kT} \tag{6.1}$$

where η is the viscosity of the fluid around the fluorophore, V_η is the hydrate volume of the fluorophore molecule, k is the Boltzmann constant and T is temperature. Fluid-phase cytoplasmic viscosity is calculated from τ_{ic} according to equation 6.1. In fact, the hydrate volume, V_η, contains the information of the characteristic spatial dimension, E_c. τ_{ic} is proportionally related to viscosity (equation 6.1) and is, according to Table 2.2 and Figure 2.4, a function of T_r and E_c as discussed above.

6.3 THE MICROTRABECULAR LATTICE AND THE CYTOSKELETON: TWO LEVELS OF ORGANIZATION OF THE INTRACELLULAR MILIEU

We may distinguish two organizational states of the intracellular milieu: the microtrabecular lattice (Porter, 1984) (Figure 6.8) and the cytoskeleton (Figure 6.1). The nature of the former is unknown, though it is presumed to consist of proteins (Clegg, 1991). On the other hand, the cytoskeleton composed of microtubules, microfilaments and intermediate filaments is well characterized (Cleveland and Mooseker, 1994; Heins and Aebi, 1994).

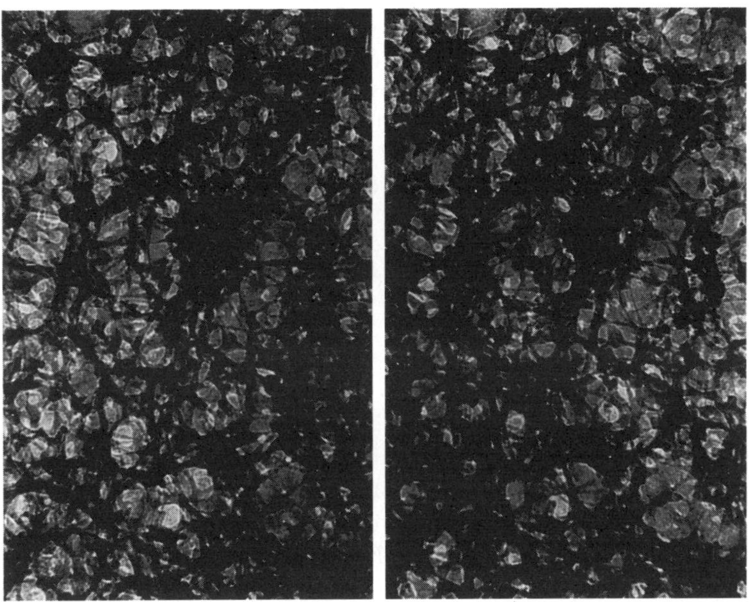

Figure 6.8 Microtrabecular lattice: stereo electron micrographs (high voltage) depicting the structure of the cytoplasmic matrix in a thin margin of a cultured newborn-rat kidney cell. Microtubules can be identified as strands of relatively uniform diameter. (× 80 000.) (Reproduced from *Journal of Cell Biology*, 1984, **99**, 3–12, by permission of The Rockefeller University Press.)

We have discussed how macromolecular crowding may have consequences upon enzyme catalysis. Its effects could be mediated by the amount of water available and mechanisms already considered in sections 6.1.1 and 6.1.3. Under these conditions, if an enzyme partitions in the interphase between the adsorbed water and the macromolecule, cell dehydration will presumably leave the enzymatic activity unaltered. However, if the enzyme distributes in the 'soluble' phase, an increase in concentration is expected due to a decrease in the amount of water.

Cytoplasm has the consistency of a viscoelastic gel whose dynamics is ruled by reversible sol–gel transitions which induce different solvent properties of water (Clegg, 1984a) (section 6.2.1). Intracellular particles appear trapped within the gel or tethered to it with no evident Brownian motion (Luby-Phelps *et al.*, 1988). We have recently proposed that the cytoplasm of living cells can be described as fractals, of the same type as percolation clusters (Rabouille *et al.*, 1992; Aon and Cortassa, 1994) (section 6.4.1). Diffusion in gel fractal networks is slowed down since there is a reduction in the number of diffusion paths available to a random walker, or obstruction by exclusion volume effects (Luby-Phelps *et al.*, 1988). Available experimental evidence with fluorescent probes suggests that cytoplasm is intrinsically compartmentalized by spatial and temporal variation of its submicroscopic structure (Luby-Phelps *et al.*, 1988). The basis of this compartmentalization was outlined in section 6.2.1.

Cytoskeleton dynamics may play a major role in the modulation of the degree of structuring of cytoplasm. The dynamics of microtubules is by far the best understood and this is the subject that will be developed in the next section.

6.3.1 MODULATION OF THE CYTOSKELETAL NETWORK

(a) Dynamic behaviour of microtubules: dynamic instability

The dynamic behaviour of microtubules in the cell is determined at three levels: the intrinsic dynamic instability of the tubulin polymer, modification of the dynamics by interacting proteins and formation of higher-order structures containing microtubules (Gelfand and Bershadsky, 1991; Cassimeris, 1993). Microtubule assembly *in vivo* may be explained quantitatively and qualitatively from the characteristic dynamic instability of nucleated microtubules at steady state (Mitchinson and Kirschner, 1984a,b; Horio and Hotani, 1986; Bayley, 1990) and the self-regulated transcription of tubulin genes by tubulin messenger RNA (Cleveland, 1988).

Dynamic instability has been defined as 'separate growing and shrinking phases of microtubules in the same population with rather infrequent transitions between them' (Kirschner and Mitchinson, 1986). According to the concept of dynamic instability, microtubules can have substantial length fluctuations when individually seen, though the population as a whole is at steady state of assembly (for reviews, see Kirschner and Mitchinson, 1986; Mandelkow and Mandelkow, 1992, and references therein). Mitchinson and Kirschner (1984) discovered the phenomenon of dynamic instability as they tested the predictions of equilibrium assembly in a simple microtubule assembly system (Oosawa and Asakura, 1975; Erickson and O'Brien, 1992). The classical theory predicted that all microtubules should grow longer when the free tubulin concentration was above C_{ss} (steady state concentration of soluble subunits) and this was what they observed. The classical theory also predicted that microtubules should neither grow nor shrink when the tubulin concentration was approximately equal to C_{ss}. However, it was found that the microtubules continued to grow. Even more surprising was the fact that, when the free tubulin was diluted below C_{ss}, they observed that some microtubules seemed to disappear very quickly, but those remaining continued to grow (Erickson and O'Brien, 1992). Thus, microtubules appear to exist in two populations: the majority growing at an appreciable rate and a minority shrinking very rapidly to provide new subunits for growth with only a small probability that a shrinking microtubule will revert to growth and vice versa (Kirschner and Mitchinson, 1986; Horio and Hotani, 1986).

(b) Dynamic instability as a regulatory device of enzyme catalysis

We asked whether the intrinsic dynamics of microtubule polymerization–depolymerization may affect in turn the dynamics of enzymatic reactions. Intuitively, one may imagine an affirmative response to this question. However, the question must be asked: is there a plausible biochemical mechanism that explains such a coupling? The search for an answer to this question started with the study of enzyme kinetics in the presence of polymerized or non-polymerized MTP (Cortassa *et al.*, 1994a; Chapter 7). An interesting result was obtained which showed that the MAPs fraction of the MTP from rat brain was responsible for the activation effect in flux observed in the presence of MTP, through the pyruvate kinase/lactate dehydrogenase (PK/LDH) couple. The microtubular lattice was able to interact with the PK/LDH couple, favouring the formation of enzymatic states of higher oligomerization (i.e. from tetramers to pentamers). We took these experimental features into account in a model which coupled the dynamics of polymerization–depolymerization of MTP and the PK activity. This is treated in the next paragraph.

(c) Coupling microtubular protein dynamics and enzyme kinetics

We have shown that a differential modulation of metabolic fluxes may occur because of the polymeric status of MTP. Recent experimental work either *in vitro* or intracellularly showed that the fluxes sustained by hexokinase/glucose 6-phosphate dehydrogenase and pyruvate kinase/lactate dehydrogenase (PK/LDH) are affected by the presence of either polymerized or non-polymerized MTP. PK kinetics displays cooperativity which is a kind of nonlinear mechanism (Chapters 1 and 5).

(d) A model which links microtubular protein polymerization–depolymerization dynamics and enzyme kinetics

The coupling between MTP dynamics and PK activity was modelled through a system of four ordinary differential equations (ODEs) (equations 6.2–6.5).

Equations 6.6–6.8, which describe the activity of PK, were obtained by fitting the kinetic behaviour of the enzyme with respect to the concentration of its substrate phosphoenolpyruvate (PEP), in the presence of either polymerized or non-polymerized MTP. Also taken into account was the dependence of PK activity as a function of MTP concentration (Cortassa *et al.*, 1994a). PK was modelled as an allosteric enzyme whose rate equation followed a Hill equation (equation 6.6). The enzyme, in the presence of MTP, may exist as a tetramer (T) or pentamer (P), the proportion of each form depending on the polymeric status of MTP (Figure 6.9). The MAPs bound to the microtubular lattice appears to promote the formation of the P form, the V_{PK}^{max} of the P form being larger than that of the T form. Indeed, the Hill coefficient, n, also changes with the degree of self-association of the enzyme, being ruled by equation 6.8. The coupling with the polymeric status of MTP is realized through the term of pentamer formation in equation 6.7, which is proportional to the amount of polymerized MTP, C_P, 0.1 being the proportionality constant that accounts for the fact that MAPs represent 10% of total MTP (equation 6.2) (Cortassa *et al.*, 1994a).

The equations that describe the MTP polymerization–depolymerization assume that tubulin may exist in one of three forms: polymerized (C_P), or non-polymerized (C_T), bound to GTP; or non-polymerized bound to GDP (C_D) (Carlier *et al.*, 1987). A conservation equation (6.9) relates the three forms of MTP.

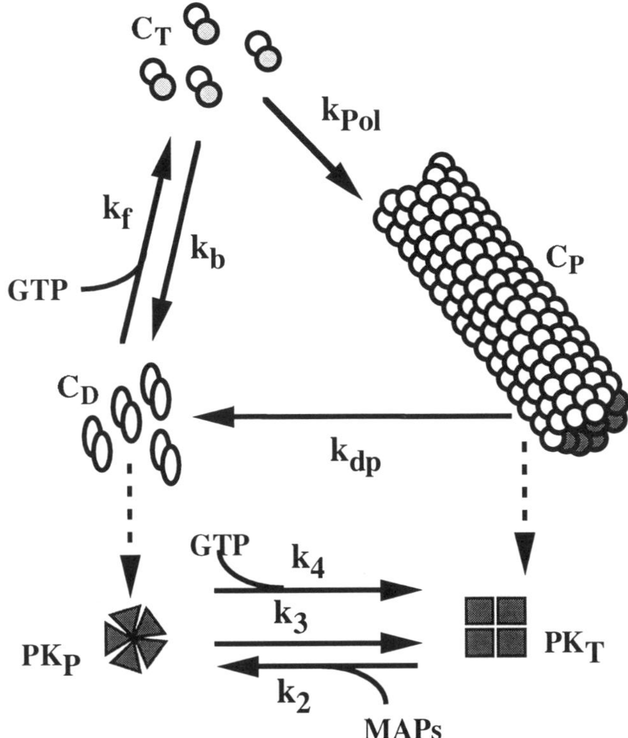

Figure 6.9 Scheme of the model which links the dynamics of assembly–disassembly of microtubular protein (MTP) and pyruvate kinase (PK) kinetics. PK, in the presence of MTP, may exist as a tetramer (PK_T) or pentamer (PK_P), the proportion of each form depending on the polymeric status of MTP, MAPs or GTP (k_4, GTP). The MAPs bound to the microtubular lattice appear to promote the formation of the pentamer (k_2, MAPs). The cycle of MTP polymerization–depolymerization assumes that tubulin may exist in one of three forms: polymerized (C_P); or non-polymerized, bound to GTP (C_T); or bound to GDP (C_D). The coupling between the kinetic behaviour of PK and the polymeric status of MTP is realized through the term of pentamer formation, PK_P, which is proportional to the amount of polymerized MTP (C_P). The dashed arrows indicate that polymerized or depolymerized forms of MTP stabilize either the tetrameric or pentameric forms of PK. (Reproduced from Aon, Cáceres and Cortassa, *Journal of Cell Biochemistry*, 1996, by permission of John Wiley & Sons, Inc.)

$$\frac{d[P]}{dt} = 0.1k_2(PK_t - [P])C_P - k_3[P] - k_4[P][GTP] \tag{6.2}$$

$$\frac{d[PEP]}{dt} = -V_{PK} - k_1(PEP_0 - [PEP]) \tag{6.3}$$

$$\frac{dC_T}{dt} = -K_{Pol}C_P^{n_1}C_T + k_f C_D[GTP] - k_b C_T(GTP_0) - [GTP]) \tag{6.4}$$

$$\frac{dC_P}{dt} = K_{Pol}C_P^{n_1}C_T - k_{dp}C_P \tag{6.5}$$

$$V_{PK} = \frac{V_{PK}^{max}[PEP]^n}{K_S^n + [PEP]^n} \tag{6.6}$$

$$V_{PK}^{max} = V_T^{max} + (V_P^{max} - V_T^{max})\frac{[P]}{PK_t} \tag{6.7}$$

$$n = 4 + \frac{[P]}{PK_t} \tag{6.8}$$

$$C_0 = C_D + C_T + C_P \tag{6.9}$$

The model simulates essential experimental results such as the kinetics of MTP polymerization and the increase in PK-catalysed fluxes as a function of MTP concentration (Figure 6.10). Figure 6.11 shows that upon variation of the rate of MTP polymerization, K_{Pol}, irreversible transitions may occur in the concentration of either polymerized or non-polymerized MTP together with PEP and the concentration of the pentameric form of PK (see Figure 1.8 for a detailed study of this point). For a relatively large range of values of the rate of polymerization, a monotonous change in the steady values of the state variables occur. However, depending on the previous history of the system (i.e. the direction of change of K_{Pol}), an irreversible jump occurs. After the jump, the state variables (e.g. the non-polymerized or polymerized MTP) remain indefinitely on the steady branch without the possibility of returning to the previous situation. The transitions in microtubular protein polymeric status also entrain irreversible changes in the PK's catalytic properties as well as of its conformational state – namely, aggregation as a pentamer. Depending on the previous history, the cytoskeleton may remain non-polymerized indefinitely or may alternate between polymerized and non-polymerized states (Figures 6.11 and 1.8). Irreversible transitions were previously described in a substrate cycle involving two antagonistic enzymes, one of them being inhibited by excess substrate (Hervagault and Canu, 1987; Schellenberg and Hervagault, 1991).

Figures 1.8 and 6.11 emphasize the multidimensional character of functional regulation in cells, i.e. the changes in steady state values of PEP versus the simultaneous variation of the total amount of MTP, C_0, and its rate of polymerization, K_{Pol}. This results in a 'bifurcation line' instead of

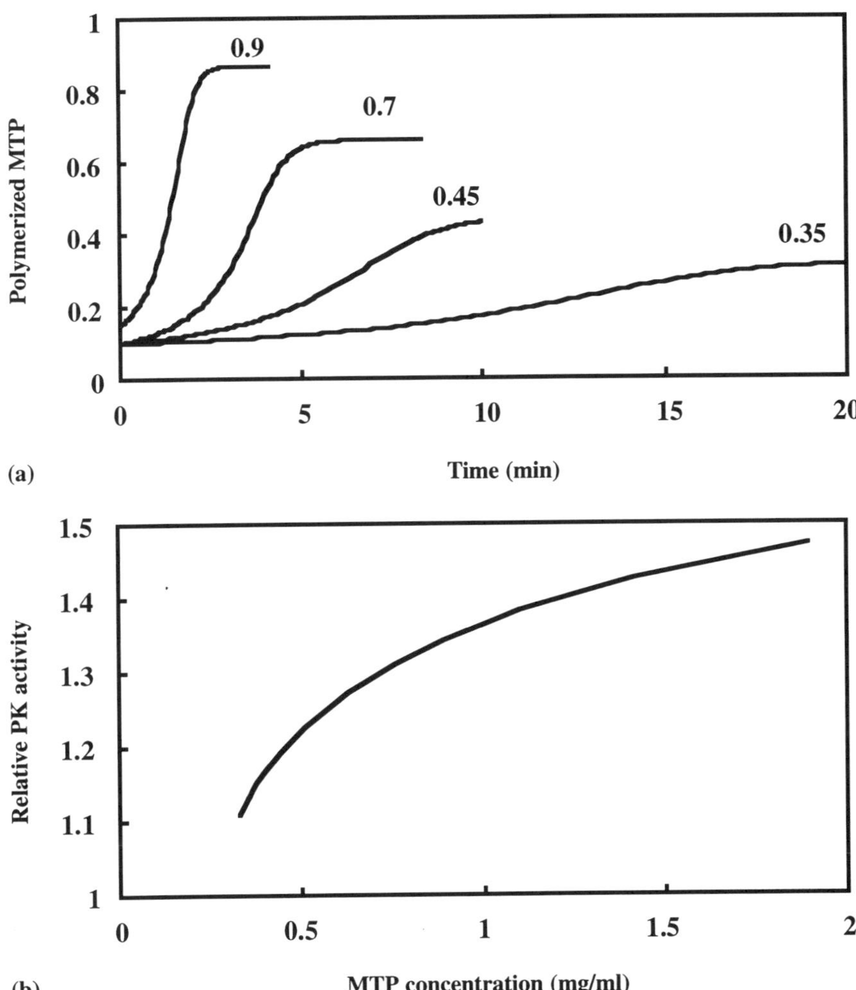

Figure 6.10 Modelling of coupled microtubular protein dynamics and enzymatic activity. The behaviour of the model depicted by ODEs 6.2–6.5 was simulated. (a) Results of the simulations of MTP polymerization kinetics at 37°C in the presence of 1 mM GTP. (b) Results obtained with the increase in the PK-catalysed fluxes in the presence of polymerized MTP. The values plotted correspond to V_{PK} (equation 6.6). The equations were numerically integrated with the following parameter values: $V_T^{max} = 14$ (mM s^{-1}); $V_P^{max} = 25$ (mM s^{-1}); $K_s^n = 0.6$ (mM^n); $K_{Pol} = 10$ (mM^{-2} s^{-1}); $k_{dp} = 2.5 \times 10^{-3}$ (s^{-1}); $k_f = 3.0$ (mM^{-1} s^{-1}); $k_b = 2.5 \times 10^{-3}$ (mM^{-1} s^{-1}); $k_1 = 1.0$ (s^{-1}); $k_2 = 10.0$ (mM^{-1} s^{-1}); $k_3 = 0.05$ (s^{-1}); $k_4 = 0.02$ (mM^{-1} s^{-1}); $PEP_0 = 10$ (mM); $GTP_0 = 1.0$ (mM); $PK_t = 0.01$ (mM); $n_1 = 2$; C_0 corresponds to the values indicated on top of each polymerization curve in (a) and the y-axis in (b). (Reproduced from Aon, Cáceres and Cortassa, 1996, *Journal of Cell Biochemistry*, by permission of John Wiley & Sons, Inc.)

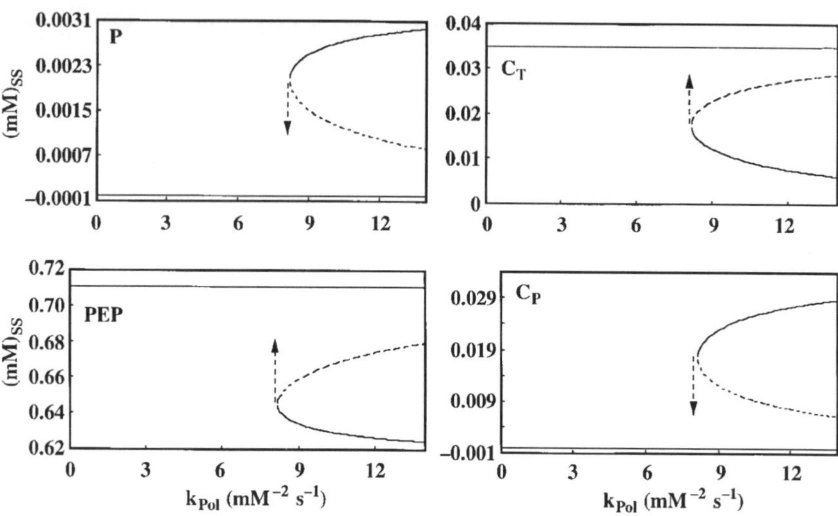

Figure 6.11 Stability analysis of a model which couples microtubule dynamics to pyruvate kinase kinetics. The steady state values of the variables of the model represented by the system of equations 6.2–6.9 were computed with AUTO (E. Doedle, 1986, Concordia University, Canada). The bifurcation behaviour presented stable (solid line) and unstable (dashed line) branches with limit points. The diagrams were obtained with the following parameter values: $k_1 = 1.0$ s^{-1}; $k_2 = 10$ mM^{-1} s^{-1}; $k_3 = 0.05$ s^{-1}; $k_4 = 0.02$ mM^{-1} s^{-1}; $k_f = 3.0$ mM^{-1} s^{-1}; $k_b = 2.5 \times 10^{-3}$ mM^{-1} s^{-1}; $k_{dp} = 2.5 \times 10^{-3}$ s^{-1}; $n_1 = 2$; $K_s = 0.6$ mM; $C_0 = 0.035$ mM; $PEP_0 = 10$ mM; $PK_t = 0.01$ mM; $GTP_0 = 1.0$ mM; $[GTP] = 0.9$ mM; $V_P^{max} = 25$ mMs^{-1}; $V_T^{max} = 14$ mMs^{-1}. The steady states (ss) of PK pentamer (P), PEP, polymerized MTP (C_P) and non-polymerized GTP-bound MTP (C_T) were obtained as a function of the rate constant of tubulin polymerization (K_{Pol}, mM^{-2} s^{-1}). Arrows indicate the sense of the irreversible transition.

'bifurcation points'; the former resulting from the concomitant balance of two parameters. In fact, the variations of both parameters, C_0 and K_{Pol}, may result from the 'lumping' of several physiological processes which, in turn, entrain different functional properties. For instance, actin and tubulin polymerization are sensitive to cytoplasmic Ca^{2+} and phosphoinositides levels (Janmey, 1994) and pH changes (Suprenant, 1991). These two physiological variables will affect K_{Pol}. In the same vein, C_0 may be modulated by changes in gene expression (Cleveland, 1988).

As the amount of protein increases, Figure 1.8 shows that either the irreversible transitions in PK activity (displayed by the level of the substrate phosphoenolpyruvate, PEP) or the level of polymerized protein, C_P, (not shown) occur at lower rates of polymerization, K_{Pol}. The transitions in MTP polymeric status also entrain irreversible changes in the catalytic properties of PK as well as of its conformational state, namely aggregation as a pentamer (Aon *et al.*, 1996a).

Overall, the kinetic behaviour of PK may be in one of two branches of activity that will render two different levels of PEP and perhaps a different metabolic status to the cell, if spread to the whole cellular field. The latter may happen if the dynamics occurs in the cytoplasmic percolation cluster above the percolation threshold.

(e) An interpretation of microtubular dynamic instability as bistable irreversible transitions

The dynamic behaviour of microtubules in the cell is determined in part by the intrinsic dynamic instability of the tubulin polymer (Mitchinson and Kirshner, 1984b; Horio and Hotani, 1986; Kirshner and Mitchinson, 1986; Bayley, 1990; Gelfand and Bershadsky, 1991; Erickson and O'Brien, 1992; Mandelkow and Mandelkow, 1992).

According to the definition of dynamic instability, the separation between growing and shrinking phases of microtubules may be explained by irreversible transitions in the bistable behaviour of polymerizing–depolymerizing MTP (Figure 6.11). Those irreversible transitions are in agreement with the 'rather infrequent transitions' between both phases of microtubules (Kischner and Mitchinson, 1986). This behaviour is significant for developing systems since it may be associated with cell commitment to a particular fate. From this viewpoint, it is remarkable that irreversible transitions appear in cytoskeleton components since their structural states may determine macroscopic or global cellular properties, as discussed in Chapter 2.

Taking into account the kinetics shown by PK in the presence of polymerized or non-polymerized MTP either *in vitro* or intracellularly (Cortassa *et al.*, 1994a), we demonstrate that it may show bistable dynamics entrained by MTP dynamics (Figure 6.10). The latter evidence strongly suggests that the dynamic instability of microtubules *in vivo* may well have consequences in the dynamics of metabolism. Moreover, previous work indicated that coupling of bistable dynamics with a spatial macroscopic cue such as a gradient may lead to the appearance of complex spatial banding patterns (Aon *et al.*, 1989b).

(f) Modulation of intracellular concentration of tubulin and dependence of the microtubule length on tubulin concentration

It appears that a feedback control mechanism for tubulin synthesis by the level of unpolymerized tubulin exists in the cell operating through mRNA stability (Ben-Ze'ev, 1986). Modulation of intracellular levels of tubulin through the stability of mRNAs is a mechanism known as autoregulated instability (Cleveland, 1988).

In vitro experiments demonstrated that the average length at which a growing microtubule starts shrinking depends upon the concentration of assembly-competent tubulin in the system. If the dependence of length upon concentration also applies to the *in vivo* situation, then regulation of intracellular tubulin levels becomes important. Studies of quantitative control of tubulin expression in yeast suggest that narrow steady state levels of tubulin are not necessary to produce appropriate organization of assembled microtubules. Cells with 50%, or less, of normal levels of tubulin could still function normally. However, too much tubulin is deleterious and even small amounts of undimerized β-tubulin are lethal (Solomon, 1991).

According to the dynamic instability explanation, two important parameters describe the behaviour of microtubule ends: catastrophe frequency and rescue frequency. The reduction of tubulin concentration increases the frequency of catastrophes and decreases the frequency of rescues, i.e. the MTP dynamics will approach the limit points in Figure 6.11.

(g) Exchange between microtubules and the soluble tubulin pool

The microtubule dynamics in living cells has been assessed through photobleaching experiments. The results of these studies showed a rapid exchange of subunits between microtubules and the soluble tubulin pool. The fluorescence recovery was close to 300 s for cytoplasmic microtubules in cultured cells and even faster for mitotic microtubules (Gelfand and Bershadsky, 1991). Mitotic microtubules appear, in general, as significantly more dynamic than microtubules in interphase cells. The majority of interphase microtubules are dynamic, but 10–20% of cytoplasmic microtubules are exchanged at a much lower rate. In particular, kinetochore microtubules undergo large changes in stability during mitosis, becoming five-fold more stable in anaphase with respect to the metaphase stage of the mitotic cycle (Zhai *et al.*, 1995).

6.4 A NON-EUCLIDEAN CYTOSTRUCTURE

Under the term **fractal**, coined by Mandelbrot (1982), are grouped different patterns which follow a geometry previously conceived as aberrant because it is indescribable from either mathematical or geometrical viewpoints. On the basis of a quantification of the fractal dimension, D, in micrographs of microtrabecular lattice, quick-frozen and deep-etched electron microscopy of axoplasm (Hirokawa, 1991) or immunofluorescent images of the cytoskeleton through intermediate filaments and microtubules, we have proposed that the groundplan of living cells can be described as fractals of the same type as percolation clusters (Feder, 1988) (Figure 2.9). One of the proposed definitions of fractals that contains its essential feature of self-similarity is that of Mandelbrot, quoted in Feder (1988): 'A fractal is a shape made of parts similar to the whole in some way.' In fact, fractal objects look the same whatever the scale, i.e. they remain invariant at several length scales. Fractals such as Cantor sets, Koch snowflakes, the Sierpinski gasket and Mandelbrot–Given curves are constructed by applying a recursive rule, i.e. a generator on an initiator (Mandelbrot, 1982; Feder, 1988). For instance, in the Sierpinski gasket the initiator is a filled triangle and the generator eliminates a central triangle. The fractal dimension, D, characterizes fractal objects whose value does not correspond to those of the familiar dimensions 1, 2 or 3 of euclidean geometry, i.e. lines, surfaces or volumes, respectively, but to an intermediate one. The latter feature raises the possibility that the complex forms shown by cell architecture can be explained by iteration of an invariant pattern that spans several length scales. Principles of fractal geometry have been shown to apply also for pulmonary airways and in blood vessels (Weibel, 1991; for a review see Cross, 1994). Fractal geometry has become a new tool for quantitative microscopy as a more valid way of measuring dimensions of complex irregular objects than integer-dimensional geometries such as euclidean geometry (Cross, 1994).

We have demonstrated elsewhere that biological polymers such as (glyco)proteins and polysaccharides can organize as fractals (Rabouille *et al.*, 1992; Figure 2.6). An ovomucin gel forms a fibrillar macromolecular lattice or hydrogel with definite solvent properties and water holding. We have already seen that the rheological properties of such hydrogels are modified by perturbing hydrophobic or electrostatic interactions which probably reflect changes in the geometry of the lattice (Rabouille *et al.*, 1989, 1990). Bound water and polymer interactions are likely to participate in microscopic selection of different macroscopic morphological features. These diverse macroscopic morphologies correspond to different fractal dimensions (D) (Rabouille *et al.*, 1992). The changes can be elicited by affecting bound water and polymer interactions. Short-range interactions (e.g. van der Waals, electrostatic, hydrophobic, hydrogen bonds) are likely to participate in microscopic selection of different morphological features at the macroscopic level. These morphologies arise from a lattice with different occupational probabilities and geometry which result in changes in the fractal dimension, D.

The finding that cytoskeleton components may be fractally organized as percolation clusters raises the interesting possibility that the groundplan of living cells might behave according to the principles of fractal geometry.

6.4.1 PERCOLATION CLUSTERS AND THE FRACTAL STRUCTURE OF THE CYTOPLASM OF LIVING CELLS

A **percolation lattice** is a space where diffusion is restricted to certain directions or, in other words, where randomness is frozen (Figure 2.9). The term was coined because of its resemblance to the behaviour of coffee in a percolator (Feder, 1988). A **percolation cluster** is the ensemble of pores or 'sites' in the lattice connected to a chosen centre of injection of a fluid which will only invade or 'wet' another pore that is directly connected to that pore in the lattice through capillary channels or bonds (Feder, 1988; Stauffer and Aharony, 1994). The cluster that spans the lattice is called the **spanning cluster** or percolation cluster.

The most remarkable feature of percolation processes is the existence of a **percolation threshold**, P_c, below which the spreading process is confined to a finite region (Feder, 1988). Near the critical threshold, P_c, as the number of wet pores in the lattice is increased, the probability of a wet pore belonging to the largest cluster increases dramatically. The percolation process undergoes a transition from a state of local connectedness to one where the connections extend indefinitely (Feder, 1988; Chapter 2). Thus, a remarkable result is that below the percolation threshold, P_c, the cluster behaves as locally connected, while above P_c the connection extends indefinitely.

A consequence of the assumption that the cellular cytoplasm behaves according to the principles of percolation processes is that local cytoplasmic behaviour, when subjected to fluctuations or perturbations, may extend and globally impose that behaviour to very remote regions in the cellular cytoplasm. This is a remarkable property since it would imply that, above P_c, cytoplasmic activities may show coherent behaviour, i.e. transitions from local (microscopic) to global (macroscopic). This coherence might be induced by fluctuations in local dynamics of biological processes – for example, enzymatic fluxes, waves of second messengers or ions (Aon and Cortassa, 1994; Chapter 7).

According to the view of the cytoplasm as a percolation cluster, one can easily imagine the following picture: metabolites, hormones, second messengers and ions would percolate through the macromolecular network with the striking possibility that the processes (or the targets) in which they are locally involved become connected and synchronous to similar processes occurring far remote in the cytoplasm or targets distributed farther away from each other.

(a) Interpretation as a percolation threshold of the lethal dose (IC_{50}) for cell proliferation inhibitory drugs

Another consequence of the cytoplasm behaving as a percolation cluster might be that the effect of exogenous substances (e.g. inhibitors) affecting cellular widespread cytoplasmic elements, such as the cytoskeleton, should exhibit threshold behaviour. More explicitly, we are suggesting that the lethal dose operationally defined as the concentration at which inhibitors or pharmacos exert 50% of their maximum effect may be interpreted as the percolation threshold in the fractal cytoplasmic lattice. Percolation fractal lattices, then, provide a conceptual framework for the interpretation of existing experimental evidence related to cell treatment with several microtubule disrupting agents.

Vinca alkaloids such as vincristine, vinblastine, vindesine, vinepidine and vinrosidine induced arrest of cell proliferation in HeLa cells in a concentration-dependent manner (Jordan *et al.*, 1991). At concentrations of vinblastine, vincristine, vindesine and vinepidine higher than those at which cell proliferation arrest was effective, all four derivatives were able to inhibit net tubulin addition at the assembly ends of bovine brain microtubules at steady state. Although all referred alkaloids exhibited similar uptake rates, the most potent proliferation inhibitors (vinblastine and vincristine) accumulated to significantly higher concentrations (Jordan *et al.*, 1991). Both inhibitory properties of the *Vinca* alkaloids, i.e. mitotic arrest and microtubule assembly, exhibited threshold concentration behaviour in their effect (Figure 6.12) (Jordan *et al.*, 1985; Jordan *et al.*, 1991).

If the lethal dose may be interpreted as a percolation threshold for microtubule disrupting agents, then a significant effect on morphology of the cytoskeleton should be apparent with increasing alkaloid concentration. Cell proliferation arrest induced by *Vinca* alkaloids at the lowest effective concentration occurred without a detectable loss of spindle organization or reduction in the mass or, apparently, in the number of microtubules. However, as the alkaloid concentration was raised the metaphase spindles showed different morphology, together with a different chromosomal organization (Figure 6.13). An increase in the length and number of the astral microtubules occurred along with shortening of the pole-to-pole distance and a severe disruption of the

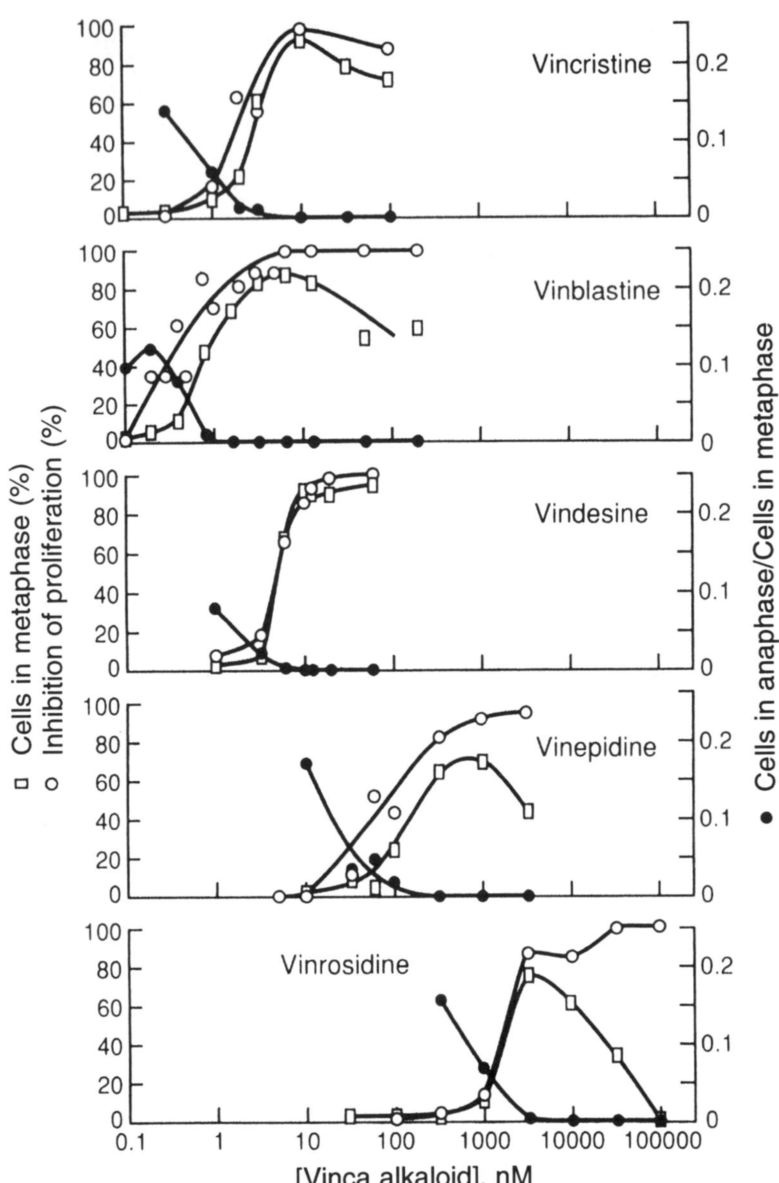

Figure 6.12 Concentration dependence for inhibition of cell proliferation, frequency of metaphase arrest and inhibition of the metaphase/anaphase transition, after incubation of HeLa cells with *Vinca* alkaloids. The percentage inhibition of cell proliferation (open circles) was calculated for each concentration of *Vinca* congener by determining the increase in cell number that occurred during 18–20 h of continuous exposure to drug compared with the increase in cell number in a parallel culture without drug. (Reproduced from Jordan, Thrower and Wilson, 1991, by permission of the American Association for Cancer Research, Inc.)

spindle organization. Spindles collapsed into monopolar configurations and the chromosomes aggregated into a ball usually associated with a star-shaped monoaster of long microtubules or as microtubules depolymerized with short residual microtubule fragments or possibly abnormal polymeric forms of tubulin (Figure 6.13) (Jordan *et al.*, 1991). At sufficiently high concentrations of all five derivatives, total microtubule depolymerization occurred.

Thus, the fractal nature of cellular cytoplasm may be further supported by the existence of a percolation threshold as could be judged from the threshold behaviour exhibited by, for example, *Vinca* alkaloids with respect to inhibition of cell proliferation.

(b) A rheological interpretation of models of cytoplasmic structure

The tensegrity model (section 6.10) describes cytoplasm as a pre-stressed molecular continuum of independent struts and tensile elements. Following application of mechanical stress, cytoskeleton stiffness and apparent viscosity change in parallel because of the postulated molecular continuum. Changes in cytoplasmic rheology (i.e. viscosity) result from filament alignment or interfilamental friction. Altering the global architecture of the cell changes the force balance of the cytoskeleton that in turn elicits changes in its stiffness and apparent viscosity (Wang and Ingber, 1994).

The cytoplasmic supramolecular network rheologically affects sol–gel transitions whose connectivity would certainly depend on the concentration of assembling filaments, the strength of their interaction or the number of cross-linkers, e.g. MAPs (Forgacs and Newman, 1994). Besides, geometrical arrangements certainly influence the connectivity of the gel network. As already stressed (section 6.4.1), a common ingredient in these systems is that, at a given value of certain characteristic parameter, a macroscopic connected cluster forms in the system – for example, a cluster of connected polymer molecules in gel formation. When such a macroscopic connected cluster is formed, the physical properties of the system can change drastically. In the case of sol–gel transition, the polymer solution has liquid properties before the gel forms, that is, a well-defined value of the viscosity, whereas in the gel phase, the viscosity is practically infinite and the system has well-defined elastic properties (Forgacs and Newman, 1994). The critical parameter at which these transitions happen is the percolation threshold that for gel formation may be the concentration of the polymers in solution.

The storage shear modulus, G, in polymer networks becomes linearly dependent on the fraction of all units that participate directly in cross-links, α, after a critical value, α_c, of the latter (Nossal, 1988). Beyond α_c, an infinite three-dimensional network, the **gel cluster**, will form. As α

increases, a greater proportion of all structural units is connected to the gel cluster (Nossal, 1988).

To summarize, a clear link exists between the behaviour of a fluid in a percolation supramolecular network, and the rheological behaviour of the molecular network itself. Otherwise stated, at the percolation threshold, P_c, the dynamics of the (bio)chemical reactions might attain not only a global cytoplasmic influence but also the rheological character

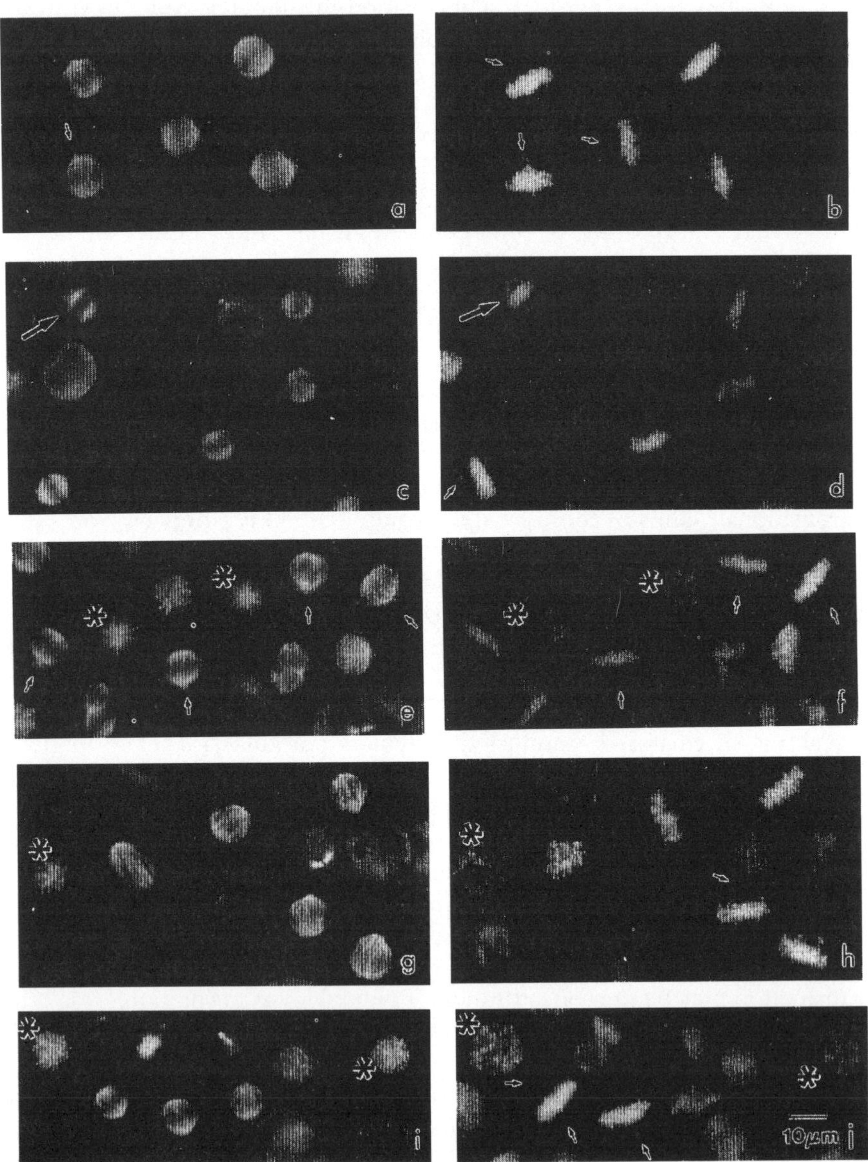

of the polymer network, i.e. it may affect a sol–gel transition. This is of utmost importance since it integrates, in a unified and coherent scheme, rheological and biochemical evidence with the self-organizing properties of nonlinear kinetic schemes. The latter is a potential source of self-emergent, qualitatively new spatio-temporal behaviours, i.e. bistability, oscillations, thresholds, chaos or waves of chemicals (section 6.10).

6.5 TOPOLOGY AND REACTION RATES

6.5.1 CATALYSIS IN HETEROGENEOUS FRACTAL MEDIA

The enhanced performance of catalysts in heterogeneous fractal media may be understood through the simultaneous operation of two mechanisms. The first one is based on the shredded topology exhibited by percolation clusters. For reactions in heterogeneous media, the larger the interface, the faster are the reactions, i.e. the rate per unit surface is constant. This is known as **Wenzel's law** (quoted in Kopelman, 1988).

Bimolecular reactions in shredded topologies, at equal length of rims, proceed faster than in connected topological arrangements (Figure 6.14). This violates Wenzel's law. It has been proposed that the increase in reactivity per unit surface may be due to segregation of reactants in unstirred batch reactions. Segregation hampers encounters between two species. A higher segregation is expected on the connected topology (Figure 6.14, top) (Kopelman, 1988). This explains why the reaction proceeds three times faster than for the disjointed topology (Figure 6.14, bottom) than for the connected one. Fractal-like kinetics is 'compact' kinetics, i.e. the random walkers mostly 'oscillate' around their original positions (Kopelman, 1988). 'Compact kinetics' precludes high segregation.

Figure 6.13 (see facing page) Microtubule and chromosome organization in cells in the presence of *Vinca* alkaloids. Cells were incubated for 18–20 h with *Vinca* alkaloid congeners at approximately the lowest concentration of each congener that inhibited cell proliferation. Left: antitubulin immunofluorescence; right: DAPI stain of chromosomes and chromatin in the same cells. (a,b) 2 nM vincristine; (c,d) 0.8 nM vinblastine; (e,f) 6 nM vindesine; (g,h) 60 nM vinepidine; (i,j) 1000 nM vinrosidine. Coverslip cultures parallel to those of Figure 6.10 were fixed, stained and analysed by immunofluorescence. The spindles of many inhibited cells were bipolar (types I and II, small and large arrows, respectively) though a few had collapsed to a monopolar star-like arrangement of microtubules surrounding a ball of chromatin (type III, asterisks in e–j). Bipolar spindles in drug-treated cells often had prominent tufts of astral microtubules (arrows in a, c and e); these were often associated with chromosomes that were not included in the compact metaphase plates but, rather, were found near the spindle poles (arrows in b, d, f, h and j). (× 800.) (Reproduced from Jordan, Thrower and Wilson, 1991, by permission of the American Association for Cancer Research, Inc.)

Figure 6.14 Catalysis in disjointed and shredded topologies. In both topologies, there is an equal length of rims. A chemical reaction (e.g. A + B → AB) results in a lower reaction rate for the connected catalyst (top) compared with the disjointed one (bottom) (see text for further discussion). (Reprinted with permission from Kopelman, *Science*, **241**, 1620–1626, copyright 1988 American Association for the Advancement of Science.)

Catalytic islands on non-catalytic supports are examples of 'dust'. Regardless the fractal nature of 'dust', the result is anomalous fractal-like kinetics, i.e. with time-dependent rate constants. Any surface diffusion-controlled reaction that occurs on clusters or islands is expected to be fractal-like, i.e. anomalously large values of the reaction order X (Kopelman, 1988).

The second mechanism deals with the transitions from two dimensions to three dimensions, or vice versa, occurring in fractal objects, i.e. a surface may fractally evolve into a volume, or a volume may by fractal folding attain the properties of a surface (Sernetz *et al.*, 1985). The latter is crucial since it has long been known that the probability of a random walker (e.g. a substrate) finding its target (e.g. enzyme) is higher in two dimensions than in three (Polya, 1921, quoted in McCloskey and Poo, 1986). Substrate or effector, and enzyme or target, may find each other in the same two-dimensional space, which increases the encounter probability between them.

The catalytic properties of enzymes at the cellular level are expected to be influenced by the fractal nature of the cellular cytoplasm. In fact, a percolation cluster is a highly shredded, disjointed object with a higher reactivity per unit surface than in a connected geometry (Figure 6.14 and Plate 6). The increase in probability of a substrate finding its target is equivalent to the increase in reactivity when local reduction of dimensions occurs in fractally folded spaces. In other words, increase in reactivity will occur when enzymes (or targets) and substrates (or effectors) coexist in the same topological dimension.

There is experimental evidence that the latter could be the case in enzyme-catalysed reactions which happen in the presence of polymerized or non-polymerized MTP (Cortassa *et al.*, 1994a; Cortassa and Aon, 1994b). The following parameters concerning either the dynamics of cytoskeleton or enzymatic reactions, are amenable to intracellular regulation:

- concentration of cytoskeletal protein;
- local concentration of substrates, enzymes or effectors;
- polymeric status of cytoskeleton components.

The regulatory effects exerted by these three parameters, either *in vitro* or intracellularly, have been described (Chapter 7). The results indicate a concentration-dependent activating effect of MTP on fluxes through coupled enzymatic systems, which reinforces the idea that a polymer such as microtubules, which has been shown to be able to structure the space in a fractal way (Figure 6.1; Rabouille *et al.*, 1992), may also improve the catalytic performance of enzymes.

We have noticed the predominance of higher oligomeric forms of higher activity of PK, an allosteric enzyme in the presence of MTP. This explains the increase in fluxes registered in the PK/LDH couple (Cortassa *et al.*, 1994a). The drastic decrease in transition time, τ, of PK/LDH kinetics as the concentration of polymeric MTP increases, is in agreement with higher oligomeric forms of enzymes.

6.5.2 IMPLICATIONS OF FRACTAL KINETICS

Assuming a fractal percolation nature of the cytoplasm and that enzyme-catalysed reactions are influenced by the organization of the cytomatrix we asked next how *in vivo* enzyme kinetics would be modified because of medium heterogeneity in which metabolic reactions occur. In the Michaelis–Menten formalism a reaction rate is expressed as a combination of elementary steps governed by mass action kinetics where the order of the reaction represents the molecularity of those steps. As already pointed out, the main assumption underlying this kind of analysis is that reactions occur in dilute solutions in a homogeneous space. Savageau has explored the kinetic consequences of reactions occurring in dimensionally restricted spaces, such as a surface, a channel or dispersed clusters (Savageau, 1995). Kopelman (1988) had already predicted for those non-euclidean geometries an increase in the order of the reaction. In such fractal geometries, the order of the reaction does not correspond to the molecularity of the elementary steps but may reflect in some cases the fractal dimension of the surface where a reaction occurs (Savageau, 1995).

To formalize these new concepts, Savageau (1995) derived a fractal rate law, whose expression is as follows:

$$\left(\frac{[S]}{K_f}\right)^{gs} = \left(\frac{v_P}{v_m}\right)\left(1 - \frac{v_P}{v_m}\right)^{-ge} \tag{6.10}$$

where

$$K_f = \left(\frac{k_2 + k_{-1}}{k_1} \right)^{1/gs} \left(\frac{V_m}{k_2} \right)^{(1-ge)/gs} \tag{6.11}$$

According to equation 6.10 additional kinetic parameters appear, namely the fractal Michaelis constant, K_f, which is to a certain extent equivalent to the Michaelis constant of the traditional derivation; the kinetic order for the substrate, gs, or the enzyme, ge, and an effective K_M, proportional to K_f, standing for the substrate concentration at half the maximal velocity. One important difference between traditional Michaelian kinetics and fractal formalism is that the fractal Michaelis constant and the effective K_M, as well as the kinetic order of the enzyme, do depend on enzyme concentration. This feature may be a good criterion to distinguish between fractal and allosteric kinetics, since both will exhibit sigmoidal kinetics. Also the classical Hill plots obtained from the $\log[v_p/(V_m - v_p)]$ vs $\log[S]$ correspond to a straight line in the case of allosteric kinetics, but for fractal kinetics they yield two straight lines with different slopes that change one into the other at $\log[K_f]$.

It is well known that *in vitro* kinetic laws and parameters do not agree with *in vivo* measured metabolic fluxes (Wright and Kelly, 1981; Cortassa and Aon, 1994b). Fractal kinetics has been put forward as a formalism that will allow a better correspondence between the *in vivo* behaviour of metabolic networks with the experimental data on enzyme kinetics *in situ* (Savageau, 1995). Up to now the adjustment of *in vitro* kinetics to *in vivo* measured or estimated fluxes has been performed on a V_{max} basis. However, the largest improvement in catalytic efficiency *in vivo* is also to be expected because of the large decrease in the effective K_M with respect to the *in vitro* data, accompanied by changes in the kinetic order (Kopelman, 1988; Savageau, 1995).

6.6 THE DYNAMICS OF THE GROUNDPLAN OF LIVING CELLS

In general, biological systems are capable of existing in different steady states under identical conditions. Such systems could pass from one steady state to another under the influence of transient modifications of the environment and these transitions might be either reversible or irreversible (Delbruck, 1949). The concept of **dynamic organization** relates to the shift between attractors of the dynamic system, through structural instability (Chapters 1 and 2).

In particular, the assembly–disassembly processes of actin, microtubules and intermediate filaments belong to the sort of physiological processes that are able to shift between different attractors.

The intracellular space may be dynamically structured following the dynamics of organization of cytoskeleton components. Since living cells are thermodynamically open systems operating far from equilibrium, the structures arising from polymerization–depolymerization of, for example, actin or tubulin, with expenditure of energy (ATP or GTP, respectively), may be described as spatio-temporal self-organized structures, i.e. dissipative structures (Nicolis and Prigogine, 1977). These have been observed in physico-chemical, subcellular, cellular and enzymatic systems (Chapters 3–5; for reviews and references therein, see: Hess and Boiteux, 1971; Goldbeter and Caplan, 1976; Winfree, 1980; Olsen and Degn, 1985; Aon and Cortassa, 1993). The striking patterns formed by microtubules during mitosis of eukaryotic cells (Alberts *et al.*, 1989) are a clear manifestation of their dynamically self-organized nature.

The energy expenditure of actin filaments and microtubules assembly–disassembly is far from negligible since apparently a great deal of the ATP hydrolysed by the living cell is used to maintain the dynamic state of actin (Korn *et al.*, 1987; Carlier, 1992).

When whole populations of microtubules are synchronized, the assembly–disassembly processes are seen as oscillations – for example, in light scattering or turbidimetry (Mandelkow and Mandelkow, 1992). These oscillations are supposed to be the macroscopic manifestation of the dynamic instability under conditions of synchronous polymerization and catastrophe of individual microtubules (Gelfand and Bershadsky, 1991).

The energy source that powers the constant activity of dynamic instability is almost certainly the hydrolysis of GTP. The tubulin dimer binds two molecules of guanine nucleotide. One of these is always GTP, is not exchangeable with nucleotides in solution, and is not hydrolysed during assembly. Oscillations in microtubule assembly driven by GTP provide, then, an unequivocal proof of the dissipatively self-organized nature of the patterns of microtubules originating from assembly–disassembly processes. Oscillations dampen macroscopically before GTP exhaustion, probably due to GDP accumulation and loss of synchrony (Mandelkow and Mandelkow, 1992). Self-organization of microtubules is obtained through the coupling of GTP hydrolysis with the assembly of microtubular protein occurring in open systems continuously dissipating GTP. The source of nonlinear behaviour is given by the threshold behaviour exhibited by the assembly of microtubules with respect to the protein concentration.

Oscillating populations of microtubules show spatial patterns including linear or circular waves that assemble and disassemble with the same time period of minutes as in the temporal oscillations, and with wavelengths of a few millimetres, and velocities of $1\,mm\,min^{-1}$, respectively (Plate 5) (Mandelkow *et al.*, 1989). Even stationary microtubules can generate spatial patterns – probably of microtubules aligned in parallel, similar to nematic liquid crystals (Hitt *et al.*, 1990). These stationary patterns of microtubules may be driven by the gravitational field (Tabony and Job, 1992) with a concomitant reduction of symmetry in the sense of the gravitational field. Microtubular striped patterns might also depend upon having a sufficient supply of hydrolysable GTP (Tabony and Job, 1992).

6.7 THE LINK BETWEEN THE FRACTAL NATURE OF THE CYTOSKELETON AND THE SPATIO-TEMPORAL SCALING SHOWN BY PHYSIOLOGICAL PROCESSES

The subcellular processes taking place at different levels of organization inside the fractal percolation cytostructure show spatial as well as temporal scaling. The characteristic space, E_c, and relaxation time, T_r, contain dynamic information about physiological processes at different levels of organization (Chapter 2). T_r and E_c are the result of integration in time and space of the inherent autonomous dynamics of (sub)cellular processes involved in cell physiology (Aon and Cortassa, 1993). At instabilities of their dynamics, subcellular processes bifurcate, exhibiting transitions between levels of organization. In fact, microtubular protein which organizes as fractal percolation lattices (Rabouille *et al.*, 1992; Aon and Cortassa, 1994) exhibits self-organization in the spatio-temporal 'window' of the transition point (Aon and Cortassa, 1993). Microtubules may reach lengths of micrometres and assemble–disassemble in the temporal domain of minutes. In this spatio-temporal 'window', transitions may give rise to new functional properties appearing as a new sort of macroscopic coherence – for example, (a)symmetry changes, sophisticated dynamics or evolution toward a new steady state. Moreover, in percolation lattices at the threshold level of occupancy of the lattice, i.e. the percolation threshold, there are transitions between local and global connectedness.

The considerations of the preceding paragraph indicate that the passage from microscopic to macroscopic functional coherence, achieved through self-organization by the assembly–disassembly of cytoskeleton components, clearly correlates self-similarity in fractal structures and spatio-temporal scaling exhibited by physiological processes at distinct levels of organization. We have interpreted the change in the slope of the allometric relationship as the spatio-temporal 'window' (micrometres and minutes) at which macroscopic coherence emerges in cellular processes (Chapter 2). This macroscopic coherence at the cellular level might be chiefly driven by processes exhibiting conspicuous increments of their spatial coordinates, i.e. appearance of spatial patterns. In other words, the temporal relaxation of the dynamics of biological processes is restricted in the long range, further from the 'transition point', when spatial patterns could emerge. This restriction from a dynamic point of view would ensure that 'essential' dynamic variables remain bounded (Kauffman, 1989). The latter is in agreement with synchronization properties since for similar processes occurring wider apart in a spatial sense, they will be able to settle their dynamics by fixing their variables to a bounded (probably small) range of variation. This notion also supports homeostatic properties of the integrated dynamic system. Environmental parameters could bring about generalized cellular responses that in turn induce cells to stop or resume division, or choose between divergent developmental paths (Chapter 9).

6.8 THERMODYNAMIC ASPECTS OF FRACTALS AND SELF-ORGANIZATION

To relate the dynamics of biological processes to fractal properties of living systems which exhibit spatio-temporal scaling, we need to link self-organizing principles to the generation of fractal objects. The process of generation of fractal organization from iteration of an invariant pattern that spans several length scales may be self-organized under non-equilibrium thermokinetic conditions of growth (Ben-Jacob and Garik, 1990; Hill, 1990; Rabouille *et al.*, 1992). One of the most accepted hypotheses about the generation of fractals is **diffusion-limited aggregation** (DLA). Simply stated, the self-similarity of fractal objects whose shapes are made of parts similar to the whole may arise from self-organizing principles such as those participating in the organization of cytoskeleton components (section 6.2). This feature has been put in evidence by the formation of fractals in an ovomucin gel (Plate 6). At high rates of evaporation the pattern formation did not occur. This suggests that only coupled events (dehydration and solute precipitation), occurring under far from equilibrium growth conditions and involving nonlinear kinetics, lead to pattern formation. The patterns are in fact quasi-crystals that grow under far from equilibrium conditions with a final stable state (Rabouille *et al.*, 1992).

It has been proposed that the cytoskeleton is organized in a fractal way (Aon and Cortassa, 1994). Certainly, polymerization of microtubules might be described by reaction (aggregation)–diffusion equations:

$$\frac{dC_P}{dt} = K_{Pol}C_T C_P^{n1} + D\frac{\partial^2 C_T}{\partial x^2} \tag{6.12}$$

where C_P is the concentration of GTP-bound polymerized tubulin, C_T is the concentration of GTP-bound non-polymerized tubulin, K_{Pol} is the rate constant of tubulin polymerization, n_1 is the number of tubulin monomers in polymerization nuclei and D is the diffusion constant of tubulin monomers in solution.

In equation 6.10, it is possible to identify two coupled processes: reaction (the first term on the right-hand side) and diffusion (the second term on the right-hand side). This type of reaction is able to exhibit spatial structures such as waves, as has been shown for an immobilized bienzymatic system (Cortassa *et al.*, 1990b). Waves of polymerizing–depolymerizing microtubules have been experimentally demonstrated in a petri dish (Mandelkow *et al.*, 1989) or in microtubular solutions (Tabony, 1994). The fluxes of each participating process (reaction and diffusion) are related to the driving forces (free-energy difference between reactant and products and the chemical potential gradient of diffusing species) through a relation that may be expressed as a Taylor series of the form $J = f(\Delta G)$. Even considering only the linear (up to first order) terms of the Taylor series, the resulting phenomenological equation is non-symmetric because of the kinetic nonlinearity in the reaction term, i.e. the $C_P^{n_1}$ factor. The asymmetry in the matrix of phenomenological coefficients is a necessary condition for a system to exhibit self-organization, even one exhibiting linear relations between fluxes and forces but displaying kinetic nonlinearities (Cortassa *et al.*, 1991). Occurrence of processes under far from equilibrium conditions implies that the coupled processes, in this particular case diffusion and reaction (aggregation), exhibit a non-symmetric matrix of phenomenological coefficients in the equations relating the rate of the process to their driving forces.

In the case of percolation clusters, nonlinearity is provided by the threshold behaviour of occupancy of the lattice that occurs at the transition between local to global connectedness (Feder, 1988; Aon and Cortassa, 1994). A necessary and sufficient condition to allow the occurrence of self-organization is to have coupled processes operating under far from equilibrium conditions, without symmetry in the matrix of phenomenological coefficients. In the particular case of the cytoskeleton, its organization is driven by the dissipation of metabolic energy in the form of GTP high energy transfer bonds.

Synthetically, cytoskeleton components may self-organize following the principles of fractal geometry.

6.9 DYNAMIC ORGANIZATION AND CONNECTEDNESS IN SUPRAMOLECULAR NETWORKS SUPPORTING CATALYSIS

The problem of how to build complex systems of many interacting parts is that alterations in the behaviour of one small subsystem will propagate throughout the system without breaking it into uncoupled subsystems. Such a problem was addressed by Ashby (1960) and revisited by Kauffman (1989). This observation is straightforwardly related to the concept of dynamic organization developed in Chapter 2.

On the one hand, we defined dynamic organization as the evolution of the spatio-temporal organization of biological processes between successive levels of organization. The evolution of the dynamics of a biological process from one behaviour into another is achieved through instabilities (bifurcations) which give rise to the emergence of collective behaviour – for example, of molecules, enzyme activity, cell-to-cell interaction, spatio-temporally coherent (Chapters 1 and 2).

On the other hand, we have suggested that the cytoplasm of living cells might be organized as a percolation cluster. The randomness of a fluid spreading through a fractally organized medium, as is the case with a percolation cluster, differentiates from the familiar diffusion processes. In the latter, randomness is essentially given by the random walks of the fluid particles – an example is the irregular thermal motion of molecules in a liquid. However, in a percolation process, randomness is frozen into the medium itself (Feder, 1988). As already noted, the most remarkable feature of percolation processes is the existence of a percolation threshold, P_c, below which the spreading process is confined to a finite region.

Several reported data indicate that, in patchy environments, rigidly deterministic unstable dynamics at the local level together with limited diffusion toward adjacent patches may give rise to an amazing range of complex spatio-temporal patterns such as spiral waves or spatially chaotic variation (Hassell *et al.*, 1991).

We have emphasized that if the cellular cytoplasm behaves as a percolation cluster, it should exhibit percolation thresholds. The cellular cytoplasm may be considered as a patchy environment whose functionality under certain conditions may be more modular (islands, or patches, functionally isolated or with weak interactions between them) than global. Most of our research in this chapter has been to describe or approach the general principles and mechanisms in cellular systems through which the spatio-temporal organization of several simultaneous processes at different levels of organization may be achieved. The very fact that deterministic systems whose dynamics may be become locally unstable and diffusively spread to adjacent patches may give rise to spatio-temporal chaos or complex spatial patterns such as spiral waves or 'crystal lattices' (Hassell *et al.*, 1991; Cross and Hohenberg, 1994) suggests that:

- a chaotic spatial pattern does not necessarily imply chaotic dynamics;
- the appearance of random fluctuations or Brownian motion of particles in the cytoplasm arises inevitably from randomness but may

also be given by deterministic dynamics in local regions or patches, weakly or strongly coupled to other adjacent regions.

In this respect let us consider the local dynamics of deterministic systems. It may be supposed that the controlled variables of a dynamic system change their values according to variations in their input over some range. If a variable is unchanging in its value as its inputs change their values, then that fixed behaviour of the variable blocks the propagation of alterations through it to other variables downstream of that variable. Each element is frozen in its forced value and leaves behind functionally isolated islands – or finite regions of the cytoplasm below the percolation threshold (Kauffman, 1989; Aon and Cortassa, 1994). Functional isolation allows parameter changes to alter the behaviour of one isolated cluster of variables without propagating changes to all other variables in the system. The existence of isolated clusters of variables is called **modularity**. Under the conditions described above a fluctuating functional modularity is achieved then, rather than structural modularity, by building systems from fully independent subsystems. Thus, **functional modularity** may be viewed as dynamic functional subsystems whose variables conform to clusters of variables with partially constant responses that do not propagate their changes to other subsystems. This is a crucial concept since in a complex network of connected subsystems, as a living cell is supposed to be, percolation thresholds will determine under which conditions large connected webs of elements or dynamic subsystems will form, as studied in two-dimensional and three-dimensional lattices with nearest neighbour coupling. In a percolation lattice, when the probability of occupation, P, is larger than the critical value, P_c, then the dynamic behaviour of the network breaks up into a state in which connections extend indefinitely. P_c, the critical value of P, marks a kind of 'phase transition' in the behaviour of such a dynamic system.

6.10 MODELS OF CYTOPLASMIC STRUCTURE AND FUNCTION

In the last two decades or so, ideas concerning the organization and functional properties of cell cytoplasm have grown around two main concepts. The first one is that the cytoplasm is a crowded (Minton, 1983; Garner and Burg, 1994) though organized protein crystal (Fulton, 1982; Clegg, 1984a,b). The second one deals with the fact that metabolism and the gene expression machinery are not indifferent to that organization; on the contrary, they are deeply influenced by it (Penman *et al.*, 1981; Pienta and Hoover, 1994; Cortassa *et al.*, 1994a). Overall, experimental evidence increasingly favours the idea that metabolism occurs strongly associated to the dynamic cellular scaffolds (e.g. Welch, 1977; Clegg,

1984a,b, 1992; Masters, 1984; Ingber, 1993; Cortassa *et al.*, 1994a) and that these insoluble matrices and their dynamics in turn deeply influence the dynamics of chemical reactions (Clegg, 1984b, 1992; Ingber, 1993; Ingber *et al.*, 1994; Aon and Cortassa, 1994; Cortassa and Aon, 1994b; Pienta and Hoover, 1994).

At present, within the framework of the 'structured view' of the cytoplasm (Fulton, 1982; Clegg, 1984a; Luby-Phelps *et al.*, 1988), there are at least two distinguishable models concerning cell function. These are the tensegrity model (Ingber, 1993) and the fractal percolation model (Aon and Cortassa, 1994). There are two main aspects of these models that may be compared: the mechanism of coherence and how this coherence is transduced into functional properties.

By 'mechanism of coherence' we mean how the passage from microscopic to macroscopic order in cell function is effected, i.e. how the activity or architecture of millions of molecules or supramolecular structures becomes synchronized in space and time in an apparently 'purposeful', functional way. Transduction is a second problem tightly linked to that of coherence; it mainly deals with the mechanisms through which cells transduce environmental stimuli (e.g. hormones, neurotransmitters) into differential gene expression, metabolism and cellular energetics. On this basis let us briefly compare the two models.

6.10.1 THE TENSEGRITY MODEL

For Ingber's model (1993), coherence is given by changes in the 'tensional integration' of components of the cytoskeleton and its transduction is essentially conformational (Ingber, 1993; Wang *et al.*, 1993). In essence, the tensegrity model proposes that the entire cytoskeleton responds to stress as a single, tensionally integrated (tensegrity) structure. Experimental evidence in support of this model came from experiments of stress–strain relationships measured with magnetic microbeads attached to the surface of living endothelial cells (Wang *et al.*, 1993). The mechanical stress applied to the surface of cells with a twisting device showed that the cytoskeleton response to the applied stress was a property of the integrated system. The partial disruption of the cytoskeleton with depolymerizing agents such as cytochalasin D or nocodazole did not completely suppress the cytoskeleton stiffening, i.e. a large resistance to mechanical deformation was evidenced at high levels of applied stress (Wang *et al.*, 1993). Using a tensegrity model to interpret the data, the stiffness–stress (force) linear response of the cytoskeleton of living cells could be simulated and it was further shown that, when the force was increased, the mechanically interdependent structural elements rearranged without topological disruption or loss of tensional continuity.

Thus, a molecular continuum of mechanically interdependent struts and tensile elements structurally stabilizes by global rearrangement rather than local deformation in response to stress (Ingber, 1993; Ingber *et al.*, 1994). In this light, the cytoskeleton response to applied stress appeared as a property of the integrated system and not a characteristic of any individual part (Wang *et al.*, 1993).

According to the tensegrity model, coherence would be given by the use of tensegrity architecture by cells, providing a mechanism to distribute mechanical stresses to regulatory molecules and enzymes that are immobilized on or interact with cytoskeletal and nuclear scaffolds and thereby to integrate cell structure and function (Ingber, 1993; Ingber *et al.*, 1994). Mechanical stress in turn is transduced into biochemical function through conformational changes, i.e. a mechanochemical transduction.

6.10.2 THE FRACTAL PERCOLATION MODEL

The fractal model of cellular cytoplasm proposes that the cytoskeleton is organized according to the principles of fractal geometry. Iteration of an invariant pattern that spans several length scales might describe the complex forms shown by cell architecture (Aon and Cortassa, 1994) (Figure 6.1). From a dynamic perspective, the main feature of percolation clusters is given by the existence of a percolation threshold for which a percolation process undergoes a transition from local to global connectedness (Feder, 1988; Aon and Cortassa, 1994).

In the fractal model, coherence arises at instability points in the dynamics of cellular processes further from the 'percolation threshold', and transduction may be realized by multiple pathways (Aon and Cortassa, 1993, 1994; Cortassa and Aon, 1994b). A main transduction process in cell function is enhanced catalytic performance. In heterogeneous fractal media transduction may be achieved through two mechanisms:

- less segregation of reactants participating in bimolecular reactions occurring in disjointed fractal topologies compared with connected topologies;
- increase of the encounter probability between substrate or effector, and enzyme or target, by their coexistence in the same topological dimension (2D).

6.10.3 (MACRO)MOLECULAR CROWDING

Another view of cytoplasmic structure emphasizes the 'macromolecular crowding' properties of the cytoplasm. According to this view, cytoplasmic organization is based on the fact that cells contain a large variety of macromolecules, largely proteins, which taken together will occupy a substantial fraction of the total cytoplasmic volume, called **volume occupied** or **crowded** (Minton and Wilf, 1981; Minton, 1983; reviewed in Garner and Burg, 1994). Crowding may also affect the topological arrangement of polymers inside the cell.

The view of the cytoplasm as a crowded object is not excluded by either tensegrity or fractal percolation models. For instance, crowding may alter the average degree of self-association of an enzyme, in turn changing its average catalytic activity (Minton, 1983, 1992; Cortassa *et al.*, 1994a). This effect was shown to be specifically induced by microtubular protein or actin in the pyruvate kinase/lactate dehydrogenase or hexokinase/glucose 6-phosphate dehydrogenase enzymatic couples (Figures 7.8–7.12). Furthermore, the supramolecular organization of microtubular protein could entrain the dynamics of enzymatic reactions through mechanisms of self-association–dissociation of enzymes toward higher or lower oligomeric forms with higher or lower levels of activity, respectively (Cortassa *et al.*, 1994a). Experimental evidence suggests that the same enzyme may be differentially modulated at different locations in the cytoplasm, depending on the cytoskeletal protein (and their associated proteins) present at that location (Figures 7.9–7.11). According to this proposal, it can be postulated that a precise biochemical coupling exists between the highly sophisticated mechanisms by which cells dynamically rearrange their (supra)molecular architecture (Figure 7.7) and the dynamics of biochemical reactions taking place concomitantly.

This is a relevant matter since the information presently available strongly suggests that the cytoskeleton is one of the endogenous determinants of cell form, and that its (supra)molecular asymmetries are involved in the development of, for example, neuronal polarity (Solomon, 1981; Cáceres *et al.*, 1984; Ferreira and Cáceres, 1989). Furthermore, during recent years it has become increasingly apparent that there are distinct cytoplasmic domains in neurons which are intimately associated with the ability of these cells to create, say, microtubule subsets that differ in molecular composition and spatio-temporal distribution (Cáceres *et al.*, 1984; Ferreira and Cáceres, 1989). Thus, a clear link is established between cell polarity, form and the underlying dynamics of cellular biochemistry, by the proposal attributing a topologically driven modulation of metabolism to the heterogeneous cytoplasmic distribution of cytoskeletal proteins (Aon *et al.*, 1996). Considering the entrainment of enzymatic reactions by the dynamics of cellular architecture, a whole set of observations is clarified. In fact, actin and tubulin, as major structural proteins in all eukaryote cells, change their distribution along the cell cycle. In the budding yeast *Saccharomyces*, the cell-cycle dependent patterns of microtubule and actin localization are well studied (Lew and Reed, 1993; Chant, 1994). Reported experimental data show that cytoskeleton components continuously assemble–disassemble along the length of the axon of cultured neurons or dramatically rearrange its spatial distribution in response to growth factors (Hirokawa, 1991; Tanaka and Kirschner, 1991; Morfini *et al.*, 1994; DiTella *et al.*, 1994). In many plant or animal cells, the interplay of intracellular ionic gradients (e.g. H^+, Ca^{2+}) may in turn regulate the polymeric status as well as localization of cytoskeleton components, since cytoskeletal assembly is sensitive to ionic conditions (Kropf, 1994). F-actin and microtubule assembly are faster at acidic and alkaline pH, respectively (Sampath and Pollard, 1991; Suprenant, 1991).

A prediction of the fractal model of cytoplasm is that a cell, by regulating its geometry or architecture, in turn regulates the level of its percolation threshold, P_c. In percolation lattices transitions between local to global connectedness happen at the threshold level of occupancy of the lattice. Different P_c levels would determine when the local level of a messenger or the product of an enzymatic reaction may extend to the whole cellular field if it spreads in a domain belonging to the spanning cluster.

The more adequate models for cellular functional coherence will be those in which structure and dynamics are more worthily combined. In shaping those models we should not forget that cellular processes, from the conformational change in the α-helix of a protein up to cell division, scale in space and time. By this we mean that, for instance, processes such as enzymatic reactions or cell division occur within characteristic spatial coordinates and when perturbed they will relax with a characteristic relaxation time (Aon and Cortassa, 1993). Microtubular protein which organizes as fractal percolation lattices (Rabouille *et al.*, 1992; Aon and Cortassa, 1994) exhibits self-organization in the spatio-temporal 'window' (micrometres and minutes) in which microtubules assemble–disassemble (Kirschner and Mitchison, 1986; Erickson and O'Brien, 1992) and regulate chains of enzymatic reactions (Cortassa *et al.*, 1994a; Cortassa and Aon, 1994b).

From these considerations, it has been proposed that the passage from microscopic to macroscopic functional coherence, achieved through self-organization by the assembly–disassembly of cytoskeleton components, clearly correlates self-similarity in fractal cytostructures and the spatio-temporal scaling exhibited by physiological processes at distinct levels of organization (Chapter 2). We have been able to show that one of the simplest self-organized phenomena, i.e. bistability or multiple stationary states, may occur in the coupled dynamics of MTP and a biochemical reaction as a function of two relevant cellular variables, C_0 and K_{Pol} (Figures 1.8 and 6.10). These two variables in turn regulate the geometry of the lattice of MTP and through this the percolation threshold, P_c, and the fractal dimension, D (Feder, 1988; Rabouille *et al.*, 1992). This reinforces the idea that polymerization–depolymerization of cytoskeleton components may be involved in the passage from microscopic to macroscopic order at the cellular level. Besides, the existence of abrupt transitions at limit points in the dynamic behaviour of MTP polymeric status, together with PK activity, further suggests a link between P_c and self-organization of cytoskeleton components along with their involvement in the synchronization of cellular activities farther apart in the cytoplasm of living cells.

Spatio-temporal regulation of glycosis and oxidative phosphorylation *in vivo*

7

The aim of this chapter is to present some recent advances in the *in vivo* control and regulation of glycolysis and oxidative phosphorylation with special emphasis on yeast and tumour cells. These insights are analysed in the conceptual framework of dynamic organization in living cellular systems, focusing on the interaction (coupling) between two major catabolic processes occurring in two different cellular compartments. Old and new experimental data are revised in the light of powerful new technologies (e.g. NMR, confocal microscopy) and quantitative techniques combined with mathematical modelling. They are described under the general topic of dynamic organization which emerges from multiple interactions between compartments and processes inside cells. Those compartments may be of structural origin – for example, plasma membrane delineating the cell's boundaries, or mitochondrial/ cytoplasmic membranes – or functional ones such as the alternating association–dissociation of enzymes to subcellular structures (e.g. mitochondria, cytoskeleton) with different kinetic properties in each state. Two novel regulatory mechanisms at the level of the cytoskeleton may add a new dimension to the *in vivo* physiological properties of cells. One of these mechanisms concerns the regulation of metabolic fluxes by polymerization–depolymerization of microtubular protein in a concentration dependent manner. The other implies the differential regulation of enzymatic fluxes sustained by the heterogeneous distribution of the major cytoskeleton constituents (actin and microtubular protein). The consequences of the fractal organization of the cytoplasm as a percolation cluster on enzyme catalysis are explored. One main suggestion prompted by the modulatory power of the polymeric status and concentration of cytoskeleton components is that it could function as an intracellular mechanism of synchronization between microscopic (local) and macroscopic (global) processes.

7.1 SUBCELLULAR DYNAMICS

In cellular function, subcellular dynamics largely represents a transdisciplinary field of confluence between biochemistry, physiology and biomathematics. This transdisciplinary field takes advantage of technologies which deal with cellular and subcellular spatio-temporal organization – for example, confocal microscopy, NMR, flow cytometry, patch-clamp techniques. In addition, biomathematics provides through mathematical modelling the possibility of analysing transient and steady states of the dynamic behaviour of biological systems. This is required for the full description of the regulatory role of different biochemical devices under defined physiological conditions, e.g. allosteric properties, hexokinase (HK) ambiquity, transcriptional control or the recently described polymerization–depolymerization of cytoskeleton components (Cortassa *et al.*, 1994a). Thermodynamically, a cell appears as a system far from equilibrium, attaining different steady states characterized by free-energy dissipation. Accordingly, subcellular dynamics may be viewed as a flux of energy mainly in the form of redox equivalents and high energy transfer bonds transduced from substrate catabolism. Different forms of energy drive cellular processes and functions regulating their dynamics and that of the exchanges between compartments of either functional or structural origin. Biochemically, the flux of energy is essentially characterized by redox and phosphorylation potentials which measure the energy status of an organism faced with particular environmental conditions (Chapter 8). A flux of energy may also be understood as the exchange between different forms of energy stores. In addition to phosphorylation and redox potentials, other forms of energy are the electrochemical potentials of ions and solutes, the build-up by electron transport processes (e.g. oxidative phosphorylation or photosynthesis) and stores of macromolecules like glycogen; these are in turn expended or used as driving forces for metabolite exchange across plasma membrane or between organelles and the cytoplasm. Through these energy transducing events, prokaryotic or eukaryotic cells are readily able to accommodate to changes in environmental conditions. These changes result in coupled modifications at the level of intermediary metabolism, DNA supercoiling and likely gene expression (Cortassa and Aon, 1993b, 1994b; Chapter 12).

Glycolysis in either of its two known routes (Embden Meyerhoff or Entner Doudoroff) is one of the main catabolic pathways employed by prokaryotic or eukaryotic cells to obtain energy and redox equivalents from substrates together with the pentose phosphate shunt, the tricarboxylic acid cycle and oxidative phosphorylation in non-photosynthetic organisms. In photosynthetic organisms, the cells are additionally able to convert the energy coming from sunlight into energy directly available for cellular work (electrochemical gradients or ATP).

Even though glycolysis and oxidative phosphorylation are among the most widely explored biochemical pathways, the spatio-temporal regulation of their interactions *in vivo* is far from being completely understood. It is germane for this chapter to distinguish between the meanings of 'control' and 'regulation'. **Control** means that if a change in the level of X occurs, a change in the rate of a process or the amount of, say, a metabolite will follow. However, this does not necessarily imply that amounts or rates are regulated by X *in vivo* ; it will only regulate if a change in X actually occurs in a physiological process (Brown, 1992). Accordingly, the term **regulation** will be used preferentially since we will be dealing with *in vivo* data, albeit 'control' may be also used, depending on the context.

7.2 STEADY STATE CONTROL

7.2.1 ANALYSIS OF GLYCOLYTIC AND OXIDATIVE PHOSPHORYLATION FLUXES IN CHEMOSTAT CULTURES OF *SACCHAROMYCES CEREVISIAE* UNDER CARBON LIMITATION

The quantitative characterization of the physiological and metabolic behaviour of yeast in continuous cultures has improved our knowledge of the regulatory machinery of glycolysis and its relationships to other pathways such as fermentation and the tricarboxylic acid (TCA) cycle.

In carbon-limited chemostat cultures of *S. cerevisiae* (yeast) a characteristic behaviour is exhibited as the dilution rate (D), i.e. the growth rate, μ (D = μ), is systematically varied. Respiratory metabolism is observed at low D whereas respiro-fermentative behaviour characterizes high D. At values of D below a critical one (D_c), a linear relationship is evidenced between the rate of oxygen consumption and the growth rate, μ (Rieger *et al.*, 1983). At values of D > D_c, the rate of respiration became independent of μ. This experimental observation was interpreted as limited respiratory capacity (Kappeli, 1986). Above D_c, the steady state levels of ethanol increased while the biomass yield decreased. This was taken as evidence of respiro-fermentative metabolism. According to the concept of limited respiratory capacity (Kappeli, 1986), saturation of respiration is the primary cause of the onset of respiro-fermentative glucose metabolism. Mechanistically, aerobic ethanol formation appears to arise from an overflow reaction at the branch point between ethanol formation and the TCA cycle which results from a limited respiratory capacity of the cells. Alexander and Jeffries (1990), comparing the data obtained by Beck and von Meyenburg (1968), Barford and Hall (1981) and Rieger *et al.* (1983), concluded that repression of respiration or the so-called Crabtree effect occurred with *S. cerevisiae* H1022 regardless of adaptation or medium culture. Respiratory repression appears to be common in yeasts of the genus *Saccharomyces*, as may be inferred from the data of Beck and von Meyenburg (1968) and Kappeli *et al.* (1985).

Based on the data obtained by Rieger *et al.* (1983) in continuous cultures of *S. cerevisiae*, we performed a quantitative analysis at the physiological level to find out how the rates of ethanol production and biomass synthesis are regulated. A main finding was that the specific fluxes of glucose and oxygen consumption (q_{glc}, q_{O_2}) and the specific fluxes of carbon dioxide, ethanol and biomass production (q_{CO_2}, q_{EtOH} and μ) were independent of the initial concentration of glucose in the chemostat (30, 10 and 5 g per litre of glucose) (Table 7.1). In addition, q_{glc} was observed to vary non-proportionally with the rate of glucose input into the chemostat. Since q_{CO_2} and q_{EtOH} were found to be proportional to q_{glc} we chose it as a 'parameter' to which the microoganism appears to adjust its catabolic fluxes instead of the glucose input into the chemostat, set by the experimenter. In fact in chemostat cultures the rate of substrate input into the vessel as well as the growth rate of the microorganism ($D = \mu$) are parameters fixed by the experimenter; thus it is trivial to infer that μ control resides in the rate of substrate provision by the pump.

Table 7.1 shows the existing proportionality between the specific flux of oxygen consumption, q_{O_2}, and μ which was noticed only during the respiratory mode of glucose breakdown. At $D > D_c$, q_{O_2} was constant, whereas an analysis of the absolute flux of oxygen (i.e. mmol O_2 h^{-1}) revealed that the constancy of q_{O_2} was the result of a correlative decrease in the oxygen uptake rate and biomass production. This was in contrast with the specific glucose uptake rate (q_{glc}) whose increase was due to a constant glucose uptake rate and a decreased rate of biomass synthesis. If the oxygen uptake decreases with μ while that of glucose is kept constant, then biomass synthesis is regulated by the oxygen uptake and ethanol production by the glucose uptake.

Table 7.1 Physiological behaviour of carbon limited chemostat cultures of *Saccharomyces cerevisiae* (reproduced from Cortassa and Aon, 1994b, by permission of Academic Press)

Inlet glucose (gl^{-1})	*D* (h^{-1})	*Glc$_{IN}$* $(mmolh^{-1})$	*Biomass* (gl^{-1})	q_{glc} $(mmolh^{-1}g^{-1})$	q_{O_2} $(mmolh^{-1}g^{-1})$	q_{CO_2} $(mmolh^{-1}g^{-1})$	q_{EtOH} $(mmolh^{-1}g^{-1})$
30	0.2	33.33	14.2	2.34	5.7	6.1	–
30	0.29	48.33	14.2	3.40	8.0	8.5	–
30	0.32	53.33	8.5	6.11	8.1	14.0	4.99
30	0.36	60.00	5.7	9.54	8.0	21.2	12.36
30	0.41	68.33	3.4	14.73	8.0	30.0	26.47
10	0.2	11.11	4.7	2.36	6.0	6.4	–
10	0.29	16.11	4.9	3.28	7.8	8.4	–
10	0.32	17.78	2.7	6.12	8.1	13.6	5.15
10	0.36	20.00	1.6	9.88	8.0	21.1	12.23
10	0.41	22.78	1.0	13.67	7.8	28.9	19.61
5	0.2	5.56	2.4	2.31	6.0	6.4	–
5	0.29	8.06	2.4	3.35	8.0	8.6	–
5	0.32	8.89	1.3	6.02	8.2	14.0	5.35
5	0.36	10.00	0.8	9.25	8.2	20.2	10.76
5	0.41	11.39	0.3	15.19	7.5	26.1	20.80

7.2.2 MODELLING AND METABOLIC CONTROL ANALYSIS (MCA) OF THE STEADY STATE BEHAVIOUR OF GLYCOLYSIS, THE ETHANOL PRODUCTION BRANCH AND THE TCA CYCLE: CARBON, NITROGEN OR PHOSPHATE LIMITATIONS

The steady states achieved by *S. cerevisiae* in chemostat cultures were simulated in a model comprising 10 ODEs. The set of ODEs described the temporal evolution (dynamics) of 10 metabolites of glycolysis as a function of the individual rate laws determined from the enzyme kinetics *in vitro*. The V_{max} in each rate equation was considered an adjustable parameter (Cortassa and Aon, 1994a). Three criteria were chosen to perform a parameter optimization:

- the system of differential equations should attain a steady state;
- the flux through glycolysis should be equal to the experimentally measured flux of glucose consumption;
- the metabolite levels should meet the experimentally determined steady state concentrations.

The equations and parameters used to perform this calculation have been presented in Cortassa and Aon (1994a). The following section shows selected results obtained from applying the matrix method of MCA (Sauro *et al.*, 1987; Westerhoff and Kell, 1987; Appendix 7), which was used to elucidate the control of glycolysis in *S. cerevisiae* based on experimental data of continuous culture obtained by Franco *et al.* (1984) (Cortassa and Aon, 1993b, 1994a).

7.2.3 CONTROL OF THE GLYCOLYTIC FLUX AND ETHANOL PRODUCTION BRANCH IN CHEMOSTAT CULTURES OF *S. CEREVISIAE*

Two different models were tried with glucose uptake as a main difference (Galazzo and Bailey, 1990; Cortassa and Aon, 1994a). Glycolysis flux control was shared mainly by glucose uptake (flux control coefficient = 1.01 to 1.043) and ATPase (flux control coefficient = –0.0089 to –0.0306) steps, in the model with substrate uptake dependent on ATP and extracellular glucose (Cortassa and Aon, 1994a). Additionally, TCA and ADH controlled the flux with values ranging within 0.0044 to 0.0144 and –0.0069 to –0.0144, respectively (Figure 7.1). The negative control coefficients exerted by ATPase and ADH are consistent with a depletion of cellular ATP; the ADH control was through competition with the ATP-producing branch of TCA. The decrease of intracellular ATP would slow down the glycolytic flux because of the ATP dependence of sugar transport kinetics. The magnitude of the flux control coefficients increased at the critical dilution rate in continuous cultures limited by carbon, nitrogen or phosphate, though the absolute values of the coefficients were larger under phosphate limitation at all dilution rates analysed (Figure 7.1). Despite quantitative differences in the values of control coefficients, the same reaction steps controlled the flux independently of the type of metabolic glucose breakdown, i.e. respiratory (at D < 0.16 h^{-1}) or respiro-fermentative (D > 0.16 h^{-1}) (Figures 7.1 and 8.8).

The control over the ratio of fluxes at the branch point to the TCA cycle and alcoholic fermentation pathways was shared by all four flux rate-controlling steps (Cortassa and Aon, 1994b). Irrespective of the nutrient limitation, a significant increase in the amount of control exerted on the flux ratio by the branch to ethanol formation happened as the dilution rate increased, which is consistent with the triggering of respiro-fermentative metabolism (Franco *et al.*, 1984; Kapelli, 1986).

7.2.4 WHAT CONTROLS GLYCOLYSIS FLUX *IN VIVO*?

The MCA of metabolic studies in chemostat cultures of *S. cerevisiae* suggested that sugar uptake has a chief role in the steady state regulation of the glycolytic flux and the flux ratio at the branch towards ethanol production and the TCA cycle (Cortassa and Aon, 1994a). This

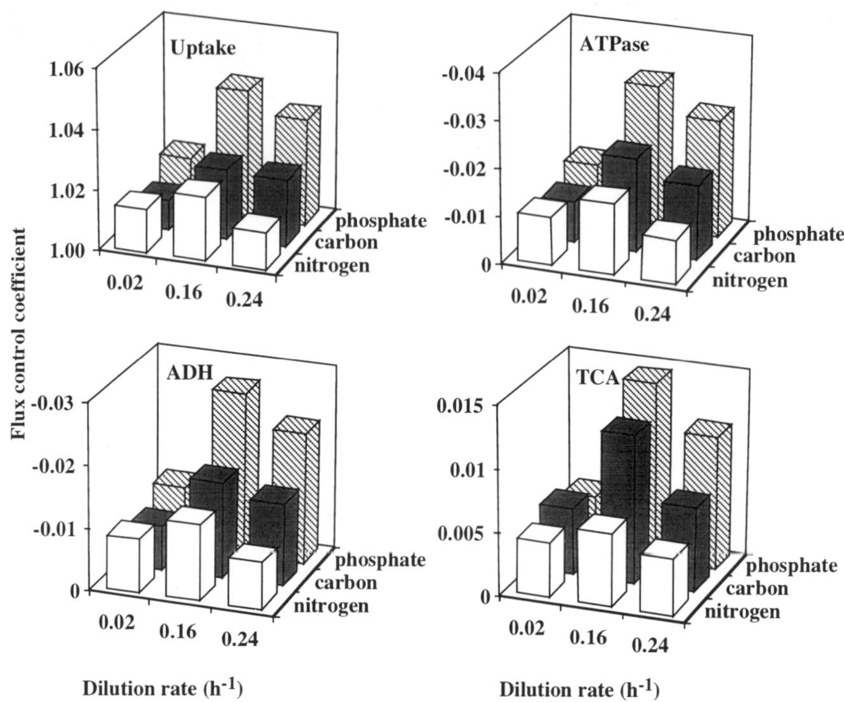

Figure 7.1 Flux control coefficients of the glycolytic pathway for chemostats run under carbon, nitrogen or phosphate limitations. The flux control coefficients were obtained by inversion of the matrix of elasticity coefficients (equation 7.21 in Appendix 7B) based on model I, as described in Cortassa and Aon (1994a). (From *Enzyme and Microbial Technology*, **16**, 761–770, reprinted with permission.)

suggests a close link between uptake and the flux to ethanol production. Glucose transport had been suggested as one of the possible controlling steps of glycolytic flux in *S. cerevisiae* (Gancedo and Serrano, 1989).

ATPase, TCA cycle and ADH-catalysed steps were the others sharing the control of the flux, when the sugar uptake kinetics explicitly depends on intracellular ATP concentration. This picture did not change either for carbon, nitrogen or phosphate limited continuous cultures of *S. cerevisiae*. In the steady states analysed by Franco *et al.* (1984) and our calculations, the rate-controlling steps of the flux also controlled the branch flux ratio at the level of pyruvate in the glycolytic pathway. These results appear to be in agreement with those theoretically and experimentally reported by Barford (1990) for several yeasts in the sense of the main role played by the cell's rate of sugar uptake. Specifically, the model used by Barford to explain the experimental results obtained either in batch or in continuous cultures proposes that, given fixed rates of transport of sugar into the cell and transport of respiratory intermediates into mitochondria, if the former is larger than the latter, a saturated specific oxygen rate profile is observed together with ethanol excretion; if both are of the same order then a derepressed specific oxygen uptake rate appears and little or no ethanol excretion (Barford, 1990).

At present, the steady state control of glycolysis *in vivo* is open to four possibilities: sugar uptake; allosteric regulation; glycolytic enzymes supercomplex; or distributed control over many enzymes in the pathway.

An analysis of transcription and translation of glycolytic enzymes at different dilution rates in glucose-limited continuous cultures of *S. cerevisiae* revealed that glycolysis flux is not regulated at transcriptional and translational levels according to measurements of mRNA levels and enzyme activities *in vitro* (phosphoglucose isomerase, phospho-fructokinase, HK I or II) (Sierkstra *et al.*, 1992). These authors concluded that the glycolytic flux is regulated mainly by allosteric interactions when the growth rate is increased. A wider interpretation could be that flux control is realized by the action of effectors (either allosteric or not) that modify enzyme activities in one of the four possibilites discussed above.

Several genes coding for glycolytic enzymes have been cloned into plasmid vectors and overexpressed in *S. cerevisiae* (Heinisch, 1986; Schaaf *et al.*, 1989). Their effect has been mainly investigated by monitoring ethanol production with respect to the wild type strain in batch cultures. The results obtained show that there are no significant differences between the transformants that overexpressed several or individual glycolytic enzymes and wild type strains as could be judged from ethanol production in batch cultures (Schaaf *et al.*, 1989). These results have been interpreted as being due to a distributed control over many enzymes rather than a 'rate-limiting step'. The growth rates described in Schaaf *et al.* (1989) are on average 0.31 h^{-1} which, in chemostats under carbon limitation, corresponded to flux control by the uptake counterbalanced by the ATP-consuming and producing processes (Cortassa and Aon, 1993b, 1994a). Therefore, we suggest a rather different interpretation from that offered by Schaaf *et al.* (1989): if glucose uptake is one of the main rate-controlling steps of the glycolytic flux of *S. cerevisiae* growing in batch as well as in chemostat cultures, then this explains the unsuccessful attempts to increase the ethanol fermentation flux in batch cultures by overexpressing enzymes of the glycolytic pathway further down the glucose uptake system at least under carbon limitation. Our main result allows the prediction that changing the nutrient limitation will not change the control. Cloning and up-modulation of genes involved in sugar uptake would confirm this conclusion.

We realize that, according to the third possibility posed by us, the results of Schaaf *et al.* (1989) may be differently interpreted if one assumes that glycolytic enzymes are part of a supercomplex. Under those conditions, up-modulated glycolytic enzymes would not integrate to the enzymatic complex, leaving the flux unchanged. The possibility of glycolytic enzymes functioning as a macromolecular complex could be an alternative explanation of the observed failure to change the flux by cloning of glycolytic enzymes in multicopy vectors and transformation of yeast. Under those conditions, either the substrate catabolized would be preferentially channelled through the glycolytic complex or the overexpressed enzymes would not integrate to that complex (Westerhoff *et al.*, 1991). Under both conditions the flux will remain unchanged.

Niederberger *et al.* (1992) showed that the flux to tryptophan could only be increased by the simultaneous overexpression of five enzymes ($\sim$24-fold with respect to the wild type) related to the synthetic pathway of that amino acid. The results were interpreted as an indication of distributed control among several steps, i.e. each enzymatic step sharing a small control over the flux, as in Schaaf *et al.* (1989).

Overall, at least with the glycolytic flux control in *S. cerevisiae*, the question remains open as to whether the overexpression of the glucose transporter will give an increase in the flux. Provisionally, it raises another explanation for the lack of response in the flux when up-modulation in several yeast glycolytic enzymes is performed.

7.3 REGULATION OF TRANSIENT AND STEADY STATE BEHAVIOUR

7.3.1 REGULATION OF GLYCOLYSIS DYNAMICS *IN VIVO* THROUGH INTERACTIVE COUPLING BETWEEN CYTOPLASMIC AND MITOCHONDRIAL COMPARTMENTS

Glycolysis and oxidative phosphorylation occur in two different subcellular compartments: cytosolic and mitochondrial, respectively. In recent years it has become apparent that, *in vivo*, the interrelationships between those compartments may be diverse, dynamic and with multiple sites of regulation (Aon *et al.*, 1991; Jones *et al.*, 1992). More precisely, the regulation of the glycolytic flux may be dynamically achieved through cytoplasmic ATP and NADH levels, which are in turn regulated by mitochondrial functions. Increased mitochondrial respiration and ATP synthesis are obtained in many tissues when ATP cellular demands increase (Brown, 1992, and references therein). Under conditions closer to thermodynamic equilibrium, respiration and ATP synthesis are mainly regulated by the phosphorylation potential and the NADH/NAD$^+$ ratio. Farther from equilibrium, other steps share the control such as the respiratory chain, the proton leak, the ATP synthase and the adenine nucleotide carrier (Brown, 1992).

We have already argued about the tight dependence of the branch toward alcoholic fermentation and the TCA cycle on glucose uptake in *S. cerevisiae*. In this section other aspects of the multidimensional interactions between glycolysis and mitochondrial functions will be stressed.

(a) Interactions between glycolysis and oxidative phosphorylation in yeast and tumour cells

As a result of the introduction of new non-invasive technologies (NMR, confocal scanning microscopy) a richer and more complex picture of the *in vivo* mutual regulation between glycolysis and oxidative phosphorylation is emerging. Particularly interesting is the potential ability of several steps to regulate glycolysis dynamics *in vivo* in concert with several mitochondrial functions (Aon *et al.*, 1991). We do not know yet how the cell 'mixes' those controls at the cellular level.

The rate of glucose utilization by yeast cells depends on whether they are glucose-repressed or derepressed. For glucose-repressed yeast cells, the rate of glucose utilization does not depend on oxygen levels, i.e. it is the same aerobically and anaerobically. However, in derepressed cells under aerobic conditions the rate of glucose utilization is about two-fold lower than anaerobically (Den Hollander *et al.*, 1986). At present, the latter is interpreted as depending upon the higher levels of expression of respiratory and gluconeogenic with respect to glycolytic enzymes under derepressed conditions (Gancedo, 1992).

A long time ago Lynen (1963) observed that the decreased uptake of glucose under aerobic conditions must be in some way related to oxidative phosphorylation). This was suggested by experiments with the uncoupler 2,4-dinitrophenol (DNP). It was shown by Lynen that aerobic glucose consumption in the presence of DNP increased to the values observed under anaerobic conditions. After the use of cyanide to block respiration of a vigorously aerated suspension of yeast cells in sugar solution, glucose-6-phosphate (G6P) changed in parallel with ATP, i.e. dropped rapidly after respiration had been blocked (Lynen, 1963). Lynen interpreted this observation as that, in yeast, mitochondrial ATP can be used for the initial phosphorylation of sugar. This finding has been confirmed by other authors (Bustamante and Pedersen, 1977; Pedersen, 1978; Arora and Pedersen, 1988). Unlike the evolution of G6P, fructose-1,6-bisphosphate (FBP) did not change immediately after cyanide poisoning. DNP initiated the same series of changes as cyanide in hexose phosphates, inorganic phosphate and pyruvate. Following DNP treatment, G6P dropped whereas FBP was first maintained at a constant level or sometimes even decreased a little and then rose. Lynen further suggested that this would mean that mitochondrial ATP can phosphorylate glucose by HK but has little or no action on further phosphorylation of hexose phosphate to form FBP. He concluded that probably the two enzymes were located at different sites in the cell.

Since the studies of Warburg (1930), aerobic glycolysis has been directly implicated as a main catabolic route of tumour cells (Dills, 1993). Under glutamine limiting conditions, energy production in normal proliferating and tumour cells mainly relies on glycolysis (Eigenbrodt *et al.*, 1985). Warburg's hypothesis suggested that cancer cells may have an impaired respiratory capacity resulting in elevated rates of glycolysis (quoted in Pedersen, 1978). In fact, most tumour cells have higher rates of both anaerobic and aerobic glycolysis than normal cells and it seems clear that transformed cells usually derive an appreciable amount of energy from glycolysis, even in the presence of oxygen (Gillies, 1981). Early observations of Crabtree (1929) pointed in a similar direction concerning the partial inhibition of respiration in fast-growing tumour cells by physiological concentrations of glucose. The Pasteur effect predicts that glycolysis would be inhibited in actively respiring cells. A collection of data presented by Gillies (1981) showed that tumour cells exhibit a normal Pasteur effect when measured in absolute terms. On a percentage basis, glycolysis is much less inhibited by respiration in tumour cells than in normal cells.

An elevation of the cytoplasmic concentration of free Ca^{2+} may be involved in the switch of cellular catabolism from respiratory to glycolytic. It has been shown that Ehrlich ascites mitochondria are able to accumulate large quantities of Ca^{2+} in an energy-dependent way and that this accumulation results in an impairment of several energy coupling processes (Teplova *et al.*, 1993; Evtodienko *et al.*, 1994). Apparently, the mitochondrial accumulation of Ca^{2+} results in: suppression of respiration in state III but not in state IV, nor under uncoupled conditions; inhibition of H^+ production following ADP addition; and prevention of $\Delta\psi$ decrease in the presence of ADP or glucose (Evtodienko *et al.*, 1994). This is in sharp contrast to the effect of Ca^{2+} on coupled respiration in most normal, non-neoplastic, tissues.

Overall metabolism in hepatomas revealed an imbalance in the reciprocal regulation of the enzymatic activities involved in synthetic and catabolic pathways. Concerning carbohydrate metabolism, increased activities of glycolytic enzymes (e.g. HK, PFK, PK) and pentose phosphate production, together with a decrease in gluconeogenic ones, have been observed (Weber, 1983; Dills, 1993) (Chapter 9). Rapidly growing tumours exhibit a capacity to utilize more glucose than do slowly growing ones and many normal tissues. It is believed that imbalances in pyrimidine and purine as well as carbohydrate metabolisms confer selective reproductive advantage to cancer cells (Weber, 1983). Highly glycolytic tumours *in vitro* rely more on glycolysis for ATP than do the slow and intermediate growth-rate hepatomas. In highly glycolytic tumours, as much as 60% of the cells' ATP supply may be provided by glycolysis. The slowly growing and intermediate growth rate hepatomas most likely obtain the bulk of their ATP from mitochondrial reactions, whereas rapidly growing tumours obtain significant amounts from both sources, mitochondrial oxidative phosphorylation being the major source of ATP (Pedersen, 1978). It has been shown that mitochondrial-bound HK in highly glycolytic tumours (AS-30D hepatoma cells) has preferred access to mitochondrially-generated ATP (Arora and Pedersen, 1988). It is likely that the source of ATP for bound HK is derived preferentially from the ATP synthase residing within the inner mitochondrial membrane compartment rather than from cytosol. The next section will analyse quantitatively the consequences of HK ambiquity on glycolytic dynamics.

Alternatively, tumour cells are able to consume glutamine as one of the main substrates for energy generation in such cells (Matsuno, 1987; Baggetto, 1992). Glutamine metabolism is strongly linked to mitochondrial energetics since enzymes related to the processing of this amino acid are located in mitochondria (Matsuno, 1987; Hugo *et al.*, 1992). Reduction of glutamine concentrations at saturating glucose concentrations further enhances the aerobic glycolytic rate (Eigenbrodt *et al.*, 1985). Under some conditions, the addition of glutamine with glucose stimulates lactate production from glucose. Ehrlich ascites tumour cells oxidize glutamine to CO_2 at a more elevated rate than any other amino acid, while it has also been shown that glutaminase activity of different rat hepatomas is linearly correlated with the growth rate and the degree of malignancy of tumour cells (Baggetto, 1992).

(b) Regulatory effects of mitochondrial function on glycolytic dynamics suggested by a simulation study

An important role is played by adenine and nicotinamide nucleotide levels in the dynamic behaviour of glycolysis. During transitions from aerobic to oxygen-limited or anaerobic conditions, the balance of both redox (NAD(P)H/NAD(P)) and phosphorylation (ATP/ADP) couples is very important in the *in vivo* regulation of glycolysis dynamics (Aon *et al.*, 1991; Verdoni *et al.*, 1992). Beauvoit *et al.* (1993) investigated to what extent mitochondrial functions, such as those related to the regulation of cytosolic ATP and NADH levels, in turn regulate glycolysis dynamics *in vivo*. From studies on resting yeast cells grown aerobically on lactate, the highest ATP/ADP ratios and NADH levels were obtained in the presence of ethanol alone, followed by those in the presence of glucose + ethanol and glucose alone, respectively. Beauvoit *et al.* (1993) concluded that the response of the respiratory rate to the ATP/ADP ratio was dependent on the redox potential.

The use of the drug almitrine in rat hepatocytes decreased the ATP and increased the ADP concentration resulting in lower ATP/ADP ratios in both cytosolic and mitochondrial compartments (Leverve *et al.*, 1994). The glycolytic flux was stimulated while the gluconeogenic flux was inhibited by the drug when hepatocytes consumed glycerol as carbon substrate. In turn, almitrine did not affect the respiratory activity of either isolated mitochondria or hepatocytes. Leverve *et al.* (1994) showed that stimulation of pyruvate kinase was probably mediated by the sensitivity of this enzyme to the ATP/ADP ratio. These results further emphasize the importance of the phosphorylation potential balance in the dynamic interaction between mitochondrial, glycolytic and gluconeogenic pathways.

A model introduced by Sel'kov (1975) and further developed by Cortassa *et al.* (1990b) and Aon *et al.* (1991), including interactions with mitochondrial activity, was used. It was first suggested and demonstrated that built-in ATP stoichiometry of glycolysis in the model is the autocatalytic mechanism source of, for example, oscillatory dynamics (Chapter 3). This autocatalytic mechanism was able to sustain sophisticated dynamics such as sinusoidal, spike and square-type oscillations along with complex oscillations exhibiting period doubling (Figure 7.2). Most of this periodic behaviour could be experimentally corroborated in intact yeast cells (Figures 7.3 and 3.6). In addition, oscillations reported in the literature for experiments with yeast extracts could be simulated (Cortassa *et al.*, 1990a and unpublished). These results argued in favor of stoichiometric mechanisms also being able to describe the *in vivo* regulation of glycolysis dynamics. Moreover, the built-in stoichiometrically based autocatalytic mechanisms of metabolic pathways could be the source of regulation of even sophisticated dynamics. Complex dynamics has already been shown for the allosteric properties of PFK finely modulated by several activators and inhibitors (Boiteux *et al.*, 1975; Sols, 1981).

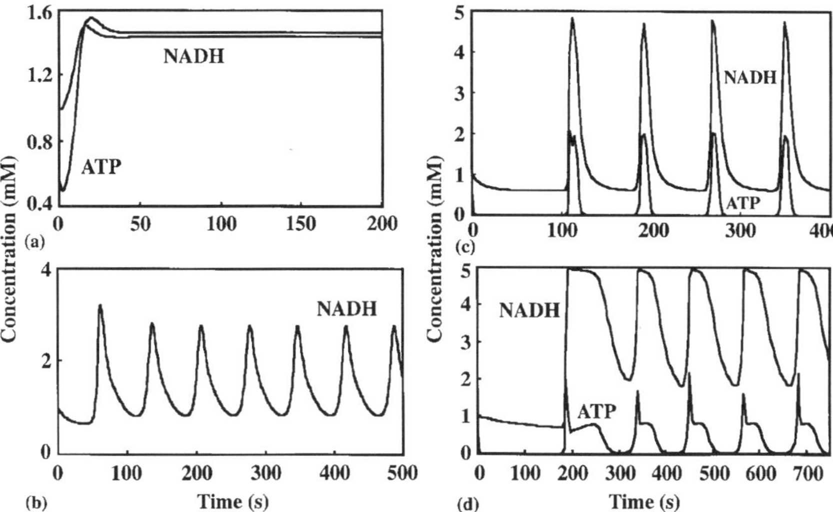

Figure 7.2 Complex dynamic behaviour shown by a purely stoichiometric model of glycolysis: model simulations of the temporal evolution of NADH and ATP for variable rates of substrate phosphorylation and NADH reoxidation. Simulations of the model described in Aon *et al.* (1991) were carried out with the following parameter values: V_{in} (mM s^{-1}) = 0.12 (a,c,d), 0.15 (b); k_1 (mM^{-1} s^{-1}) = 0.3 (a), 0.05 (b), 0.1 (c), 0.03 (d); k_3 (mM^{-3} s^{-1}) = 0.1 (a,b,d), 0.3 (c); k_5 (s^{-1}) = 0.1 (a,b,d), 1.0 (c); k_7 (mM^{-1} s^{-1}) = 0.1 (a,b,c), 0.01 (d); k_9 (s^{-1}) = 0.05 (a,b,d), 0.3 (c); τ = 0.83; d_p = 0.941 (a), 1.771 (b), 4.826 (c), 2.986 (d). The kinetic parameters of the proton pump were: K_M = 2 mM; V_M = 0.5 mM s^{-1}. The total nucleotide (C_A) and phosphate pool (P_T) were both 10 mM while the pyridine nucleotide pool (C_N) was 5 mM. The initial values (mM) of the state variables were: S_1 = 0.157; ATP = 0.552; I = 0.0031; Pyr = 0.394; NADH = 1.0. The model was numerically integrated on a PC-60 III Commodore using the program SCoP with an Adams method (Duke University, 1989). The analysis of the stability and bifurcation properties of the ODEs system described in Aon *et al.* (1991) was performed with AUTO (E. Doedle, 1986, Concordia University, Canada). (Reproduced from Cortassa and Aon, 1994b, by permission of Academic Press.)

Theoretical predictions derived from the model were subjected to experimental verification and vice versa, i.e. experimentally observed behaviour was simulated by the model. Stability analysis allowed the prediction that either an increase in the work load represented by mitochondria towards the glycolytic ATP synthesizing machinery, or a decrease or increase in the influx of glucose into the system, could enhance the tendency of glycolysis to oscillate. The former prediction suggests that elimination of mitochondrial oxidative phosphorylation, so that the entire work load of producing the ATP needed for intracellular

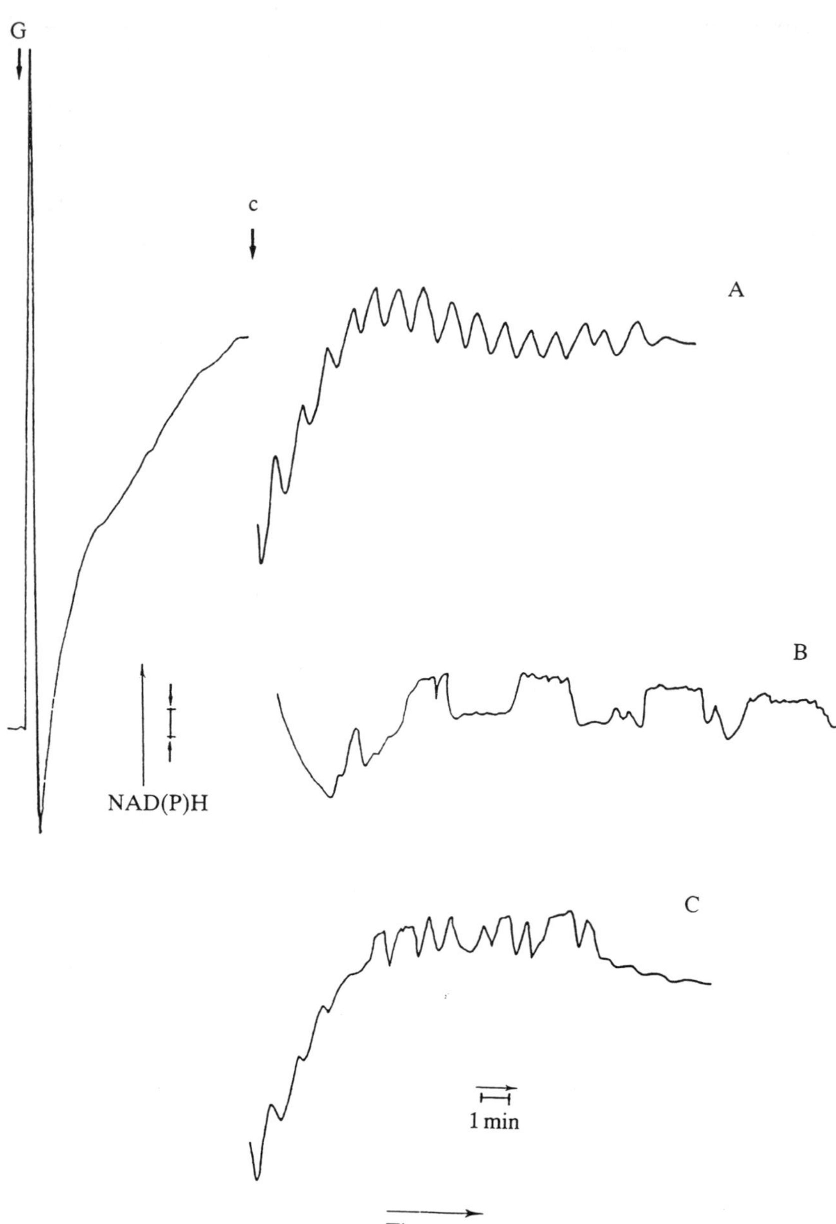
G
c
A
B
C
NAD(P)H
1 min
Time

free-energy transduction befalls glycolysis, will cause oscillatory behaviour. This should be noticeable at moderate work loads as relaxation towards a steady state through damped oscillations. At extreme work loads, this might lead to sustained oscillations (Cortassa *et al.*, 1991).

The activity of NADH reoxidizing processes could also be affecting the dynamics of glycolysis. The plausibility of model predictions was investigated by affecting in different ways the cytosolic levels of ATP and NADH and looking at the appearance (approximating a Hopf bifurcation) or disappearance (departing from a Hopf bifurcation) of glycolytic oscillations (Aon *et al.*, 1991).

At present it cannot be excluded that the effects of ATP/ADP and cellular redox states may run in part through allostericity of PFK and in part through the autocatalytic nature of the pathway.

(c) Mitochondrial processes involved in the decrease of the levels of cytosolic ATP that affects the dynamics of glycolysis

The glycolytic flux decreases as the ATP/ADP ratio increases. Glucose addition on aerobiosis sustains a low phosphate potential but with high ATP turnover (Den Hollander *et al.*, 1986; Beauvoit *et al.*, 1993). A drain of cytosolic ATP mediated by mitochondrial functions is probably a trigger of qualitative changes in glycolysis dynamics (Aon *et al.*, 1991). Ethanol as a substrate induces higher phosphate and redox potentials than glucose (Cortassa *et al.*, 1995). Accordingly, resting *S. cerevisiae* cells grown under derepressed conditions induced a low glycolytic flux together with a large respiratory flux when supplied with glucose + ethanol (Beauvoit *et al.*, 1993).

An important observation in the pioneering studies of Chance *et al.* (1964) was that inhibition of oxidative phosphorylation triggered yeast glycolytic oscillations. When cells were impaired in mitochondrial

Figure 7.3 (see facing page) Induction of square-type cytoplasmic NAD(P)H glycolytic oscillations by progressively inhibiting alcohol dehydrogenase with pyrazole. A suspension of yeast cells (10^7 cells ml^{-1}) grown, harvested and starved as described in Aon *et al.* (1991) was preincubated for 5 min in the absence (upper trace) or in the presence of pyrazole, 0.1 mM (middle trace) or 0.01 mM (lower trace), and pulsed with 20 mM glucose in the cuvette of a fluorimeter at 25°C with stirring followed by 10 mM KCN after 3–5 min. The fluorescence scale is 1.5 µM NAD(P)H. (Reproduced from Aon *et al.*, 1991, by permission of The Company of Biologists Ltd.)

respiration, by either respiratory inhibitors or mutants in the bc1 complex (ubiquinol:cytochrome c reductase) (Berden *et al.*, 1988) a pulse of glucose did give rise to glycolytic oscillations (Aon *et al.*, 1991). Experiments performed in the presence of bongkrekic acid, an inhibitor of the adenine nucleotide translocator in living cells (Lumbach *et al.*, 1970; Gbelska *et al.*, 1983), affected the oscillatory mechanism. This effect was apparently exerted through the control of the cytoplasmic–mitochondrial ATP/ADP exchange under conditions in which yeast cell respiration remained the same despite preincubation with BKA. Additional results obtained in the presence of FCCP, or FCCP + BKA, indicated the existence of at least two regulatory levels of cytoplasmic ATP pool by mitochondria: the proton motive force ($\Delta\mu_{H^+}$) and the adenine nucleotide translocator (Aon *et al.*, 1991). Overall, the results obtained strongly suggested that glycolysis dynamics in starved yeast cells operates very close to a Hopf bifurcation, which is approached when the cytosolic ATP decreases.

Glucose transport or phosphorylation and mitochondrial functions apparently interact through a regulatory feedback in yeast cells (Aon *et al.*, 1990b). This regulatory feedback may operate at the level of HK ambiquity (section 7.2.2). In yeast, a tight coupling was shown to exist between glucose transport and phosphorylation (Fiechter *et al.*, 1981) as well as glycolysis dynamics. It is likely that the decrease in cytosolic ATP levels triggers oscillations by a decrease in the rate of substrate phosphorylation at sufficiently low rates of substrate uptake (Cortassa and Aon, 1994b). Experimental evidence in favour of this possibility was obtained by studying the oscillatory behaviour along the growth curve of *S. cerevisiae* strain X2180 in batch culture. The oscillatory behaviour persisted throughout the exponential phase of growth. With starved cells collected at equal cell densities in the exponential phase of growth, it could be verified that, inside the range of glucose uptake corresponding to the oscillatory domain, higher glucose uptakes were coincident with higher amplitudes in the oscillations (Cortassa and Aon, 1994b). It has been shown that undamped oscillations of intracellular NAD(P)H exist close to the shift from glucose to ethanol consumption in batch cultures of *S. cerevisiae*; the oscillations died when glucose was exhausted from the medium (Richard *et al.*, 1993).

(d) Cytoplasmic NADH reoxidation also affects glycolytic dynamics

A correlation between a continuous reduction of the NADH pool and de-energization of mitochondrial membranes was provided by the use of mutants in the bc1 complex and uncoupling agents of mitochondrial activity, such as FCCP. Two experimental observations suggested that de-energization of mitochondria correlates with low rates of NADH reoxidation:

- a low glycolytic flux and a tendency of respiratory mutants to increase their intracellular pool of NAD(P)H after successive pulses of glucose;
- a similar pattern of behaviour shown by the wild type strain when subjected to a pulse of glucose after preincubation with FCCP (Aon *et al.*, 1991).

When mitochondrial membranes were uncoupled, KCN-induced oscillations quickly damped or exhibited the tendency to disappear as shown by FCCP treatment and respiratory mutants. These results further suggested that the rate of NADH reoxidation by mitochondrial dehydrogenases may be proton motive force ($\Delta\mu_{H^+}$) regulated. When glycolysis dynamics approached the oscillatory region at high ATP load, i.e. high cytoplasmic drain of ATP, the rate of NADH reoxidation was able to trigger oscillations. Transient changes in the rate of NADH reoxidation by mitochondria triggered oscillations in model simulations (Aon and Cortassa, unpublished data).

At high ATP drains, the rate of cytoplasmic NADH reoxidation by mitochondrial dehydrogenase (Bruinenberg *et al.*, 1985; Alexander and Jeffries, 1990) and alcohol dehydrogenases provides another regulatory level of the dynamics of glycolysis (Aon *et al.*, 1991). This additional regulatory level influences not only the approach of glycolysis to the oscillatory region but also, once inside, it further regulates the shape of the oscillations (sinusoidal to square type, spikes: Figure 7.2) without affecting respiration.

(e) Mitochondrial subcellular organization and its possible role in the regulation of glycolysis and oxidative phosphorylation dynamics *in vivo*

Mitochondrial DNA staining of a bc1 complex mutant revealed that mitochondria were differently organized at the subcellular level, resembling 'aggregated' structures rather than a network or 'mitochondrion'. Interestingly, the same mutant exhibited uncoupled mitochondria (Aon *et al.*, 1991; Aon, unpublished data). These observations suggested a link between the magnitude of the proton motive force of mitochondria and their subcellular organization (Skulachev, 1990). In starved cells of a wild type yeast strain metabolizing glucose, the mitochondrial network exhibited dynamic changes in organization as shown by video image analysis (Figure 7.4). The temporal evolution of the 'bleaching' of the fluorescent probe at the mitochondrial membrane (Figure 7.4) may be due to local de-energization of the mitochondrial membrane. Apparently, the 'bleaching' effect was not due to rupture of mitochondrial integrity, as could be verified by electron microscopy and double labelling of yeast cells with dyes (DAPI, DASPMI: Bereiter-Hahn, 1976; Bereiter-Hahn *et al.*, 1983) staining nucleic acids and mitochondrial membranes (Aon, unpublished data).

Mitochondrial morphological reorganization *in vivo*, as judged from fluorescence quantitation of video films of yeast cells, was shown to be sensitive to genetic background as well as inhibitors which affect the respiratory electron transport (KCN), the mitochondrial electrochemical

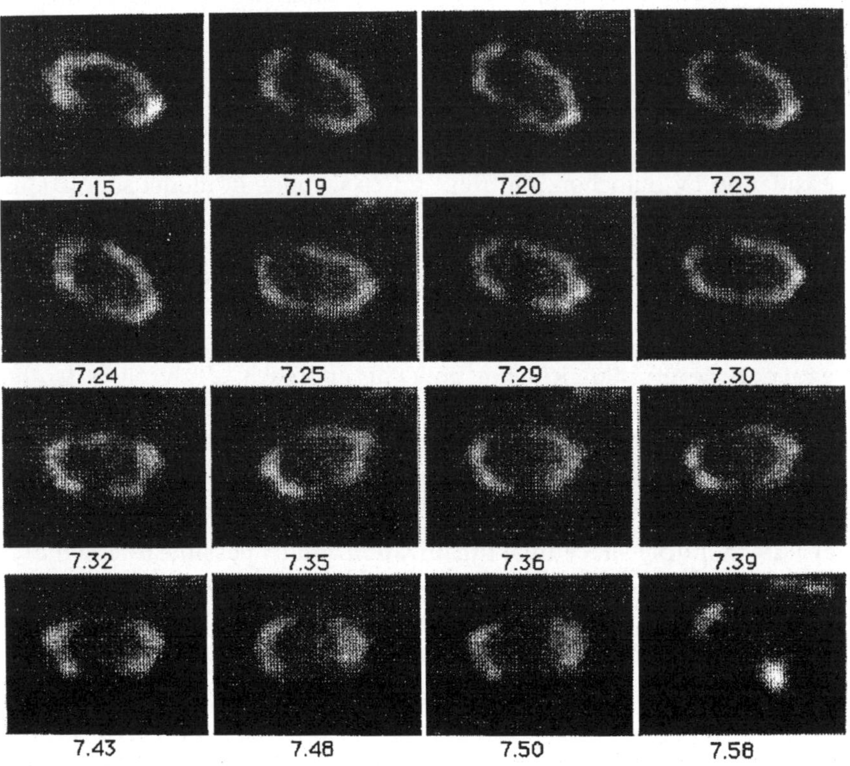

Figure 7.4 Video film (excitation at 340 nm, emission at 450 nm) of a yeast cell metabolizing glucose (20mM) of the strain X2180 of *Saccharomyces cerevisiae* grown, harvested, starved and stained with the vital dye DASPMI (as described in Aon *et al.*, 1992). Pictures were taken from video films at the intervals indicated below (e.g. 7.15 = 7 min 15 s) after glucose addition and subjected to image processing as described in Aon *et al.* (1991). (Reproduced from Cortassa and Aon, 1994b, by permission of Academic Press.)

gradients (FCCP) or the adenine nucleotide translocator (bongkrekic acid). The genetic background influence was deduced from strains that were either haploid or diploid or mutants *petite* (Genta and Aon, unpublished data).

These local effects could have a wide range of effects on glycolytic dynamics – for example, when cells are subjected to nutritional shortage. Simulation studies suggested that short pulses (10 s) of membrane de-energization, i.e. through lowering of the transmembrane electric potential, triggered glycolytic oscillations (Aon and Cortassa, unpublished data). In the mathematical model presented in section 7.3.2 we were able to describe oscillatory dynamics at low values of the association constant of HK to mitochondria, or at intermediate values of the dissociation constant of the enzyme. This observation raises the possibility that local changes in mitochondrial membrane energization displace glycolytic dynamics into or out of the region of periodic behaviour through the effect on the association–dissociation constants of the enzyme.

The putative link between the dynamics of mitochondrial subcellular organization and its effect(s) on the dynamic behaviour of glycolysis deserves further consideration for the future.

7.3.2 THE ROLE OF HEXOKINASE AMBIQUITY IN SUBCELLULAR DYNAMICS

HK ambiquity appears as an additional level of regulatory interactions taking into account that glycolysis and oxidative phosphorylation happen in two different compartments. From *in vitro* studies it was postulated that HK ambiquity might play an important role in the regulation of the glycolytic flux *in vivo* (Wilson, 1980). The levels of mitochondrial-bound HK in tumour cells displaying high rates of glycolysis greatly exceeded those of normal tissues (Arora and Pedersen, 1988). Not only is HK elevated in many rapidly growing tumours, but also a large fraction of the total enzymatic activity (40–60%) is frequently bound to the mitochondrial fraction. Some workers have presented evidence indicating that mitochondrially bound HK may be involved in limiting (or inhibiting) respiration when Ehrlich ascites tumour cells are utilizing glucose (Crabtree effect). The continuous phosphorylation of glucose by ATP (involving mitochondrially bound HK) may reduce the phosphate level available for oxidative phosphorylation and therefore prevent maximal state III respiration rates from being obtained.

It is well known that HK is much less sensitive to product inhibition by G6P when bound to mitochondria than when free in solution. It has been shown that G6P has a dual role with respect to HK, i.e. it inhibits bound HK to a lesser extent and modulates the binding of the enzyme to mitochondria. These two facts, when combined with the finding that tumour mitochondrial HK has preferred access to mitochondrially generated ATP, provide an explanation for high rates of G6P synthesis (13 mg G6P h^{-1} mg^{-1} mitochondrial protein) and modulation of G6P intracellular level (Arora and Pedersen, 1988). Such rates are unlikely to be achieved in normal cell types where the level of HK, both total and mitochondrial-bound, is much lower than in tumours.

The association of HK isozyme I to mitochondria (70%) has been ascertained quantitatively in astrocytes *in vivo* by confocal microscopy (Lynch *et al.*, 1991). Alteration in HK isozyme I binding to mitochondria in the presence of a glycolysis inhibitor 2-deoxyglucose was taken as evidence for modulation of HK binding to mitochondria by the rate of glycolysis.

Following particular conditions of G6P concentration, pH and ionic strength, HK may be released from the mitochondria into the cytosol. The kinetic properties of bound HK differ from those of the 'free' enzyme. The bound form exhibits a lower K_m for ATP and a decreased susceptibility to inhibition by G6P than does the soluble form. These observations ascertain the role of the 'ambiquity' of HK as a general coupling device for the simultaneous regulation of glycolysis and oxidative phosphorylation in normal as well as in tumour cells. This results in a potentially rich regulatory picture to which the enzyme itself may contribute.

(a) Transient behaviour of glycolysis and oxidative phosphorylation in tumour cells

Gumaa and McLean (1969) reported short-term regulation experiments (3 to 30 s) in Ehrlich ascites tumour cells after glucose addition. Changes in the concentration of intermediates of the glycolytic pathway and of adenine nucleotides were measured following glucose addition. A high and constant rate of ATP utilization was exhibited during the first 15 s, declining sharply after 20 s. In the period from 20 to 40 s, the respiration rate almost doubled, thereafter falling to almost 20% of the maximum rate value attained. Following the addition of glucose, ADP rose and ATP fell, and there was a transition from state IV to III due to increased ADP. The evidence then suggested that during the initial period after glucose addition there is a direct utilization by mitochondria of the ADP generated by the HK reaction. Initially, ATP generated by glycolysis did not make any contribution, as shown by virtually unchanged levels of FBP and lactate in the lapse of 5 s. Gumaa and McLean concluded that during the first 5 s the regeneration of ATP can be almost entirely attributed to the activity of mitochondrial oxidative phosphorylation. At 10–15 s after glucose addition, the relative contribution of the glycolytic pathway and mitochondrial systems were almost equal and there was a burst of lactate production; the mitochondrial contribution to ATP formation remained linear at that time. At 20–30 s after glucose addition the rate of both ATP utilization and ATP regeneration decreased. The overall contribution of the two compartments to ATP formation throughout the 30 s period was equally distributed between glycolysis and oxidative phosphorylation. Gumaa and McLean (1969) further suggested that the relative contribution of glycolysis and oxidative phosphorylation could be regulated by a dynamic association–dissociation of HK from mitochondria.

Three lines of evidence support the HK↔HK* interconversion:

- The phosphorylation of glucose increased linearly during the first 15 s in the presence of inhibitory concentrations of G6P (0.5 mM), suggesting that HK was in the mitochondrially bound form (HK*).
- There was a dramatic triggering of HK inhibition by G6P at 15–20 s, with a small modification of the amounts of intracellular G6P or ADP suggesting that HK was cytoplasmic (HK) and susceptible to inhibition by G6P.
- The half-time release of the enzyme from mitochondria by G6P was of the same order of the onset as inhibition of glucose phosphorylation.

Additionally, a burst of glycolytic ATP production (5–15 s) together with a sigmoidal accumulation of FBP were described (Gumaa and McLean, 1969).

(b) Regulation of the glycolytic flux by hexokinase ambiquity: study of transients and steady state regulation

Taking into account the facts described above, we developed a mathematical model (Figure 7.5) to answer the following questions.

- Can the dynamic compartmentation of HK between mitochondria and cytoplasm account for the transient behaviour of glycolytic metabolites following a glucose pulse?
- Does HK ambiquity influence the control of the glycolytic flux under steady state conditions?

The behaviour of the model was tested for transient and steady states as a function of the constants of association or dissociation of the enzyme to mitochondria.

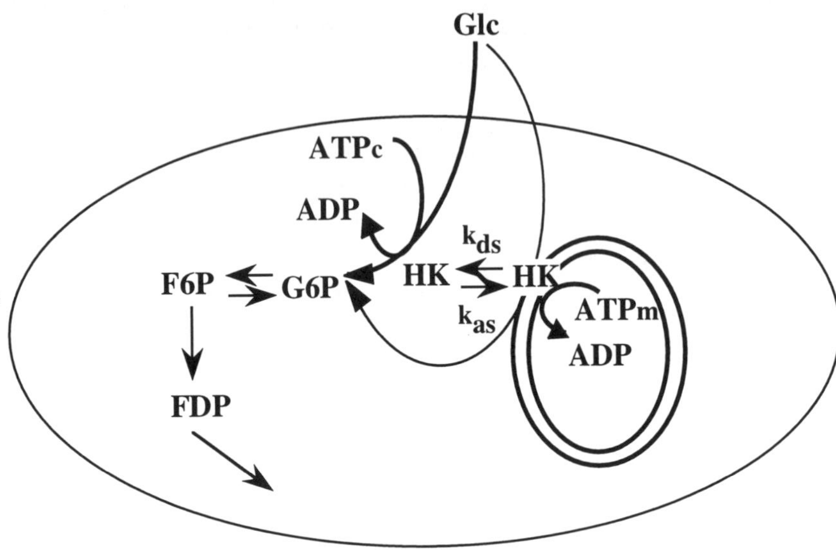

Figure 7.5 Scheme of a glycolytic model which takes into account hexokinase (HK) ambiquity. The main features of this model are that the ATP pool is split into two – one mitochondrial (ATP_m) and the other cytoplasmic (ATP_c) – and that HK coexists under two forms, one bound (to mitochondria) and the other free (or cytoplasmic) whose proportion is governed by the association–dissociation dynamics (equations 7.7, 7.14, 7.15).

(c) A mathematical model of glycolysis that takes into account the dynamic compartmentation of HK

Transient behaviour

The model is described by a set of five nonlinear ODEs (7.3–7.7) and three conservation equations (7.8–7.10) (Appendix 7). It simulates the initial burst in glucose 6-phosphate (G6P) with a concomitant slower increase in FBP during the first 20–100 s. Coincidently with existing experimental evidence, glucose phosphorylation increases linearly and rapidly during the first 15 s after glucose addition (Figure 7.5). The G6P burst could be regulated by increasing the relative amount of mitochondrial-bound HK (HK*), i.e. a higher association constant to mitochondria (K_{as}). The G6P burst is clearly higher when HK is more 'mitochondrial' than 'cytoplasmic' (Figure 7.6a,b).

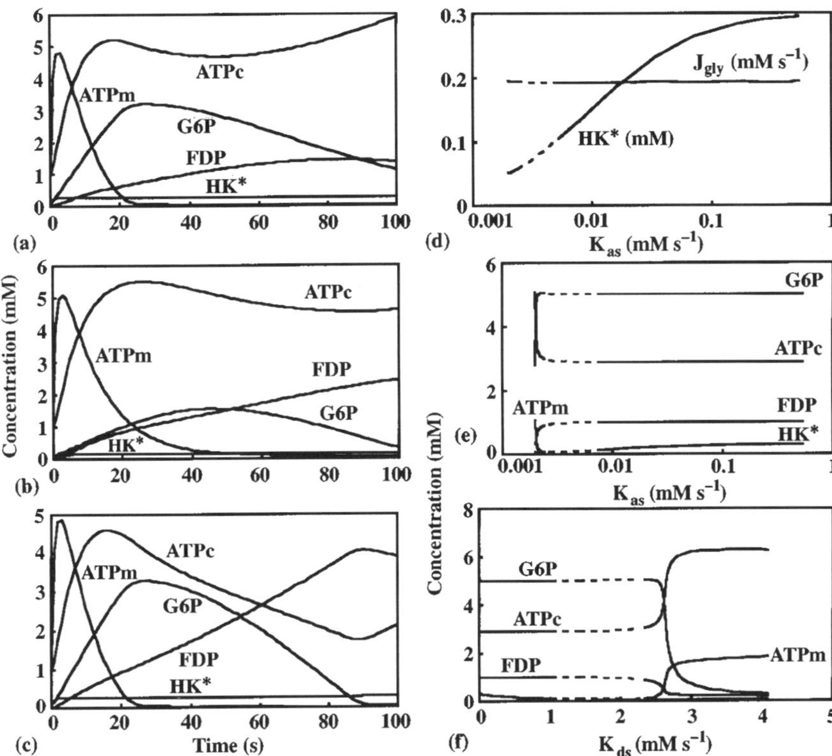

Figure 7.6 Simulation studies of the influence of HK ambiquity in (a–c) transient and (d–f) steady state behaviour of glycolytic flux and its intermediates. The model developed in Appendix 7A was numerically integrated to simulate its dynamic behaviour with (a–c) SCoP (Duke University, 1989) or (d–f) a stability analysis performed with AUTO (Doedle, 1986, Concordia University). The following parameters were used: (a) k_1 (s⁻¹) = 1.0; k_2 (s⁻¹) = 5 × 10⁻³; Glc (mM) = 1.5; HK_t (mM) = 0.3; C_A (mM) = 10; K_{mg} (mM) = 0.1; K_{cg} (mM) = 0.1; K_{mat} (mM) = 0.1; K_{cat} (mM) = 0.1; V_{max} (mM s⁻¹) = 0.1; K_{g6} (mM) = 0.1; K_{atp} = 0.1; n = 2; k_4 (mM⁻¹ s⁻¹) = 0.1; k_6 (s⁻¹) = 0.1; P_t (mM) = 10; k_{as} (s⁻¹) = 1.0 (a,c), 0.1 (b); k_{ds} (s⁻¹) = 0.1; K_{as} (mM) = 0.1; k_3 (mM⁻¹ s⁻¹) = 0.1 (a,c), 0.05 (b); k_9 (s⁻¹) = 0.05. The same parameter values were used to obtain a steady state and solve as a function of a bifurcation parameter. The initial values (mM) of the state variables were: G6P = 0.381; ATP_m = 5.116 × 10⁻³; ATP_c = 5.405; FBP = 2.007; HK* = 0.278. The ODEs (equations 7.3–7.7) were solved at the steady state as a function of the bifurcation parameters indicated in (d)–(f). The model was numerically integrated as described for Figure 7.2. (Reproduced from Cortassa and Aon, 1994b, by permission of Academic Press.)

The initial burst (first 10 s) of G6P is largely accounted for by mitochondrial ATP, while the subsequent rise of G6P is sustained by cytoplasmic ATP. The initial linear rise in mitochondrial phosphorylation of ATP (5–10 s) levels off to one fifth of that value at 20 s. This result agrees with the observation of Gumaa and McLean (1969). These results suggest direct utilization by HK* of the ATP from oxidative phosphorylation. Figure 7.6c shows that although the initial burst in G6P is higher than that of FBP, the latter makes an overshoot later. From the transient behaviour shown in Figure 7.6c it is clear that the G6P overshoot is built up on mitochondrial ATP while FBP slowly and linearly accumulates coincidently with G6P and ATP_c consumption. FBP accumulation stimulates the 'lower' part of glycolysis, explaining the ATP_c rise after G6P reaches a minimum and FBP a maximum.

Synthetically, the transient, short-term behaviour of the model based on the two forms of HK (mitochondrial-bound and unbound) satisfactorily describes the changes in the amounts of glycolytic intermediates after a pulse of glucose to tumour cells.

Steady state behaviour

The picture of the short-term regulation of glycolytic dynamics by HK ambiquity is rather different at the steady state. The main result shows that the steady state behaviour of glycolysis remained unchanged for a large range of values of the association constant (K_{as} = 0.001–1.0 mM s^{-1}) to mitochondria, the net result of which was an increase in the bound form of the enzyme (Figure 7.6d). We have systematically studied the effect of the association constant of HK on the stationary levels of glycolytic intermediates. We could not observe any effect of K_{as} on the steady behaviour of either ATP_c, ATP_m, FBP, G6P or bound HK except for a very narrow range at low values of K_{as} ($\sim$0.004–0.0045 mM s^{-1}), where an appreciable increase in the steady levels of G6P and FBP was noticed. A decrease in ATP_c and ATP_m was also evident (Figure 7.6e).

The rate of dissociation of HK from the mitochondrial membrane (K_{dis}) only influenced the steady levels of glycolytic intermediates for a small range of variation of that parameter (K_{dis} = 2.5–2.6 mM s^{-1}) whereas it did not exert any influence for most of the range studied (K_{dis} = 0.0–4.5 mM s^{-1}). G6P and the cytoplasmic ATP pool were the mainly affected variables (Figure 7.6f).

Overall, it is likely that HK ambiquity influences the transient behaviour of glycolytic flux and intermediates. On the other hand, the steady state flux of glycolysis was not affected even for large changes in the amount of the mitochondrial-bound form of the kinase. Both kinetic constants of association–dissociation of the kinase to the mitochondrial compartment did exert their influence on the appearance (or disappearance) of periodic behaviour (dashed lines in Figure 7.6).

7.3.3 THE DYNAMIC COUPLING CONCEPT AND ITS RELATIONSHIP TO THE INTERACTION OF CYTOPLASMIC AND MITOCHONDRIAL COMPARTMENTS *IN VIVO*

Coupling was defined in Chapter 1 in a broad sense, as being how two processes are related through a common intermediate. Given two coupled dynamic processes, a dynamic change of a variable in one of the processes that affects the intermediate that couples them (e.g. ATP) will influence the steady state behaviour of the second process, i.e. transition from asymptotic or relaxed oscillations to sustained oscillations.

Regulation of glycolysis dynamics *in vivo* through dynamic coupling between processes taking part in cytoplasmic and mitochondrial compartments is likely to be achieved through the cytoplasmic availability of ATP and NAD(P)H pools. The latter has been shown to operate at several steps, according to existing experimental evidence (Aon *et al.*, 1991; Beauvoit *et al.*, 1993): the mitochondrial proton motive force through the regulation of the ATP synthetic or hydrolytic fluxes; the adenine nucleotide translocator, presumably through the regulation of the rate of cytoplasmic–mitochondrial ATP/ADP exchange; and the mitochondrial dehydrogenases by regulating the rate of NADH reoxidation. By increasing the cytoplasmic load of ATP, the mitochondrial ATPase subsystem of yeast (in turn regulated by proton motive force) concomitantly changed the dynamic properties of glycolysis, i.e. transition from a stable branch of asymptotic steady states through a Hopf bifurcation to an unstable branch of oscillatory steady states.

In addition to the short-term regulatory properties of G6P brought about by HK ambiquity, long-term responses that compromise several metabolic pathways should also be considered. Depletion of essential amino acids in mammalian cells initiates the so-called pleiotypic response, which implies the inhibition of several processes such as glucose uptake and the synthesis of nucleic acid and protein (Rabinovitz, 1992). In Ehrlich ascites tumor cells a major block is imposed on glycolysis and glucose uptake by G6P accumulation following amino acid deficiency, probably due to inhibition of PFK by uncharged tRNA (Rabinovitz, 1992). In mammalian cells G6P is a potent inhibitor of HK.

It has been shown that ATP induces a proton-permeability pathway in isolated mitochondria which would play the role of an energy-dissipating reaction (Prieto *et al.*, 1992). These authors proposed that the activation of that mitochondrial proton-conducting pathway by dissipating energy would contribute to the redox balance of yeast cells during aerobic growth through acceleration of NAD(P)H reoxidation.

Dynamic ambiguity of HK was able to regulate the transients induced by, for example, glucose pulses in tumor cells, though not the steady state control of the glycolytic flux (Figure 7.6). The latter point was also shown with the approach of metabolic control analysis (MCA). In fact, HK ambiguity did not influence either the quantitative values of flux control coefficients or rate-controlling steps when taken into account in the matrix approach of MCA (Aon *et al.*, 1996b). One important result of the mutual regulation of glycolysis and mitochondria through HK ambiguity is that it appears as a biochemical device more able to regulate transients than to control steady state behaviour. Under those conditions, HK ambiguity behaves as a regulatory mechanism exerting feedback on glucose phosphorylation through mitochondrial pools of ATP.

7.3.4 MODULATION OF GLYCOLYTIC FLUX BY STRUCTURAL POLYMERS: A POSSIBLE *IN VIVO* REGULATORY MECHANISM

Several workers had been interested in the elucidation of enzyme binding properties with actin, its effect on enzyme kinetic properties, and the interaction between enzymes in supramolecular aggregates with its influence upon transition times in consecutive catalytic steps (Masters, 1984; Clarke *et al.*, 1985; Luther and Lee, 1986). Subsequent work showed that the microcompartmentation of aldolase and glyceraldehyde-3-phosphate dehydrogenase in four different cell types was associated with the actin cytoskeleton and that both enzymes existed in a soluble as well as in a structure-bound form (Minaschek *et al.*, 1992). Changes in kinetic properties of glycolytic enzymes have been reported upon binding to actin filaments (Masters, 1984; Luther and Lee, 1986; Keleti and Ovádi, 1988; Keleti *et al.*, 1989). PFK is able to interact with microtubules and tubulin, its activity being inhibited by increasing amounts of tubulin and microtubules (Lehotzky *et al.*, 1993). The reversible association of GAPDH with F-actin has been implied in the increase of glycolytic activity and its integration with cell motility and metastatic potential. A close correlation was observed between GAPDH expression (mRNA abundance) and both cell motility and metastatic potential of rat prostatic adenocarcinoma (Epner *et al.*, 1993).

It may be conceived that the assembly–disassembly of cytoskeleton components may play the role of a macroscopic, coherent inducer of changes in systemic properties such as cellular fluxes (Aon and Cortassa,

1993). Supraregulatory effects of microtubular protein assembly–disassembly on fluxes through coupled enzymatic reactions of carbon metabolism, both *in vitro* and in a permeabilized yeast cell system, have been described (Cortassa *et al.*, 1994a). A concentration-dependent stimulatory effect in the presence of polymerized or non-polymerized microtubular protein on metabolic fluxes sustained by enzymes of carbon metabolism has been shown (Figure 6.3).

(a) The groundplan of living cells: heterogeneous distribution and organization of cytoskeletal proteins drive differential modulation of metabolic fluxes

At present, it is widely accepted that the organization inside a cell is that of a structured cytoplasm or a protein crystal with 40% of water (Figure 6.1) (Fulton, 1982; Clegg, 1984b) (Chapter 6). Actin filaments and microtubules – major components of the cellular cytoskeleton in all eukaryote cells – dynamically assemble–disassemble with expenditure of energy (Carlier, 1992; Gelfand and Bershadsky, 1991). Actin and tubulin distribution changes during the cell cycle (Lew and Reed, 1993; Chant, 1994). In the budding yeast *Saccharomyces* the cell cycle dependent patterns of microtubule and actin localizations are well studied. The yeast actin cytoskeleton, in immunofluorescence light microscopy (Adams and Pringle, 1984; Kilmartin and Adams, 1984; Novick and Botstein, 1985; Lew and Reed, 1993) or immunoelectron microscopy (Mulholland *et al.*, 1994), appears as cytoplasmic cables and cortical patches asymmetrically organized along the axis of cell growth. Reported experimental data show that microtubules and actin filaments continuously assemble–disassemble along the length of the axon of cultured neurons.

Hypothetically, the cellular cytoplasm has been characterized as a random fractal of the percolation cluster type (Aon and Cortassa, 1994; Cortassa and Aon, 1994b). A similar idea has been presented by Forgacs (1995). In multicellular or unicellular organisms, in a developing embryo or an organ, or microorganisms facing environmental changes or competition with other species, cells must give a coherent answer to different stimuli (e.g. hormones, neurotransmitters, substrates). Among those coherent responses, cells must divide, differentiate, migrate or redirect metabolic fluxes. A main biological problem extensively treated throughout this book is how the cell orchestrates a coherent spatio-temporal response in the (hypothetical) random fractal of its internal milieu.

Based on original experimental evidence, we develop here the idea that cytoplasmic spatial segregation of cytoskeleton components (actin and microtubular protein) may provide a plausible biochemical mechanism of the spatio-temporal organization of cellular biochemistry (Figure 7.7). More specifically, the nature of the cytoskeleton components together with its concentration, polymeric status (assembled or unassembled) or topological arrangement, are main targets of cellular regulation. The topologically driven modulation of metabolic fluxes by cytoskeleton components may exert their regulatory effects through the local activity of enzyme(s), or concentration of substrates, enzymes or effectors, inducing different oligomeric forms of allosteric enzymes (Roberts and Somero, 1987; Lehotzky *et al.*, 1993; Cortassa *et al.*, 1994a; Marmillot *et al.*, 1994), as well as by modulating the level of percolation thresholds (Feder, 1988; Aon and Cortassa, 1994).

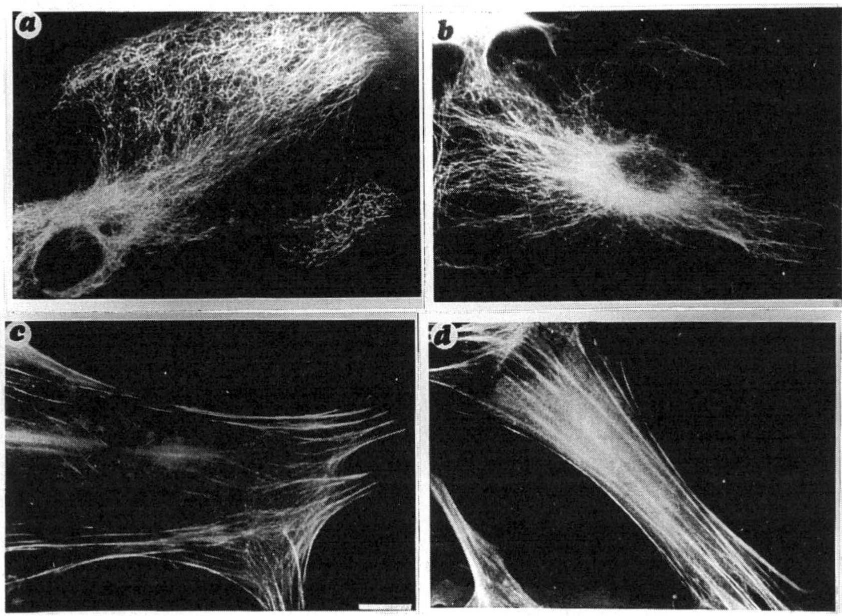

Figure 7.7 Heterogeneous distribution of cytoskeletal proteins at the cellular level. Micrographs showing the distribution of (a,b) microtubules and (c,d) actin filaments in (a,c) control and (b,d) kinesin-heavy chain (KHC) antisense-treated astrocytes. Under both conditions microtubules and microfilaments extend throughout the cell cytoplasm. Microtubules were labelled with an antibody against tyrosinated α-tubulin, and actin filaments were labelled with fluorescein-phalloidin. Bar, 10 μm. (Reproduced from *Journal of Cell Biology*, 1994, **127**, 1021–1039, by permission of The Rockefeller University Press.)

(b) Effect of microtubular protein (MTP) on enzymatic fluxes catalysed by glycolytic enzymes: the role of MTP concentration and its polymeric status

The catalytic activity of enzymes from representative reactions of carbon catabolism was assayed at different MTP concentrations in either its polymerized or non-polymerized state. Non-polymerized MTP increased both the flux through HK/G6PDH and PK/LDH enzymatic systems relative to controls in the absence of MTP (Figure 7.8a,b).

$$\text{Glucose} + \text{ATP} \xrightarrow{\;HK\;} \text{Glucose 6 P} + \text{ADP}$$
$$\text{Glucose 6 P} + \text{NADP}^+ \xrightarrow{\;G6PDH\;} 6 \text{ phospho} - \text{gluconate} + \text{NADPH}$$

$$(7.1)$$

$$\text{Phosphoenolpyruvate} + \text{ADP} \xrightarrow{\;PK\;} \text{Pyruvate} + \text{ATP}$$
$$\text{Pyruvate} + \text{NADH} \xrightarrow{\;LDH\;} \text{Lactate} + \text{NAD}^+$$

$$(7.2)$$

Other polymers could not elicit the effect on enzymatic fluxes observed with MTP. The effect of neutral polymers (glycogen) or the inert synthetic PEG 8000 on enzymatic fluxes was tested in the PK/LDH and HK/G6PDH reactions. The results obtained were consistent with the flux regulation effect induced by polymerized brain MTP being specific and unlikely to be due to locally high concentrations of polymers (Cortassa *et al.*, 1994).

In the presence of polymerized MTP, the flux through HK/G6PDH exhibits a lower increase and flux saturation for MTP concentrations higher than 0.8 mg ml^{-1} (Figure 7.8a). On the other hand, the presence of microtubules provoked a linear increase in the flux of the PK/LDH coupled reaction (Figure 7.8b). The change in kinetic behaviour was also reflected by a drastic decrease in transition time, τ (i.e. the time it takes the coupled enzymatic reaction to achieve the steady state condition) from ~50 s to 20 s at MTP concentrations higher than the critical concentration for polymerization. The change in kinetic parameters of PK, namely V_{max} and n, was also reflected by the decrease in τ which appears to be induced by an interaction of PK/LDH enzymes with polymerized MTP (Cortassa *et al.*, 1994a).

The effect of the presence of polymerized and non-polymerized MTP on metabolic fluxes was further studied in another two enzymatic systems related to carbon (PFK/PK/LDH) or nitrogen (GDH) metabolism. In the presence of polymerized MTP, the fluxes in both enzymatic systems increased but for different ranges of MTP concentration (Figure 7.8c,d). A linear increase in flux up to 40% through the PFK/PK/LDH system was observed with respect to the control for a concentration range of MTP from 0.2 to 1.2 mg ml^{-1} (Figure 7.8c). The increase in flux

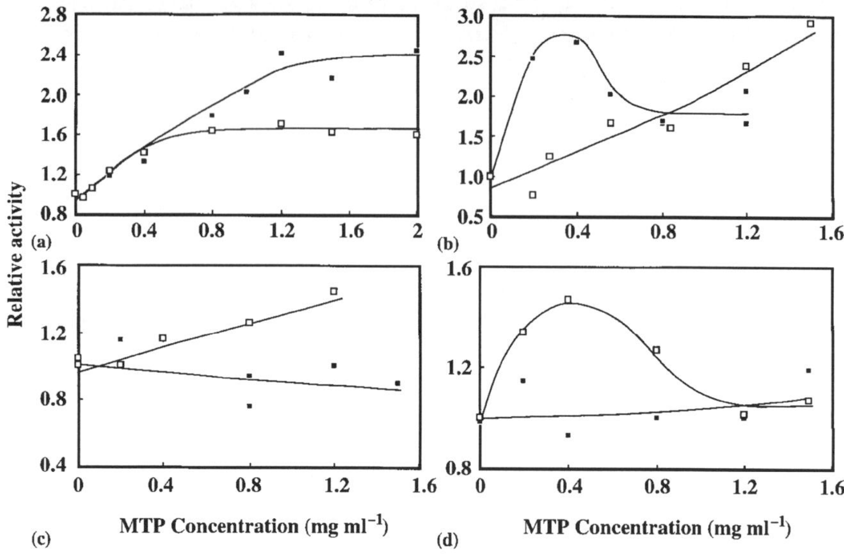

Figure 7.8 Effect of the presence of microtubular protein in its polymerized or non-polymerized state on enzymatic fluxes. (a) Activity of the enzyme couple HK/G6PDH is plotted as a function of MTP concentration with respect to the activity of the couple in the absence of MTP. The activity measurements in the presence of polymerized MTP were referred to a control assayed in the presence of 2 mM GTP. The increase in relative activity was referred to the control in the absence of MTP: activity 1.0. (b–d) Activities of the systems PK/LDH, PFK/PK/LDH and GDH, respectively. Filled and open squares indicate activity in the presence of non-polymerized and polymerized MTP, respectively. The conditions of activity assay were as described in Cortassa *et al.* (1994a). Each point represents the mean of duplicates from two independent experiments. The lines joining the experimental points were drawn by eye. (Panels a,b are from Cortassa, Cáceres and Aon, *Journal of Cellular Biochemistry*, 1994, reprinted by permission of John Wiley & Sons, Inc.)

through the GDH-catalysed reaction was of the same order of magnitude as in the PFK/PK/LDH reactions but the maximal stimulation was achieved at a lower concentration value of 0.4 mg ml^{-1} of MTP. At MTP concentrations higher than 0.4 mg ml^{-1}, the increase in flux was annihilated (Figure 7.8d).

Non-polymerized MTP showed almost no effect upon fluxes in both enzymatic systems. For PFK/PK/LDH a slight decrease in flux was noticed (Figure 7.8c). In the case of GDH, the nucleotide GTP is a strong inhibitor. Fluxes three-fold higher (3.92 ± 0.12 versus 1.3 ± 0.14 μmol

NAD min^{-1}) were measured in the absence and in the presence of GTP. However, this inhibition does not account for the MTP concentration dependence of GDH activity stimulation observed in the presence of polymer (Figure 7.8d).

An insight into the observed mechanism of activation was obtained from plots of enzymic rates versus substrate concentration at fixed MTP concentration. The substrate concentrations of the limiting enzyme (i.e. NADP or PEP for HK/G6PDH or PK/LDH couples, respectively) were varied. These two enzymatic couples were chosen because they were paradigmatic in the sense of their kinetics (hyperbolic or sigmoidal) or interactive abilities with MTP.

The presence of 1 mg ml^{-1} of polymerized or non-polymerized brain MTP provoked changes in kinetic parameters (K_M and V_{max}) of the assayed G6PDH coupled to hexokinase. The dependence of the HK/G6PDH flux on MTP concentration suggested that the soluble components of MTP could account for the activation of the flux through these enzymes. This hypothesis was tested by mathematical modelling (Cortassa *et al.*, 1994a). The simulation results reproduced the experimental data, supporting the idea that the MTP not engaged in the microtubule structure is activating G6PDH.

The results obtained with the coupled PK/LDH system as a function of PEP showed that the presence of polymerized brain MTP induced an increase in cooperation as measured by n (Hill coefficient) (Figure 7.9b). Furthermore, non-polymerized MTP induced an even higher increase in cooperativity and a 100% increase in V_{max} (Figure 7.9a). The change in kinetic parameters of PK could be mechanistically induced by interaction of the enzyme(s) with MAPs (see next section). Mathematical modelling explained the flux activation induced by MTP in the PK/LDH couple by a displacement of the equilibrium of association of the allosteric enzyme by stabilization of the higher oligomeric states coincident with higher activity. That the same mechanism might be at work intracellularly was suggested by the insensitivity *in situ* of PK's flux to FBP. On the other hand, since no changes in the $K_{0.5}$ of PK were registered it could be inferred that the mechanism of activation by MTP through self-association of the enzyme differs from the activation elicited by an allosteric effector, such as FBP.

MAPs-mediated increase in flux

Part of the activity of PK/LDH could be located in the pellet of polymerized MTP. Apparently, MAPs present in the preparation of

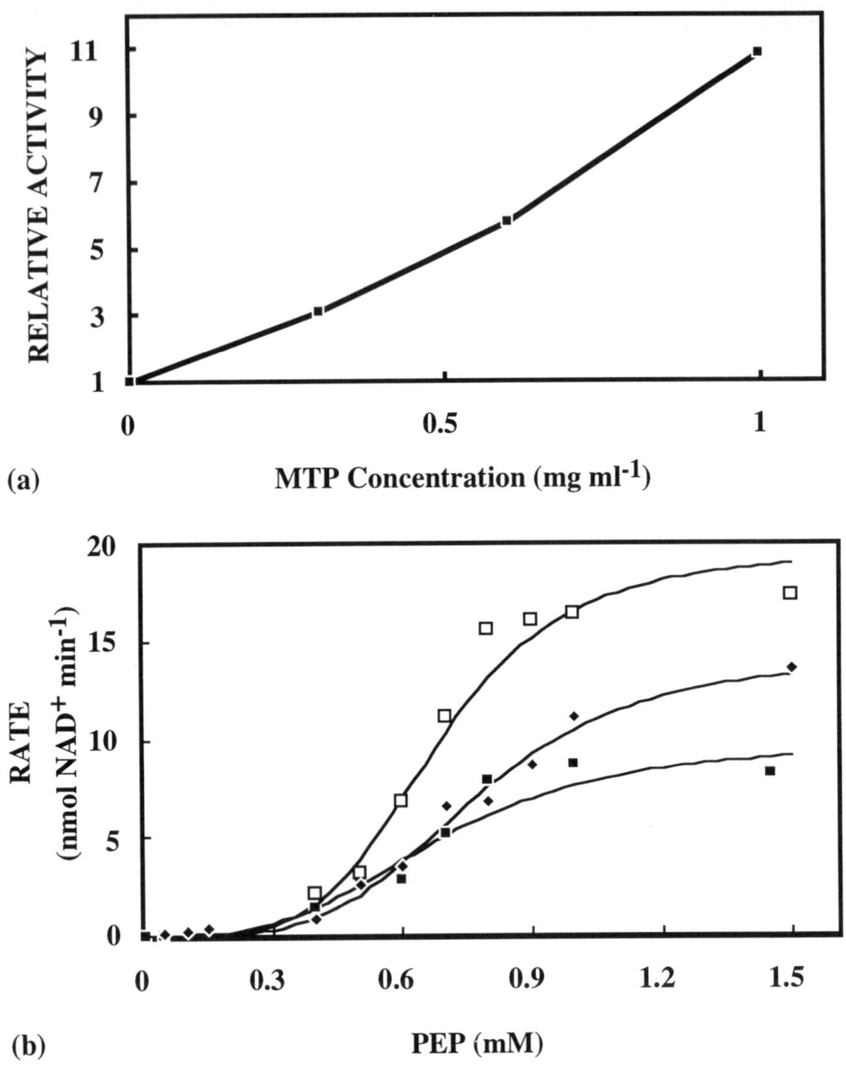

Figure 7.9 (a) Increase of PK activity and (b) effect on kinetic parameters as a function of varying concentrations of polymerized or non-polymerized microtubular protein (MTP). (a) The activity of the enzyme couple PK/LDH is plotted as a function of MTP concentration with respect to the activity of the couple in the absence of MTP. The activity measurements in the presence of non-polymerized MTP were assayed as described in Figure 7.8. (b) In the y-axis the initial rates of enzymatic fluxes in the absence of MTP (filled squares) or presence of polymerized (filled diamonds) or non-polymerized MTP (open squares) (0.4 mg ml⁻¹ MTP) are plotted as a function of PEP. The continuous lines represent the fitting to the experimental data. (Reproduced from Aon, Cáceres and Cortassa, *Journal of Cellular Biochemistry*, 1996a, by permission of John Wiley & Sons, Inc.)

MTP were able to induce the increase in flux independently of tubulin (Cortassa *et al.*, 1994a) (Figure 7.10). Taken together, both facts suggested that interaction and activation of PK in the presence of the MTP lattice was MAPs-mediated. The activating effect exerted by MAPs was lost when the MAPs fraction was boiled for 5 min. We further

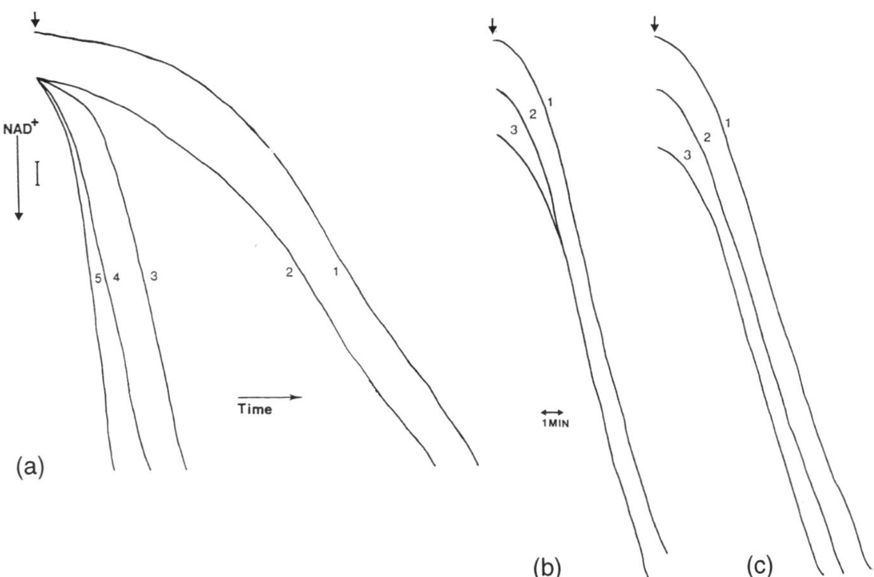

Figure 7.10 The activation of PK by MTP appears to be due to thermolabile microtubule-associated proteins (MAPs). The MTP fraction was separated into tubulin and MAPs by passage through phosphocellulose as described in Cortassa *et al.* (1994a). The MTP fraction was heated at 100°C for 10 min and centrifuged at 10 000 g for 20 min and the supernatant (the thermostable fraction of MAPs) was assayed with the PK/LDH couple. (a) The activation effect is due to the (total) MAPs fraction (trace 4: 0.11 mg ml⁻¹) and not to tubulin (trace 2: 0.6 mg ml⁻¹). The activating effect exerted by MAPs was lost when the MTP fraction was boiled for 5 min (identical to trace 2). Trace 1 shows the control activity without either MTP or MAPs. The MTP concentrations assayed in traces 3 and 5 were 0.3 and 0.6 mg ml⁻¹, respectively. (b,c) The MAPs-mediated activation was dependent upon the thermolabile fraction of MAPs by incubating either MTP or MAPs for 1 h in the presence of monoclonal antibodies (from bovine brain) against the main components of thermostable MAPs: MAP2 (traces 2) and tau (traces 3). The increase in PK activity still remained as in the control in the absence of antibodies (traces 1) when assayed in the presence of MTP (b) or the MAPs fraction (c). (Panel (a) is reproduced from Aon, Cáceres and Cortassa, *Journal of Cellular Biochemistry*, 1996a, by permission of John Wiley & Sons, Inc.)

assessed that the MAPs-mediated activation was dependent upon the thermolabile fraction of MAPs by incubating in the presence of monoclonal antibodies against MAP2 and tau from bovine brain. Figure 7.10 shows that the activity increase of PK in the MAPs fraction, incubated with either antibody, still remained with respect to the control.

Actin

An opposite effect on enzyme kinetics was obtained with F-actin polymerized in the presence of 5% PEG, with respect to that observed with MTP. Each enzymatic couple was inhibited to a different extent by increasing concentrations of actin (Figure 7.11a). Actin slightly inhibited (10%) the flux through the HK/G6PDH couple, and significantly (ca. 40%) at 0.4 mg ml^{-1} (or higher) of PK/LDH (Figure 7.11a). The putative effect of molecular crowding induced by PEG may be ruled out since controls were run in parallel in the presence of 5% PEG but in the absence of actin. None of the two enzymatic couples tested was able to interact with the actin lattice even after 2 h centrifugation at 40 000 g. The whole activity was recovered in the supernatants.

Non-polymerized G-actin behaved similarly to the control (Figure 7.11b); the cooperativity, n, and the $K_{0.5}$ of the reaction were 8.2 ± 0.3 and 0.62 ± 0.034 mM in both cases, whereas the V_{max} values were 11.2 ± 1.5 and 13.7 ± 1.8 nmol min^{-1}, respectively. The high cooperativity

Figure 7.11 (see facing page) Effect of F-actin and G-actin on PK/LDH and HK/G6PDH sustained fluxes and kinetic parameters. (a) Decrease of PK/LDH (filled square) and HK/G6PDH (open square) fluxes and effect on kinetic parameters of PK in the presence of (b) G-actin or (c) F-actin polymerized for 30 min or 120 min in the presence of 7% PEG. G-actin (a, 0.25–1.0 mg ml^{-1}; b,c, 0.4 mg ml^{-1}) was polymerized as in Suzuki *et al.* (1989) in the presence of the enzymes HK/G6PDH and PK/LDH, except for glucose (former couple) or NADH/PEP (latter couple) which were added after completion of the polymerization step. Both polymerization and enzymatic activity were monitored as described in Cortassa *et al.* (1994a). The lines drawn under the experimental points are the best-fit curves according to a generalized Hill equation (below) for the PK/LDH activity (b,c). The fitting procedure was performed with a Levenberg–Marquardt algorithm and with Enzfitter by R.J. Leatherbarrow (Elsevier, Biosoft). Both fitting procedures comprise non-regression analysis of the data according to the following equation:

$$V = \frac{V_{max}[PEP]^n}{(K_{0.5})^n + [PEP]^n}$$

(Reproduced from Aon, Cáceres and Cortassa, *Journal of Cellular Biochemistry*, 1996a, by permission of John Wiley & Sons, Inc.)

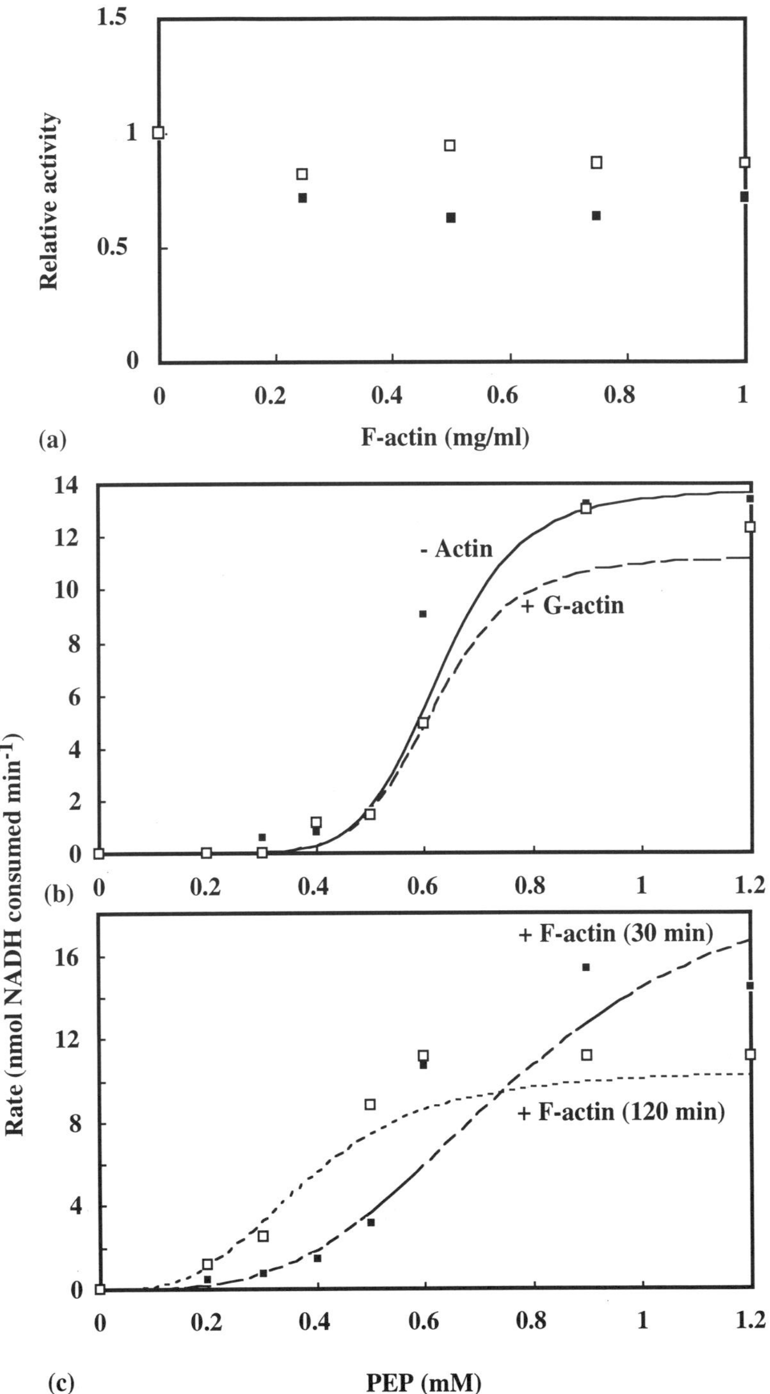
1.5
1
0.5
0
Relative activity
0   0.2   0.4   0.6   0.8   1
F-actin (mg/ml)
(a)
14
12
10
8
6
4
2
0
- Actin
+ G-actin
0   0.2   0.4   0.6   0.8   1   1.2
(b)
Rate (nmol NADH consumed min⁻¹)
+ F-actin (30 min)
16
12
8
4
0
+ F-actin (120 min)
0   0.2   0.4   0.6   0.8   1   1.2
PEP (mM)
(c)

exhibited by PK was probably induced by PEG since our previous data in the absence of PEG gave n values of 3.3 (Cortassa *et al.*, 1994a).

The kinetics of PK/LDH showed differences depending on the presence of inhibitory concentrations of F-actin (0.4 mg ml^{-1}) with different topological arrangements induced by 7% PEG (Figure 7.11c). Depending on the polymerization time, either random filaments (30 min) or bundles (120 min) of F-actin are induced in the presence of 7% poly(ethylene glycol) (PEG) (Suzuki *et al.*, 1989). A dramatic decrease in cooperativity, with respect to the control in the absence of actin, was observed in the presence of both actin arrangements (from n = 8.2 ± 0.3 to 3.3 ± 0.3) along with almost unchanged $K_{0.5}$ for random actin filaments ($K_{0.5}$ = 0.78 ± 0.27) and half $K_{0.5}$ (= 0.38 ± 0.023) for actin bundles (Figure 7.11c). Also dramatic was the difference in V_{max} values between the two topological arrangements (20.6 ± 3.02 in comparison with 10.5 ± 0.8 nmol min^{-1} in random or bundle filaments, respectively) (Figure 7.11c).

The results presented indicate that the inhibitory effects on the fluxes catalysed by PK/LDH are due to a significant decrease in cooperativity as well as maximum velocity of PK.

(c) Sensitivity of metabolic fluxes to the intracellular polymeric status of cytoskeleton components

We have ascertained the plausibility of the regulatory role of cytoskeleton components by measuring the same metabolic fluxes in permeabilized yeast cells assayed *in vitro*. The yeast *Saccharomyces* contains tubulin and actin in its cytoskeleton (Solomon, 1991). In order to test if enzymatic fluxes were responsive *in situ* to microtubule disassembly, yeast cells were permeabilized in the presence of nocodazole (Jacobs *et al.*, 1988; Solomon, 1991). In the HK/G6PDH couple, nocodazole promoted a slight increase in both K_M and V_{max} with respect to permeabilized control cells (Figure 7.12a). PK increased the *in situ* reaction flux while retaining its allosteric properties (Figure 7.12b). Comparing the data obtained *in vitro* and *in situ*, there is a qualitative resemblance in the kinetic behaviour shown by PK in the presence of non-polymerized MTP and its intracellular equivalent in the presence of nocodazole (compare Figures 7.9 and 7.12b). Essentially, PK kinetics changed in the same sense of increase in V_{max} as the *in vitro* data.

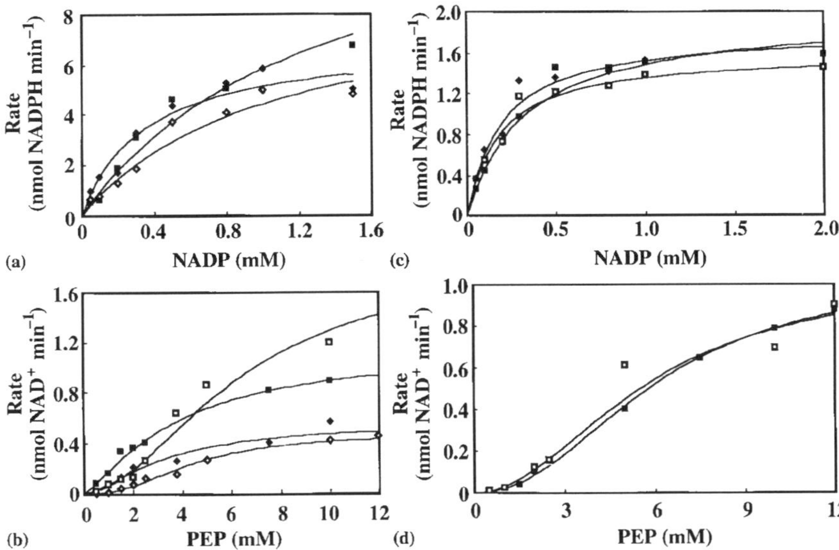

Figure 7.12 Intracellular effect of compounds which are known to affect the cytoskeleton. Intracellular fluxes of HK (a,c) and PK (b,d) were measured in *Saccharomyces cerevisiae* strain A364A yeast cells which were permeabilized as described in Cortassa *et al.* (1994a). (a,b) The kinetic behaviour of both enzymatic systems in the absence (open diamond) or presence of compounds known to stabilize (1 μM taxol, filled diamond) or depolymerize (15 μg ml⁻¹ nocodazole, filled and open squares) microtubules. (c,d) The effect of a drug known to depolymerize (0.3 μg ml⁻¹ cytochalasin B: CB) actin in mammalian cells but not in yeast cells: control cells (filled diamond) or control cells with DMSO (filled squares), or cells treated with CB (open squares). The permeabilized cells were used as the enzymatic source and assayed *in situ* as described in Cortassa *et al.* (1994a). The continuous lines represent the best fit to the experimental data which for the Michaelis–Menten kinetics of (a) was performed according to the following equation:

$$V = \frac{V_{max}[\text{NADP}]}{K_m + [\text{NADP}]}$$

For the case of allosteric kinetics, the fitting was ruled by the equation for Figure 7.11. (Panels a,b reprinted from Cortassa, Cáceres and Aon, *Journal of Cellular Biochemistry*, 1994, by permission of John Wiley & Sons, Inc.)

Compounds such as taxol or cytochalasin B which are known to stabilize microtubules or depolymerize actin in mammalian cells but not in yeast cells (Solomon, 1991; Barnes *et al.*, 1992; Cortassa *et al.*, 1994a) did not affect the kinetics of either PK or HK in permeabilized yeast cells in comparison with the controls (Figure 7.12c,d). This provides evidence for the reliability of intracellular measurements of enzymatic activities, principally with respect to unspecific effects.

Correlative changes of the organization of the actin macromolecular network and glycolytic activity have been described (Bereiter-Hahn *et al.*, 1995). G1 blocked heart endothelial cells showed actin stress fibres, whereas after release of blockage peripheral lamellae and microspikes appeared concomitantly with increased lactic acid production. Similar changes were induced by phorbol myristate acetate treatment (Bereiter Hahn *et al.*, 1995). These results also support the plausibility of changes in metabolism dynamics following rearrangements of cytoskeleton components.

(d) Local and global dynamics of biochemical reactions in percolation clusters

Biologically speaking, what could be the link between self-organizing fractal properties of the groundplan of living cells and the generalized coherent responses that cells give to different environmental cues or stimuli? By exhibiting coherence, we mean that the spatio-temporal coordination of a cell's metabolism, energetics and gene expression (Cortassa and Aon, 1994b) induces, for example, the stopping or starting of cell division or differentiation, or the resumption, slowing down or acceleration of growth, when environmental parameters change (Chapters 6 and 9).

The dynamics of chemical reactions could be influenced by the fractal nature of the cytoplasm through two postulated mechanisms. The first is the 'compact kinetics' expected to happen in highly shredded, disjointed objects such as percolation clusters (Kopelman, 1988). The second deals with transitions from three dimensions to two and through this the substrate or effector, and enzyme or target, find each other in the same topological dimension (2D), which increases the probability of encounter and foreseen reactivity (Chapter 6).

Because of the property of global connectedness exhibited by percolation clusters further from the percolation threshold (Figure 6.14) fluctuations in the local dynamics of biological processes (e.g. enzymatic fluxes, waves of second messengers or ions) may extend to very remote cytoplasmic regions. In that way, subcellular processes taking place in functional compartments of cytoplasmic macromolecular networks may become connected and synchronous to similar processes occurring farther away in the cytoplasm or targets distributed far from each other (Aon and Cortassa, 1994). This mechanism may be relevant to cell signalling induced by second messenger waves. Spatial waves of redox substances or cAMP are exhibited by chemical reactions taking place far from equilibrium in either artificial (e.g. the Belousov–Zhabotinskii reaction) or biological systems, respectively (Figures 3.3 and 5.5). The rise in intracellular levels of calcium is an apparent transduction mechanism of extracellular stimuli, including hormones and neurotransmitters (Berridge *et al.*, 1988; Lechleiter *et al.*, 1991). Physiological examples of Ca^{2+} waves are the sperm–egg fusion, cell–cell signalling in monolayers of cultured astrocytes triggered by the neurotransmitter glutamate and synchronization of large cell assemblies of ciliated epithelial cells, vascular endothelial cells and hepatocytes (Meyer, 1991, and references therein) (Plate 2).

Messenger waves may be generated by a series of diffusion and amplification steps. Positive feedback (autocatalysis) and cooperativity are the main nonlinear mechanisms for bistable amplifiers with an intrinsic threshold (Meyer, 1991). The amplifier has been modelled as consisting of a threshold concentration, C_{thr}, and a delay time, τ_α. The amplifier is inactive for messenger concentrations below C_{thr}, whereas above C_{thr} the amplifier only remains inactive for the delay time, τ_α, after which an instantaneous increase of the messenger concentration by the amplifier ensues to a maximal level, C_{max} (Meyer, 1991; for another derivation of the profile equation see Sneyd and Kalachev, 1994) (Figure 7.13). A local (spatially speaking) instability in the dynamics of the amplifier mediated by any of the nonlinearities mentioned may change the local concentration of messenger. Depending on the history of the system's dynamics, the chemical concentration may abruptly increase or drop. If such an increase in the local concentration of messenger occurs in a site belonging to the spanning cluster, i.e. above the percolation threshold (which is a function of the geometry of the lattice: Rabouille *et al.*, 1992), it may globally extend to cytoplasmic regions far away from the site where the initial perturbation of the dynamics arose. The synchronization effect may be explained by the fact that percolation of the messenger through the spanning cluster would provoke the perturbation of other amplifiers by increasing the messenger concentration above the threshold C_{thr} and triggering similar effects in neighbouring sites (Figure 7.13). Summing up, the local dynamics of an amplifier may extend to the whole cellular field if it belongs to the spanning cluster.

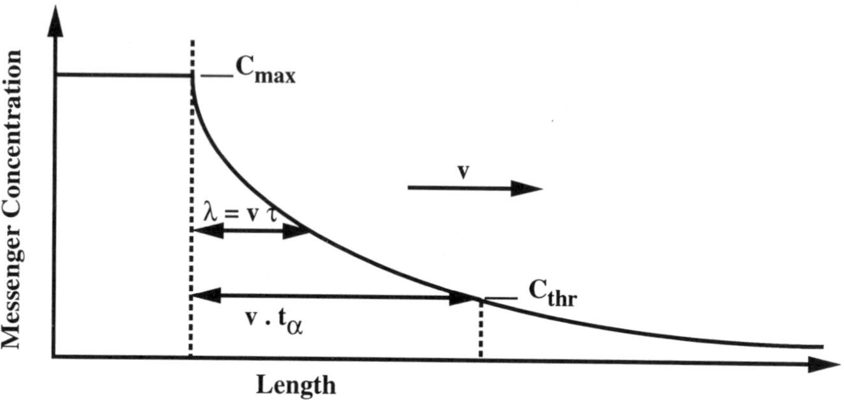

Figure 7.13 Parameters related to the propagation of messenger waves. Experimentally observed properties of a messenger wave are the length of the gradient (λ), the local rise time of messenger concentration (τ) and the wave velocity (v). The model amplifier is inactive for a delay time (t_α) after the threshold concentration (C_{thr}) is reached. The exponential profile of the gradient corresponds to the solution for λ. (Reproduced from Meyer, 1991, by permission of Cell Press.)

(e) Actin and tubulin percolation lattices

According to the proposal of the cytoplasmic organization as a percolation cluster, the following question may be formulated. For enzymatic catalysis, what could be the consequences of the cytoplasmic segregation of polymers which arrange according to different fractal dimension, D (Figure 7.7)? An answer to this question may be deduced from the fact that the percolation threshold is a function of the geometry of the lattice (Feder, 1988; Rabouille *et al.*, 1992), i.e. for a regular two-dimensional quadratic lattice it is 0.593 and for a triangular lattice it is 0.5 (Feder, 1988). This is relevant because of the property of global connectedness exhibited by percolation clusters further from the percolation threshold, P_c (Aon and Cortassa, 1994 and unpublished). Cytoskeleton components which arrange as fractal percolation lattices exhibit self-organization in the spatio-temporal domain of micrometres and minutes, after which new functional properties emerge according to the scaling shown by physiological processes (Chapter 2). We have called that spatio-temporal domain the 'transition point' at which (sub)cellular processes may exhibit transitions from local to global coherence (Aon and Cortassa, 1993). On the other hand, percolation lattices exhibit transitions between local and global connectedness at their threshold level of occupancy, i.e. the percolation threshold. On this basis, we have suggested that the percolation threshold of the fractal cytostructure is related to the scaling of physiological processes at the transition point in the spatio-temporal domain for which cytoskeleton lattices self-organize.

We previously suggested that if an increase in the local concentration of messenger occurs at a site belonging to the spanning cluster, i.e. above the percolation threshold, it may globally extend to cytoplasmic regions far away from the site where the initial perturbation of the dynamics arose. More specifically, a local (in a spatial sense) instability in the dynamics of '**amplifiers**', i.e. chemical reactions exhibiting intrinsic nonlinearities such as allostery or autocatalysis, may change the local concentration of messenger or ions. Depending on the history of the system's dynamics, i.e. the 'amplifier' was 'feeling' the local increase or decrease in the concentration of an effector, the chemical concentration may abruptly increase or drop (Figures 1.7, 6.9 and 7.13) (Aon and Cortassa, unpublished).

Overall, by regulating its geometry or architecture a cell in turn regulates the level of its percolation threshold, P_c. Different P_c levels would determine when the local level of a messenger or the dynamics of an enzymatic reaction (e.g. its product) may extend to the whole cellular field if it belongs to the spanning cluster. This proposal links the highly sophisticated mechanisms by which cells rearrange their architecture (Figure 7.7) and the dynamics of biochemical reactions taking place inside them (Figures 1.8 and 6.9).

7.4 OUTLOOK

Several main new features emerge from *in vivo* studies of the spatio-temporal regulation of glycolysis dynamics and oxidative phosphorylation. Those novel aspects concern the control of the glycolytic flux and branching to alcoholic fermentation by substrate uptake, the putative effect of short-term mitochondrial subcellular reorganization, and the regulatory effects of the polymeric status and concentration of microtubular protein or actin on enzymatic fluxes. The *in vivo* manifold dynamic regulation of phosphorylation, redox potentials and hexokinase ambiquity have been extensively described.

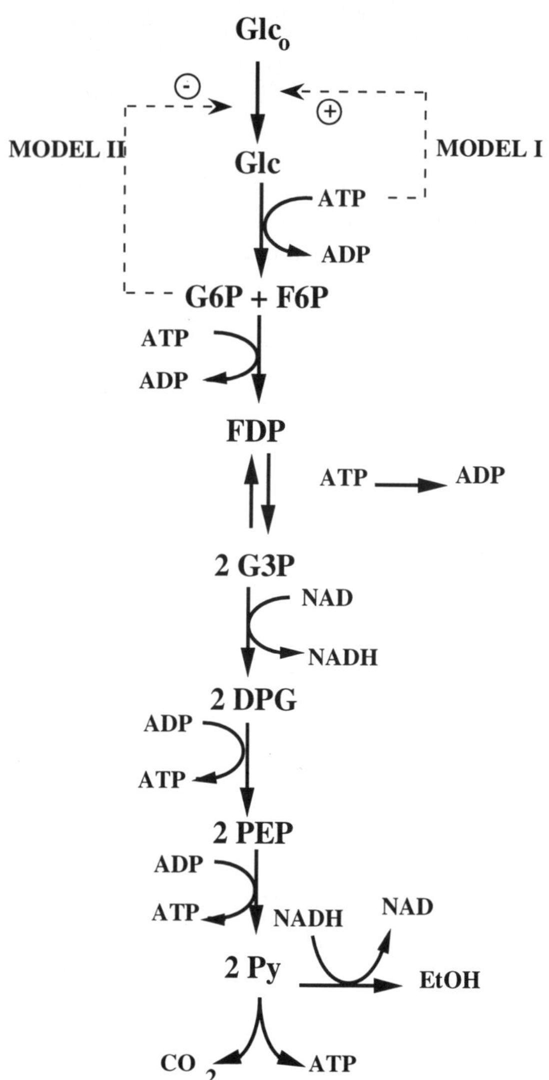

Figure 7.14 Scheme of the glycolytic pathway model used for interpretation and control analysis with MCA of ethanol production by *Sacchromyces cerevisiae* in chemostats run under carbon, nitrogen or phosphate limitations. (Reprinted, with permission, from *Enzyme and Microbial Technology*, **16**, 761–770.)

The concept of dynamic coupling is proposed in quantifying the relaxation of two or more coupled processes following perturbation. In this way, it may be possible to discern a temporal causality, i.e. which processes may regulate others according to the temporal hierarchy of their occurrence.

The microtubular network could be involved in the general intracellular signal transduction pathway of environmental stimuli. Microtubular and actin lattices viewed as nonlinear, dissipative systems have the potential ability to self-organize macroscopically under far from equilibrium conditions. A topologically driven modulation of metabolic fluxes is proposed which is different on thermodynamic and dynamic grounds from 'topodynamic regulation' (Kaprelyants, 1988), the driven force of which was proposed to be thermal diffusion. In our case, nonlinear dynamics of biological systems at some distance from thermodynamic equilibrium is the basis of self-organized, emergent properties at transitions from microscopic to macroscopic phenomena at the cytoplasmic level (Chapter 6).

Our emphasis is on the cytoarchitecture and not on the catalysts. In fact, the supramolecular organization of cellular architecture can in turn entrain the dynamics of enzymatic reactions through mechanisms of association–dissociation of enzymes toward higher or lower oligomeric forms with higher or lower activities, respectively (Figures 1.8 and 6.11). According to this proposal, it can be postulated that a precise biochemical coupling exists between the highly sophisticated mechanisms by which cells dynamically rearrange their (supra)molecular architecture (Figure 7.7) and the dynamics of biochemical reactions taking place concomitantly. Our modelling results simulate how equilibrium displacements between different oligomeric states of enzymes may well be a main mechanism through which structural polymers exert their modulatory role (Figure 6.19).

How the cell 'mixes' or puts on or off controls and regulatory effectors or mechanisms under different physiological and environmental conditions against different genetic backgrounds is a main avenue for systematic research in the near future.

APPENDIX 7A MODELLING HEXOKINASE AMBIQUITY

The model of glycolysis used takes into detailed account only the first steps of glycolysis, up to the production of FBP. The steps of glycolysis downstream from FBP were lumped. Another feature of the modelling concerns the splitting of the ATP pool into two: cytoplasmic (ATP_c) and mitochondrial (ATP_m), with the underlying assumption that when hexokinase is mitochondrially bound (HK*) it preferentially uses the ATP pool provided by oxidative phosphorylation (ATP_m). The soluble hexokinase (HK) uses the ATP_c pool.

A differential equation accounting for HK* depends on the association (K_{as}) and dissociation (K_{dis}) constants of the enzyme to the mitochondrial compartment. K_{dis} depends on the concentration of G6P, a fact supported by substantial experimental evidence (Tuttle and Wilson, 1970).

The model is described by a set of five nonlinear ODEs (7.3–7.7) and three conservation equations (7.8-7.10) (Figure 7.5):

$$\frac{d[G6P]}{dt} = V_b + V_f - V_p \tag{7.3}$$

$$\frac{d[ATP_m]}{dt} = V_{op} - V_b - V_{tr} \tag{7.4}$$

$$\frac{d[ATP_c]}{dt} = V_{tr} + V_{1gly} - V_f - V_p - V_c \tag{7.5}$$

$$\frac{d[FDP]}{dt} = V_p - V_{1gly} \tag{7.6}$$

$$\frac{d[HK^*]}{dt} = V_{e1} - V_{e2} \tag{7.7}$$

$$C_A = [ATP_m] + [ATP_c] + [ADP] \tag{7.8}$$

$$P_t = P_i + 2[FDP] + [ATP_m] + [ATP_c] + [G6P] \tag{7.9}$$

$$HK_t = HK + HK^* \tag{7.10}$$

with the following rate expressions:
- Bound hexokinase (HK*):

$$V_b = \frac{k_1[ATP_m][Glc][HK^*]}{(K_{mg} + [Glc])(K_{mat} + ATP_m)} \tag{7.11}$$

- Soluble hexokinase (HK):

$$V_f = \frac{k_2[ATP_c][Glc][HK_t - HK^*]}{(K_{cg} + [Glc])(K_{cat} + ATP_c)} \tag{7.12}$$

- Phosphofructokinase:

$$V_p = \frac{V_{max}[ATP_c][G6P]^2}{([K_{g6}]^2 + [G6P]^2)(K_{atp} + ATP_c)}$$
(7.13)

- Association of HK to mitochondria:

$$V_{e_1} = k_{as}([HK_t - HK]$$
(7.14)

- Dissociation of HK to mitochondria:

$$V_{e_2} = \frac{k_{ds}[HK][G6P]}{(K_{as} + [G6P])}$$
(7.15)

- Oxidative phosphorylation:

$$V_{op} = k_4(C_A - [ATP_m] - [ATP_c])(Pt - 2[FDP] - [G6P] - [ATP_m] - [ATP_c])$$
(7.16)

- ATP transport from mitochondria to cytoplasm:

$$V_{tr} = k_6[ATP_m]$$
(7.17)

- Non-glycolytic ATP-consuming processes:

$$V_c = k_9[ATP_c]$$
(7.18)

- 'Lower' glycolysis:

$$V_{1gly} = 4k_3[FDP](C_A - [ATP_m] - [ATP_c])(P_t - 2[FDP] - [G6P] - [ATP_m] - [ATP_c])$$
(7.19)

APPENDIX 7B THE MATRIX METHOD OF METABOLIC CONTROL ANALYSIS (MCA) FOR THE CALCULATION OF FLUX, METABOLITE AND BRANCH CONTROL COEFFICIENTS AS APPLIED TO GLYCOLYSIS AND THE BRANCH TO ETHANOLIC FERMENTATION AND THE TCA CYCLE

The matrix method is based on the connectivity property of MCA (Chapter 8):

$$EC = I \quad ; \quad C = E^{-1}$$
(7.20)

where E is the matrix of elasticity coefficients, C is the matrix of control coefficients and I is the identity matrix that contains ones along its main diagonal and zeros elsewhere.

For the Embden–Meyerhoff pathway and the branches to ethanol production and TCA cycle, 10 reaction steps are involved and eight metabolites are considered as variables (Figure 7.14). The matrix of elasticity coefficients E should be of range 10:

$$
E=\begin{bmatrix}
1 & 1 & 1 & 1 & 1 & 1 & 1 & 1 & 1 & 1 \\
0 & \epsilon_{Glc}^{HK} & 0 & 0 & 0 & 0 & 0 & 0 & 0 & 0 \\
\epsilon_{G6P}^{IN} & 0 & \epsilon_{G6P}^{PFK} & 0 & 0 & 0 & 0 & 0 & 0 & 0 \\
0 & 0 & 0 & \epsilon_{FDP}^{ALD} & 0 & 0 & \epsilon_{FDP}^{PK} & 0 & 0 & 0 \\
0 & 0 & 0 & \epsilon_{G3P}^{ALD} & \epsilon_{G3P}^{GAPD} & 0 & 0 & 0 & 0 & 0 \\
0 & 0 & 0 & 0 & 0 & \epsilon_{DPG}^{PGK} & 0 & 0 & 0 & 0 \\
0 & 0 & 0 & 0 & 0 & 0 & \epsilon_{PEP}^{PK} & 0 & 0 & 0 \\
0 & 0 & 0 & 0 & 0 & 0 & 0 & \epsilon_{Py}^{ADH} & \epsilon_{Py}^{TCA} & 0 \\
\epsilon_{ATP}^{IN} & \epsilon_{ATP}^{HK} & \epsilon_{ATP}^{PFK} & 0 & \epsilon_{ATP}^{GAPD} & 0 & \epsilon_{ATP}^{PK} & 0 & 0 & \epsilon_{ATP}^{ATsa} \\
0 & 0 & 0 & 0 & 0 & 0 & 0 & V_{TCA}^{r} & -V_{ADH}^{r} & V_{ATsa}^{r}*
\end{bmatrix}
\qquad (7.21)
$$

$$
*V_{ATsa}^{r} = \frac{n_1 V_{ADH} - n_2 V_{TCA}}{V_{ATsa}}; \quad \text{with } n_2 = 4 \text{ and } n_1 = 0
$$

$$
C=\begin{bmatrix}
C_{IN}^{Jgly} & -C_{IN}^{Glc} & -C_{IN}^{G6P} & -C_{IN}^{FDP} & -C_{IN}^{G3P} & -C_{IN}^{DPG} & -C_{IN}^{PEP} & -C_{IN}^{Py} & -C_{IN}^{ATP} & C_{IN}^{Jr} \\
C_{HK}^{Jgly} & -C_{HK}^{Glc} & -C_{HK}^{G6P} & -C_{HK}^{FDP} & -C_{HK}^{G3P} & -C_{HK}^{DPG} & -C_{HK}^{PEP} & -C_{HK}^{Py} & -C_{HK}^{ATP} & C_{HK}^{Jr} \\
C_{PFK}^{Jgly} & -C_{PFK}^{Glc} & -C_{PFK}^{G6P} & -C_{PFK}^{FDP} & -C_{PFK}^{G3P} & -C_{PFK}^{DPG} & -C_{PFK}^{PEP} & -C_{PFK}^{Py} & -C_{PFK}^{ATP} & C_{PFK}^{Jr} \\
C_{ALD}^{Jgly} & -C_{ALD}^{Glc} & -C_{ALD}^{G6P} & -C_{ALD}^{FDP} & -C_{ALD}^{G3P} & -C_{ALD}^{DPG} & -C_{ALD}^{PEP} & -C_{ALD}^{Py} & -C_{ALD}^{ATP} & C_{ALD}^{Jr} \\
C_{GAPD}^{Jgly} & -C_{GAPD}^{Glc} & -C_{GAPD}^{G6P} & -C_{GAPD}^{FDP} & -C_{GAPD}^{G3P} & -C_{GAPD}^{DPG} & -C_{GAPD}^{PEP} & -C_{GAPD}^{Py} & -C_{GAPD}^{ATP} & C_{GAPD}^{Jr} \\
C_{PGK}^{Jgly} & -C_{PGK}^{Glc} & -C_{PGK}^{G6P} & -C_{PGK}^{FDP} & -C_{PGK}^{G3P} & -C_{PGK}^{DPG} & -C_{PGK}^{PEP} & -C_{PGK}^{Py} & -C_{PGK}^{ATP} & C_{PGK}^{Jr} \\
C_{PK}^{Jgly} & -C_{PK}^{Glc} & -C_{PK}^{G6P} & -C_{PK}^{FDP} & -C_{PK}^{G3P} & -C_{PK}^{DPG} & -C_{PK}^{PEP} & -C_{PK}^{Py} & -C_{PK}^{ATP} & C_{PK}^{Jr} \\
C_{ADH}^{Jgly} & -C_{ADH}^{Glc} & -C_{ADH}^{G6P} & -C_{ADH}^{FDP} & -C_{ADH}^{G3P} & -C_{ADH}^{DPG} & -C_{ADH}^{PEP} & -C_{ADH}^{Py} & -C_{ADH}^{ATP} & C_{ADH}^{Jr} \\
C_{TCA}^{Jgly} & -C_{TCA}^{Glc} & -C_{TCA}^{G6P} & -C_{TCA}^{FDP} & -C_{TCA}^{G3P} & -C_{TCA}^{DPG} & -C_{TCA}^{PEP} & -C_{TCA}^{Py} & -C_{TCA}^{ATP} & C_{TCA}^{Jr} \\
C_{ATPse}^{Jgly} & -C_{ATPse}^{Glc} & -C_{ATPse}^{G6P} & -C_{ATPse}^{FDP} & -C_{ATPse}^{G3P} & -C_{ATPse}^{DPG} & -C_{ATPse}^{PEP} & -C_{ATPse}^{Py} & -C_{ATPse}^{ATP} & C_{ATPse}^{Jr}
\end{bmatrix}
\qquad (7.22)
$$

The inversion of matrix E results in C, which provides the flux control coefficients in its first column, the concentration control coefficients in the second to ninth columns and the flux ratio control coefficients at the branch (in the case of branched pathways) in the last column of the matrix. See Cortassa and Aon (1994a), for the numerical values of the elasticity coefficients used to calculate the control coefficients for each steady state.

The numbers in the last column of the matrix correspond to the control coefficient of the enzyme i on the flux ratio, defined according to the following expressions:

$$C_{E_i}^{J_r} = \frac{d\ln(J_r)}{d\ln(E_i)} \tag{7.23}$$

where

$$J_r = \frac{J_1}{J_2} = \frac{J_{TCA}}{J_{ADH}} \tag{7.24}$$

$$d\ln(J_r) = d\ln(J_1) - d\ln(J_2) \tag{7.25}$$

with $J_1 = J_{TCA}$ and $J_2 = J_{ADH}$, the flux through the branch to the TCA cycle and ethanol production, respectively.

Assuming that the control coefficients $C_{E_i}^{J_{TCA}}$ and $C_{E_i}^{J_{ADH}}$ are both positive, the sign of $C_{E_i}^{J_r}$ would reflect the relative weight of the control that E_i exerts on J_{TCA} and J_{ADH}, i.e. if negative it means that $C_{E_i}^{J_{TCA}}$ is lower than $C_{E_i}^{J_{ADH}}$.

The matrix calculations were performed with the software Maple (Waterloo Maple Software, Waterloo, Ontario)

About the dynamic behaviour of microorganisms

8

8.1 THE PHASE SPACE AND THE CHEMOSTAT

The use of chemostat or continuous cultures to study a microorganism at the steady state provides a rigorous experimental approach for the quantitative evaluation of the microorganism's physiology and metabolism. In addition, chemostat cultures allow definition of the phase of behaviour which suits a purpose that we may fix ourselves – for example, maximum substrate consumption and output fluxes of metabolic by-products of interest.

The description of a particular network of reactions or processes implies the characterization of all possible kinds of steady state that may be attained by this particular network in the whole parametric space (Chapter 1) and the specification of how a living system containing these networks will behave under defined environmental conditions. In order to achieve this aim we should additionally know the structure of control of a metabolic network for a finite range of parameters and not only for a small neighbourhood around a steady state. When enough experimental data are available about the kinetics of any metabolic pathway, a mathematical model can be formulated and its dynamic behaviour analysed according to dynamic bifurcation theory (DBT) (section 1.4). DBT provides the possibility of analysing the steady states of a system continuously, as a function of a bifurcation parameter (Doedle, 1986; Abraham, 1987; Aon *et al.*, 1991). When plotted, the latter renders a bifurcation diagram (Chapter 1). This is the rationale employed in the numerical approach proposed as the basis of a biothermokinetic method which combines metabolic control analysis (MCA) and DBT to explore quantitatively and predict new (un)desirable behaviours in metabolic pathways and microorganisms (Aon *et al.*, 1996b; see also Appendix 8A). The behaviour of the steady state fluxes as a function of a parameter in the chemostat is, conceptually, the same as a bifurcation diagram (Figure 8.1).

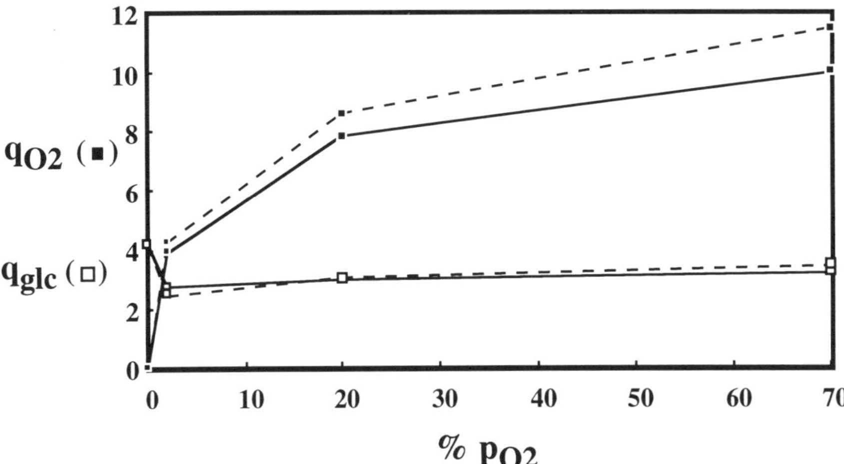

Figure 8.1 Steady state specific fluxes of oxygen and glucose consumption as a function of p_{O_2} in continuous cultures of *Pseudomonas mendocina* growing in synthetic medium. *P. mendocina* was grown ($D = 0.25$ h^{-1}) in synthetic medium in a chemostat at 30°C, pH 7.0, in the presence of 3 g l^{-1} of glucose and subjected to aerobic→microaerophilic (plain lines) and vice versa (dashed lines) transitions in p_{O_2}. Each point is the mean of duplicates from two independent cultures and experiments; q_{O_2} and q_{glc} are in mmol h^{-1} g^{-1} dw. (From Verdoni, Aon and Lebeault, 1992, by permission of the American Society for Microbiology, ASM Press.)

The biothermokinetic method provides a straightforward means of quantitation in both kinetic and thermodynamic terms. The link between the analysis of the dynamic behaviour of a biological system based on DBT and the control in terms of MCA is met through the first derivative of the double logarithmic plot of bifurcation diagrams whose slope directly shows flux or metabolite control coefficients. In summary, the slope of the log–log form of the bifurcation diagram allows graphical and ready estimation of the extent of control exerted on the flux or on metabolite concentrations by a bifurcation parameter under study (Figure 8.1; Appendix 8A).

The biothermokinetic analysis also gives information about the structure of control of a network for finite ranges of parameters, either in the physiological domain or outside it. The contribution of DBT to the present approach is to offer the possibility of analysing the type and stability of steady states (i.e. asymptotic, oscillatory or multiple steady states) for finite ranges of parameters. Thus, this approach overcomes some of the limitations of MCA when applied to metabolic pathways with potential sources of nonlinear behaviour such as feedback or feed-forward, and allows analysis of the performance of biochemical systems in energetic terms. Overall, the biothermokinetic method may be used for the prediction and quantitation of fluxes and performance in, say, biochemical or microbiological systems (Chapter 11) (Aon *et al.*, 1996b).

If metabolic pathways are considered as energy transducers, control coefficients of thermodynamic efficiency may be calculated for different parameters.

8.2 FUNCTIONAL SELF-SIMILARITY AT DIFFERENT LEVELS OF DESCRIPTION OF LIVING SYSTEMS

The chemical reaction for the synthesis of a microorganism is a complex one. The following chemical equation represents the amount of carbon, NADPH, NADH, NH_4^+ and CO_2 required or produced during the synthesis of 1 g of yeast cell biomass from glucose as carbon source:

$$7.4C_6H_{12}O_6 + 7.2NH_3 + 7.9NADPH + 14.5NAD^+ \rightarrow$$
$$9.9C_4H_{7.5}O_{1.7}N_{0.73} + 4.7CO_2 + 7.9NADP + 14.5NADH + 18H_2O \quad (8.1)$$

The arrow which indicates the direction of the reaction 'hides' a complex network of around 1000 chemical reactions occurring inside a cell; for example, bacteria (Bailey and Ollis, 1977; Stouthamer and Van Verseveld, 1987), yeast (Cortassa *et al.*, 1995) or mammalian.

The ensemble of processes occurring in a whole network, or in a part or in only one chemical reaction taking place in it, can be described by a similar phenomenological equation. The Monod and Michaelis–Menten formalisms are mathematically isomorphic equations, i.e. they describe similar functional relationships between dependent and independent variables, though at different levels of description. In the case of microbial growth, the dependent variable is the growth rate of a microorganism, μ (the result of the whole metabolic network of cells in a reactor), as a function of μ_{max} and the substrate concentration (independent variable), whereas the Michaelis–Menten equation represents the rate of an enzyme-catalysed chemical reaction as a function of V_{max} and the substrate concentration:

$$\mu = \frac{\mu_{max}[S]}{K_s + [S]} \quad (8.2)$$

Equation 8.2 is the usual form of the Monod equation used extensively by microbiologists to describe phenomenologically the growth of microbial cultures. As mentioned above, the Monod equation is mathematically isomorphic to the Michaelis–Menten formalism that with slight modifications of notation, i.e. V_{max} instead of μ_{max} and K_M instead of K_S, is in turn the phenomenological description adopted by biochemists to characterize a 'Michaelian' enzyme. Figure 8.2 plots the type of functional dependence described by equation 8.2.

The fact that widely different levels of organization (i.e. a microbial culture or a single chemical reaction) may be described with the same phenomenological law stresses the existence of a functional self-similarity (Chapter 2) between biological processes operating at different levels of organization. Independently of mechanistic details, the functional dependence (described by equation 8.2) existing between variables and parameters is isomorphic at two widely differing spatio-temporal coordinates (see Chapter 2). This isomorphic functional transformation is the basis of the concept of functional self-similarity.

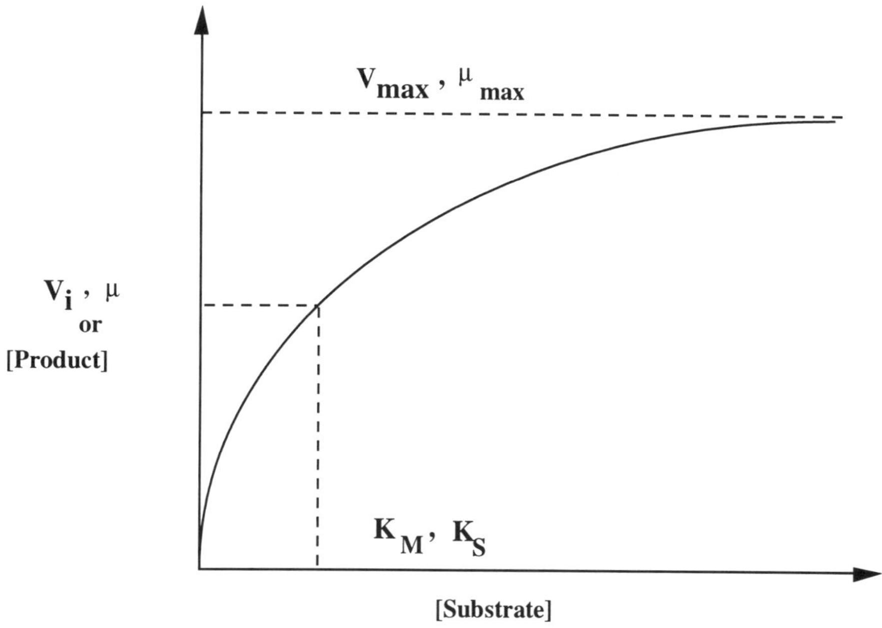

Figure 8.2 Functional self-similarity. The Monod or Michaelis–Menten equations phenomenologically describe similar dependences between the growth rate of a microbe culture (μ) or initial velocities (v_i) of an enzymatic reaction versus the substrate concentration. Both the enzyme and the microbial population represent widely distant levels of organization. However, independently of mechanistic details, the functional relationship existing between variables and parameters is isomorphic for either an enzymatic reaction or the complex chemical reaction to synthesize a microorganism. This isomorphic functional transformation is the basis of the concept of functional self-similarity (Chapter 2).

The concept of functional self-similarity is analogous to geometrical self-similarity, which is one of the main characteristics of fractals (Chapter 6).

8.3 BLACK AND GREY BOXES: THE DESCRIPTION OF THE METABOLIC BEHAVIOUR OF MICROORGANISMS

The description of a complex network of chemical reactions may be achieved at different levels of description or detail. Schematically, there are two levels of description: black and grey boxes. In the approach called black box, only the input(s) and output(s) of the system (e.g. microorganisms) are specified (Roels, 1983). If we progressively improve our descriptive power by increasing the knowledge about what is occurring inside the box (e.g. mechanisms, reactions), it then becomes a grey box. The grey level of the box will be lighter depending upon how deep is our knowledge about the physiological and dynamic condition of the system under study (Figure 8.3).

The dynamic states of cultures of microorganisms are essentially of two types: transient or steady. The main difference between both is their time dependence (Figure 1.1). At the steady state, in turn, two types of behaviour may be described: non-expanding (steady state) and expanding (balanced growth). The former are obtained in chemostat cultures, whereas the latter correspond to batch logarithmic growth.

8.3.1 STEADY AND BALANCED GROWTH

Growth is balanced when the specific rate of change of all metabolic variables (concentration or total mass) is constant (Barford *et al.*, 1982):

$$\frac{1}{x_{iAV}}\frac{\Delta x_i}{\Delta t} = \text{constant} \tag{8.3}$$

where Δx_i = change in x_i during time Δt; x_{iAV} = average value of x_i during time Δt.

A steady state is a sort of balanced growth but with the further requirement that the overall rate of change of any metabolic variable or biomass be zero:

$$\frac{dx_i}{dt} = 0 \tag{8.4}$$

We have applied or developed for ourselves different quantitative methods at different levels of description to both types of dynamic behaviour that are described in the titles of the following sections.

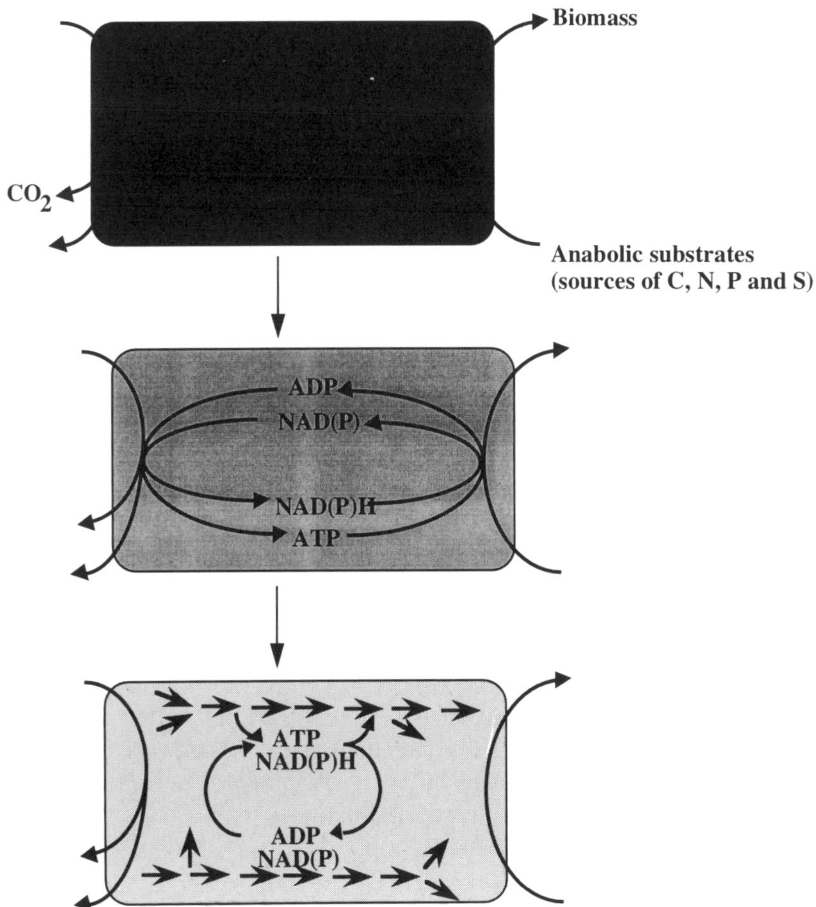

Figure 8.3 Black and grey boxes: levels of description of the metabolic behaviour of microorganisms. The black box (top), grey box (middle) and light-grey box (bottom) show successively more detailed descriptions of the microbial metabolism concerning biochemical networks and cellular energetics. The black box description takes into account only relationships between inputs (i.e. substrates) and outputs (i.e. biomass, CO_2). With more detailed descriptions of the black box, a rigorous correspondence may be established between 'macroscopic' variables such as biomass or product yields and the biochemistry and physiology of cells. In this vein, we developed a method to quantitate instantaneously the fluxes through the main anabolic pathways during yeast growth (Figure 8.4 and section 8.3.2). This method was further applied to describe the metabolic and energetic status of yeast cells mitotically or meiotically cycling (Chapter 11).

8.3.2 QUANTITATION OF FLUXES OF CARBON, PHOSPHORYLATION AND REDOX INTERMEDIATES DURING BALANCED GROWTH OF *SACCHAROMYCES CEREVISIAE*

One main conceptual advantage in the description of physiological or metabolic behaviours of microorganisms is that by subjecting them to a stressful situation not all the pathways of the whole biochemical network will be equally active under that particular situation. Under balanced growth conditions, pathway utilization (for example, glyoxylate or tricarboxylic acid cycles and glycolysis or gluconeogenesis) may be quantified according to the required flux of precursors for biosynthesis as well as the minimal catabolic fluxes needed to fulfil the ATP and redox equivalent requirements to sustain growth (Holms, 1986). We developed a method that allows us to quantitate the instantaneous fluxes of anabolic precursors needed for the synthesis of 1 g of biomass for cultures growing steadily (logarithmically) (Cortassa *et al.*, 1995).

The basis for a flux evaluation was laid down by the knowledge about yeast chemical composition and the metabolic pathways used by this microorganism (Fraenkel, 1982; Jones and Fink, 1982; Bruinenberg *et al.*, 1983). According to the method developed, we open the cell's black box and learn about the dynamics of its metabolic and energetic machinery by quantifying anabolic fluxes when yeast are growing on minimal medium on various carbon substrates (glucose, glycerol, lactate, pyruvate, acetate or ethanol). The fluxes through the central amphibolic pathways were calculated from the total required amount of a given carbon intermediate multiplied by the growth rate (Figure 8.4) (Holms, 1986; Cortassa *et al.*, 1995).

Figure 8.4 (see facing page) Metabolic pathways leading from glucose to the key intermediary metabolites (a), and anabolic fluxes through the central amphibolic pathways (b). (a) Glucose as carbon source is converted to key intermediary metabolites (shaded background) through the central amphibolic pathways. The main catabolic routes of degradation of the substrate are also indicated. Enzymes catalysing the indicated steps are shown in italics. (b) The estimated fluxes (in ovals) during growth on glucose were calculated from the sum of the intermediate requirements multiplied by the growth rate. The fluxes refer to the rate of intermediate disappearance. Fluxes are given in mmol h^{-1} g^{-1} dw. The calculation of fluxes and their experimental corroboration have also been performed for yeast batch growth on glycerol, pyruvate, lactate, acetate or ethanol. (From Cortassa, Aon and Aon, 1995, *Biotechnology and Bioengineering*, copyright John Wiley & Sons, Inc.)

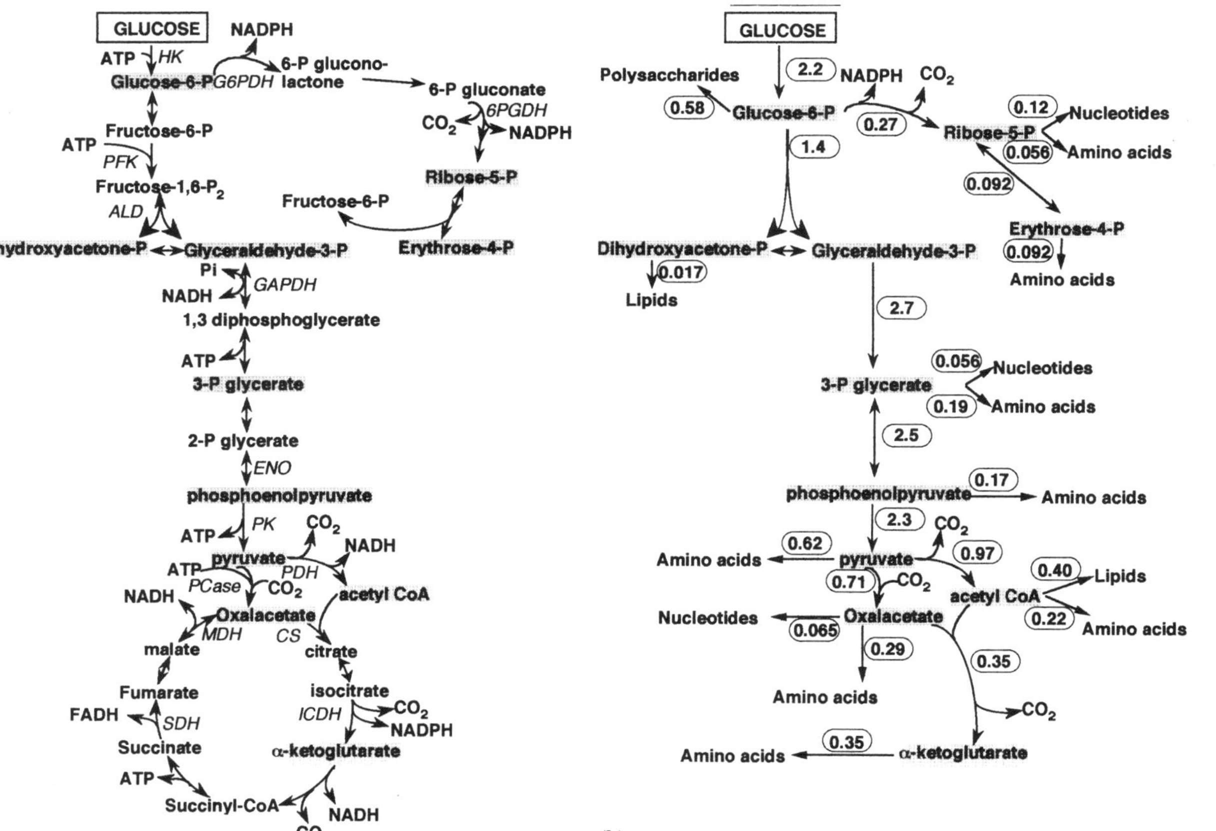

GLUCOSE
ATP
HK
Glucose-6-P
G6PDH
NADPH
6-P glucono-lactone
6-P gluconate
6PGDH
CO2
NADPH
Fructose-6-P
ATP
PFK
Fructose-1,6-P2
ALD
Ribose-5-P
Fructose-6-P
Dihydroxyacetone-P
Glyceraldehyde-3-P
Erythrose-4-P
Pi
NADH
GAPDH
1,3 diphosphoglycerate
ATP
3-P glycerate
2-P glycerate
ENO
phosphoenolpyruvate
ATP
PK
CO2
NADH
pyruvate
ATP
PCase
PDH
CO2
acetyl CoA
NADH
Oxalacetate
MDH
CS
malate
citrate
Fumarate
isocitrate
FADH
SDH
ICDH
CO2
NADPH
Succinate
α-ketoglutarate
ATP
Succinyl-CoA
NADH
CO2
(a)

GLUCOSE
Polysaccharides
0.58
Glucose-6-P
2.2
NADPH
CO2
0.27
Ribose-5-P
0.12
Nucleotides
0.056
Amino acids
0.092
Erythrose-4-P
0.092
Amino acids
1.4
Dihydroxyacetone-P
Glyceraldehyde-3-P
0.017
Lipids
2.7
3-P glycerate
0.056
Nucleotides
0.19
Amino acids
2.5
phosphoenolpyruvate
0.17
Amino acids
2.3
CO2
0.62
Amino acids
pyruvate
0.97
0.71
CO2
acetyl CoA
0.40
Lipids
0.22
Amino acids
Nucleotides
0.065
Oxalacetate
0.29
0.35
Amino acids
CO2
0.35
Amino acids
α-ketoglutarate
(b)

Substrate features such as the degree of reduction with respect to the biomass, the enthalpy of combustion and the metabolic pathways that are activated to assimilate the carbon source drive the cell to a different energetic status. In particular, *S. cerevisiae* displays a variety of growth rates on different carbon sources which cannot be attributed solely to the difference in their thermodynamic properties – for example, the heat of combustion of these carbon substrates. For instance, carbon and energy balances are expected to vary together with fluxes through the central amphibolic pathways. The rate at which metabolic pathways should work in order to sustain a certain growth rate will also depend on the nature of the carbon source. In fact, the nature of the assimilation pathway of each carbon source influences the flux values (Figure 8.4). However, substrates sharing most metabolic pathways, such as ethanol and acetate, despite changes in the macromolecular composition – namely, carbohydrate content (34 ± 1% and 21 ± 3%, respectively) – did not show large variations in the overall fluxes through the main amphibolic pathways. For instance, in order to supply anabolic precursors to sustain growth rates in the range of 0.16 h^{-1} to 0.205 h^{-1}, similar large fluxes through acetyl CoA synthase were required by acetate (4.2 mmol h^{-1} g^{-1} dw) or ethanol (5.2 mmol h^{-1} g^{-1} dw) (Cortassa *et al.*, 1995).

The V_{max} activities of key enzymes of the main amphibolic pathways measured in permeabilized yeast cells allowed qualitative confirmation of the operation of those pathways for all substrates and were consistent on most substrates with the estimated fluxes required to sustain growth.

The methodology presented can be applied to other eukaryotic organisms of known chemical composition.

8.3.3 PHYSIOLOGICAL PERFORMANCE OF YEAST CELLS GROWING ON DIFFERENT CARBON SOURCES

The catabolic flux experimentally determined as the difference of the total carbon consumption minus the calculated anabolic flux was much larger than the flux required to replenish precursors for macromolecular monomers (Table 8.1, columns 5 and 7, and Figure 8.4). Experimentally, for glucose, we determined a substrate consumption flux of 10.6 ± 0.8 mmol h^{-1} g^{-1} dw from which only 2.2 mmol h^{-1} g^{-1} dw (or 21%) would be necessary to supply carbon compounds at a growth rate of 0.307 h^{-1} (Table 8.1, columns 2 and 6). The catabolic flux so determined was 3.8-fold higher than the anabolic flux required to sustain growth at a rate of 0.307 h^{-1} (2.2 mmol h^{-1} g^{-1} dw: Figure 8.4; Table 8.1, columns 2 and 6). The required catabolic carbon flux for a fully coupled yeast cell of the composition shown in Table 8.1 at a P/O ratio of one is 0.69 mmol h^{-1} g^{-1}

dw for growth on glucose (the latter value was obtained from the sum of columns 4 and 5 in Table 8.1). The experimental results would imply that 79% of the carbon flux consumed is either oxidized or fermented, compared with a theoretical catabolic flux of 0.69 mmol h^{-1} g^{-1} dw (or 24%) of the total flux of carbon required, i.e. 2.9 mmol h^{-1} g^{-1} dw for a fully coupled yeast growing on glucose (P/O = 1.0: Table 8.1, column 7).

In the case of acetate, 55% of the carbon flux is catabolized (theoretical value 4.2 mmol h^{-1} g^{-1} dw to sustain anabolic precursors supply compared with the experimental 9.4 $\pm$ 0.6 mmol h^{-1} g^{-1} dw) compared with the theoretical 43% needed when yeast metabolism is fully coupled to produce biomass (P/O = 2; 3.1 mmol h^{-1} g^{-1} dw compared with 7.3 mmol h^{-1} g^{-1} dw: Table 8.1, columns 5 and 7).

We computed the theoretical Y_{ATP}^{max} assuming fully coupled metabolism, i.e. that all the energy and precursors produced in catabolism are directed toward biomass synthesis. The required flux of phosphorylation energy takes into account the anabolic demand plus the energy required to sustain catabolic fluxes such as those needed to direct carbon through the pentose pathway (PP) to produce NADPH. The ATP flux (usually defined as q_{ATP}) was in part met by NADH and FADH oxidation produced together with the monomers of macromolecules. An increase in P/O implied an increase in Y_{ATP}^{max} (Table 8.1).

For gluconeogenic substrates, catabolism relied largely on the functioning of the TCA cycle and respiratory chain while growth on glucose adds catabolism through fermentative pathways. For instance, when 90% of the carbon catabolized through glycolysis is directed to ethanol fermentation, the Y_S decreases significantly from 103 to 60 g dw mol^{-1} of glucose without a significant variation in the Y_{ATP}^{max}. This is accompanied by a 3.4-fold increase in the flux directed through glycolysis.

On the other hand, much lower Y_{ATP}^{max} values were calculated if the oxidation of NADH was not taken into account, i.e. 24 compared with 36 g dw mol^{-1} ATP for growth on glucose and 19 compared with 39 g dw mol^{-1} ATP for growth on glycerol. In pyruvate, as it is more oxidized than biomass, and lactate, which is converted to pyruvate via ferricytochrome c-dependent lactate dehydrogenase, the ATP requirement and consequently the Y_{ATP}^{max} varied slightly with NADH oxidation being considered, or not, and with different P/O ratios (Table 8.1, column 8).

A yield index (YI) may be evaluated from the ratio between the substrate flux required to sustain a given growth rate and the experimentally measured flux of substrate consumption (Table 8.1, last column). The comparison between the YI obtained with different carbon sources revealed that carbon substrates with the shortest chains (ethanol

Table 8.1 Fluxes and physiological parameters of yeast cells growing on different carbon sources in minimal medium

Carbon source	Growth rate[a] (h⁻¹)	P/O ratio	Fluxes[b]				$Y_{ATP}^{max\,c}$	Theor Y_S^d	Exptl Y_S^e	Yield index[f]
			PP pathway	Oxidative catabolism	Anabolism	Total carbon				
Glucose	0.307±	g	0.049	0.90	2.2	3.2	24	95	28±3	0.30
	0.005	1	0.10	0.59		2.9	36	103		0.27
		2	0.17	0.18		2.6	76	117		0.24
Glycerol	0.18±	g	0.0067	1.2	2.5	3.7	19	47	15±2	0.33
	0.02	1	0.11	0.57		3.2	39	55		0.28
Pyruvate	0.15±	g	0.00	2.2	2.3	4.5	13	33	10±2	0.32
	0.02	1	0.00	2.1		4.4	14	33		0.31
		2	0.12	1.3		3.7	14	40		0.26
		3	0.20	0.85		3.3	15	44		0.23
Lactate	0.156±	g	0.081	1.9	2.5	4.5	13	34	31±4	0.92
	0.004	1	0.099	1.8		4.5	14	35		0.90
		2	0.23	1.1		3.8	15	41		0.77
		3	0.29	0.68		3.5	15	45		0.71
Ethanol	0.205±	g	0.00	5.7	5.2	11	8.9	19	31±2	>1
	0.002	1	0.00	3.6		8.8	14	23		>1
		2	0.00	1.1		6.4	26	32		0.84
Acetate	0.16±	g	0.00	6.3	4.2	10	8.7	16	17±3	>1
	0.01	1	0.00	5.5		9.6	10	17		>1
		2	0.00	3.1		7.3	11	23		0.77
		3	0.00	1.6		5.8	13	28		0.62

(a) Experimental growth rate (μ)±SEM used for calculation of the flux through the PP pathway, oxidative catabolism and the total (catabolic plus anabolic) flux of substrate. It is equivalent to *ln2*/(doubling time).

(b) Expressed as mmol carbon substrate h⁻¹g⁻¹dw.

(c) Y_{ATP}^{max} is the theoretical yield calculated from the known requirements of ATP for synthesis, polymerization and transport (g yeast dry weight per mol ATP).

(d) Molar growth yield defined as the amount of biomass (g yeast dw) synthesized from 1 mol of carbon substrate.

(e) Experimentally determined molar growth yield for CH1211 strain of S. *cerevisiae* growing in minimal medium on different carbon substrates (g dw mol⁻¹ carbon substrate ± SEM).

(f) Yield index defined as the ratio of the theoretical substrate flux required to sustain the experimental growth rate to the experimental flux of substrate consumption. It may also be obtained from the ratio exptl Y_S (column 10) over theor Y_S (column 9) since the latter are inversely proportional to the substrate consumption rates.

(g) Calculation performed without taking into account the oxidation of NADH produced in anabolism by the mitochondrial electron transport chain.

Data reproduced from Cortassa, Aon and Aon (1995) *Biotechnology and Bioengineering*, John Wiley & Sons, Inc.

and acetate) and lactate were more efficiently utilized to produce yeast cell mass, exhibiting YI of 0.84, 0.77 and 0.77, respectively, for $P/O = 2$ (Table 8.1). This index enabled the validation of the calculated fluxes and molar growth yield by comparison with experimental data.

8.4 BIOENGINEERING OF BIOCATALYSTS FOR BIOTRANSFORMATION: A TRANSDISCIPLINARY APPROACH

The near exhaustion of natural resources is imposing on our present civilization an immediate need to recycle wastes rather than dispose of them (Odum, 1989) and to remedy pollution of water, soil and air by industrial activity. Bioremediation, bioaccumulation and biodegradation are several ways, with different aims, of performing those tasks that are assembled in a complex multidisciplinary field called environmental biotechnology (Litchfield, 1991; Alexander, 1994) (Figure 8.5). Bioremediation is appropriate where no toxic by-products will be formed; it is economical and there are microorganisms which can degrade the contaminants (Lichtfield, 1991; Alexander, 1994). The possibility of toxic by-product formation depends on the metabolic pathways used by the microorganisms to degrade the wastes.

The main goals of disciplines such as genetic ecology (Goldstein, 1991) or microbial ecology (Lynch and Hobbie, 1988) are to exploit the existing genetic potential in natural populations of microorganisms and to study how environmental factors affect the abundance and function of genes. Fortunately, our biosphere posseses an enormous catabolic potential, mainly in microorganisms that already exist in several ecological niches with diverse either 'natural' selection pressures (temperature, salinity, pH) or 'artificial' ones established by contaminants that are themselves in polluted environments (Lowe *et al.*, 1993). The basis of the broad and highly multidisciplinary fields of metabolic and cellular engineering (less than 20 years old) has now been established (for reviews, see Bailey, 1991, and Cameron and Tong, 1993). At present, metabolic and cellular engineering are both defined as the purposeful modification of intermediary metabolism or cell properties using recombinant DNA techniques, include the following categories (Cameron and Tong, 1993):

- improved production of chemicals already produced by the host organism;

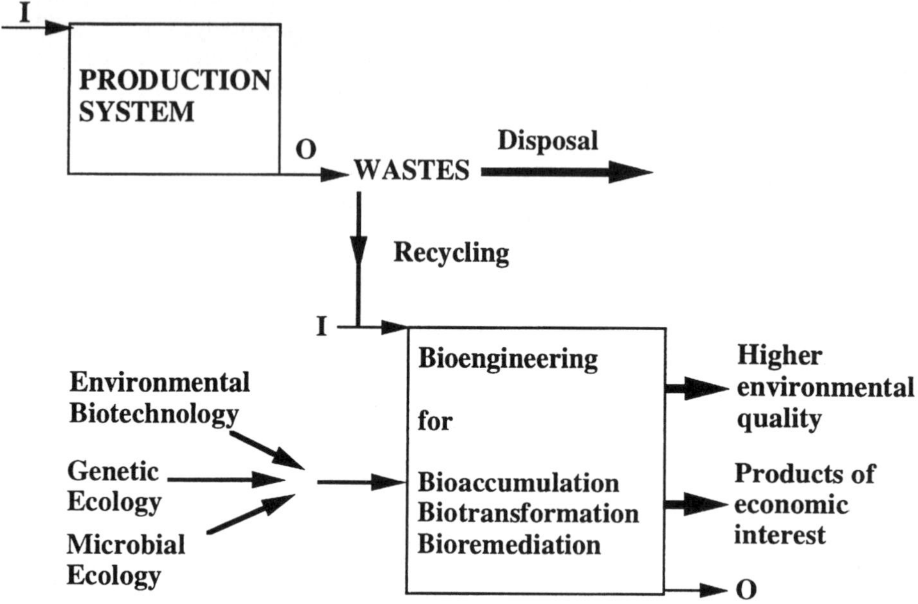

Figure 8.5 Environmental biotechnology in the integrated production–waste-recycling system. Environmental biotechnology participates in a complex multidisciplinary field in which it interacts strongly with genetic and microbial ecology, for the metabolic (or cellular) engineering of microorganisms for biotransformation or bioremediation. The fact that wastes are recycled, rather than disposed of, introduces the suggestion that bioaccumulation, bioremediation or biotransformation may be oriented toward higher environmental quality together with the obtention of products of economic interest from naturally existing or engineered microorganisms.

- extended substrate range for growth and product formation;
- addition of new catabolic activities for degradation of toxic chemicals;
- production of chemicals new to the host organism;
- modification of cell properties.

We have proposed the optimization of specific biotransformation processes by quantitatively acting on the microorganism itself rather than on the process, with the double purpose of achieving higher performance of products of economic interest and higher environmental quality. The possibility of redirecting substrate fluxes to microbial products or biomass may be achieved either through modification of environmental parameters or by acting on the microorganism itself (Figure 8.6) (Aon *et al.*, 1996b). The originality of the present approach is that it integrates several disciplines in a purposeful scheme – namely, microbial physiology and bioenergetics, thermodynamics and enzyme kinetics, biomathematics and biochemistry, genetics and molecular biology, Thus, it will be called a **transdisciplinary approach** (TDA).

The general scheme of the TDA approach for bioengineering of biocatalysts is of an iterative nature and may be outlined as follows.

1. Physiological and bioenergetic studies in continuous culture. The expected outcome will be the determination of the phases of interesting behaviour according to the aim of the bioengineering.
2. Metabolic studies, mathematical modelling and metabolic control analysis. Determination of intracellular concentrations of metabolites and enzyme rate laws. The determination of the elasticity coefficients and their array in a matricial form is the outcome of this step.
3. Inversion of the matrix of elasticity coefficients. Such inversion renders a matrix of control coefficients both for flux and metabolite concentrations. Thus, this step allows the identification of the rate-controlling steps of a flux or metabolite level in a metabolic pathway.
4. Genetic engineering. Gene overexpression or up-modulation of the enzymes which control the flux. The outcome of this step is a modified microorganism optimized for a specific biotransformation process.

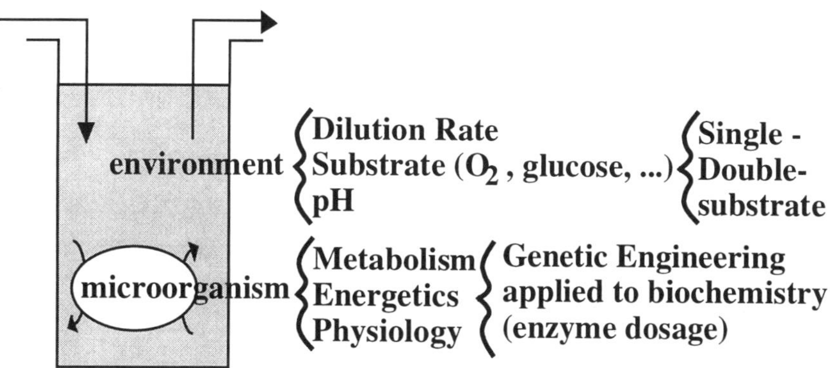

Figure 8.6 The microorganism as a target for bioengineering at metabolic, energetic and physiological levels in chemostat cultures. (From Aon, Cortassa and Verdoni, in *Gas, Oil and Environmental Biotechnology*, V, Institute of Gas Technology, Chicago; in press.)

5. This step iterates the first step. The engineered microorganism should be assayed under the physiological conditions defined in that first step. The assay should allow the evaluation of the improvement achieved in the biotransformation process with the use of the engineered microorganism.

The following sections describe in detail the first three steps as applied to either prokaryotic or eukaryotic microorganisms.

8.5 BIOENERGETICS AND THE GROWTH RATE OF MICROORGANISMS

It seems obvious that there is a relationship between the growth rate of a microorganism and its physiology and bioenergetics; however, the nature of that relationship is not trivial. One aspect of that association was specified by Linton and Stephenson (1978) and Andersen and von Meyenburg (1980). It consisted of a correlation that was found to exist between the carbon and energy content of any given organic substrate as defined by its heat of combustion and the growth yield (Linton and Stephenson, 1978) (Figure 8.7a). It was shown that growth yields are directly proportional to the heat of combustion of organic substrates over the range 2.5–11.0 kcal g^{-1} carbon substrate. These authors went further to suggest that when the heat of combustion of a given organic compound is less than 11.0 kcal g^{-1} substrate carbon, the energy generated during the biological oxidation of the compound is insufficient to reduce all the available carbon in the molecule to the level of cell carbon, i.e. growth of the microorganism is basically energy-limited. Another finding was that, apparently, there is a maximum limit to the growth yield (1.43 g dw g^{-1} substrate carbon) which corresponded to 68% of carbon conversion efficiency by assuming a bacterial carbon content of 48% (Linton and Stephenson, 1978).

Andersen and von Meyenburg (1980) varied the growth rate of *E. coli* in batch cultures with different substrates and found that the growth yield increased with the growth rate. However, that increase in growth yield was correlated with a constant specific rate of oxygen consumption which led them to suggest that aerobic growth of *E. coli* in batch cultures is limited by the rate of respiration and the concomitant rate of ATP generation by oxidative phosphorylation (Figure 8.7b).

According to the data of Hempfling and Mainzer (1975) for *E. coli* B and Tempest and Neijssel (1984) for *Klebsiella aerogenes* NCTC 418, it could be shown that when the growth rate was varied through the dilution rate in chemostat cultures (not by the nature of the organic substrate), the rate of respiration increased with growth rate (i.e. dilution rate) (see also review by Marr, 1991). In *E. coli*, except for succinate, the growth yield was constant and the oxygen maintenance was widely different: the higher the heat of combustion of the substrate, the higher was the energy of maintenance) (Hempfling and Mainzer, 1975; Marr, 1991).

When bacteria are grown at a number of different growth rates by adding different carbon sources or different supplements of amino acids, vitamins or nucleosides, the macromolecular composition of the cells changes (Schaechter *et al.*, 1958). The DNA amount per cell varies over a four-fold range (Cooper, 1991).

Bioenergetically, higher growth yields are obtained upon increasing growth rates at maximum rates of respiration if the changes in growth rate are induced by the carbon substrate. Under these conditions, the relative macromolecular content per bacterial cell increases with growth rate (Cooper, 1991). Approximately constant growth yields with increasing specific rates of respiration were obtained with increasing growth rates in chemostat cultures. In the former case, the rate of provision of ATP by oxidative phosphorylation appears to limit growth (Andersen and von Meyenburg, 1980), while in the latter precursor metabolites may control growth rate (Marr, 1991).

Low cell proliferation rates are apparently related to high metabolic uncoupling and high rates of metabolite production (Stouthamer, 1979; Kell, 1987; Linton, 1990). Bioenergetically, then, it appears that the highest rates of metabolite production occur in organisms possessing low growth yields and the highest capacity to dissipate metabolic energy. Microorganisms displaying low proliferation rates and metabolic uncoupling (i.e. low degree of coupling between anabolism and catabolism) apparently show the highest capacity to dissipate metabolic free energy through high rates of metabolite production (Tempest and Neijssel, 1984, 1992; Kell, 1987; Linton, 1990; Verdoni *et al.*, 1990; Aon *et al.*, 1995).

8.6 PHYSIOLOGICAL AND BIOENERGETIC STUDIES IN CONTINUOUS CULTURES: ENVIRONMENTALLY INDUCED REDIRECTION OF METABOLIC FLUXES

The regulation of the degree of coupling between catabolic and anabolic fluxes will in turn modulate the amount of flux redirection toward microbial products (Chapter 11). Microbial products may be divided into two groups: fermentation products, and proteins and polysaccharides.

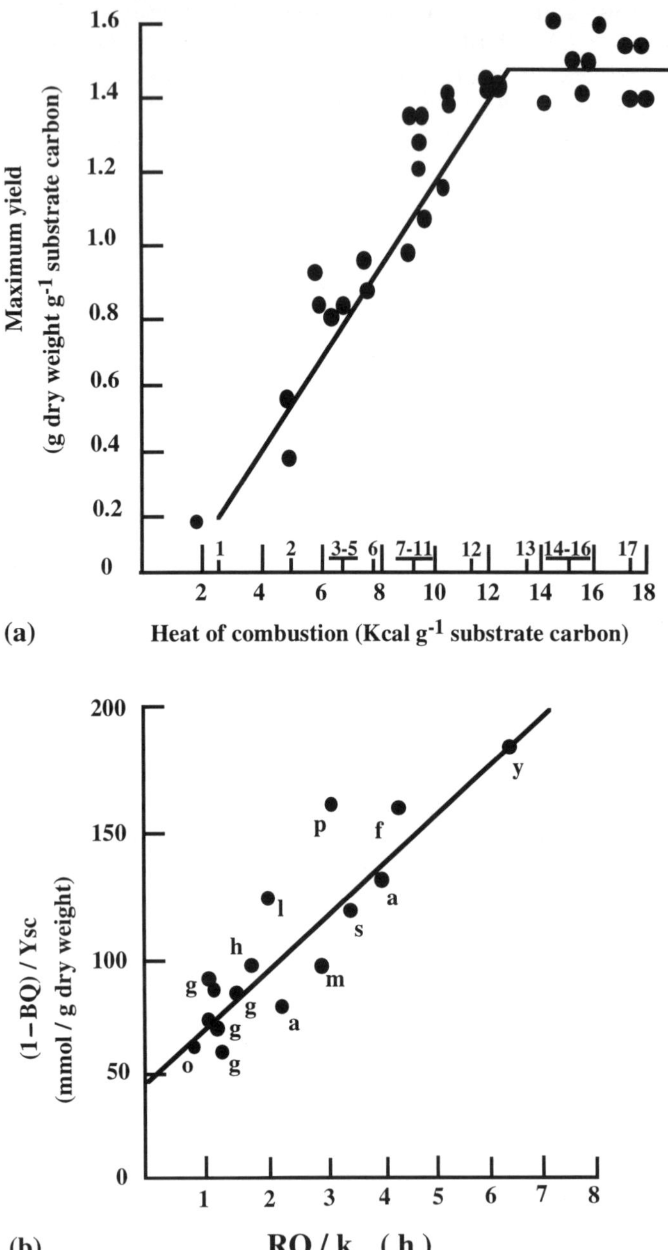

1.6
1.4
1.2
1.0
0.8
0.6
0.4
0.2
0
Maximum yield
(g dry weight g⁻¹ substrate carbon)
1
2
3-5
6
7-11
12
13
14-16
17
2   4   6   8   10   12   14   16   18
Heat of combustion (Kcal g⁻¹ substrate carbon)
(a)
200
150
100
50
0
(1 – BQ) / Ysc
(mmol / g dry weight)
y
p
f
l
a
s
h
m
g
g
a
g
o
g
1   2   3   4   5   6   7   8
RQ / k   (h)
(b)

Redirection of fluxes toward fermentation products may be generated as a result of alteration in the redox and phosphorylation potentials (Aon *et al.*, 1991; Verdoni *et al.*, 1992; Aon *et al.*, 1996b). During transitions from aerobic to oxygen-limited or anaerobic conditions, the correct balance of both redox (NADH/NAD; NADPH/NADP) and phosphorylation (ATP/ADP) couples (Senior *et al.*, 1972; Kell *et al.*, 1989; Anderson and Dawes, 1990) are germane to the question of the excretion of fermentation products. Let us exemplify the latter with the behaviour of *Pseudomonas mendocina* during growth. In continuous culture, *P. mendocina* growing in synthetic medium induced a fermentative metabolism under microaerophilic conditions with excretion of ethanol as well as lactic, formic and acetic acids. From a redox point of view, the fermentation of glucose was balanced by lactate alone or by a 50:50 mixture of ethanol plus acetate (Verdoni *et al.*, 1992). The amounts of fermentation product detected in the extracellular medium suggested that *P. mendocina* attained the redox balance by lactic acid and ethanol production rather than equimolar amounts of ethanol and acetate. Under those microaerophilic conditions, *P. mendocina* could redirect not only catabolic pathways but induced new anabolic ones such as uronic acids or alginate synthesis (Verdoni *et al.*, 1992). In fact, around 50% of the carbon was found in the extracellular medium as uronic acids.

Figure 8.7 (see facing page) Growth yields in relation to the energy content (a) and growth rates (b) in various organic substrates. (a) The maximum growth yields observed in a wide range of organic substrates for various microorganisms growing in batch and continuous culture are plotted against the heat of combustion for various heterotrophic growth substrates: 1, oxalate; 2, formate; 3, citrate; 4, malate; 5, fumarate; 6, succinate; 7, acetate; 8, benzoate; 9, glucose; 10, phenylacetic acid; 11, mannitol; 12, glycerol; 13, ethanol; 14, propane; 15, methanol; 16, ethane; 17, methane. (Reproduced from Linton and Stephenson, 1978, by permission of Elsevier Science.) (b) Corrected reciprocal growth yield (1 − BQ/Y_{sc}) plotted against RQ/k values for batch cultures growing on various carbon sources. Letters indicate carbon sources as follows: a, acetate; f, fumarate; g, glucose; h, galactose; l, lactate; m, mannose; o, glycerol; p, pyruvate; s, succinate; y, glycolate. The line drawn in the figure was obtained by linear regression analysis, excluding the values from the lactate, pyruvate and mannose cultures ($r = 0.97$; slope $= 19$ mmol h^{-1} g^{-1} dw; y-intercept $= 47$ mmol g^{-1} dw). Y_{sc} is the ratio of the cell mass formed at the end of the exponential growth phase to the molar amount of substrate carbon added to the culture; BQ is the ratio of the specific rate of by-product carbon production over the specific rate of substrate carbon consumption; RQ is the respiratory quotient. (Reproduced from Andersen and von Meyenburg, 1980, by permission of the American Society for Microbiology, ASM Press.)

According to bioenergetic calculations, a Y_{ATP}^{max} of 25 g dw mol^{-1} ATP was obtained, which is consistent with the theoretical maximum expected for *E. coli* growing on glucose and mineral salts (Stouthamer, 1979). Apparently, an increase in the stoichiometry of phosphorylation (P/O ratio change from 1 to 3) allowed the alginate pathway to become a net ATP-yielding synthetic process (Jarman and Pace, 1984).

Another microorganism such as *Azotobacter beijerincki* (an aerobe that catabolizes glucose via the Entner–Doudoroff pathway like Pseudomonads) accumulates intracellularly up to 50% of its dry weight as polyhydroxybutyrate (PHB) under oxygen limitation (Senior *et al.*, 1972; Anderson and Dawes, 1990). The PHB synthetic pathway functions as a store of carbon and energy and as an electron sink under oxygen limitation.

Saccharomyces cerevisiae, under conditions of oxidative metabolism at low glucose concentrations, utilized most of the ATP generated by catabolism to synthesize biomass (high degree of coupling) (Beck and von Meyenburg, 1968; Fiechter *et al.*, 1981; Bruinenberg *et al.*, 1983; Rieger *et al.*, 1983; Kappeli, 1986; Gancedo and Serrano, 1989; Alexander and Jeffries, 1990). In the absence of oxygen or when the cells display respiro-fermentative metabolism (Figure 8.8), since a great deal of the ATP generated by catabolism is not used for growth, the microorganism is said to be uncoupled (Stouthammer, 1979; Tempest and Neijssel, 1984; Verdoni *et al.*, 1990). Under those conditions ethanol is the main catabolic product (Figure 8.8).

Thus, by choosing the environmental conditions (sugar concentration, type of sugar, oxygen availability), it is possible to compel a kind of sugar catabolism to favour the quick consumption of sugars (e.g. hexoses, pentoses) derived from, say, agricultural waste products, to produce some microbial product of interest. For instance, it is possible in yeasts, by adjusting the degree of coupling between catabolism and anabolism, to redirect metabolic fluxes for the production of ethanol or biomass (Figure 8.8). One way to further adjust the degree of coupling – for example, to achieve maximal uncoupling – is to modify the most rate-controlling steps of a catabolic pathway to get higher sugar catabolism. This is the subject of the following sections.

8.6.1 METABOLIC CONTROL ANALYSIS AND MICROBIAL PHYSIOLOGY

Environmentally induced redirection of metabolic fluxes was quantitatively integrated as a first exploratory step to bioengineering with a transdisciplinary approach (Aon *et al.*, 1996b) (section 8.4). Indeed, metabolism is a network of reactions that is much too complicated to

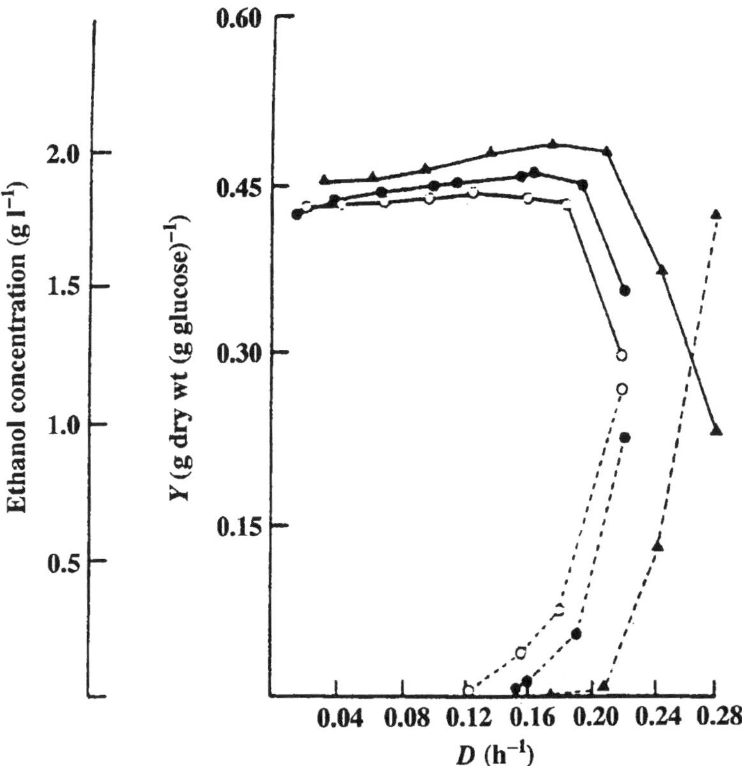

Figure 8.8 Physiological and metabolic variables of *S. cerevisiae* under carbon (filled triangle), phosphate (filled circle) or nitrogen (open circle) limitations at different growth rates in chemostat cultures. The triggering of fermentative metabolism at high dilution rates occurred at lower growth rates in nitrogen or phosphate limitations with respect to carbon limitation. The continuous and dashed lines indicate growth yield and ethanol concentration, respectively. (Reproduced from Franco, Smith and Berry, 1984, by permission of the Society for General Microbiology.)

address in its full complexity. MCA was applied to two sorts of data or 'grey boxes', each with different levels of grey, to approach the elucidation of the pathway(s) or step(s) that hamper(s) an increase in the desired fluxes (Figure 8.3 and below). In the 'grey dark' box (Figure 8.3), metabolic networks are considered to be constituted by 'blocks' of reactions – such as glycolysis, tricarboxylic acid cycle, oxidative phosphorylation, biomass synthesis – lumped together and related to each other by mass exchanges. The 'raw material' of this first quantification comprises physiological data obtained from chemostats such as the fluxes corresponding to the specific rates of glucose $[q_{glc}]$ and O_2 $[q_{O_2}]$ consumption, and the specific rates of production of, say, carbon dioxide $[q_{CO_2}]$ or organic acids, attained at each steady state, as well as the amount of biomass produced.

We present examples of analysis performed with MCA on the quantification of the control exerted by oxygen levels and glucose consumption $[q_{glc}]$ upon catabolic and anabolic fluxes in *P. mendocina* and *S. cerevisiae*, respectively, growing in continuous cultures.

8.6.2 REDIRECTION OF CATABOLIC AND ANABOLIC FLUXES DURING GRADUAL AEROBIC TO OXYGEN-LIMITED TRANSITIONS OF CHEMOSTAT CULTURES OF *PSEUDOMONAS MENDOCINA*: QUANTIFICATION OF THE CONTROL EXERTED BY O_2 ON CATABOLIC AND ANABOLIC FLUXES

Several metabolic and physiological adaptive changes were shown by *P. mendocina* when subjected to energetic stress induced by anaerobiosis and absence of electron acceptors (Verdoni *et al.*, 1990). In batch cultures, the pattern of growth of *P. mendocina* in rich medium with glucose excess and absence of electron acceptors was accompanied by qualitative and quantitative differences in the proton motive force, mixed-acid fermentation, growth arrest and metabolic uncoupling, i.e. a large fraction of the ATP produced by catabolic processes was not directly used for biomass synthesis (Chapters 11 and 12).

When the aerobic to oxygen-limited transitions where gradually performed in chemostat cultures of *P. mendocina* in synthetic medium, at least three phases in the physiological behaviour could be distinguished and characterized by different control coefficients of the p_{O_2} on catabolic and anabolic fluxes (Verdoni *et al.*, 1992) (Figure 8.1, Table 8.2, Appendix 8A). The aerobic to oxygen-limited transition and vice versa were induced through a decrease or an increase of the p_{O_2} (Figure 8.1, plain and dashed lines, respectively). First, at high p_{O_2} when the culture was glucose-limited the catabolism was oxidative (RQ = 1.0) and the levels of oxygen mainly controlled catabolic fluxes. At lower p_{O_2}, oxygen controlled both catabolic and anabolic fluxes (Table 8.2). When nitrogen was flushed into the chemostat, microaerophilic conditions were induced. Under these conditions uronic acids (or alginate) and products from a mixed-type fermentative metabolism were excreted. The production of alginate becomes a net ATP-yielding process at high P/O ratios, allowing the microorganism to achieve a high yield of biomass on ATP (Y_{ATP}^{max}) as high as 25 g dw mol^{-1} ATP, which corresponds to the theoretical maximum for *E. coli* cells.

Table 8.2 Flux and concentration control coefficients by oxygen (calculated as in Appendix A1) of *Pseudomonas mendocina* growth in continuous culture at 30°C, pH 7.0 and $D=0.25$ h^{-1} (data reproduced from Verdoni, Aon and Lebeault, 1992, by permission of the American Society for Microbiology, ASM Press)

pO_2 (%)	$C^{qCO_2}_{pO_2}$	$C^{qO_2}_{pO_2}$	$C^{X}_{pO_2}$	$C^{qglc}_{pO_2}$
70	0.22	0.19	<0.01	0.07
20	0.32	0.30	0.24	0.04
2	0.51	0.52	0.23	−0.08
0.01	0.54	0.53	0.27	−0.10
2	0.31	0.31	0.17	0.10
20	0.26	0.23	−0.09	0.10
70				

8.7 CONTROL OF METABOLIC FLUXES IN LOWER EUKARYOTIC AND PROKARYOTIC CELLS

The theorems of MCA in Appendix 8A are the basis of the matrix method (Sauro *et al.*, 1987; Westerhoff and Kell, 1987) that was applied to elucidate the rate-controlling steps of the glycolytic flux in *S. cerevisiae* growing in chemostat cultures in defined medium under different nutrient limitations, namely carbon, nitrogen or phosphate. The results of these studies are discussed in Chapter 7. The most conspicuous results of such an analysis indicate that, irrespective of the nutrient limitation, a significant increase in the amount of control exerted by the lumped steps, pyruvate decarboxylase–alcohol dehydrogenase, on the flux ratio at the branch point to ethanol formation happened as the dilution rate increased, which coincides with the triggering of respiro-fermentative metabolism (Franco *et al.*, 1984; Kapelli, 1986; Cortassa and Aon, 1994a). The fact that sugar uptake also controlled the flux distribution towards ethanol production and the TCA cycle suggested a close link between the uptake and the flux to ethanol production. This notion agrees with theoretical and experimental data reported by Barford (1990) for several yeasts, pointing to the main role played by the cell's rate of sugar uptake.

Sugar transport appears to be a main rate-controlling step of catabolism in yeast. Initially, Bisson and Fraenkel (1983) postulated the existence of low ($K_m = 20$ mM) and high ($K_m = 1$ mM) affinity hexose transport systems for *S. cerevisiae* (reviewed by Lagunas, 1993; and Weusthuis *et al.*, 1994). The understanding of the conditions under which high- and low-affinity transport systems activate or inactivate becomes physiologically relevant. The low-affinity glucose uptake appears to be constitutively expressed, whereas the high-affinity one depends on culture conditions

(Bisson and Fraenkel, 1984; Lagunas, 1993). The high-affinity transport system was low in the presence of high glucose concentrations (100 mM) as well as during prolonged incubation in the stationary phase of batch growth (Bisson and Fraenkel, 1984). Higher levels of uptake were found on low concentrations of glucose and in the presence of gluconeogenic substrates at normal concentrations. These results suggested that some component of the high-affinity uptake is repressible by glucose. In fact, a derepression up to 10-fold of the high-affinity glucose uptake could be shown following a shift from 100 mM to 5 mM glucose (Bisson and Fraenkel, 1984; Lagunas, 1993). It is significant that the high-affinity glucose transport had been glucose-repressed in all fermentative yeasts but not in respiratory yeasts (Does and Bisson, 1989). Fermentative yeasts such as brewers's and baker's yeast are glucose sensitive, so that ethanol excretion ensues when sugar is supplied in excess (Fiechter and Gmunder, 1989). Repressibility of high-affinity uptake did not take place in non-fermentative yeast strains such as *Pichia heedii* and *Yarrowia lipolytica* (Does and Bisson, 1989). *P. heedii* and *Y. lipolytica* did not lose high-affinity uptake when grown on high glucose. In obligate respirative yeast, the specific glucose uptake q_s is strongly correlated to the specific oxygen uptake rate q_{O_2}. For these yeasts, if respiration rate is low, glucose uptake is low; glucose surplus is not excreted as ethanol.

When *Agrobacterium radiobacter* NCIB 11883 was grown in ammonia-limited continuous culture at low dilution rate and with glucose as the sole carbon source, it produced an extracellular succinoglucan polysaccharide and transported glucose using the same periplasmic glucose-binding proteins as during glucose-limited growth (Cornish *et al.*, 1988). Using an inhibitor of glucose transport (6-chloro-6-deoxy-D-glucose), they showed that the latter was a major rate-controlling step for succinoglucan production in *A. radiobacter*. Both the rates of glucose uptake and succinoglucan production were inhibited almost identically in cells which had been grown under ammonium limitation, to yield a flux control coefficient >0.8.

Reported experimental data show that growth of prokaryotic cells obtained from lactose-limited or glucose-limited continuous cultures is limited by the uptake capacity of glucose via the lactose permease or phosphotransferase system (PTS), respectively (Herbert and Kornberg, 1976). In *E. coli*, at values of $\mu > 0.2\,h^{-1}$, a strict correlation was found to exist between the rates of growth and of glucose utilization. A control coefficient of 1.0 could be calculated in the range $0.2 < \mu < 0.5\,h^{-1}$ from the data of Herbert and Kornberg (1976). On the other hand, cells harvested at an exponential growth rate, $\mu = 0.4\,h^{-1}$, showed more than twice the rate of glucose utilization of cells harvested at $\mu = 0.1\,h^{-1}$

(Herbert and Kornberg, 1976). The latter suggested that the glucose PTS activity is intimately related to the rate at which cells utilize glucose for growth (Herbert and Kornberg, 1976). The control exerted by enzyme II^{glc} of glucose metabolism in *E. coli* was low on the growth rate and glucose oxidation in batch cultures, and high for uptake and phosphorylation of the glucose analogue α-methyl glucoside (control coefficients = 0.55 and 0.65, respectively) (Ruijter, 1992; Ruijter *et al.*, 1992).

8.7.1 CONTROL OF THE DEGREE OF METABOLIC UNCOUPLING OF *S. CEREVISIAE* GROWING IN CHEMOSTAT CULTURES

In order to quantify the control exerted by q_{glc} upon oxygen, carbon dioxide and ethanol fluxes in *S. cerevisiae* growing aerobically in chemostat cultures, we have recalculated from the data of Rieger *et al.* (1983) the control exerted by the specific rate of glucose consumption (parameter) on the oxygen, carbon dioxide and ethanol fluxes (variables). The procedure employed was similar to that used previously for chemostat cultures of *P. mendocina* to calculate flux or concentration control coefficients (Appendices 8A and 8B for details).

It is timely to remark that the q_{glc} was chosen as a parameter because the values of fluxes obtained were independent of the initial concentration of glucose in the chemostat (30, 10 and 5 g l^{-1} glucose; Chapter 7 and Table 7.1) (Rieger *et al.*, 1983) but it did vary according to the rate of glucose consumption by the cells. This allows us to suggest that the parameter that the microbe is sensing is the q_{glc} and not the glucose input into the chemostat, set by the experimenter.

As expected, at $D < D_c$, q_{glc} exerted control on q_{CO_2} from the TCA cycle, since the respiratory quotient (RQ) is equal to 1. Meanwhile at $D > D_c$ the control exerted by q_{glc} was mainly on the q_{CO_2} arising from alcoholic fermentation (RQ = 1.7 to 3.6).

The physiological approach on which the MCA analysis is based only allows 'coarse' predictions about the way in which the metabolic machinery of the microorganism should be manipulated to optimize alcoholic fermentation. When the aim is to redirect catabolic or anabolic fluxes, manipulating environmental conditions is not enough for quantitative modification of a desirable flux toward a product of interest.

A useful result of the 'coarse' approach, however, is that under conditions of high growth rates and high glucose concentration, the amount of ethanol produced in chemostat cultures of *S. cerevisiae* appears to be maximal (Figure 8.8), thus pointing to the physiological condition in which the rate-controlling steps of alcoholic fermentation should be determined.

8.8 QUANTITATIVE PREDICTION OF INTERESTING BEHAVIOURS IN METABOLIC PATHWAYS

When the aim is the production of a metabolite, the analysis of a network through the biothermokinetic method allows precision about not only which steps and how much they should be modified to achieve higher yields of the desired product but also in what phase of behaviour (i.e. parametric domain characterized by a structure of control) the engineering should be performed.

The description of a particular network of reactions or processes implies the characterization of all possible kinds of steady states that may be attained by this particular network and the specification of how a living system containing these networks will behave under defined environmental conditions. On the basis that dynamic bifurcation theory (DBT) makes possible the analysis of the steady states of a system continuously as a function of a bifurcation parameter, we have proposed a biothermokinetic numerical approach in order to explore quantitatively and predict new behaviours in metabolic pathways and microorganisms (Figure 8.1).

Analysis of the behaviour of several nonlinear systems indicated that the structure of control is variable, i.e. flux and concentration control coefficients change as a function of bifurcation parameters (Aon *et al.*, 1996b). It is then possible to define phases according to the sign and value of the control coefficient. By graphically inspecting the slopes in Figure 8.9, phases with high, low, nil or negative control by the same parameter may be observed. Structure of control is then defined as the distribution of, say, flux and concentration control coefficients, in different phases of the bifurcation diagram.

It may also be noted that along extensive steady state regions the glucose (Glc) concentration is insensitive to an increase in the bifurcation parameter (Figure 8.9b). These extensive regions show that the control exerted by one parameter is nil – namely, the steady state value of either metabolites or fluxes is constant over a large range of values of the bifurcation parameter (Figure 8.9). This provides an interpretation of the concept of homeostasis, which is characterized by nil or very low control coefficients.

In addition, different phases of behaviour according to the structure of control reveal different thermodynamic performance and may be characterized thereby (Figure 8.10). Thus, the thermodynamic analysis defines in a concomitant way what kind of performance in, say, efficiency terms may be expected from engineering a certain phase of

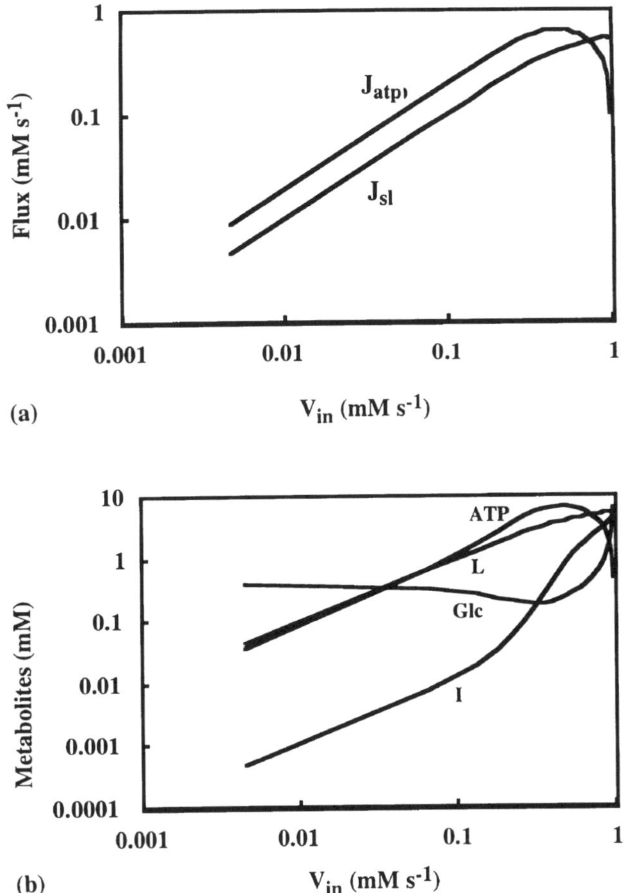

Figure 8.9 Double logarithmic plots of bifurcation diagrams of (a) the steady state fluxes and (b) metabolite concentrations as a function of the substrate input (V_{in}) in a stoichiometric model of glycolysis. The steady state branches of equations 1.21–1.24 and their stability were computed numerically with AUTO (E. Doedle, 1986, Concordia University). The logarithmic plot of the bifurcation diagrams gives the control exerted by the bifurcation parameter under study since the control coefficients are depicted by the slope of such plots. The bifurcation diagrams were obtained with the following parameter values: k_1 (mM^{-1} s^{-1}) = 0.3; k_3 (mM^{-2} s^{-1}) = k_5 (s^{-1}) = 0.1; k_9 (s^{-1}) = 0.05; k_P (mM s^{-1}) = 0.391. The kinetic parameters of the proton pump were: K_M = 2 mM; V_M = 0.5 mM s^{-1}. The total nucleotide (C_A) and phosphate pool (P_t) were both 10 mM. (From Aon, Cortassa and Verdoni, in *Gas, Oil and Environmental Biotechnology*, V, Institute of Gas Technology, Chicago; in press.)

behaviour of the metabolic converter (Appendix 8A). All considerations related to the calculation of the kinetic control structure (e.g. flux and concentration control coefficients) also hold for the thermodynamic analysis (Figures 8.9 and 8.10). In an *a posteriori* analysis of the steady state behaviour of a system, the thermodynamic functions such as dissipation and efficiency are completely defined by the steady state levels of the metabolites (Chapter 1) (Aon *et al.*, 1990a; Cortassa *et al.*, 1990; Aon and Cortassa, 1991). This approach was used to predict the appearance of glycolytic oscillations (Aon *et al.*, 1991) and to quantify the energetic performance of glycolysis in its interaction with mitochondrial functions in starved yeast cell suspensions (Figure 8.10) (Aon *et al.*, 1991; reviewed in Cortassa and Aon, 1994b).

To summarize, the different phases in the control diagram are characterized by different thermodynamic performances and may be quantified thereby. Additionally, measurements of thermodynamic performance are of great interest in bioengineering for predicting in a quantitative way how a certain parametric change will shift the performance of an energy converter in terms of efficiency.

8.9 DEPENDENCE OF THERMODYNAMIC EFFICIENCY ON THE TOPOLOGY OF METABOLIC PATHWAYS AS WELL AS THEIR BUILT-IN KINETIC NONLINEARITIES

Two types of energy converters differing in the mechanism source of nonlinear kinetics but with essentially the same topology, i.e. linear chain reaction, were compared. The sources of nonlinear kinetics were cooperativity as for the allosteric enzyme phosphofructokinase (Goldbeter and Dupont, 1990) and the built-in stoichiometric autocatalysis of ATP production (Sel'kov, 1975; Cortassa *et al.*, 1990a), both for glycolysis. The allosteric cooperative mechanism has been shown to be the primary source of glycolytic oscillations in yeast extracts whereas stoichiometric autocatalysis, first proposed by Sel'kov (1975) and further developed by Cortassa and collaborators (1990a, 1991), has been successfully applied to modelling of glycolytic oscillations *in vivo* (Aon *et al.*, 1991, 1992; Cortassa and Aon, 1994a,b) (Chapters 3 and 7).

In addition, a third type of energy converter with a different and more complex topology was studied in comparison with the other two. This energy converter possesses a circular chain reaction with a branch and a covalent cycle which is involved in the feeding of substrate to the reaction cycle. Essentially, the system exhibited the same sort of nonlinearity, i.e. stoichiometric autocatalysis (Cortassa *et al.*, 1990a, 1991; Aon and Cortassa, 1991) between circular and linear topology (Chapter 12: Figure 12.8 and Appendix 12B).

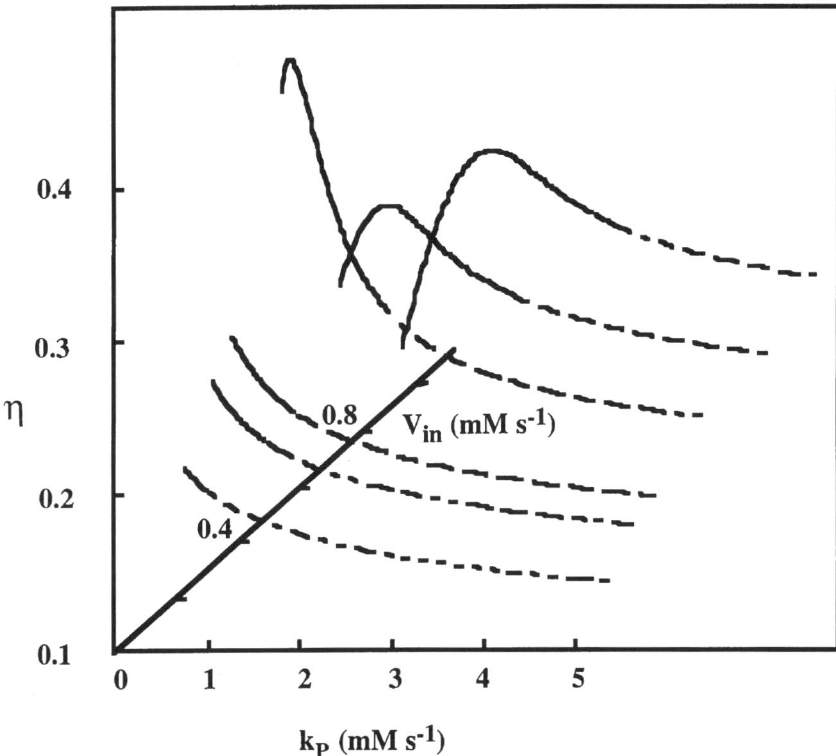

Figure 8.10 Three-dimensional representation of the bifurcation properties of a stoichiometric model of glycolysis and its interaction with mitochondrial functions. The behaviour of the thermodynamic efficiency, η, was studied for different values of V_{in} depicted in the z-axis as a function of the ATP load (k_P). The following parameter were used: $k_1 = 0.0949$ mM⁻¹ s⁻¹; $k_3 = 0.1$ mM⁻² s⁻¹; $k_5 = 0.1$ s⁻¹; $k_9 = 0.05$ s⁻¹; $k_7 = 0.1$ s⁻¹. C_A and P_t were 10 mM and K_M for the ATPase activity was 2 mM. ΔG_a and ΔG_s were 7.3 and 47 kcal mol⁻¹ and RT 0.596 kcal mol⁻¹. The continuous and dashed lines correspond to asymptotically stable and oscillatory unstable steady state parametric regions, respectively, with Hopf bifurcations in the limit between both regions. (From Aon, Cortassa and Verdoni, in *Gas, Oil and Environmental Biotechnology*, V, Institute of Gas Technology, Chicago; in press.)

The three converters were analysed for possible energetic advantage under oscillatory dynamics as compared with asymptotic steady state behaviour. Based on DBT and a numerical method employed by AUTO, we analysed the steady states continuously as a function of a bifurcation parameter. We chose the rate of uptake of the substrate as a bifurcation parameter for the three converters. The results obtained were essentially different. Figure 8.11 shows the stable (plain line) and unstable (dashed line) steady state branches of thermodynamic efficiency, η, as a function of the substrate input rate (V_{in}). The stable and unstable branches are delimited by Hopf bifurcations. The calculations of η based on equations 1.10 and 1.33 (Chapter 1, Appendix 1A) showed that the oscillatory regime allows the declining bias of η to reverse as the system approaches the Hopf bifurcation point (Figure 8.11a). The explanation for the reversal of the declining tendency of thermodynamic efficiency, as glycolysis approaches a Hopf bifurcation point, is contained in Figure 8.11b, which shows the phase plane relations of the fluxes with the input force in a limit cycle. Maxima in glycolytic ATP synthesis (J_a) and catabolic (J_{sl}) fluxes occurred simultaneously, whereas ΔG_s, the chemical potential of glucose conversion into ethanol, approached a minimum (in absolute value). This small phase shift resulted in a lower dissipation of catabolic free energy than the dissipation obtained in the case in which maxima in fluxes coincided with maxima in catabolic free energy. Thus, the instantaneous phase relations of forces and flows are relevant to the general time average outcome of the converter since the gain in efficiency is accomplished through a phase shift between fluxes and the catabolic force.

In the allosteric model, the oscillatory regime did not reverse the declining bias of thermodynamic efficiency as the system approaches the Hopf bifurcation point (Figure 8.11d). Efficiency decreased in the case of the ring-shaped topology depicting the PTS transport dynamics coupled to glycolysis. Since the latter primarily contains the same source of nonlinear kinetics, i.e. stoichiometric autocatalysis, it further shows that

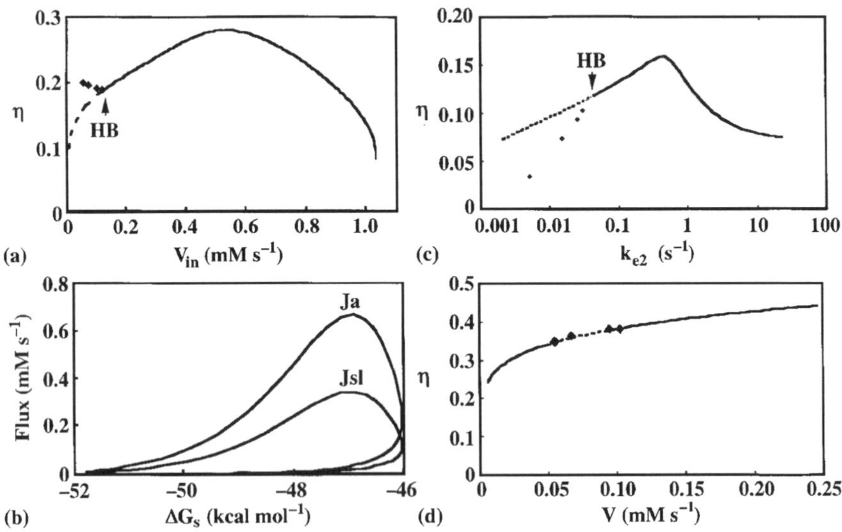

the topology of the metabolic pathway also has an influence on the putative energetic advantage exerted by periodic dynamics (Cortassa *et al.*, 1990a; Aon and Cortassa, 1991).

One possible consequence of such different results obtained with models of metabolic pathways showing dissimilar topologies and sources of nonlinear kinetics is that biological systems are able to display very

Figure 8.11 (see facing page) Thermodynamic performance of metabolic pathways with different topologies and feedback mechanisms as a function of the rate of substrate uptake by the cell. The putative advantage of oscillatory dynamics on thermodynamic performance was investigated on three different metabolic pathways. (a,b) A stoichiometric model of glycolysis in which the nonlinear mechanism is the autocatalytic feedback exerted on glucose phosphorylation by the stoichiometry of the ATP production of the anaerobic functioning of the glycolytic pathway. The glycolytic pathway is also an example of a linear-branched topology. (a) The stability analysis as a function of the substrate input V_{in} provided the steady state concentrations of the metabolites from which the thermodynamic functions could be calculated. The kinetic parameters used were: $k_P = 0.7516$ mM s^{-1}; $k_1 = 0.0949$ mM^{-1} s^{-1}; $k_3 = 0.1$ mM^{-2} s^{-1}; $k_5 = 0.1$ s^{-1}; $k_9 = 0.05$ s^{-1}; $k_7 = 0.1$ s^{-1}. The values of the constants C_A and P_t were both 10 mM, and K_M for the ATPase activity was 2 mM. Values for ΔG_a and ΔG_s were 7.3 and 47 kcal mol^{-1} and RT 0.596 kcal mol^{-1}. (b) Phase space relations of the fluxes (J_a, ATP synthesis; J_{sl}, glucose degradation) with the input force (ΔG_s, the chemical potential of glucose conversion) in a limit cycle, which is included to explain the reversion of the declining tendency of thermodynamic efficiency in (a), as glycolysis approaches a Hopf bifurcation (HB). The parameters correspond to those described in (a) for the lowest V_{in} values represented by diamonds. (c) A model of the phosphotransferase (PTS) system in bacteria that represents a global circular topology including a covalent cycle and branches. The parameter k_{e2} represents the rate at which the phosphate group of the covalent cycle A–A~P is transferred to glucose when transported into the cell (Figure 12.8). Thermodynamic efficiency was analysed as a function of k_{e2} and the following parameters were used: $k_{e1} = 0.9$ s^{-1}; $k_1 = 0.3$ mM^{-1} s^{-1}; $k_3 = 0.1$ mM^{-2} s^{-1}; $k_5 = k_7 = 0.1$ s^{-1}; $k_9 = 0.05$ s^{-1}; $\tau\, d_p V_{max} = 0.4141$; $K_g = 1.0$ mM; $K_{pe} = 0.1$ mM; $K_{pk} = 0.5$ mM; $K_p = 1.0$. The kinetic parameters of the proton pump were: $K_M = 2$ mM; $V_M = 0.5$ mM s^{-1}. The total nucleotide (C_A) and phosphate pool (P_T) were both 10 mM, whereas the total sum of substrate A involved in the covalent modification cycle (TA) was 1 mM. (d) An allosteric model described in Goldbeter and Dupont (1990). The simulations were performed with the following parameter values: $k_1 = 2.9093$ mM s^{-1}; $k_2 = 0.15$ s^{-1}; $k_3 = 0.1$ mM s^{-1}; $k_s = 3.0$ mM; $k_{adp} = 5.0$ mM; $k_p = 2.7088 \times 10^{-1}$ mM; $C_n = 10.0$ mM; n = 2; L = 5000. Thermodynamic efficiency was studied as a function of V, the rate of substrate input. For selected points (diamonds in a,c,d) in the oscillatory domain (dashed lines) the simulation of the temporal evolution of the system enabled the computation of the thermodynamic efficiency, η, in order to examine the energetic advantage of oscillatory dynamics according to equations 1.32 and 1.33 in all models represented in (a), (c) and (d). In all figures, the plain and dashed lines represent stable and unstable steady states, respectively.

diverse mechanisms of adaptation by using all or part of them under different physiological situations. Far from showing model-dependent conclusions, these results further exemplify the enormous diversity of dynamic mechanisms and, consequently, responses that microorganisms or cells utilize to compete in natural environments.

8.10 MICROBIAL GROWTH IN NATURAL ENVIRONMENTS

8.10.1 SUBSTRATE UPTAKE REGULATION AS A MICROBIAL STRATEGY FOR SURVIVAL

Expenditure of energy from a limited store to concentrate the substrate from the environment would be at the expense of growth. Organisms that follow that pattern of behaviour have favoured concentrative sugar uptake systems in their evolution instead of low affinity-facilitated diffusion systems for monosaccharides (K_M of 2–10 $\times$ 10^{-3} M for glucose). Consequently their habitats are limited to environments with high sugar concentrations. From an industrial standpoint, it is this combination of properties of the sugar uptake and utilization system (low affinity, high velocity, low efficiency) that makes these organisms valuable for the rapid conversion of sugars at high concentrations to metabolic products rather than to cell material (Romano, 1986).

Knowledge about the 'average' maintenance energy for the calculation of the growth rate or generation time of soil populations is uncertain. Laboratory calculations of around 0.4 h^{-1} correspond to microorganisms growing in luxuriant nutrient media but figures as low as 3 $\times$ 10^{-5} h^{-1} correspond to dormant organisms. The uncertainties about estimation of maintenance energies for microorganisms in nature increase when we consider r- or K-strategists. K-strategists might be considered to have very low maintenance energies. The physiological features of K-strategists are moderate growth rate and yield efficiency for substrate utilization. These microorganisms use more resistant (e.g. polymeric) substrates and behave as moderately nutrient demanding. On the other hand, r-strategists utilize rapid growth rate and high yield efficiency, giving maximum colonization with the use of readily available substrates and high nutrient demands. The population dynamics of r-strategists (exploitation) shows explosive density-independent, crash-through substrate depletion or opportunistic grazing. On the contrary, K-strategists exhibit relatively damped, density-dependent control by interspecific competition and selective grazing (Lynch and Hobbie, 1988).

The ecological strategy adopted by an organism may depend on the resource (substrate) quality and the climatic severity. For example, an organism normally regarded as an r-strategist on root-derived carbon in summer might become a K-strategist on lignocellulosic substrates in the cold of winter.

Low thermodynamic efficiencies, optimal with respect to growth rate, have been described for microorganisms growing on either highly reduced substrates or more oxidized ones than the biomass (Westerhoff *et al.*, 1983). Seemingly, evolutionary pressure has worked on the fastest growing microorganisms: in an ecosystem containing a population of similar microorganisms, the fastest one will outgrow its more slowly growing competitors. However, if selection would have worked on an intermittent scenario for substrate availability, i.e. alternate periods of excess substrate and famine, efficiency becomes a selective pressure for survival (Westerhoff *et al.*, 1983).

8.10.2 UPTAKE DETERMINES THE SORT OF INTERACTION BETWEEN COMPETING MICROBIAL POPULATIONS

Microbial cells nearly always occur within complex interacting communities. In complex ecosystems containing a diversity of organisms the metabolic activity of any individual organism is likely to have an effect on its near neighbour. Two competing microbial populations may achieve different types of interactions depending on their ability to support advantageous substrate uptake with respect to each other. Commensalism, amensalism, parasitism, predation and mutualism are the main categories of microbe–microbe interactions (Figure 8.12). The following system of ODEs (8.5–8.10) is able to describe most of these interactions:

$$\frac{dX_1}{dt} = \mu_1 X_1 \tag{8.5}$$

$$\frac{dX_2}{dt} = \mu_2 X_2 \tag{8.6}$$

$$\frac{dS}{dt} = -k_1 X_1 - k_2 X_2 \tag{8.7}$$

$$\mu_1 = \frac{\mu_1^{max}[S]}{K_s^1 + [S]} \tag{8.8}$$

$$\mu_2 = \frac{\mu_2^{max}[S]}{K_s^2 + [S]} \tag{8.9}$$

$$L = 2.42[0.4(S_0 - S) - 0.00987(X_1 + X_2)] \tag{8.10}$$

Figure 8.12 shows the temporal evolution of biomass production of two microbial populations of *Lactobacillus bulgaricus* and *Streptococcus thermophilus*. We could verify that the sort of microbial interaction which will result is dependent upon the potential for substrate uptake by each microbe (Aon, Cortassa and Manca de Nadra, unpublished data). For very different uptake potentials, the species with the slower uptake is either overgrown by the other in mixed cultures, i.e. amensalism, or does not change its behaviour with respect to pure cultures, i.e. commensalism (Figure 8.12b). When the potentials for substrate uptake are closely similar but not equal, mutualism arises, i.e. both species increase their growth rates when mixed in comparison with pure cultures (Figure 8.12a).

8.10.3 SURVIVAL OF MICROORGANISMS IN NATURAL ENVIRONMENTS

An insufficient amount of energy is available in natural ecosystems and so most microbes exist in a state of starvation that has been called 'starvation survival' (Morita, 1988). The concept of maintenance energy arose to describe maintenance reactions which include turnover of cell macromolecules, osmotic work and motility (Pirt, 1975; Chesbro, 1988). Apparently, maintenance energy requirements in most cases consume far

Figure 8.12 (see facing page) Types of interaction between two microbial populations: (a) mutualism; (b) commensalism. The interacting species are *Lactobacillus bulgaricus* (*L.b.*) (filled symbols) and *Streptococcus thermophilus* (*S.t.*) (open symbols) in pure and mixed cultures. The culture medium contains 1% glucose, 1% yeast extract, 1.5% peptone, 1% tryptone and 0.1% Tween 80 at pH 6.8 (Amoroso, 1990). The model is described by equations 8.5–8.7; the points correspond to the simulations performed with the model. Equations 8.5–8.7 were numerically integrated on a PC-60 III Commodore using the program SCoP with an Adams method (Duke University, 1989). The simulations of the mutualism type of microbial interaction were performed with the following parameters: $k_1 = 0.24$ (*L.b.* pure), 0.055 (*L.b.* mixed); $k_2 = 0.2$ (*S.t.* pure), 0.045 (*S.t.* mixed); $K_S^1 = 0.02$ (*L.b.*); $K_S^2 = 0.03$ (*S.t.* pure), 0.002 (*S.t.* mixed); $\mu_1^{max} = 0.333$ (*L.b.* pure), 0.417 (*L.b.* mixed); $\mu_2^{max} = 0.2$ (*S.t.* pure), 0.338 (*S.t.* mixed). The initial values of X_1 (*L.b.*) and X_2 (*S.t.*) in pure and mixed cultures were: 2.91 (*L.b.* pure), 2.97 (*S.t.* pure) and 4.13 (*L.b.*) + 3.54 (*S.t.*) in mixed cultures. The simulations of the commensalism type of microbial interaction were performed with the following parameters: $k_1 = 0.1$ (*L.b.* pure), 0.09 (*L.b.* mixed); $k_2 = 0.25$ (*S.t.* pure), 0.09 (*S.t.* mixed); $K_S^1 = 0.015$ (*L.b.* pure), 5×10^{-4} (*L.b.* mixed); $K_S^2 = 0.02$ (*S.t.* pure), 0.03 (*S.t.* mixed); $\mu_1^{max} = 0.4$ (*L.b.* pure), 0.475 (*L.b.* mixed); $\mu_2^{max} = 0.2$ (*S.t.* pure), 0.555 (*S.t.* mixed). The initial values of X_1 (*L.b.*) and X_2 (*S.t.*) in pure and mixed cultures were: 2.98 (*L.b.* pure), 2.44 (*S.t.* pure) and 2.2 (*L.b.*) + 1.8 (*S.t.*) in mixed cultures.

more energy than is available in the soil substrate. Thus, very low maintenance energy is required by microorganisms that exhibit metabolic arrest, permitting their survival for extremely long periods (Morita, 1988). Presumably, cells under starvation survival in different natural habitats (soil or aquatic) retain the functionality of their genome, enzymes and uptake systems that may be resynthesized after resuscitation, albeit longer lag phases would be required for cells subjected to longer periods of starvation (Morita, 1988).

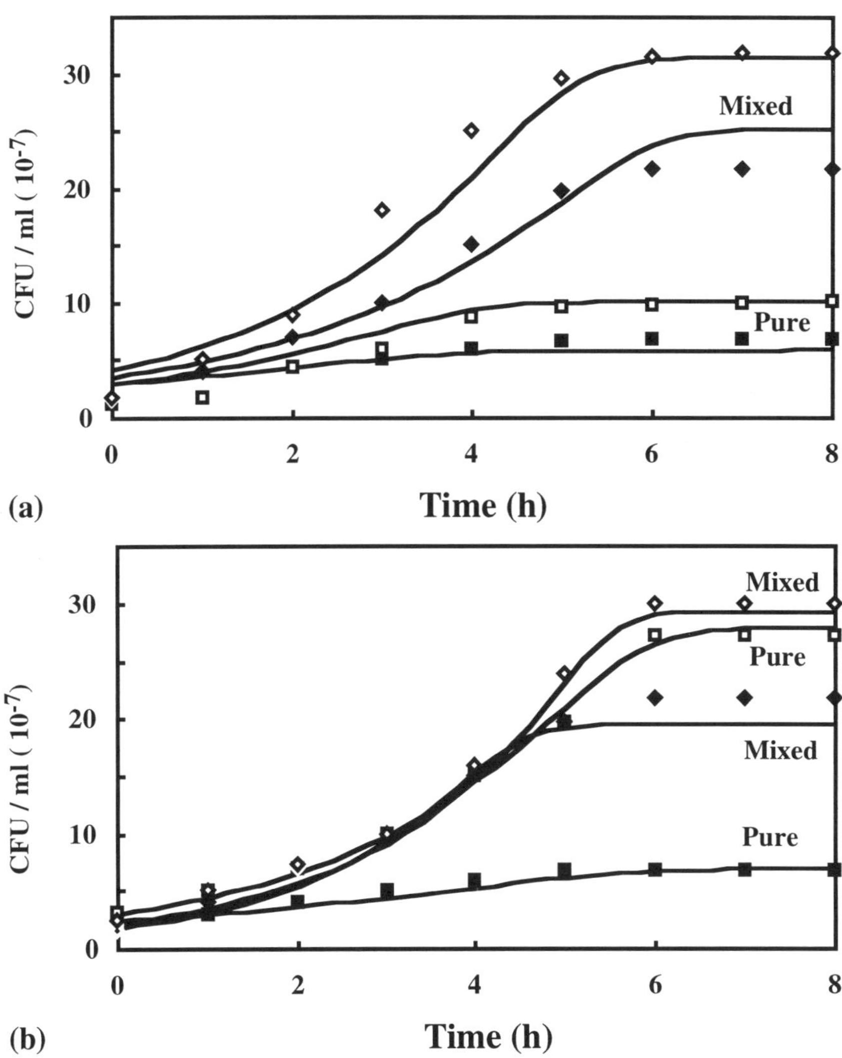

Microbial growth rate will become substrate dependent at some critical concentration which corresponds to that of subsaturation of the uptake system. At that moment, chronic starvation starts, during which it has been shown that the major part of the cell's energy budget will be dedicated to error restraint of protein translation (Chesbro, 1988). In *E. coli*, protein mistranslation (proofreading and editing of macromolecular synthesis) can increase to more than 100 (one wrong amino acid each 1000 right incorporations versus one in 100).

8.11 A PERSPECTIVE FOR A TRANSDISCIPLINARY APPROACH

A transdisciplinary attitude, i.e. beyond but through disciplines, is essential for the study of biological complexity. Complex problems such as microbial dynamic behaviour, environmental biotechnology, or even those posed by cognition or cellular function, involve several disciplines. This is a real handicap towards a realistic and successful approach to those riddles. To overcome this obstacle we have proposed a transdisciplinary approach (TDA). Let us say that TDA is the approach which supports this book. One main feature of this approach is that it may be carried out by a reduced number of non-specialists. This can be accomplished by stressing:

- the need for agreement on concepts rather than mastering of a discipline;
- the ability to develop a common language to communicate ('interphase language') rather than technical abilities;
- emphasis on knowledge of principles and utilities (i.e. the power to answer specific questions) of techniques and methods;
- the 'migration' of concepts and useful metaphors between disciplines.

This final point is the essence of TDA, by creating 'interphase domains' of consensus concepts or metaphors for a common language. Transdisciplinarity also relies on a tight dependence between phenomenology (i.e. empirical qualitative data) and quantitation. Biomathematics, through modelling, provides the quantitative and heuristic tools that are unavoidable at the time of interpreting and working with the living complexity.

APPENDIX 8A THE BIOTHERMOKINETIC METHOD

8A.1 THEORETICAL BACKGROUND

This method combines metabolic control analysis (MCA) and dynamic bifurcation theory (DBT) to quantitate biochemical networks fully, metabolically and energetically, and to predict and characterize new behaviours. The theoretical backgrounds are presented next.

8A1.1 MCA and determination of the rate-controlling steps in metabolic fluxes

MCA was developed in order to understand quantitatively what limits (and to what extent) the flux through a metabolic pathway. This rationale gives us the possibility of overcoming the restrictions that hamper an increase in the flux.

To deal with metabolic networks of arbitrary complexity, MCA defines several types of coefficient – namely, control and elasticity coefficients, which reflect global and local properties of the network, respectively (Kacser and Burns, 1973; Heinrich and Rappoport, 1974; Kacser and Porteus, 1987):

$$C_X^J = \left(\frac{\frac{dJ}{J}}{\frac{d[X]}{[X]}} \right)_{ss} = \left(\frac{d(lnJ)}{d(ln[X])} \right)_{ss} \tag{8.11}$$

$$C_X^M = \left(\frac{\frac{dM}{M}}{\frac{d[X]}{[X]}} \right)_{ss} = \left(\frac{d(lnM)}{d(ln[X])} \right)_{ss} \tag{8.12}$$

$$\varepsilon_M^{v_i} = \left(\frac{\frac{dv_i}{v_i}}{\frac{d[M]}{[M]}} \right)_{ss} = \left(\frac{d(lnv_i)}{d(ln[M])} \right)_{ss} \tag{8.13}$$

where C_X^J is the flux (J) control coefficient (section 8.6.2) and X is a parameter involved in the metabolic pathway, such as an enzyme concentration or the oxygen partial pressure; C_X^M is the concentration control coefficient for the metabolite M. These two types of coefficient reflect global properties of the system. The third type of coefficient, $\varepsilon_M^{v_i}$, is the elasticity coefficient that measures the dependence of the rate of step i (v_i) on the concentration of metabolite M. The latter reflects a local property that depends only on the rate law governing the enzyme kinetics of the step under consideration and the metabolite(s) concentration.

Several theorems constitute the main body of MCA. We will take advantage of these theorems to derive the control of ethanol production by *S. cerevisiae* in continuous cultures (see below). The **summation theorem** states that the sum of all flux control coefficients, with respect to the activities of each of the enzymatic steps involved in the metabolic pathway being considered, is equal to unity (Kacser and Burns, 1973; Kacser and Porteus, 1987):

$$\sum_i C_{X_i}^J = 1 \tag{8.14}$$

The sum of all metabolite control coefficients with respect to the activities of each of the enzymatic steps is equal to zero (Kacser and Burns, 1973; Kacser and Porteus, 1987):

$$\sum_i C_{X_i}^M = 0 \tag{8.15}$$

where X_i is the activity of enzyme i.

The **connectivity theorem** states that the sum over all products of the metabolite control coefficients with respect to the activity of the enzyme catalysing step i, multiplied by the elasticity coefficient of the same enzyme, is equal to unity:

$$\sum_i C_{X_i}^M \varepsilon_M^{V_i} = 1 \tag{8.16}$$

These theorems are the basis of the matrix method applied in Chapter 7 (Appendix 7B) (Sauro *et al.*, 1987; Westerhoff and Kell, 1987).

8AI.2 Applying MCA to microbial physiology

The chemostat allows us to monitor steady state fluxes as a function of an external parameter (e.g. oxygen). The fluxes corresponding to the specific rates of consumption or production of, say, glucose $[q_{glc}]$, oxygen $[q_{O_2}]$ and carbon dioxide $[q_{CO_2}]$ attained at each steady state by a microorganism may be monitored, as well as the amount of biomass produced (Verdoni *et al.*, 1992) (Figure 8.1). In order to quantify the control exerted by a specific external parameter on catabolic and anabolic fluxes during growth (Verdoni *et al.*, 1992), MCA may be applied (Heinrich and Rappoport, 1974; Sauro *et al.*, 1987; Westerhoff and Kell, 1987).

In applying MCA, the following assumptions were made:
- The steady states attained are asymptotically stable and without discontinuities between them (e.g. multiple stationary states).
- In analysing *P. mendocina* physiological behaviour in chemostat cultures, the natural logarithm of the steady state values existing between two successive p_{O_2} (e.g. 70% to 20% transition) follows a straight line.

MCA defines control coefficients that characterize global properties of the system. Flux and metabolite control coefficients are defined as the fractional change in flux, J, or metabolite concentration, M, as a result of an infinitesimal change in an effector X according to equations 8.11 and 8.12.

In equation 8.11, C_X^J is the control coefficient of the effector X; J is the flux through the pathway and ss denotes the steady state at which the system relaxes after a perturbation (e.g. changes in p_{O_2}). We have studied the control exerted by the effector p_{O_2} on q_{O_2}, q_{CO_2} and q_{glc}, and the amount of biomass, x, stands for metabolite concentration, M (biomass is considered as the final product of the whole metabolic network). Thus $C_{O_2}^{q_{O_2}}$, $C_{O_2}^{q_{CO_2}}$ and $C_{O_2}^{q_{glc}}$ are the flux control coefficients and $C_{O_2}^x$ is the concentration control coefficient for biomass. The control coefficients were calculated according to equation 8.11 and the following procedure: the natural logarithms of q_{glc}, q_{O_2} and q_{CO_2}, biomass concentration were plotted as a function of the natural logarithm of p_{O_2}. The values of the control coefficients are the slopes (first derivatives) of the plots described above.

8A.2 DYNAMIC BIFURCATION THEORY AND ITS APPLICATION TO DESCRIBE THE BEHAVIOUR OF METABOLIC AND ENERGY TRANSDUCING SYSTEMS

DBT allows the analysis and characterization of the qualitative dynamic behaviour of any system in terms of type and stability of steady states exhibited by that system when a parameter (called **bifurcation parameter**) is varied (Kubicek and Marek, 1983; Doedle, 1986; Aon *et al.*, 1991). The parameter values for which the stability properties change (e.g. from stable to unstable steady states) are called **bifurcation points**. A visualization of the ensemble of steady states and the localization of the bifurcation points (or limit points for the bistability case, where sudden transitions between stable branches of asymptotic steady states occur) may be obtained with DBT. This type of analysis is of great help since complex metabolic networks may exhibit various dynamic behaviours apart from asymptotic steady states, such as overshoots, bistability, oscillations and chaos (Chapter 1) (Kubicek and Marek, 1983; Abraham, 1987).

We applied DBT based on the numerical analysis of ODEs that was performed with AUTO (Doedle, 1986), a software designed for numerical analysis of nonlinear systems of the form:

$$\frac{d[v]}{dt} = f(v,k) \qquad v, f \in \mathfrak{R} \qquad (8.17)$$

where υ stands for the system variable(s) and k denotes the free parameter(s). The numerical procedure used by AUTO is based on an analysis of the Jacobian matrix of the system with a Newton chord method and the following are computed:

- branches of solutions of stable or unstable steady states, either asymptotic or periodic;
- evaluation of the eigenvalues and accordingly the stability of such branches;
- location of bifurcation points, limit points, Hopf bifurcation points and computation of the bifurcating branches (Doedle, 1986).

AUTO starts the search of solutions of a system of ODEs from a steady state of the system. Systematically varying a parameter (a bifurcation parameter), it will render the steady state values of the state variables of the system, e.g. the steady state metabolite concentrations.

8A.2.1 Direct calculation of the flux and metabolite control coefficients from the log–log plot of bifurcation diagrams

The analyses of the slope of the log–log form of a bifurcation diagram (i.e. the steady state fluxes and metabolite concentrations) are the control coefficient of fluxes or the metabolites, respectively. More precisely, the slope of the log–log form of the bifurcation diagram allows graphical and ready estimation of the extent of control exerted on the flux or on metabolite concentrations by the bifurcation parameter under study (Figure 8.9).

8A.2.2 Quantifying the energetic performance of a free-energy transducer

From the steady state metabolite concentrations the behaviour of, say, thermodynamic efficiency may be readily studied as a function of a bifurcation parameter (Figure 8.10). If the efficiency versus the bifurcation parameter is plotted on a log–log graph, the slope of the curve gives a control coefficient of the energy converter efficiency (Figure 8.10).

APPENDIX 8B METHODOLOGY TO DETERMINE FLUXES OF CARBON, PHOSPHORYLATION AND REDOX INTERMEDIATES DURING GROWTH OF *S. CEREVISIAE* ON DIFFERENT CARBON SOURCES

The amounts of key intermediary metabolites, ATP and redox equivalents required from the key intermediary metabolites to synthesize the macromolecular monomers of yeast cell biomass were the same, irrespective of the carbon source, except for acetate, because of the synthesis of glycine and serine (Cortassa *et al.*, 1995). However, the total amounts of phosphorylation energy, redox equivalents and CO_2 required to synthesize biomass from each substrate were very different because of the diversity in the demand to produce each of the key intermediary metabolites from the various carbon sources.

The numbers next to arrows in Figure 8.4 correspond to anabolic fluxes. When the minimal catabolic fluxes needed to sustain growth were computed, such fluxes varied quantitatively if the NADH (and FADH) produced in anabolism was, or was not, considered to be oxidized at the level of the mitochondrial electron transport chain generating ATP. The net demand of ATP varied according to the number of proton translocation sites in the electron transport chain during NADH oxidation (P/O ratio). In other words, the amount of substrate that should be completely oxidized to fulfil the ATP demand for biosynthesis varied with the P/O ratio (Table 8.1). In addition, as the complete oxidation of the substrate involves TCA cycle function and NADPH generation (oxidative catabolism: Table 8.1), the catabolic utilization of the PP pathway in a cyclic functioning mode (generating NADPH with the concomitant oxidation of the substrate to CO_2) (Cortassa *et al.*, 1995) differed if we consider P/O ratios of 1, 2 or 3 (Table 8.1). The PP pathway had to be used to complete the demand of NADPH at P/O ratios of 2 for pyruvate and lactate (Table 8.1). The PP pathway fluxes required to sustain a constant supply of intermediates for the synthesis of amino acids and nucleotides were low compared with the flux through the gluconeogenic pathway (10% to 15%). In the case of glucose, for a P/O ratio of 2.0 the catabolic flux required to satisfy the demand of NADPH is comparable to the total flux of substrate oxidized through the TCA cycle (Table 8.1). Much larger fluxes through the oxidative branch of the PP pathway would be required if the isocitrate dehydrogenase used NAD preferentially as acceptor and if the malic enzyme was not operative. These two steps contributed a large fraction, if not all, of the NADPH required for biosynthesis that would otherwise have to be supplied by the PP pathway functioning in its cyclic mode. An increased flux through the oxidative branch of the PP pathway would lead to an increase in carbon and ATP fluxes and decrease in Y_s an Y_{ATP}, respectively.

Dynamics of cell growth and division   9

9.1 SPATIO-TEMPORAL COORDINATION OF GENE EXPRESSION, ENERGETICS AND METABOLISM: ITS SIGNIFICANCE FOR CELL GROWTH AND DIVISION

The intracellular spatio-temporal coordination of gene expression, energetics and intermediary metabolism is at the heart of cell growth, proliferation, differentiation modes and functional adaptation to environmental changes.

Two main paths may be taken in approaching the functional spatio-temporal coordination at cellular and tissue levels: one structural and the other dynamic. From an equilibrium thermodynamics viewpoint both are incompatible. However, this is not the case for far from equilibrium systems, which cellular systems are. In these, transitions between different dynamic regimes (Nicolis and Prigogine, 1977) and levels of organization (Aon and Cortassa, 1993) occur around bifurcations at unstable points of their dynamics. Under such conditions, biological processes become self-organized and millions of molecules, or subcellular structures or even cells, achieve spatio-temporal coherence which is congruous with biological function. In other words (according to the thermokinetics of systems far from equilibrium) structure and dynamics, space and time, consistently and harmoniously arise one from the other.

According to the structural viewpoint, the tissue matrix of the cell provides the harnessing mechanisms to direct the energy derived from chemical reactions to specific functions (Pienta and Hoover, 1994; Cortassa *et al.*, 1994a; Cortassa and Aon, 1994b). The tissue matrix consists of the dynamic linkages between the skeletal networks of the nucleus (the nuclear matrix), the cytoplasm (the cytoskeleton) and the extracellular environment (the extracellular matrix) (Pienta and Hoover,

1994). The tissue matrix provides the spatio-temporal coordination for successful gene expression by the cell. This theme, already developed in Chapter 6 and extensively discussed in Chapter 10, is here focused from the perspective of cell growth and division.

According to the dynamic prospect, structural organization of, say, the tissue matrix is maintained by continuous free-energy dissipation. Such dissipation of free energy to maintain a stationary state, or to evolve toward different dynamic regimes, renders biological processes globally irreversible. In fact, these steady states correspond to distinct cellular functional properties (Figure 1.7). Mutual interaction between environment and the genetic background results in a spatio-temporal coordination of changes in metabolic fluxes implying enzyme activity modulation by effectors or catalyst concentration.

Most of the time, microorganisms respond to changes in environmental conditions by varying patterns of gene expression. In some cases, these variations involve adaptive responses of intracellular catabolic pathways. For instance, in the presence of utilizable carbon or nitrogen substrates, or available oxygen or other electron acceptors, an increased rate of synthesis of enzymes involved in their metabolism is generally observed (Harder and Dijkhuizen, 1983; Smith and Neidhardt, 1983a,b). Gene expression may affect cell energy dissipation through the following main mechanisms: changes in stoichiometry of metabolic pathways, concentration of enzymes, or enzyme alteration of kinetic parameters by, say, mutations. Chapter 11 gives a quantitative analysis, based on metabolic control analysis and dynamic bifurcation theory, of the reciprocal influence of genetic background and environment on metabolic fluxes.

Intracellular spatio-temporal coordination depends on linear, cyclical or branched (or combinations between them) enzymatic reaction pathways. Furthermore, Chapter 11 shows that the structure of control may be changed by gene expression depending on pathway topology. In fact, differential gene expression feeds back on the rate-controlling steps of different pathways not only through modulation of the enzyme amount but also through topology. In the dynamic view of the cell as a dissipative system, supra-regulatory control over synthetic and degradative fluxes exerted by the length of the cell cycle (in turn greatly influenced by the physiological condition) might induce changes in its own structure of control (i.e. rate-controlling steps) and changes in fluxes by gene up- or down-regulation (Chapter 11).

In bacteria and yeast, catabolite-regulated gene expression is an essential mechanism for adjusting to changes in nutrient availability (Entian, 1986; Carlson, 1987a,b; Thevelein, 1992). In higher plants, carbon metabolite regulation of gene expression plays a fundamentally important role in maintaining an economical balance between supply and demand of biomolecules in various organs or tissues (Sheen, 1990).

Transitions affected by, for example, mitotic cycling or environmental perturbations or signals trigger a generalized cellular response that may, say, stop cell division or change a cell's fate. Experience shows that the cellular response is coherent in nature and evidenced at higher levels of organization (i.e. at supramolecular–cellular dimensions and characteristic times of several minutes or hours when changes in growth rates happen; Chapter 2). Coherence is expressed at different levels of organization comprising different cellular activities or functions. To illustrate the latter, Figure 9.1 shows the simultaneous increase in the rate of lactate production following a pH increase in *Dictyostelium discoideum*. The destiny of *D. discoideum* cells as stalk or spore cells is affected by low extracellular (Gross *et al.*, 1983; Oohata, 1992) or intracellular pH (Aerts *et al.*, 1987) or increase in propionate concentration (Figure 9.2).

This chapter assembles reported experimental evidence, with special emphasis on the interrelationships between cellular metabolism, energetics and gene expression.

9.1.1 PHYSIOLOGY OF METABOLIC TRANSITIONS: RESPONSES OF CELL GROWTH AND DIVISION FOLLOWING NUTRIENT STARVATION

(a) Starvation and flux imbalance

Starvation is one of the most common stresses encountered by living organisms. It is estimated that much of the microbial biomass in the world exists under nutrient-depleted conditions (Werner-Washburne *et al.*, 1993, and references therein).

Microorganisms, such as the yeast *Saccharomyces cerevisiae*, respond to starvation by ceasing growth and entering a non-proliferating state referred to as **stationary phase** or G_0. Major structural changes do occur in cells as a result of starvation and are accompanied by changes in metabolism and physiology. Examination of starved cells using light microscopy reveals changes in cell morphology. *E. coli* cells become much smaller and almost spherical when they enter stationary phase (Siegele and Kolter, 1992). The surface properties of starved cells are also different from those of growing cells. Changes in the fatty acid composition of cell membranes have been observed during starvation of several species. In *E. coli* there is a conversion of all unsaturated membrane fatty acids to the cyclopropyl derivatives as cells enter stationary phase (Siegele and Kolter, 1992).

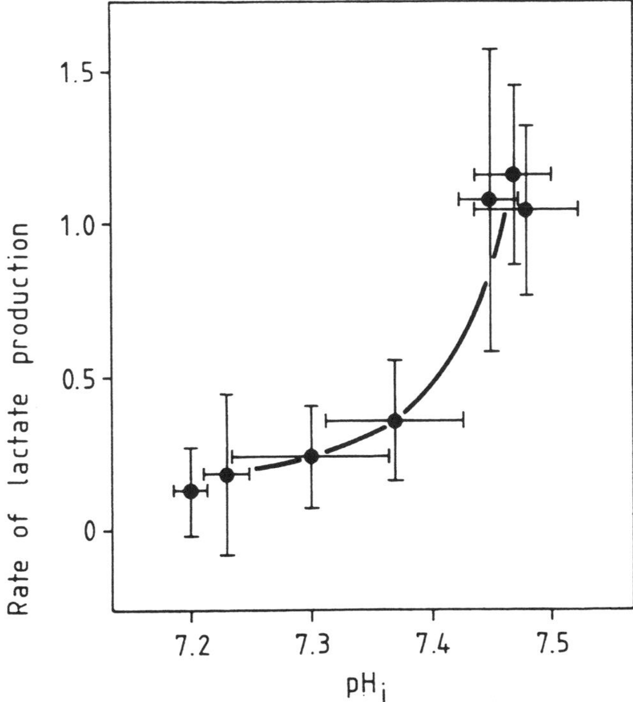

Figure 9.1 The relationship between lactate production and pH_i during the cell cycle of *Dictyostelium discoideum*. Synchronously dividing cell populations of *D. discoideum* were obtained by resuspending stationary-phase cells into fresh medium. After synchronization, determinations of pH_i and the aerobic rate of lactate production started at regular 2–2.5 h intervals. The graph demonstrates that if pH_i is high, lactate production is high; and if pH_i is low, lactate production is low. (Reproduced from Aerts, Durston and Moolenaar, 1986, by permission of Elsevier Science.)

The rate of protein turnover increases approximately five-fold in starved *E. coli* cells. In rapidly growing *E. coli* cells, the bulk of nucleic acids is stable. RNA stability decreases when cells enter stationary phase and 20–40% of total RNA is lost during the first few hours of starvation (Siegele and Kolter, 1992). Thus, in the starved cell, degradation of endogenous protein and RNA, mostly in the form of ribosomes, may provide a source of energy to support endogenous metabolism.

Another response of cells to starvation stress is to increase their spontaneous mutation rate under conditions where post-selection mutations accumulate. The level of tRNA modification could be one

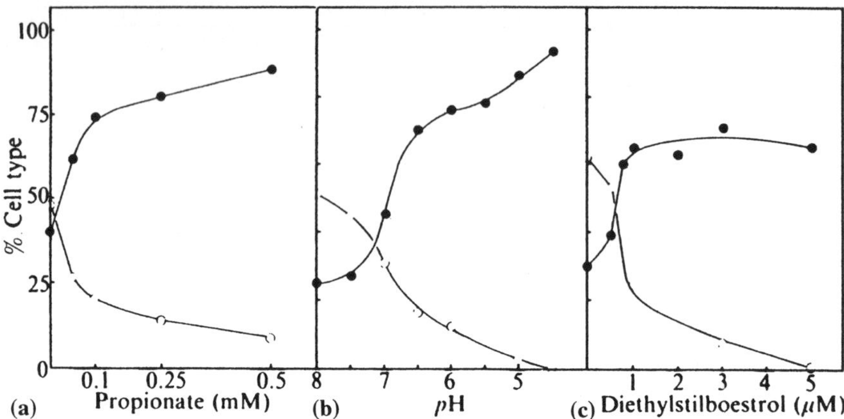

Figure 9.2 Intracellular pH control of cell differentiation in *D. discoideum*: the effects of (a) propionate, (b) extracellular pH and (c) diethylstilboestrol on stalk and spore formation. Filled and open circles correspond to stalk and spore cells, respectively. (Reprinted with permission from *Nature*, **303**, 244–245, copyright 1983 Macmillan Magazines Ltd.)

signal that triggers whatever process leads to the appearance of post-selection mutations (Siegele and Kolter, 1992).

As cells become starved their overall metabolic rate decreases, but some level of endogenous metabolism is maintained. This allows the starved cells to maintain some level of ATP (or other high-energy compounds) and the proton motive force across the membrane. One function of endogenous metabolism is to maintain the capability of transporting substrates into the cell. ATP levels and the energy charge decrease as cells go from exponential into stationary growth phase. This experimental observation, repeatedly found in starving microorganisms, may be interpreted as a mechanism allowing free-energy dissipation to decrease while retaining the potential to restore growth as environmental conditions become favourable.

(b) Synthesis of macromolecules

This section describes the dynamics of processes requiring high amounts of metabolic energy such as the synthesis of protein and nucleic acids. Stationary phase cells show a number of common physiological features irrespective of the nutritional limitation responsible for the arrest of proliferation (Boucherie, 1985). Yeast cells starved of carbon, nitrogen, sulphur or phosphorus survive for long periods and arrest at a related point in the cell cycle, suggesting that stationary phase may be identical in these cells (Werner-Washburne *et al.*, 1993, and references therein). Well in advance of glucose exhaustion, a transition phase was observed in batch cultures of *S. cerevisiae* grown on glucose, characterized by a decrease in growth rate and a progressive reduction of protein and RNA accumulation.

Overall transcription declines dramatically as yeast cells approach stationary phase. Most individual mRNA species that are abundant during the exponential phase are barely detectable after the diauxic shift (Werner-Washburne *et al.*, 1993, and references therein). Stationary-phase cells contain approximately half as much poly(A) RNA as is found in exponentially growing cells, which has been estimated to be 5% of the total RNA in exponentially growing cells (Werner-Washburne *et al.*, 1993, and references therein).

Under aerobiosis, arrest of cell proliferation in response to glucose exhaustion is known to be associated with a significant increase in the rate of synthesis of a large variety of enzymes subjected to carbon catabolite repression (Beck and von Meyenburg, 1968; Boucherie, 1985). When cells became stationary, protein accumulation was 10% of that in log-phase cells and incorporation of labelled RNA precursor was undetectable.

(c) Nitrogen, carbon, phosphate or sulphur limitations

Yeast cells enter stationary phase when starved of carbon, nitrogen, phosphorus and sulphur. Carbon limitation appears to be a complicated process when yeast cells growing in a rich medium make entry into stationary phase. This is substantiated by the preferential catabolism of glucose by fermentation, the modification of metabolism and gene expression by glucose repression, and the subsequent response to depletion of non-fermentable carbon sources (Werner-Washburne *et al.*, 1993, and references therein).

In prokaryotes a set of reactions to amino acid deprivation called the **stringent response** is characterized by inhibition of protein synthesis and reductions in transcription of ribosomal RNA (80%), mRNA (25%) and tRNA (20%). Shulman *et al.* (1977) concluded that the regulation of transcription of the three types of RNA is not coordinated. In yeast the uniform response to various modes of protein synthesis may itself trigger the inhibition of rRNA transcription that is characteristic of the stringent response. In this stringent response, therefore, we may identify two response levels: the cellular one, revealed by the decrease in growth rate, and the molecular one, for which several mechanisms have been described. However, as we do not yet know which processes were destabilized leading to the self-organized structure of the cell cycle, we cannot ascertain whether the molecular dynamics triggered by amino acid starvation is involved in the cell clock change in 'ticking' frequency, i.e. slowing down of the proliferation rate.

Cell growth and division in *S. cerevisiae* have been analysed during nitrogen starvation (Johnston *et al.*, 1977b). Although during the initial stages of starvation the cell number increased, there was little net accumulation of RNA or protein. Net accumulation of macromolecules followed when cells of the strain AG1-7 where shifted to nitrogen-free medium. A 2.5-fold increase in labelled DNA was detected but there was little increase in labelled protein or RNA. Nevertheless, continued protein synthesis was necessary for completion of DNA division cycles under nitrogen deprivation, as could be demonstrated by cycloheximide treatment that prevented further increases in cell number. The nitrogen-deprived cells exhibited extensive protein and RNA degradation. Johnston *et al.* (1977a,b) concluded that once the DNA division cycle is initiated, cells can complete division with little dependence on continued net cell growth. Since 98% of the nitrogen-deprived cells were arrested in the G_1 phase of the cell cycle upon cessation of growth, the authors further suggested that, after the 'start' event, little net growth is needed to allow cells to complete events in the DNA division cycle.

(d) Storage carbohydrates: glycogen and trehalose

The intracellular accumulations of both glycogen and trehalose vary with the growth status of yeast (Figure 9.3) and appear to be a general response to stress, including heat-shock, and limitation of sources of nitrogen, sulphur, phosphorus or carbon (Werner-Washburne *et al.*, 1993, and references therein). The intracellular concentrations of the two carbohydrates follow a different kinetics (Figure 9.3).

When cells of the strain C276 of *S. cerevisiae* growing exponentially were deprived of nitrogen, sulphur or phosphorus, they exhibited both glycogen and trehalose accumulation, albeit to different extents. Cell number increased several-fold before starved cells arrested as unbudded G_1 cells. Since these cells showed low levels of both glycogen and trehalose when growing exponentially in minimal synthetic medium, these results suggested that reserve carbohydrate accumulation is a general response to nutrient limitation (Lillie and Pringle, 1980). By monitoring glucose in the medium it was confirmed that glucose limitation was not a limiting factor in these experiments. This general observation also includes carbohydrate accumulation during sporulation in *S. cerevisiae* (Chapter 10).

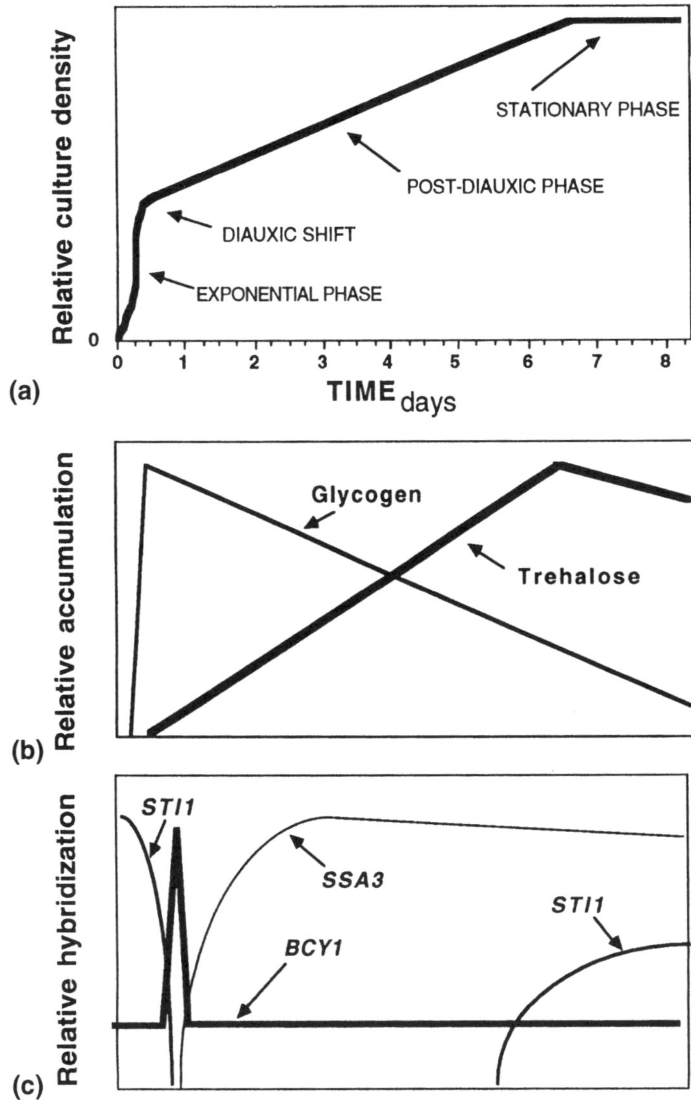

Figure 9.3 Cell division and macromolecular dynamics in wild type *Saccharomyces cerevisiae* cells grown to stationary phase in rich, glucose-based medium (YPD): (a) relative culture density of a typical growth curve measuring turbidity at 600 nm; (b) storage carbohydrate accumulation; (c) relative abundance of three mRNA species based on Northern blot hybridization. (Reproduced from Werner-Washburne *et al.*, 1993, by permission of the American Society for Microbiology, ASM Press.)

After the shift to nutrient-deprived conditions, glycogen and trehalose followed a different time course of accumulation; the synthesis of glycogen and trehalose began within the first and fourth hour after the shift, respectively. During diauxic growth, glycogen accumulation began when half of the glucose was still present in the medium and it reached a maximum shortly before the glucose in the medium was exhausted. On the contrary, trehalose accumulation did not begin until the external glucose was nearly exhausted. During growth in the absence of glucose, yeast cells contained 0.5 to 1% and 0.6% of their dry weight as glycogen and trehalose, respectively, compared with 1% to 3% in the stationary phase of growth.

9.2 COORDINATION OF CELL GROWTH AND DIVISION

In most populations of proliferating cells, the processes of growth and division occur in a coordinated fashion. It was mentioned in section 9.1.1a that yeasts respond to starvation by entering a non-proliferating state at stationary phase or G_0. On the other hand, when deprived of any one of a number of required nutrients, yeast cells arrest proliferation in the G_1 phase of the cell cycle. Under nutritional conditions satisfactory for cell proliferation, the cell begins a new mitotic cycle (Figure 9.4). Completion of the cell cycle requires the coordination of the synthesis of a variety of macromolecules, assemblies and movements. The chromosomes must be replicated, condensed, segregated and decondensed. The spindle poles must duplicate, separate and migrate to opposite ends of the nucleus (Hartwell and Kastan, 1994). Also, cells must have the capacity to arrest cell cycle progression when DNA damage is induced by unprogrammed extrinsic events such as exposure to inhibitors of DNA replication or spindle assembly or to agents that physically damage DNA (Hartwell and Kastan, 1994).

The unbudded, pre-replicative status of a stationary-phase cell is characteristic of a cell that has not yet performed 'Start'. Control of cell division in the budding yeast *S. cerevisiae* occurs within the G_1 portion of the cell cycle. Hartwell *et al.* (1973, 1974) isolated temperature-sensitive cell division cycle (cdc) mutants affected in their ability to progress through the G_1 portion of the cell cycle (Chapters 3 and 11). **Start** is defined operationally as the cell cycle step blocked by the pheromone response pathway and by conditional cdc28 mutations (Hartwell, 1974). It is believed that during Start many external and internal signals are integrated so that ultimately the cell is committed to different fates, such as proliferation, G_0 and sporulation (Figure 9.4) (Wheals, 1987; Aon *et al.*, 1995; Mónaco *et al.*, 1995).

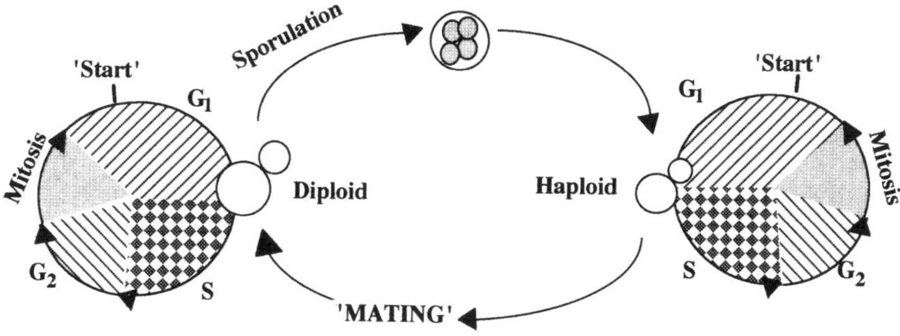

Figure 9.4 Vegetative and life cycles of *S. cerevisiae*, emphasizing the alternation of yeast cells between mitotic and meiotic cycles, each belonging to vegetative and life cycles, respectively. Yeast growth (macromolecular synthesis) and cell proliferation occur in the vegetative cycle, whereas yeast differentiation (sporulation) happens predominantly associated with degradative processes. Our research deals with metabolism and cellular energetics underlying a putative 'common decision' that in yeast could regulate the exit from the mitotic cycle to the differentiation path (Chapters 10 and 11).

Activation of the $p34^{CDC28}$ protein kinase, encoded by the *CDC28* gene, occurs as cells traverse Start and is required for the performance of this step. The $p34^{CDC28}$ protein kinase is activated by association with the G_1-cyclin regulatory proteins (Werner-Washburne *et al.*, 1993, and references therein). This activation is blocked by nutrient starvation. The coordination of macromolecular synthesis and the DNA-division cycle are thought to be achieved through a series of changes in cyclin-dependent kinases (CDKs). The active forms of the CDKs are a complex of at least two proteins, a kinase and a cyclin. In wild type cells, G_1-cyclins, which are post-translationally regulated in a cell-cycle dependent manner, are degraded as cells leave G_1 and enter S-phase and they do not accumulate again until cells have re-entered G_1. Newly synthesized cyclin subunits combine with pre-existing cdc2 subunits to form an inactive maturation promoting factor (MPF). The complex is then activated in an autocatalytic fashion (Werner-Washburne *et al.*, 1993, and references therein; Dirick and Nasmyth, 1991). When taken into account in a mathematical model, these events showed that the control system can operate in three modes: as a steady state with high MPF activity, as a spontaneous oscillator, or as an excitable switch (Tyson, 1991; Novak and Tyson, 1993).

To maintain the cell mitotically cycling requires the operation of processes able to sustain a certain rate of macromolecular synthesis and a critical size (Wheals, 1987). This is translated in biological terms by the statement that growth is rate-limiting for cell proliferation (Johnston *et al.*, 1977a; Aon *et al.*, 1995).

The substrate input rate is one of the parameters that determines the dynamic behaviour of the yeast cell cycle (Munch *et al.*, 1992b; Cortassa and Aon, 1993a, 1994b); when the substrate input decreases below a certain threshold value, the behaviour is characterized by a stationary state of a cell in G_1 (Start or G_0) (Aon *et al.*, 1995; Mónaco *et al.*, 1995).

Several cdc mutants have been shown to continue growth after the arrest of cell division at the restrictive temperature. The volume, dry weight and protein content of the average cell attained values two- to four-fold greater than those of the average cell at the permissive temperature. This result suggested that growth is fairly independent of progression of the cell through the events of the DNA-division cycle: studies with nitrogen-deprived yeast cells suggest that, after the Start event, little net growth is needed to allow cells to complete events in the DNA division cycle. Starved cells were arrested at the beginning of the DNA-division cycle as evidenced by the presence of less than 2% budded cells (Johnson *et al.*, 1977a)

Johnston and Singer (1978) observed that in cells treated with RNA synthesis inhibitors, orthophenanthroline or hydroxyquinoline (HQ), protein accumulation ceased although the rate of protein synthesis showed no significant decline during the first 100 min of treatment. On the other hand, rates of RNA synthesis displayed an exponential decline, beginning immediately upon addition of the chelating agent. Simultaneously, cells accumulated in G_1, as evidenced by an increase in the proportion of cells without buds. Furthermore, the execution point of cells treated with HQ was found to be in G_1 (Johnston and Singer, 1978) which suggested that the initial alteration in RNA metabolism was probably the major influence in blocking cells in G_1.

9.2.1 DEPENDENCE OF CELL CYCLE PROGRESSION ON PROTEIN SYNTHESIS

A dividing cell of *S. cerevisiae* is sensitive to a number of variables in its environment and reacts to unfavourable conditions by arresting cell division in G_1 (Hartwell, 1974). Accumulation of cells in G_1 may also be achieved by a change from a fermentable to a non-fermentable carbon source or vice versa. Cells that are starved for nitrogen after passing Start complete a cell cycle with less than 10% increase in cellular protein (Johnston *et al.*, 1977a). Treating cells with cycloheximide, at sufficient concentrations to inhibit protein synthesis, arrests cell division, and cells accumulate at multiple points in the cell cycle (Burke and Church, 1991). Therefore, the small amount of protein synthesis (10%) that occurs under nitrogen starvation is essential to completion of cell division. Using cdc2, cdc6, cdc7, cdc8, cdc17 and cdc21 mutants arrested at the restrictive temperature and subsequently shifted back to the permissive temperature in the presence of cycloheximide, it was possible to show that cells arrested in S-phase completed DNA synthesis but could not complete nuclear division if protein synthesis was inhibited. *S. cerevisiae* cells that arrest in the G_2 stage complete nuclear division in the absence of protein synthesis (Burke and Church, 1991). However, protein synthesis is required later in the cell cycle to complete cytokinesis and cell separation. The same effects on protein synthesis inhibition were obtained with two different drugs, trichodermin and cycloheximide (Burke and Church, 1991). It was concluded that some events, such as DNA synthesis and nuclear division, can be completed in the absence of protein synthesis.

Inhibiting protein synthesis in a random population of dividing cells causes them to arrest at one of the three steps in the cell cycle that require protein synthesis. The three steps look morphologically different: unbudded cells with a single nucleus (G_1), budded cells with an undivided nucleus (medial nuclear division) and budded cells with a divided nucleus (late nuclear division) (Burke and Church, 1991).

Several histones were synthesized almost exclusively in late G_1 and early S phases. Additionally, the rate of synthesis of eight proteins was shown to be modulated during the cell cycle; of these, five were unstable with half-lives of 10–15 min (Lorincz *et al.*, 1982).

Wild type cells of the yeast strain aS288C grown in minimal medium plus 3-amino-1,2,4-triazole (AT), an inhibitor of the sixth enzyme of histidine biosynthesis, respond to the addition of the inhibitor by increasing the specific activities of his1, his2, his3, his4A and his4C. In addition, enzymes of the arginine, lysine and tryptophan pathways derepress in response to AT inhibition. Strains carrying mutations in the *TRA3* and *TRA5* genes showed constitutively derepressed enzymes for arginine, histidine, lysine and tryptophan biosynthesis and temperature sensitivity for growth on both minimal and complete medium at 36°C. Asynchronous cultures of *tra3⁻* and *tra5⁻* strains accumulate as unbudded cells at 36°C, suggesting an early block during G_1. The *tra3* mutants synthesize very little DNA at 36°C while RNA and protein synthesis continue at 36°C for at least 5 h, albeit at a reduced rate (Wolfner *et al.*, 1975).

9.2.2 INTERDEPENDENCE OF ENZYME EXPRESSION, METABOLIC RATES AND CELL CYCLE ACTIVITY

Changes in metabolic rates influencing the level of enzyme expression are expected to occur during the cell cycle of *S. cerevisiae*. It had already been stated, long ago, that the rate of glucose consumption affects enzyme synthesis. In yeast, Beck and von Meyenburg (1968) stated for the first time that the presence of glucose *per se* does not have a regulatory effect, but the rate of glucose consumption affects enzyme synthesis. The regulation of enzyme formation would be a function of the rate of glucose consumption and subsequent steady state rates of metabolic routes, rather than of the glucose concentration. This was the first time that a dynamic property was invoked as an explanation of changing metabolism and the degree of enzyme expression (Kappeli, 1986; Kappeli and Sonnleitner, 1986). It has been shown that the enzyme pattern of *S. cerevisiae* depends on the growth rate and that the latter regulates the relative length of the budding phase and the single cell phase in the cell cycle (Chapter 11).

Yeast glycolytic genes appear to be coordinately regulated in response to, say, carbon source, for the following reasons (Moore *et al.*, 1991):

- All their gene products are required for the catabolism of glucose.
- All are efficiently expressed.
- All have very strong codon biases and optimal translation initiation regions.

A comparative study of the steady state levels of all the glycolytic mRNAs during exponential growth of *S. cerevisiae* on glucose or lactate showed that these were induced to different extents by glucose (Figure 9.5). The regulation of glycolytic mRNA levels is probably mediated at the transcriptional level. The level of each glycolytic mRNA relative to that of the actin mRNA control was calculated, and then the relative mRNA level on glucose was divided by that on lactate to give the level of glucose activation for each mRNA. *TDH* and *PGII* were not significatively induced while pyruvate decarboxylase I (*PDCI*) and phosphoglucomutase I (*PGMI*) were only moderately activated. Two peaks of maximal induction by glucose were observed at phosphofructokinase and pyruvate kinase both encoded by mRNAs that were previously shown to be coregulated at the translational level in *S. cerevisiae* (Figure 9.5). However, it is premature to suggest that the differential regulation of glycolytic mRNA levels might have a significant influence upon glycolytic flux in *S. cerevisiae* (Cortassa and Aon, 1994b) (Chapters 7 and 11).

Different enzyme patterns under various growth conditions appear to be the result of control by repression or derepression at different thresholds for different enzymes over the budding cycle (Figure 9.6). The repression of TCA enzymes at high growth rates (Polakis *et al.*, 1965; Beck and von Meyenburg, 1968; Wales *et al.*, 1980) would explain the decrease

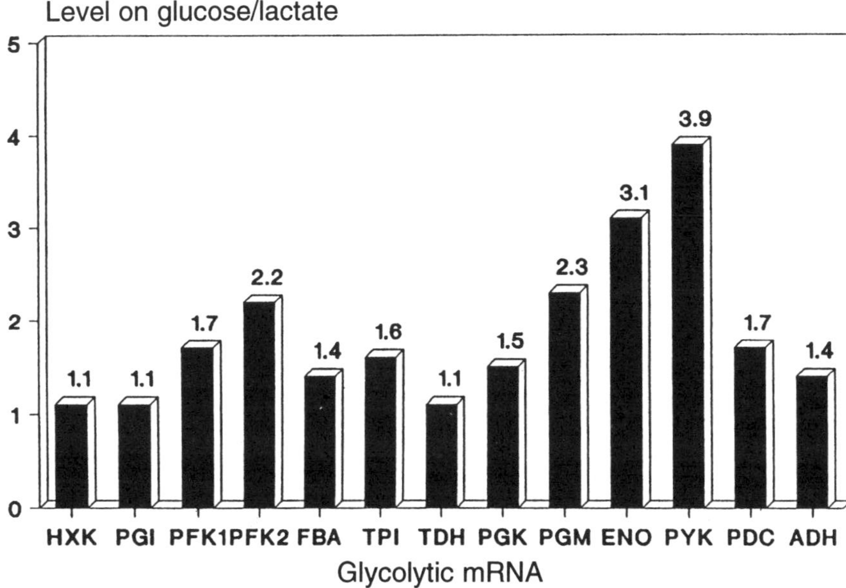

Figure 9.5 Differential regulation of yeast glycolytic genes by the carbon source. The abundance of each glycolytic mRNA (relative to the actin mRNA internal control) was compared during growth on glucose or lactate. The relative abundance on glucose was divided by the equivalent value for lactate. The mRNAs are placed in the order in which they occur on the glycolytic pathway. (Reproduced from Moore *et al.*, 1991, by permission of the American Society for Microbiology, ASM Press.)

in the oxygen uptake rate (Chapter 7). The limited respiratory capacity at high growth rates may be understood by the combined action of flux saturation at the level of pyruvate and enzyme repression. The K_M for oxidation of pyruvate by pyruvate dehydrogenase in isolated mitochondria is approximately 10-fold lower than the K_M of pyruvate decarboxylase. Therefore, when the intracellular pyruvate concentration is low, pyruvate is preferentially channelled into the TCA cycle (Van Urk *et al.*, 1989; Van Dijken *et al.*, 1993). In *S. cerevisiae* the type of catabolism exhibited by the cells has been shown to be determined not solely by the respiratory capacity, but also by the pattern of anabolic reactions. Changes of anabolic processes will lead to corresponding and well balanced changes in the activity of catabolism (Kappeli and Sonnleitner, 1986).

The activities of TCA cycle enzymes are induced when yeast cells are either grown to stationary phase on glucose-containing media or shifted

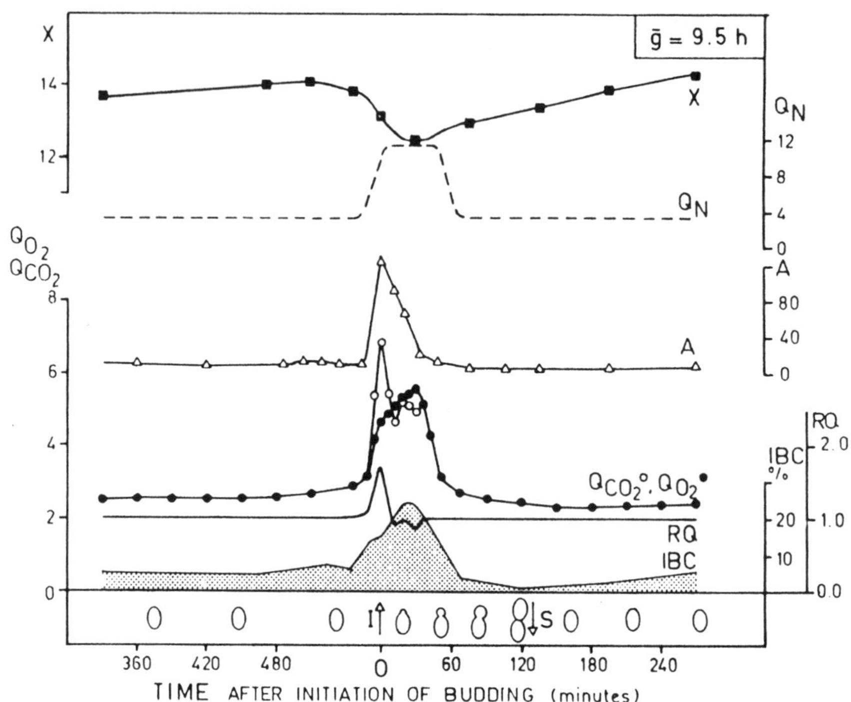

Figure 9.6 Correlative metabolic events along the cell cycle of *S. cerevisiae*, during synchronous growth under glucose limitation in continuous culture at D = 0.073 h⁻¹. X, dry weight (mg ml⁻¹); Q_{O_2} and Q_{CO_2}, specific oxygen and carbon dioxide fluxes (mmol h⁻¹ g⁻¹ dw) RQ, respiratory quotient; % IBC, percentage of initial budding cells; A, ethanol concentration; QN, specific rate of nitrogen uptake (mg h⁻¹ g⁻¹ dw); I, initiation of budding; S, scission of the daughter from the mother cell. (Reproduced from *Arch. Mikrobiol.*, von Meyenburg, 1969, by permission of Springer-Verlag.)

from high-glucose medium to derepressing conditions. During growth of *S. cerevisiae* on 1% glucose, the level of citrate synthase activity was observed to increase 16-fold, beginning late in exponential growth phase. In contrast, cells grown in medium containing ethanol–glycerol showed no reduction in enzyme levels during growth phase (Hoosein and Lewin, 1984). The increase in citrate synthase activity in cells grown on glucose was shown to occur as a result of an increase in enzyme amount due to an increase in translatable mRNA for citrate synthase. These results were interpreted as *de novo* enzyme synthesis attributable to new transcription of the citrate synthase gene or to RNA processing.

The amounts of alcohol dehydrogenase I (*ADH I*) mRNA and the rate of ADH I protein synthesis decreased upon growth in ethanol-containing medium and during growth into stationary phase (Denis *et al.*, 1983). Yeast cells grown on glucose-containing medium were transferred to medium containing ethanol; the ADH I and ADH II protein synthesis was monitored in time by immunoprecipitation with an antibody specific to ADH. ADH I protein synthesis declined in the first 2 h of growth on ethanol (non-fermentable carbon source); the proportion of ADH I protein synthesis decreased from ~6% (at 15 min after the shift) to 1% (at 3–4 h after the shift) of total protein synthesis. This decline was coincident with a decrease from 1 to 0.2% of functional *ADH I* mRNA within 3–4 h after switching the cells to ethanol-containing medium. Synthesis of *ADH I* and *ADH II* mRNA was first detected approximately 1 h after release from glucose repression. It appeared that the decline in ADH I protein synthesis after switch to growth on ethanol was not the result of a major change in either the rate of ADH I protein degradation or the turnover rate for the ADH. Essentially, the same results were obtained under conditions in which glucose became depleted naturally from the medium and diauxic growth proceeded. However, the amount of *ADH II* mRNA and its rate of protein synthesis were much greater during diauxic growth than during growth on ethanol-containing medium.

(a) Glucose catabolism during the cell cycle of *S. cerevisiae*

The fermentative ability of yeasts depends on the type of sugar, its concentration and the availability of oxygen (Berry *et al.*, 1981). *S. cerevisiae* is Crabtree positive, i.e. it shows repression of respiration, and performs an alcoholic fermentation in spite of aerobic conditions, provided that excess of glucose is present (Berry *et al.*, 1981; Postma *et al.*, 1989; Alexander and Jeffries, 1990). Aerobic-respiring (AR) yeasts such as *Candida shehatae*, *Pichia stipitis*, *C. tenuis*, *Pichia segobiensis* and *Pachysolen tannophilus* are unable to grow without oxygen (Bruinenberg *et al.*, 1983). AR yeasts such as *C. utilis* appear to have a fundamental requirement of respiration for growth, and do not repress their respiration in the presence of excess glucose. For aerobic-fermenting (AF) yeasts such as *S. cerevisiae*, respiration seems to be optional; they can grow in the absence of oxygen (adequately supplemented with ergosterol, nicotinic acid and unsaturated fatty acids) (Andreasen and Stier, 1953, 1954, quoted in Van Dijken *et al.*, 1993), and repress their respiration in favour of fermentation when in the presence of large amounts of a fermentable sugar.

The switch between pure respiratory or respiro-fermentative metabolism would induce differences in the transit of yeast through the cell cycle (Figure 9.6) (von Meyenburg, 1969; Munch *et al.*, 1992a,b; Aon *et al.*, 1995; Mónaco *et al.*, 1995). The initiation of DNA synthesis and bud appearance occur simultaneously. A shift down in dilution rate (from 0.15 to 0.07 h^{-1}) induces a rising in q_{CO_2} and q_{O_2} (the former higher than the latter) 20 min before budding onset (von Meyenburg, 1969). The sudden rise in gas metabolism is parallelled by an increase in the ethanol concentration and the respiratory quotient (RQ = 1.7) (Figure 9.6). These changes in metabolic rates are associated with bud emergence, and immediately afterwards (during G_2) ethanol is consumed with a concomitant decrease of the RQ. The slight dry weight increase during the single cell phase and its subsequent decrease at the initiation of budding indicate accumulation of some cell constituents and their fast consumption (von Meyenburg, 1969). Kuenzi and Fiechter (1972) showed that trehalose and glycogen are partly degraded during the budding process, which probably explains the short increase of the RQ to fermentative values. The increase in the rate of ATP generation (Q_{ATP}) enables the cells to form the new bud, nucleic acids and proteins at a higher velocity over the budding period. Von Meyenburg (1969) concluded that even under excess glucose there is a timing of energy generation at the expense of reserves that appears to function over the cell cycle for the control of the overall growth process of the individual cell (Figure 9.6).

Yeast cells may be committed to start cell cycle activity by changes in substrate uptake (Munch *et al.*, 1992a,b). Experiments performed with either spontaneously or forced synchronized cultures of baker's yeast showed that the synchrony of the culture could be maintained by weak substrate pulses (Munch *et al.*, 1992a). That the oscillations in several physiological parameters corresponded to cell cycle activity was confirmed by flow cytometry DNA measurements in both forced and spontaneously synchronous cultures (Munch *et al.*, 1992a,b).

After mobilization of stored carbohydrates in the budding yeast population, these cells have to refill their carbohydrate storage. It has been verified that the higher the dilution rate, the less is the amount of incorporated storage material together with a reduced G_1 phase (Chapter 11). Consequently, the amount of cells ready to start a new cell cycle is smaller when the forcing frequency in synchronous cultures increases. Carbon dioxide production is also lower with the increase in forcing frequency due to the decreased mobilization of storage material (Munch *et al.*, 1992a).

9.2.3 GENE EXPRESSION AND METABOLIC COORDINATION IN RESPONSE TO HORMONE ACTION

A knowledge of transcriptional and post-transcriptional control mechanisms is essential for a full understanding of cell growth, cellular responses to hormonal stimulation and cell transformation. Such control involves the activation of transcription of specific mRNAs, as well as the initiation of translation of specific proteins, some of which (such as the nuclear regulation proteins) may have a very short lifespan. There is evidence that compartmentalization of polysomes into free and membrane-bound polysomes may be in turn a mechanism of compartmentalization of mRNAs; for example, the actin mRNA was found associated with both free and cytoskeletally bound polysomes through the microfilament system (Vedeler *et al.*, 1991).

Insulin stimulation of transcription of L-type pyruvate kinase and fatty-acid synthase needs ongoing cellular protein synthesis. Post-transcriptionally, insulin modifies mRNA stability. Glucose, insulin and glucagon have profound effects on the stability of the L-type pyruvate kinase messengers: after actinomycin D block, the half-life of the pyruvate kinase mRNA is 1 h in the presence of glucagon and 24 h for insulin and glucose.

The L-type pyruvate kinase gene was controlled at the transcriptional level by glucose and insulin in adult rat hepatocytes in primary cultures (Decaux *et al.*, 1989). Insulin action on gene expression of cultured hepatocytes isolated from fasted rats requires the presence of glucose for the accumulation of aldolase B mRNA, L-type pyruvate kinase and an unidentified 5.4-kilobase mRNA (Decaux *et al.*, 1989, 1991). In contrast, maintaining the amount of albumin and transferrin mRNAs in these hepatocytes requires the presence of insulin alone, glucose having no effect by itself (Decaux *et al.*, 1991). Insulin was shown to act on gene transcription by increasing the transcription of genes coding for proteins of the glycolytic pathway through a glucose-dependent mechanism and a glucose-independent mechanism by acting on the transcription of genes coding for serum proteins (Decaux *et al.*, 1991) (for further discussion see Chapter 12).

The common denominator in all the examples in this section lies in the fact that a cellular response (such as growth and division) to, say, hormones depended on the dynamic balance of macromolecular synthesis and degradation. This is further explored from the aspect of the flux coordination hypothesis presented in Chapter 11.

9.3 REGULATION OF GENE EXPRESSION DURING METABOLIC TRANSITIONS

It is widely accepted that the choice of a developmental fate in organisms strongly depends on differential gene expression. Mammalian genome regulation essentially involves two steps; the first step is the activation of tissue-specific genes through conversion from a sequestered to exposed state of the chromosomal DNA. The second step resembles the corresponding process in bacteria, i.e. activation or inactivation of the exposed genes as a result of interaction with appropriate effector molecules (Puck and Krystosek, 1992). In unicellular bacteria, yeast or multicellular higher plants, genes coding for enzymes of metabolic pathways are subjected to regulation by the metabolic status of cells (Sheen, 1990, 1994). There is evidence that on the second step of genome expression metabolism and energetics may effectively regulate gene expression (Decaux *et al.*, 1991; Marie *et al.*, 1993; see also Chapter 11).

Glucose and acetate trigger global repression of maize photosynthetic gene transcription (reviewed in Sheen, 1994). Using chimeric genes created by fusing maize photosynthetic gene promoters and a *CAT* reporter gene, it has been shown directly that seven maize photosynthetic gene promoters are specifically and coordinately repressed by glucose and sucrose (Sheen, 1990). Glucose repression, also called catabolite repression in *E. coli* and yeast (Gancedo, 1992), or anabolite repression by sucrose and glucose (Sheen, 1990) in higher plants, are main examples of metabolic regulation of gene expression. However, major differences exist in the mechanisms of metabolic regulation in prokaryotic and eukaryotic cells. In bacteria, genes encoding enzymes with related metabolic functions are frequently arranged contiguously in the chromosome and are coregulated by short sequences upstream of the gene cluster. In yeast and higher plants, the situation is different since genes encoding related enzymes of metabolic pathways, or photosynthetic genes in higher plants, are not arranged in clusters but scattered among several chromosomes (Sheen, 1990). Glucose signalling in higher plants lacks specificity; while in mammals, glucose responses cannot be triggered by other hexoses (Sheen, 1994). Maximal gene expression responses in plant and mammalian cells are produced by glucose concentration at about 10 mM in the medium; 10-fold higher concentration of glucose is required in yeast (for a review, see Sheen, 1994).

We will now present a series of experimental results of the dynamics of microbial or plant metabolic transitions that emphasize the mutual dependence between metabolism, energetics and gene expression.

9.3.1 MECHANISMS OF DIFFERENTIAL GENE EXPRESSION AND METABOLIC FLUXES IN RESPONSE TO OXYGEN OR CARBON SOURCE REGULATION

Environmentally induced transcriptional control of gene expression appears to be a level of coordination between metabolism and gene expression. Gene expression of the class of environmentally responsive genes is apparently sensitive to the availability of substrates for growth such as oxygen, carbon source, and the phase of growth in batch cultures. One of the best documented mechanisms of gene expression regulation by environmental factors is that effected by carbon sources (Moore *et al.*, 1991; Lombardo *et al.*, 1992; Sierkstra *et al.*, 1992). The transcriptional machinery appears to integrate a coordinated response of the cell through triggering of a particular metabolic and energetic status of the intracellular environment.

(a) Oxygen regulation

In addition to differences in gene expression under steady-state aerobic or anaerobic growth conditions, cells respond to a transient decrease in oxygen tension through classes of environmentally responsive genes. Those genes encode oxygen-dependent functions, such as cytochrome subunits and oxidases and desaturases in heme, sterol and fatty acid biosynthesis, that are induced at low oxygen tension (Zitomer and Lowry, 1992).

The fumarate reductase, dimethyl sulphoxide–trimethylamine N-oxide reductase and nitrate reductase operons in *E. coli* are regulated in response to anaerobiosis and nitrate availability (Tseng *et al.*, 1994). Whereas fumarate reductase expression increased five-fold as the growth rate decreased from 0.6 to 0.12 h^{-1} during aerobic growth, little change was seen under anaerobic conditions. In contrast, growth-rate dependent expression of nitrate reductase occurred under anaerobic conditions but not under aerobic conditions (Tseng *et al.*, 1994).

The assembly of the respiratory apparatus in yeast requires the coordinate expression of many genes. To regulate cellular metabolism efficiently under aerobic or anaerobic growth, the facultative aerobe *S. cerevisiae* must differentially express a large number of genes in response to oxygen (Zitomer and Lowry, 1992). Several of these genes are heme-activated. Heme serves as the prosthetic group in the cytochromes and some oxygen-binding proteins such as catalases; heme synthesis has an absolute requirement for oxygen. Most of the nucleus-encoded genes involved in oxygen-dependent functions are known to be regulated by heme. Two categories may be distinguished (Zitomer and Lowry, 1992): those encoding respiratory functions (e.g. cytochrome subunits) and those encoding oxidative damage repair (e.g. catalase, superoxide dismutase). Another set of genes is repressed by heme and for most of these genes the utilization of oxygen in electron transport or in membrane or heme biosynthesis is involved (Zitomer and Lowry, 1992).

There are also heme-dependent transcriptional activators: heme activation protein (HAP) complexes. One of these complexes, the HAP2/3/4, which transcriptionally activates a number of genes, responds to two stimuli: heme and non-fermentable energy sources (Zitomer and Lowry, 1992). Since *S. cerevisiae* exhibits the Crabtree effect, i.e. it ferments under glucose excess even in the presence of oxygen, many respiratory genes are 10- to 20-fold repressed when grown in glucose-containing media (Gancedo, 1992). The iso-1-cytochrome c (*CYC1*) gene expresses to different levels depending upon the presence of heme and glucose (Zitomer and Lowry, 1992).

(b) Carbon source regulation

Apparently, the metabolism of glucose is crucial for triggering repression of several genes (Gancedo, 1992; Johnston and Carlson, 1992; Sheen, 1994). Studies with other hexoses and various glucose analogues suggest that glucose/hexose phosphorylation but not further metabolism is required for signalling in plants as well as in yeast (Rose *et al.*, 1991; Gancedo, 1992; Sheen, 1994). However, further glucose metabolism after phosphorylation is necessary for glucose sensing in mammalian cells (German, 1993).

Several genes have been shown to be differentially regulated by glucose at the transcriptional or post-transcriptional levels (Figure 9.5). Those genes are related to the regulation of the expression of glycolytic enzymes (Moore *et al.*, 1991; Sierkstra *et al.*, 1992), the iron–protein subunit (Ip) of succinate dehydrogenase (Lombardo *et al.*, 1992), the *ADH2* gene of the alcohol dehydrogenase II (*ADH II*) (Vallari *et al.*, 1992), or the *GDH2* gene of the NAD-dependent glutamate dehydrogenase (NAD-GDH) in *S. cerevisiae* (Coschigano *et al.*, 1991). In addition, transcriptional modulation of gene expression by glucose in liver metabolism depends on the presence of insulin. For example, the transcription of the glucokinase gene is stimulated by insulin without the aid of glucose, whereas the presence of both glucose and insulin is needed for the transcriptional activation of most glycolytic and lipogenic genes in hepatocytes (Vaulont and Kahn, 1994) (section 12.3).

The proposed mechanisms of transcriptional activation include the following.

- Positive transcriptional activators such as the RAP1 protein which binds to a specific sequence motif in the promoters of many glycolytic genes (Moore *et al.*, 1991) or the ADR1 protein that binds to an upstream activation sequence in the non-coding region of the *ADH II* gene.

- Transcriptional regulation at the level of an upstream activation site of the *GDH2* gene (Coschigano *et al.*, 1991).
- Post-transcriptional regulation of the half-life of mRNA transcripts (mRNA stability).

It has been shown that the half-life of the transcripts of the iron–protein subunit (Ip) of succinate dehydrogenase is vastly shorter in glucose-containing media than in media containing a non-fermentable carbon source (Lombardo *et al.*, 1992). In the presence of glucose, the half-life of the mRNA corresponding to the Ip gene appears to be less than 5 min; in glycerol medium, the half-life increased to 60 min or longer.

Regulated changes in mRNA stability play an important role in modulating the level of expression of many eukaryotic genes (Cleveland and Yen, 1989). It is known that transcription serves as a primary level of control in gene expression while subsequent post-transcriptional events can play important roles in regulating the final levels of a gene product. It appears that in many cases RNA degradation requires ongoing protein synthesis whose mechanistic basis is still unknown.

The RNAs encoding proto-oncogenes, such as *c-fos* and *c-myc*, and lymphokines and cytokines, are among the least stable RNAs in higher eukaryotes, with half-lives of 15–30 min (Cleveland and Yen, 1989). A common feature of this group of RNAs is the presence of a 30- to 80-base AU-rich domain in the 3′ untranslated region. The insertion of a 51-base AU-rich domain from the granulocyte-macrophage colony-stimulating factor (GM-CSF) into the 3′ untranslated region of β-globin mRNA reduces the half-life of the normally long-lived β-globin mRNA to only 30 min in transfected cells (Shaw and Kamen, 1986). It seems that those sequence determinants found within 3′ untranslated regions influence mRNA instability, and that mRNA instability is often linked to ongoing translation (Cleveland and Yen, 1989). For histones and tubulin, *in vivo* mRNA degradation has been shown to be co-translational and probably occurs through the action of a ribosome-bound nuclease.

Another example of the coordination between gene expression and metabolism is provided by alginate biosynthesis by *Pseudomonas aeruginosa*. Mucosity in *P. aeruginosa* appears to be, at least partially, under environmental control. In pulmonary infections caused by *P. aeruginosa*, initially the microorganism is non-mucoid but during the course of the infection it changes to the mucoid, alginate-producing phenotype (Deretic *et al.*, 1989). The mucoid variants are unstable with respect to alginate synthesis and rapidly become non-mucoid when cultured *in vitro* on artificial medium (Sá-Correia *et al.*, 1987). A pivotal step in alginate biosynthesis is the activation in mucoid cells of transcription of the *algD* gene encoding GDPmannose dehydrogenase.

This enzyme catalyses a key step in the alginate pathway whereby double oxidation of GDP mannose into GDP mannuronic acid, a precursor of alginate polymerization, channels the pool of sugar intermediates into alginate production (Deretic *et al.*, 1989).

In plants and *S. cerevisiae*, it has been implied that the intracellular concentration of phosphorylated sugars is involved in regulation of gene expression at the transcriptional level. Repression of genes judged by the level of transcripts in turn regulated by the rate of transcription and/or the rate of degradation of RNA species has been described (Graham *et al.*, 1994). Photosynthetic end-product repression by sucrose or glucose has been shown in protoplasts (Sheen, 1990) or cell cultures (Graham *et al.*, 1994) of higher plants. The metabolic repression of photosynthetic genes apparently overrides other forms of regulation, such as light, tissue type and developmental stage, because it is executed in young leaf cells under light. The half-life of α-amylase mRNA was less than 1 h in cells provided with sucrose, but increased to 12 h in cells starved of this sugar. Both transcriptional and post-transcriptional control mechanisms are important in the metabolic regulation of α-amylase gene expression. *De novo* synthesized proteins appear to be involved in those mechanisms since α-amylase mRNA accumulated massively when cells provided with sucrose were treated with the protein synthesis inhibitor, cycloheximide (Shen *et al.*, 1994). The expression of α-amylase and actin genes appeared to be mediated by different mechanisms.

Genes coding for enzymes involved in pathways such as the glyoxylate cycle (Graham *et al.*, 1994), i.e. malate synthase (*MS*) and isocitrate lyase (*ICL*), or photosynthetic gene promoters (Sheen, 1990) have been shown to be subjected to catabolite (by acetate) or anabolite repression (by glucose or sucrose). It has been proposed that either hexose sugars or the flux of hexose sugars into glycolysis is involved in signalling the intracellular metabolic status that correlates with changes in gene expression (Graham *et al.*, 1994). Induction of the *MS* and *ICL* gene transcription depends on the intracellular concentration of sugars falling below a critical threshold of sugar concentration and on their phosphorylation status, since glucose analogues that cannot be phosphorylated, such as 3-methylglucose, do not initiate the repression response.

In *S. cerevisiae* it has been implied that glucose repression could be initiated by an enhanced glycolytic flux instead of high concentrations of either extracellular glucose or intracellular glucose 6-phosphate (Sierkstra *et al.*, 1993). In fact, a mutant of *S. cerevisiae* carrying a deletion in the glucose-6-phosphate isomerase gene (*pgi1*) did not show glucose repression in the presence of high concentrations of extracellular glucose and intracellular glucose 6-phosphate in combination with a low glycolytic flux (Sierkstra *et al.*, 1993).

9.3.2 INTERDEPENDENCE BETWEEN ENERGY METABOLISM AND GENE EXPRESSION DURING METABOLIC TRANSITIONS IN *ESCHERICHIA COLI*

Prokaryotic cells readily accommodate to changes in environmental conditions such as oxygen level, osmolarity or nutrient availability. These changes result in modifications at the level of both intermediary metabolism and gene expression (Figure 9.7) (Smith and Neidhardt, 1983a,b; Neidhardt, 1987). Recognition of DNA sequences by specific transcription factors is the basis of the bacterial paradigm of regulation of gene expression. DNA topology through its supercoiling status also appears to be an important contributor to gene regulation involving changes in gene activity.

Increasing evidence supports the notion that supercoiling may regulate gene expression (Pruss and Drlica, 1989). The state of DNA supercoiling depends on a balance of topoisomerase activities which, it was shown *in vivo*, may affect the expression of specific genes. Apparently, DNA supercoiling provides a kind of 'conformational information' in addition to the coding information (Gellert *et al.*, 1983). It appears also that intracellular DNA is kept under some kind of configurational stress and that maintenance of that stress in the presence of DNA-relaxing topoisomerases requires the continuous activity of DNA gyrase (Gellert *et al.*, 1983).

The induction of several genes elicited by anaerobiosis (Figure 9.7) or osmotic signals depends on DNA topology – namely, the degree of DNA supercoiling (Ni Bhriain *et al.*, 1989; Pruss and Drlica, 1989; Gober and Kashket, 1989). The extent of DNA supercoiling is mostly determined by the relative levels of activity of two enzymes, topoisomerase I and DNA gyrase (Wang, 1985). The level of transcription of the genes coding for topoisomerase subunits (gyrA, gyrB) is in turn regulated by the levels of DNA supercoiling (Menzel and Gellert, 1983). DNA gyrase, a topoisomerase II, introduces negative helical turns in DNA topology. Energetically, the introduction of negative supercoils is an upstream process, i.e. DNA gyrase hydrolyses ATP to drive the unwinding of DNA (Gellert *et al.*, 1976). According to *in vitro* experiments that showed that the activity of topoisomerase II depends on the ATP/ADP ratio rather than on the ATP concentration itself (Westerhoff *et al.*, 1988), the possibility was raised that DNA supercoiling, by sensing the energy status of the cell, could play the role of an 'energy intermediate' to transduce metabolic signals into differential gene expression. Thus, on the basis of the

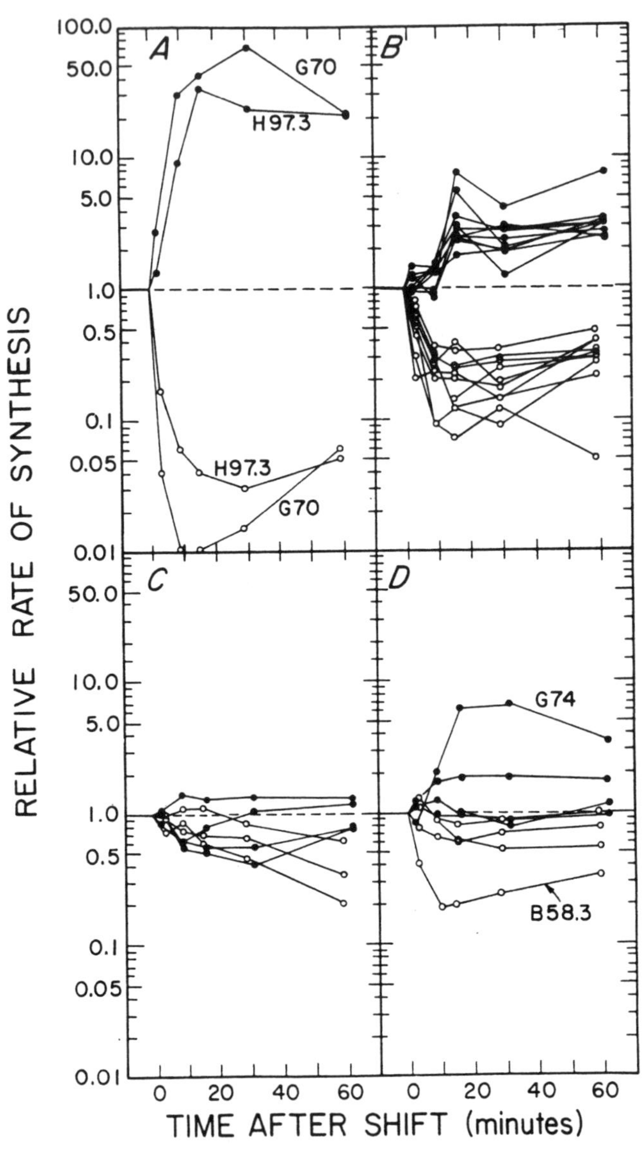

100.0
50.0
10.0
5.0
1.0
0.5
0.1
0.05
0.01
50.0
10.0
5.0
1.0
0.5
0.1
0.05
0.01
A
B
C
D
G70
H97.3
H97.3
G70
G74
B58.3
RELATIVE RATE OF SYNTHESIS
0
20
40
60
0
20
40
60
TIME AFTER SHIFT (minutes)

dependence of DNA supercoiling on the ATP/ADP ratio by experiments performed *in vitro* (Westerhoff *et al.*, 1988), and taking into account that gyrase activity depends on ATP, it was postulated that the phosphorylation potential might be one of the free-energy intermediates being transduced *in vivo* for the regulation of DNA supercoiling through gyrase activity (Westerhoff *et al.*, 1988). It then became plausible that transitions in the cellular energetic status could be coupled to changes in gene expression through DNA supercoiling (Chapter 12).

Though chromatin DNA in eukaryotic cells is topologically underwound, bulk DNA is not under superhelical tension because the supercoils are constrained by histone–DNA interactions. Positive DNA supercoiling downstream of the transcription complex is believed to decondense the chromatin fibre (Lee and Garrard, 1991). Uncoiling the negatively supercoiled nucleosomal DNA downstream of the transcriptional complex would pave the way for RNA polymerase passage. It appears that positive DNA supercoiling provides a mechanism for generating, but is not required for maintenance of a conformation in chromatin characteristic of highly transcribed genes (Lee and Garrard, 1991). Torsional stress is apparently required for transcription to occur in living cells, since circular but not linearized or nicked DNA molecules are transcriptionally active *in vivo*.

9.3.3 NITROGEN METABOLISM AND GENE EXPRESSION

The metabolism of nitrogenous compounds in *S. cerevisiae* has been reviewed extensively by Cooper (1982). *S. cerevisiae* can utilize a number

Figure 9.7 (see facing page) Proteins induced by anaerobiosis in *Escherichia coli*. Relative rates of synthesis of individual proteins at various times after a shift from aerobic to anaerobic conditions (filled circles), or after transition from anaerobic to aerobic conditions (open circles). All values were normalized to the measured preshift rate of synthesis. (a) Proteins G70 (Pyr formate-lyase, β) and H97.3; (b) proteins B15.0, C40.3, C43.1, F35.8 (F P-kinase I), F38, F43.8 (enolase), F50.5, G31.5 (only filled circles), G54.7 (Pyr kinase I) and I33.5 (glycerol P-dehydrogenase); (c) proteins A15.5, A48, B11.8 (filled circles only) and B76; (d) control proteins B46.7 (ATPase β), B58.3 (PTS, EI), F39.7 (aspartate aminotransferase) and G74 (Pyr formate-lyase, α). (Reproduced from Smith and Neidhardt, 1983, by permission of the American Society for Microbiology, ASM Press.)

of compounds as nitrogen sources – for example, allantoin, allantoate, NH_4^+ and amino acids. The glutamine synthase/glutamate synthase (GS/GOGAT) cycle is mainly involved in ammonia assimilation in higher plants or *S. cerevisiae*. Work with mutants lacking functional GDH argues in favour of multiple routes of ammonia assimilation: glutamate and ammonia can be interconverted through a system which uses GS/GOGAT in a high energy-consuming process (Cooper, 1982).

NADH- and NADPH-dependent glutamate dehydrogenase, glutamate synthase and glutamine synthase are the key enzymes of both catabolic and anabolic pathways of nitrogen conversion. All ammonia incorporated into cellular macromolecular components should be transformed through at least one of these enzymatic steps.

The interconversions of ammonia and glutamate are at the interphase between biosynthesis and degradation of nitrogenous compounds. The role of NAD-dependent glutamate dehydrogenase coded by the gene *GDH2* in *S. cerevisiae* revealed that the expression of the enzyme was subjected to coordinated regulation by nitrogen and the carbon source. High glucose concentrations inhibit the expression of *GDH2*. Under catabolite repression conditions, the presence of glutamate was able to derepress *GDH2*. At low glucose concentrations, or in the presence of gluconeogenic substrates, glutamate dehydrogenase was expressed at high levels irrespective of the presence of glutamine in the culture medium, which acts as a repressor in the presence of 1% glucose (Coschigano *et al.*, 1991).

Ammonia assimilation in isolated barley (*Hordeum vulgare* L.) chloroplasts has been studied by metabolic control analysis. The flux control coefficient for GS did not change significantly with 2-oxoglutarate concentration (from 0.58 at 5 mM 2-oxoglutarate to 0.4 at 20 mM 2-oxoglutarate (Baron *et al.*, 1994). The control exerted by GOGAT over the flux depended on the glutamine concentration. At high concentrations of glutamine the GS/GOGAT cycle exerted a major control over the flux, but was far less important at low glutamine levels (Baron *et al.*, 1994). The flux control coefficient for GOGAT decreased significantly with decreasing glutamine concentrations (from 0.76 at 20 mM glutamine to 0.19 at 10 mM glutamine). The coordinate regulation of an enzyme involved in nitrogen metabolism may be understood in terms of the requirement of carbon skeletons and ATP in order to drive the reactions that assimilate nitrogen into glutamate and glutamine (Mora, 1990).

All the examples discussed up to this point may be interpreted in dynamic terms according to the concepts developed in Chapters 1, 2 and 8. Environmental factors (pH, p_{O_2}, carbon source concentration) act as bifurcation parameters driving the dynamics of cellular systems to particular behavioural domains and eliciting, in some cases, global behavioural changes. Those changes may concern, for example, changes in cell cycle activity, metabolic transitions, or regulation of gene expression (transcriptional control by carbon or nitrogen sources). According to the type of substrate (fermentative, gluconeogenic, nitrogenous) and its concentration, different metabolic rates and patterns of gene expression are induced.

9.4 METABOLIC TRANSITIONS AND CELLULAR TRANSFORMATION

9.4.1 NEOPLASIA: A CELL CYCLE SICKNESS?

Cancer may be viewed as a disease of the cell cycle (Gerson, 1978, quoted in Gillies, 1981). Proliferative and non-proliferative states are two branches in the dynamic behaviour of cells. It could well be conceived that the transition point in the parametric control space between proliferative to resting behaviour of normal cells has indeed been shifted in tumour cells. Such a shift may be a consequence of the change in dynamics of the processes at lower levels of organization that determine the proliferative behaviour.

In the G_1 period of the cell cycle are located the so-called 'restriction' (R point: Pardee *et al.*, 1978) or 'Start' (Hartwell *et al.*, 1974) events. Depending on these events being completed, either a cell progresses through its division cycle or it is led to increase its biomass without proliferation (Figure 9.8). If the restriction (or Start) point of G_1 is passed, cells will replicate their DNA and will be committed to divide. Sensitivity to protein synthesis is one of the most important observations concerning the R point or Start (Hartwell *et al.*, 1974; Pardee *et al.*, 1978).

From observations made in normal or tumour cells, and lower or higher eukaryotes, several general features emerge. It seems clear that glycolysis is necessary for unimpeded progression of a cell through its growth cycle, at least in *S. cerevisiae*. The G_1/S transition was associated with high glucose fluxes and respiro-fermentative metabolism, whereas cells in G_1 showed low glucose fluxes and respiratory-type metabolism (Aon *et al.*, 1995). It has been observed that ethanol formation always takes place when cells enter a reproduction phase (von Meyenburg, 1969; Kappeli, 1986).

On the other hand, proliferative higher eukaryotic cells, like most tumour cells, have higher rates of both anaerobic and aerobic glycolysis than normal cells (Aisenberg, 1961). The Pasteur effect predicts that glycolysis would be inhibited in actively respiring cells. A collection of

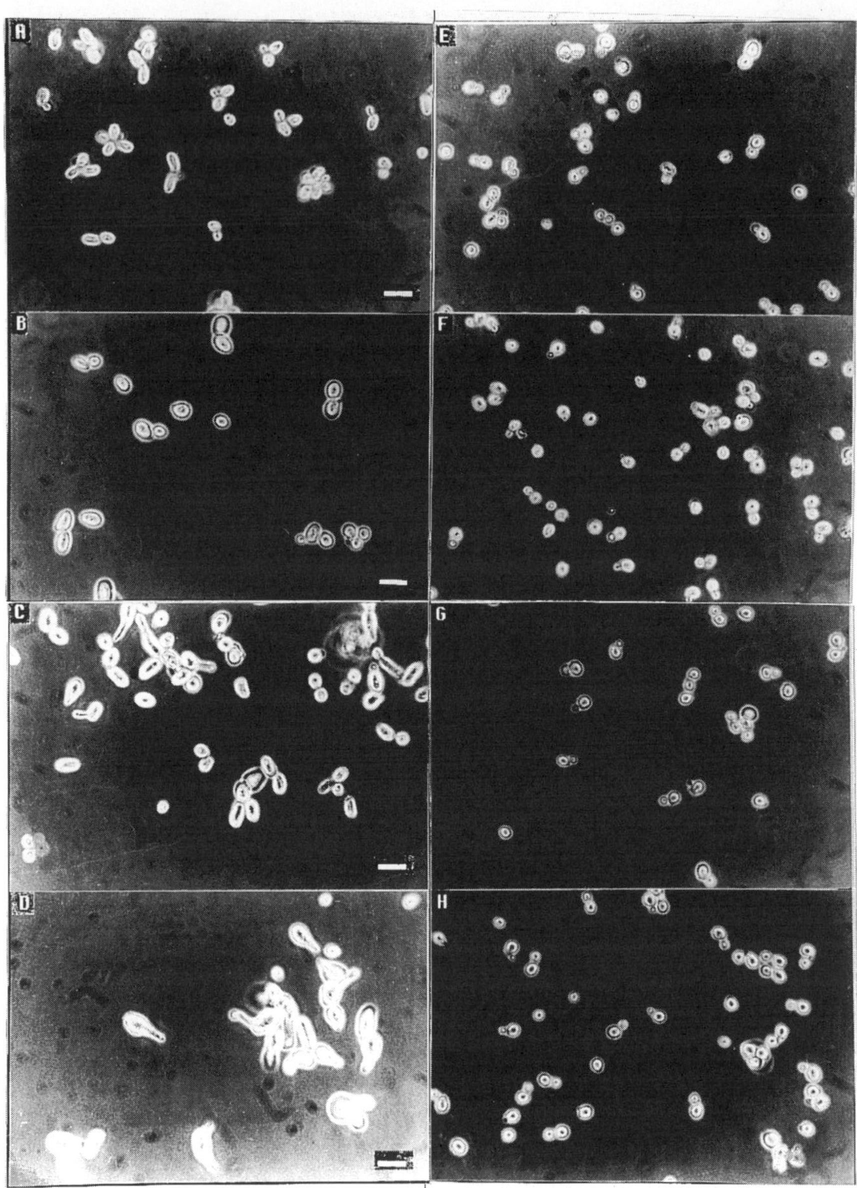

Figure 9.8 The 'Start' point of the G_1 period of the cell cycle of *S. cerevisiae*. Yeast cells were grown at 25°C in YNB-glucose medium and shifted to 37°C (restrictive temperature), as described in Aon *et al.* (1995). (A)–(D) and (E)–(H) show the *S. cerevisiae* strains cdc28 and wild type A364A, respectively, at time zero (A,E) and 1 h (B,F), 3 h (C,G) or 5 h (D,H) after the temperature transition. Cdc28 is a temperature-sensitive cell division cycle mutant of *S. cerevisiae* (A–D). Note the progressive induction of the 'shmoo' phenotype of the cdc28 strain arrested at 'Start'. Bar=5 μm.

data presented by Gillies (1981) showed that tumour cells exhibit a normal Pasteur effect when measured in absolute terms, but on a percentage basis glycolysis is much less inhibited by respiration in tumour cells than in normal cells. However, it is clear that transformed cells usually derive an appreciable amount of energy from glycolysis even in the presence of oxygen (Chapter 7).

In several systems, either prokaryotic or eukaryotic, it has been shown that intracellular pH (pH_i) changes in parallel with the increase in the rate of the glycolytic flux (Gillies, 1981; Den Hollander *et al.*, 1981). Maxima in glycolytic rates are observed in pH_i ranges that have been shown to correspond to maximum growth rate: 6.6–7.4 in *Tetrahymena* and 7.0–7.2 in *S. cerevisiae* (Gillies, 1982) (Table 10.1). In lower eukaryotes, glycolytic flux is activated during G_0 to G_1 transition in the cell cycle, concomitant to a transient alkalinization of pH_i of about 0.3 units (Gillies *et al.*, 1981). These two events – glycolytic activation and pH increase – occurred before the onset of DNA synthesis, which was followed by mitosis. The exit of synchronous yeast cell populations from G_0 to G_1 after oxygenation and refeeding with glucose for more than 30 min was correlated with an alkaline pH transient of approximately 0.3 pH units above an initial pH_i of 7.0. The intracellular alkalinization in *S. cerevisiae* resulted regardless of the starting pH_i and occurred entirely before the onset of DNA synthesis, as in *Tetrahymena* though much slower (Gillies, 1982). Mitosis began at about 130 min, which was over 1 h after the pH_i transient.

9.4.2 COORDINATED METABOLIC AND ENERGETIC CHANGES IN CELLS
STIMULATED TO PROLIFERATE

Several correlative events, triggered in mitogen-induced lymphocyte proliferation, have been well characterized. Peripheral lymphocytes exist for long periods in a vegetative (G_0) state if unstimulated. One of the earliest events to take place after addition of phytohaemagglutinin (PHA) to lymphocytes is a dramatic increase in the rate of aerobic glycolysis which precedes the onset of both protein and DNA synthesis (Hedeskov, 1968). Glucose utilization was almost doubled in lymphocytes incubated with PHA. The data indicated that an increased rate of glycolysis is necessary for the mitogen-induced proliferation of lymphocytes. An increased K^+ uptake, which is mediated in part by an increased activity of the ATPase, was also registered (Gillies, 1981, and references therein).

Cell populations of spleen lymphocytes stimulated with mitogens increased both pH_i and the rate of DNA synthesis, whereas unstimulated cells remained at a low pH_i and did not have significant rates of DNA synthesis (Gerson, 1982, quoted in Gillies, 1982).

In sea urchin eggs there was a sharp increase in the rate of protein synthesis, starting at pH 6.9 with an optimum at pH 7.4, which corresponded closely to the pH values *in vivo* where protein synthesis is activated at fertilization (Winkler, 1982, quoted in Busa and Nuccitelli, 1984). In a cell-free system a TMV product (110 000 MW) required 36 min at pH 6.9 and 18 min at pH 7.4 to be synthesized (Winkler, 1982, quoted in Busa and Nuccitelli, 1984).

9.4.3 METABOLIC TRANSITIONS AND MALIGNANCY: PLEIOTROPIC REPROGRAMMING OF PURINE, PYRIMIDINE AND CARBOHYDRATE METABOLISM

A flux imbalance is exhibited by transformation with respect to normal cells. It has been shown that a general increase in synthetic over catabolic fluxes is closely correlated with tumour malignancy (Figure 9.9) (Weber, 1983). Furthermore, enzymic and metabolic imbalance are characteristics of neoplasia. Cancer cells revealed a complete reprogramming of gene expression manifest in the imbalance of activities of key enzymes of pyrimidine *de novo* and salvage biosynthetic as well as degradative pathways. The changes in enzymic patterns were linked to either transformation or progression (Table 9.1). In the former case, enzyme activity increased in all the tumours examined, even the slowly growing ones. In the case of progression-linked changes, significant correlations with the proliferation rate of the neoplasms were observed (Weber, 1983).

Changes in enzymic patterns, combined with a general increase in synthetic over catabolic pathways, confer neoplastic cells with reproductive advantages (Figure 9.9). The enzymic and metabolic imbalance is characteristic of neoplasia. The following types of tumour showed a similar overall pattern of enzymic imbalance (Weber, 1983): chemically-induced rodent tumors (hepatomas, kidney tumors and sarcoma in rats); colon carcinomas (rats and mice); myeloma, lymphosarcoma and lymphocytic leukaemia in mice; MC-29 virus-induced hepatoma in chickens and in humans; hepatocellular and renal cell carcinomas; colon tumour xenografts; human primary lung and colon adenocarcinomas; human leukaemias and lymphomas.

In neoplastic cells the global changes at the biochemical and enzymic levels may be the result of a pleiotropic reprogramming of gene expression. The experimental data supporting this proposal are based on two important assumptions:

- that the activity of the enzyme measured, in the presence of optimum substrate and cofactor, was proportionate with the amount of enzyme;

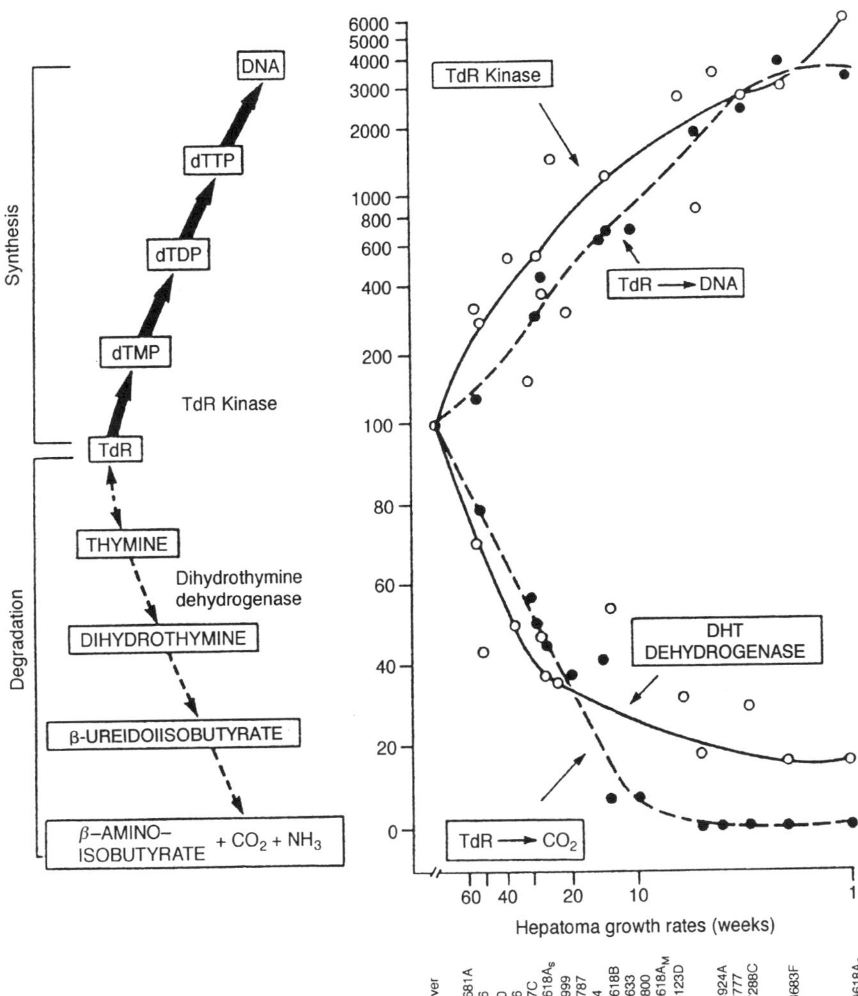

Figure 9.9 Biochemical strategy of cancer cells: unbalanced synthetic and catabolic pathways in hepatomas. Reciprocal behaviour of activities of opposing key enzymes, thymidine (TdR) kinase and dihydrothymine (DHT) dehydrogenase and synthetic and catabolic pathways of DNA by incorporation of thymidine or by its degradation to CO_2. The activities were expressed as percentages of control values in normal rat liver. (Reproduced from Weber, 1983, by permission of the American Association for Cancer Research, Inc.)

Table 9.1 Enzymic markers of transformation (the relationship of enzymic activities with transformation and progression was determined in rat hepatomas) (reproduced from Weber, 1983, by permission of the American Association for Cancer Research, Inc.)

Metabolic areas	Synthetic pathways	Degradative pathway
	Increased activity	*Decreased activity*
Pyrimidine and DNA	Carbamoyl-phosphate synthetase II	Thymidine phosphorylase
	Aspartate carbamoyltransferase	Dihydrothymine dehydrogenase
	Dihydroorotase	
	Orotate phosphoribosyltransferase	
	Orotidine-5′-phosphate decarboxylase	
	UDP kinase	
	CTP synthetase	
	Ribonucleotide reductase	
	DNA polymerase	
	Uridine–cytidine kinase	
	Deoxycytidine kinase	
	Thymidine kinase	
Purine and RNA	Amidophosphoribosyltransferase	5′-Nucleotidase
	Formylglycinamidine ribonucleotide synthetase	Inosine phosphorylase
	Adenylosuccinate synthetase	Uricase
	Adenylosuccinate lyase	Adenylate kinase
	AMP deaminase	
	AMP dehydrogenase	
	GMP synthetase	
	tRNA methylase	
	Arginyl-tRNA synthetase	
	RNA polymerase	
Ornithine	Ornithine decarboxylase	
Pentose phosphate	Glucose-6-phosphate dehydrogenase	Ornithine carbamoyl-transferase
	Transaldolase	
	PRPP synthetase	
	Decreased activity	*Increased activity*
Carbohydrate	Glucose-6-phosphatase	Hexokinase
	Fructose-1,6-diphosphatase	6-Phosphofructokinase
	Phosphoenolpyruvate carboxykinase	Pyruvate kinase
	Pyruvate carboxylase	
Membrane cAMP	Adenylate cyclase	cAMP phosphodiesterase

- that the amount of enzyme was an indicator of the gene expression.

The amount of enzyme was measured by two independent methods: kinetic and inmunological.

The metabolic characterization has been performed in a wide variety of transformed cells. On the one hand, the biochemical patterns of neoplastic cells were contrasted with those of normal (resting) liver as well as regenerating and differentiating liver. On the other hand, results were obtained with a spectrum of hepatomas of different growth rates allowing interpretation of whether the biochemical changes were due to neoplastic transformation and progression or to rapid growth rate or different degrees of differentiation. The results obtained could also be generalized to renal and colon carcinomas, chemically induced muscle sarcoma and human neoplasms (hepatocellular carcinomas, renal cell carcinoma, colon carcinoma, lung neoplasia and leukaemic cells).

In neoplastic liver cells, the exhibited pyrimidine metabolism increased enzyme levels of *de novo* biosynthesis and salvage pathways (Table 9.1). **Salvage enzymes** are defined as those providing alternative routes to produce metabolites like UMP, CTP and dTMP. On the other hand, a decrease in activities of enzymes involved in pyrimidine degradation was corroborated. Purine metabolism showed an increased potential to produce IMP and particularly GMP, GDP and GTP, and a decreased capacity to degrade purines. Existing evidence indicates that the capacity for GTP biosynthesis is up-regulated in cancer cells. Pyrimidine metabolism revealed increased activities of the synthetic enzymes and a decrease of the catabolic ones. Carbohydrate metabolism showed increased activities of glycolytic enzymes and those of the pentose phosphate pathway, with a decrease in gluconeogenic enzymes (Table 9.1).

In cancer cells, the activities of key enzymes of *de novo* biosynthesis were elevated and the salvage enzymes of *de novo* biosynthesis were also increased. In chemotherapy, in principle, the activities of these enzymes should be inhibited (Weber, 1983, 1990). An enzyme-pattern targeted chemotherapy of IMP dehydrogenase (IMPDH) activity and GTP concentration with tiazofurin was tried on end-stage leukaemia patients (Weber, 1990). Tiazofurin (2-β-D-ribofuranosylthiazole-4-carboxamide) is a C nucleoside which in sensitive cells is metabolized through the activities of a kinase and NAD pyrophosphorylase to an NAD analogue, thiazole-4-carboxamide adenine dinucleotide, the latter being the active metabolite of tiazofurin (Weber, 1990). Tiazofurin inhibits IMPDH activity. Allopurinol, administered routinely with tiazofurin to decrease uric acid formation, provided a combination chemotherapy impact, resulting in inhibition of both *de novo* and salvage pathways of GTP biosynthesis in blast cells. This rationally designed chemotherapy provided complete haematological remissions in 55% of 21 consecutively treated, unselected patients (Weber, 1990, and references therein). K562 human erythroleukaemic cells cultured in the presence of tiazofurin exhibited decreased IMPDH activity and GTP concentration. The decrease was followed by down-regulation of the c-Ki-*ras* oncogene and subsequent induced differentiation (Weber, 1990).

Overall, it appears that the enzymic and metabolic pattern of ordered transformation and progression-linked alterations in tumour cells is a manifestation of the reprogramming of gene expression. Apparently, metabolic imbalance signals the reprogramming of gene expression conferring selective advantages to cancer cells (Chapter 11).

9.4.4 STARVATION RESPONSE OF TUMOUR CELLS

Ehrlich ascites tumour cells deprived of glucose or amino acids, or in the presence of 2-deoxyglucose, showed a decrease in the rates of peptide chain initiation and elongation. These changes occurred in parallel with a decrease in the intracellular ATP concentration. Glucose deprivation affected protein synthesis by inhibition of the rate of protein synthesis per ribosome in polysomes and the rate of chain initiation. The fact that deprivation of amino acids resulted in a prompt and marked decrease in glucose utilization and in the amount of intracellular ATP was taken as further evidence for an indirect mechanism of slowing protein synthesis acting through limitation of the energy supply (Van Venrooij *et al.*, 1972).

A response of cells to starvation stress is to increase their spontaneous mutation rate under conditions in which post-selection mutations accumulate. The level of tRNA modification could be one signal that triggers whatever process leads to the appearance of post-selection mutations (Siegele and Kolter, 1992). Cairns *et al.* (1988) and Hall (1990) have pointed out that when organisms are allowed to persist on medium which permits survival but not growth, mutation appearances result from time rather than generation dependence. This explains why some calculations of mutation rates for developing tumours may be inapplicable (Strauss, 1992). The distinction between the conditions for the measurement of mutation and those required for the changes involved in carcinogenesis may provide an explanation for the high rate of cellular transformation in rodent cells as compared with mutation (Strauss, 1992).

9.5 NEOPLASIA VIEWED FROM AN INTEGRATIVE PERSPECTIVE OF CELLULAR METABOLISM, ENERGETICS AND GENE EXPRESSION

Carcinogenesis is a complex process involving at least three steps: initiation, promotion and progression. The most accepted view of carcinogenesis includes an initiation step that would correspond to the reaction of DNA with a carcinogen, resulting in a mutation which is converted into a tumour by promotion and progression (Strauss, 1992).

Nowell (1976) has presented the idea that most neoplasms have a unicellular origin and clonal growth pattern underlined by genetic instability. In other words, neoplasms develop as a clone from a single cell which somehow acquires a selective growth advantage over adjacent normal cells. Neoplastic proliferation proceeds with enhanced genetic instability associated with the expanding tumour population. Karyotypic alterations, including whole chromosome loss or gain, ploidy changes and a variety of chromosome aberrations, are common in cancer cells (Hartwell and Kastan, 1994). Mutation of genes that encode components of cell cycle checkpoints, which are believed to integrate DNA repair systems with cell cycle progression, increase genetic instability. Negative controls on cell cycle progression are exerted during development, differentiation, senescence and cell death. These negative controls may play an important role in preventing tumorigenesis (Hartwell and Kastan, 1994).

The idea that point mutation is central to carcinogenesis was supported by the close correlation observed between mutagenesis and carcinogenesis. However, two main pieces of evidence from recent research shed doubts on the latter idea:

- There is a wide class of nonmutagenic carcinogens.
- Cancers show so many genetic changes that measured mutation rates in somatic cells cannot account for the mutation rates of developing tumours (Strauss, 1992).

It has been suggested that developing tumours are similar to *E. coli* kept under selective conditions. Under those conditions mutation is time-dependent rather than replication-dependent and it continues in the absence of cell proliferation (Cairns *et al.*, 1988; Hall, 1990; Siegele and Kolter, 1992).

The picture outlined in this chapter about the dynamics and coordination of metabolism, energetics and gene expression in growing and proliferating cells reveals some common features. Transitions effected by, for example, mitotic cycling, environmental perturbations or signals such as hormones trigger a generalized cellular response associated with the arrest of cell division or change in a cell's fate. There is redirection of metabolic fluxes involving activation or inhibition of certain metabolic pathways such as protein or nucleic acid synthesis, or glycogen and trehalose accumulation. When cells stop division because of starvation or when they are stimulated to grow, metabolism reacts as a whole: changes at the level of gene expression, mutation rates, intermediary metabolism, cellular energetics (e.g. membrane energization) and eventually cell morphology may be observed (Chapter 11).

Overall, it appears that the complex changes associated with transformed or neoplastic cells involve the spatio-temporal coordination of cellular metabolism, energetics and gene expression. To achieve a more thorough understanding of such profound alterations in cellular function, it is necessary to identify a mechanism able to deal with the global coordination and transduction of environmental changes or stimuli. Thus, the question is: how is this coordination accomplished at the cellular level?

9.5.1 THE ROLE OF THE CYTOSKELETON IN THE SYSTEMIC COORDINATION OF GENE EXPRESSION, METABOLISM AND CELLULAR ENERGETICS

Cancer cells have altered cell architecture as well as altered cell function and altered energy requirements (for reviews, see Pienta and Hoover, 1994; Cortassa and Aon, 1994b). Disruption of the actin microfilament skeleton correlates with altered cell structure, gene expression and energy metabolism. In general, remodelling of cell architecture is linked to changes in energetic metabolism (Bereiter-Hahn *et al.*, 1995). The biochemical basis of such linkage may be given by the altered kinetic behaviour of, say, glycolytic enzymes (Cortassa *et al.*, 1994a) associated with the cytoskeleton (Knull and Walsh, 1992; Epner *et al.*, 1993). This evidence is in agreement with the existence of the so-called tissue matrix system.

It is well documented that many of the metabolic changes associated with growth transitions are due to differential expression of gene products. On the one hand, differential genetic expression involves not only transcription and translation but DNA topology as well; on the other hand, neoplastic transformation appears to be associated with genetic instability. The latter is reflected in improved mutation rates for cells under prolonged selective conditions. The efficiency of this phenomenon appears to rely on the genetic make-up of the cell. In chick red blood cells, fibroblasts, brain cells and red blood cell precursors, it has been demonstrated that different DNAs are hypersensitive to hydrolysis when isolated nuclei are treated with DNAase I (Weintraub and Groudine, 1976, quoted by Puck and Krystosek, 1992). It was further shown that active genes are preferentially digested by DNAase I, their sensitivity to DNAase digestion being largely tissue-specific and associated with their transcriptional potential. Applied to cancer cells, these concepts highlighted the fact that the amount of hypersensitive DNA in twelve malignant cells was greatly reduced over that found in the corresponding normal cells (Puck and Krystosek, 1992). It was further proposed that conversion of genes from the relatively resistant to the hydrolysis-sensitive state (gene exposure) was the necessary first step for gene expression (Puck and Krystosek, 1992). Nick translation experiments demonstrated that the region of distribution of exposed DNA which sets apart the normal from the malignant form of the CHOK1 cells is confined to a shell encompassing the periphery of the nucleus (Figure 9.10). All

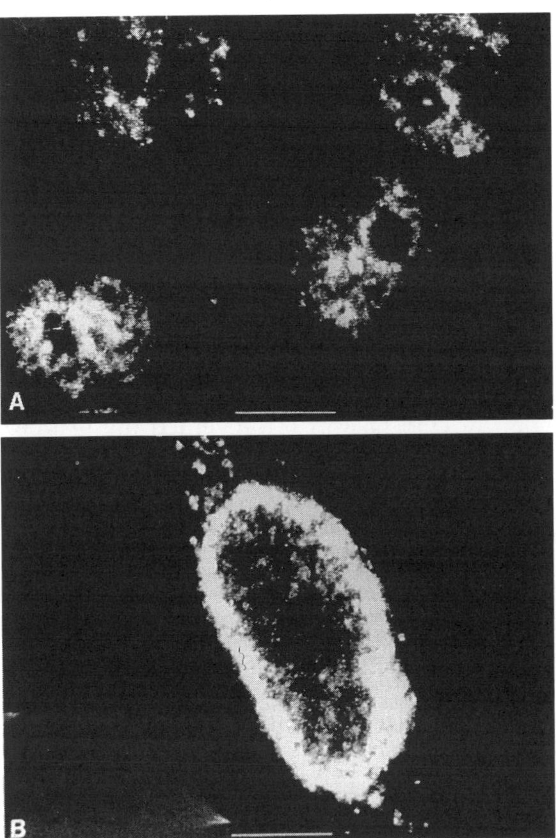

Figure 9.10 The genome exposure defect in cancer cells: differences in DNA exposure pattern revealed by nuclei of (A) a typical cancer cell, and (B) the corresponding normal or reverse-transformed cell. (A) The cell nuclei are malignant CHO-K1 cells. (B) The same cell after reverse transformation with cAMP derivatives. Both types of cell were subjected to the nick translation procedure, which permits visualization of the exposed DNA region (ie. DNase I-sensitive region) as a shell around the nucleus. Each of 12 different cancer cell lines tested behaved as in (A), while their normal or reverse-transformed counterparts resembled that in (B). Apparently, an organized cytoskeleton is necessary for the genome exposure reaction. (Reproduced from Puck and Krystosek, 1992, by permission of Academic Press.)

non-cancer cells tested thus far displayed the peripheral nuclear pattern of exposed DNA, but this was absent in several malignant ones, as shown in Figure 9.10. Approximately 30% of the genome in mammalian cells is converted from the sequestered to the exposed stage in a single action governed by effectors such as cAMP, retinoic acid and nerve growth factor, acting on transformed cells. Whether a given chromosomal region remains expanded and exposed, or condensed and sequestered, is largely determined by the nature of the protein(s) attached to key repetitive DNA sequences and the integrity of the fibre network of the cell (Puck and Krystosek, 1992). It has been postulated that the cytoskeleton or other components of the tissue matrix system, such as the nuclear matrix, may control sequestration or exposure by attachment to specific sites along the chromosomes in interphase (Puck and Krystosek, 1992).

The next step along this line of reasoning may be formulated through the following question: is there a role for the cytoskeleton in this transition from sequestered to exposed genes?

Apparently, the cell cytoskeleton is a necessary structure for producing the specific genome exposure pattern for different cell types. Of particular importance is the fact that cytoskeleton-disorganizing agents, such as colcemid or cytochalasin B, prevented the reappearance of the shell of exposed DNA in cAMP reverse-transformed cancer cells. These findings indicated a direct relationship between the cytoskeleton organization and the shell of exposed DNA around the nuclear periphery (Puck and Krystosek, 1992). Experiments have shown that all three elements of the cell cytoskeleton are involved in the reverse transformation reaction of the CHOK1 cancer cells. These are remarkable observations. Two main consequences follow:

- Genome regulation (genome exposure) may be effected through the so called tissue matrix system, which consists of linkages and interaction of the nuclear matrix, the cytoskeleton and the cell periphery (Penman *et al.*, 1981; Pienta *et al.*, 1989; Getzenberg *et al.*, 1991; Puck and Krystosek, 1992) (Chapter 10).
- Since the effects of the cytoskeleton are so diverse, the response of cells so coherent and the genome regulation so large, a mechanism involving coherence in cellular response must be invoked.

For the latter, the nature of the organization of cellular cytoplasm becomes a timely topic. The cytoplasmic organization may be analysed around two main aspects: the mechanism of coherence and how this coherence is transduced into functional properties, i.e. metabolism and gene expression (Chapters 6 and 7).

Spatio-temporal coordination of cellular energetics and metabolism during development

10

10.1 THE PROBLEM

Chapter 4 approached, from the angle of dynamic organization, the question of how millions of molecules could be spatio-temporally organized in morphogenetic fields such that form arises in developing organisms at macroscopic scales (micrometres to millimetres, i.e. the characteristic lengths over which biological form is generated). Indeed, molecules caught in the spatio-temporal window of milliseconds and nanometres are not able to create anistropy *per se* at spatial dimensions of micrometres to millimetres and minutes to hours. Even though the maternal environment is asymmetrical, it is the internal (e.g. egg's) process dynamics that, within appropriate boundaries, will distribute molecular components anisotropically. We hypothesize that only the dynamics of processes, i.e. fluxes or movements of molecular, supramolecular or subcellular structures, and not molecules themselves are able to create spatial anisotropy in a long-range scale such as that needed for developmental polarity (section 4.14).

A higher eukaryotic organism comprising a great variety of specialized parts (supramolecular structures, subcellular organelles, cells) with different functions is thought to arise from developmental decisions which follow from elementary steps. A central theme concerns the nature of control mechanisms whereby cells of common lineage diverge into alternative developmental pathways. Learning about these control mechanisms has been hampered by the fact that cells are committed long before any phenotypic sign of such restriction can be distinguished. The search for signals of a chemical nature has thrown light on the understanding of the dynamics of processes underlying developmental decisions.

Progress in molecular and cellular biology has provided a rather good phenotypic characterization of differentiated cells through molecular markers; it has even allowed an identification of the elements (mostly proteins, together with their coding genes) participating in the above-mentioned control mechanism. However, the way in which the different biological activities (mostly well characterized *in vitro*) inherently participating in developmental decisions are coordinated in space and time to give an organized whole is largely unknown. Thus stated, the problem of morphogenesis becomes one of understanding how molecules (proteins), macromolecules and supramolecular structures are functionally organized.

10.2 GENES DO NOT DETERMINE FORM DIRECTLY BUT ACT THROUGH LENGTHY CASCADES OF PLEIOTROPIC EFFECTS

Morphology and organization, like other complex phenotypes, are the outcome of interactions among the expression of multiple genes and the environment. The relationship of genes to cell form is not like that of genes to proteins. Due to phenotypic complexity, enzyme activity and fitness are not unambiguously related (Dykhuizen *et al.*, 1987; Harold, 1990). The activity of individual enzymes takes part in a vastly complex metabolic network. The main questions (Dykhuizen *et al.*, 1987) are: how do those changes in activity, nested deep in the biochemistry of cells, affect the fluxes and intermediate pools of metabolic pathways, and how, in turn, do these changes affect reproductive success? The answers to these questions are inextricably linked to the main topic of this and the following chapters, i.e. the relationship between genetics, physiology and bioenergetics in growing, proliferating or differentiating organisms.

Cell growth and proliferation may be considered as the result of a dynamically coordinated, dissipative balance of fluxes (Chapters 9 and 11) (Aon *et al.*, 1995; Cortassa *et al.*, 1995; Aon and Cortassa, 1995). According to this view, the dynamic balance of catabolic and anabolic fluxes is associated with a dissipative pattern of matter and energy. Cellular processes such as metabolic reactions or cell growth may be characterized through systemic properties such as fluxes and the Gibbs free energy of the substrates being consumed by the cell. Therefore, one of the main targets of genetic control of the rate of energy dissipation by an organism is a systemic cellular property such as metabolic fluxes. On this basis we propose that form appears to arise epigenetically, from the intrinsic, autonomous dynamics of cellular and subcellular processes (Chapter 11). From this viewpoint morphogenesis represents the end result of multiple pathways integrated in a whole: cyclic, parallel, interdigitating (Harold, 1990, 1991). Epigenetics has been mainly concerned with temporal control of gene activity, segregation of gene activities during somatic cell division, the heritability of active or inactive states and their reversibility at specific times in the life cycle. Overall, epigenetics deals with the strategy through which the information encoded by genes is unfolded during development (Waddington, 1947; Holliday, 1988).

To address the subject, we have developed a general biothermokinetic approach around the concept of dynamic organization and applied it to growing, proliferating or differentiating microorganisms with special emphasis on *Saccharomyces cerevisiae*. Based on the concept of energy dissipation as a fundamental property of all living organisms that depends explicitly on fluxes and free-energy differences, the biothermokinetic approach focuses on metabolic fluxes which are regarded as a result of the reciprocal interaction of genetic background and environment. Epigenetics, in our view, is the realm of the outcome which emerges from the interaction of genetic background with intracellular and extracellular environments. This chapter aims to quantify the targets of control of the rate of energy dissipation through metabolic fluxes, at the level of enzyme concentration or enzyme activity and structural polymers such as microtubular protein.

10.3 CELLULAR CYTOARCHITECTURE AND DEVELOPMENT

A main problem of development concerns the mechanism(s) through which a morphogen of either chemical, mechanical or electrical nature induces a change in a cell's commitment to different developmental paths. It is becoming increasingly evident that the understanding of how a cell interprets signals from a developmental field is crucially dependent on the spatio-temporal organization of its cytostructure. This emergent view is accompanied by the recognition that prominent steps of cellular metabolism related to main metabolic pathways (Welch, 1977; Clegg, 1984a,b; Srere, 1987; Srivastava and Bernhard, 1987; Keleti *et al.*, 1989) and gene expression such as transcription (Cook, 1989) and translation (Cook, 1991) occur in association with the cell's dynamic scaffolds, whether nuclear or cytoplasmic.

Mechanical (Ingber *et al.*, 1994), fractal (Aon and Cortassa, 1994), rheological (Forgacs and Newman, 1994) and molecular/macromolecular crowding (Minton, 1983; Parker, 1993; reviewed in Garner and Burg, 1994) models or views of cytoplasmic structure and function have emerged in recent years. These models are extensively discussed in Chapter 6 from two main viewpoints: the mechanisms of functional coherence and of stimuli transduction (section 6.10).

The cell's position in a tissue could be sensed by cell cytoarchitecture, since the genome itself appears to be responsive to major controlling signals such as cell shape and surface contact from intercellular and intracellular macromolecular networks (Penman *et al.*, 1981; Shinohara *et al.*, 1989). In turn, cell cytoarchitecture is under genomic control (Penman *et al.*, 1981). For the latter to be possible, the genome itself and its expression must be responsive to major controlling signals such as cell shape and surface contact from the intracellular macromolecular networks. Several studies have shown that much of the macromolecular metabolism of the cell, including DNA, RNA and protein synthesis, responds to changes in cell shape (reviewed in Pienta *et al.*, 1989). It was shown for a quiescent rat embryo fibroblast cell line that initiation of DNA synthesis could be triggered by microtubule disruption with vinblastine (Shinohara *et al.*, 1989). It was suggested from those studies that disruption of normal microtubules could elicit intracellular signals leading to cell proliferation. Because of the involvement of the nuclear matrix components in DNA rearrangements (i.e. sister chromatid exchanges at the site of DNA replication), it has been proposed that alterations in the nuclear matrix structure could contribute to genetic instability (Pienta *et al.*, 1989). Genetic instability might place errors within the genetic apparatus that may be the basis of progression and the appearance of tumour cell heterogeneity (Chapter 9).

Rheologically, the macromolecular networks which constitute the cytostructure may originate boundaries of immiscibility between adjacent cytoplasmic regions. Boundaries of immiscibility between adjacent blocks of tissue are the basis of developmental compartmentation in insects, and play an important role in segmental organization in a wide variety of species (Forgacs and Newman, 1994, and references therein). Global connectivity changes or percolation in networks of cytoskeleton fibres result in boundaries of immiscibility and related morphogenetic changes in cells (section 6.4.1).

The vast amount of DNA present in the nucleus justifies the belief in a precise three-dimensional organization (Pienta *et al.*, 1989). The nuclear matrix provides a structural framework which organizes the DNA in the nucleus and has a direct role in the regulation of gene expression (Getzenberg *et al.*, 1991). Conceptually, the nuclear matrix is viewed as the nuclear equivalent to the cytomatrix (Pienta *et al.*, 1989). Moreover, the nuclear matrix is the component of the tissue matrix system which connects the DNA structurally and functionally through the cytomatrix to the extracellular matrix. The cytoskeleton is part of a tissue matrix system which forms a structural and functional bridge from the DNA to the cell periphery and beyond to the extracellular matrix as well as other cells (Pienta and Hoover, 1994). Overall, the cytoarchitecture consists of linkages and interactions that interlock the nucleus, the cytoplasm and the cell periphery.

The cytoskeleton, as a main component of the intracellular macromolecular architecture, appears to depend on cellular energetics. In fact, ATP-dependent regulation of cytoplasmic microtubule disassembly has been described (Bershadsky and Gelfand, 1981). Depletion of the cellular ATP pool in cultured mouse fibroblasts with various inhibitors of energy metabolism led to inhibition of the microtubule disassembly induced by colcemid or vinblastine (Bershadsky and Gelfand, 1981). Glucose catabolism partially restored the intracellular ATP levels, thus abolishing the inhibition of microtubule disassembly. Although this effect was specific for ATP, it was ruled to be due to not a direct effect of the nucleotide but an indirect one, presumably related to tubulin phosphorylation by microtubule-associated protein kinase required for microtubule disassembly.

In *Candida albicans*, pH-induced morphological transitions may be cytoskeleton-mediated. Hyphal growth in a salt–glucose medium containing 4% calf serum at pH 7 at 37°C continued for 24 h and could be reverted to yeast cells at pH 4. Although microtubules were distributed in the cytoplasm in a similar fashion as during hyphal growth, long microfilaments disappeared within 30 min after the change of external pH.

10.4 BIOTHERMOKINETICS OF GENE EXPRESSION AND THE SWITCH BETWEEN DEVELOPMENTAL PATHWAYS

Chapter 9 stressed that the intracellular spatio-temporal coordination of gene expression, energetics and intermediary metabolism is at the heart of a cell's functional adaptation to growth, proliferation, differentiation or environmental changes. In this sense, the dynamics of cellular energetics – including its continuous free-energy dissipation – allows functional biological processes to evolve toward different steady states that probably correspond to different cellular functional states. The main point is that spatio-temporal coordination results from mutual interaction between environment and genetic background (see also Chapters 9 and 11). Gene expression may affect cell energy dissipation through the following main mechanisms: changes in the stoichiometry of metabolic pathways, enzyme concentration, or enzyme alteration of kinetic parameters by, say, mutations.

Growth, proliferation and the life cycle of microorganisms strongly depend on interaction between events occurring at different levels of organization (Lloyd *et al.*, 1982; Aon and Cortassa, 1993). Metabolism includes processes with relaxation times in the order of a few seconds to minutes. In the domain of fast-relaxing processes are the effects of activation or inhibition of enzyme kinetics, whereas slow relaxation is found in transcription and translation of proteins. Interdependence between processes occurring in different spatio-temporal domains is manifest through the mutual influence of differential gene expression and metabolic rates and cellular energetics. In Chapter 11 this idea is pursued further to show in a quantitative way how this mutual dependence is achieved from a dynamic perspective. We also propose a hypothesis of dynamic nature founded on biochemistry, physiology, bioenergetics and kinetics, to explain the interdependence between a microorganism's physiology, energetics and biochemistry during cellular growth, proliferation and differentiation.

The metabolic state of an organism is somehow involved in the switch between developmental pathways. The cell's metabolic status regulates gene expression through changes in steady-state levels of transcripts of enzymes belonging to, for example, the glyoxylate cycle during plant development (Graham *et al.*, 1994). The data available emphasize the importance of the metabolic and energetic status of the organism in the choice of developmental paths. Let us assume that a cell that undertakes a developmental path is subjected to a sort of metabolic transition. During metabolic transitions the status of free-energy intermediates (phosphorylation and redox potentials, solute gradients, proton motive force, pH_i) changes both qualitatively and quantitatively.

It is now well established that the decision to culminate rather than to begin or continue migration in *Dictyostelium discoideum* is a response to a reduction in the concentration of the unprotonated form of the weak base ammonia in aggregates (Gross, 1994). Early aggregates of *D. discoideum* consist of a random mixture of pre-stalk and pre-spore cells. Pre-stalk cells subsequently sort out to the apex (and base) of the aggregate during tip formation, leaving the body of the aggregate occupied by a majority of pre-spore cells (Gross, 1994). Intracellular pH (pH_i) appears to be involved in the choice of differentiation pathways in *D. discoideum*. In populations of developing *D. discoideum* containing 75–80% pre-spore cells, a pH_i value was detected of about 0.2 pH units higher than in populations containing 50% pre-spore cells (Figure 9.2). Abolition of changes in pH_i by means of a weak acid or base prevent the regulation process (Gross *et al.*, 1983; Aerts *et al.*, 1985, 1987). The action of ammonia in maintaining the migratory slug phase of cell aggregates had been shown to depend on penetration by the uncharged form of the weak base (NH_3), since its effectiveness was a function of the pH of the medium (for a review, see Gross, 1994). Recent evidence suggests that ammonia acts not by raising cytosolic pH but by raising the pH of some component of the extensive acidic vesicle system (for a review, see Gross, 1994).

Walsh and Koshland (1985) described complete cessation of proliferation when acetate was added to *Escherichia coli* cells that overproduced (50-fold) citrate synthase. These authors showed that even if cells do not proliferate they continue to metabolize glucose. Therefore, uncoupled metabolism linked to the proliferation arrest is likely to occur. Proliferation arrest is, in most cases, a condition for cell differentiation. At this point the subject converges with the study of the control of cell growth and division presented in Chapter 9.

10.4.1 INTEGRATION OF METABOLISM IN SOROCARPS OF *DICTYOSTELIUM DISCOIDEUM*: DYNAMIC SYSTEM ANALYSIS OF METABOLISM *IN VIVO*

The life cycle of the amoeba *D. discoideum* is characterized by a free-living vegetative phase in the presence of enough food. However, under starvation conditions the amoebas aggregate and differentiate into one of two cell types. Differentiation is completed with the formation of a fruiting body composed of a cellulose stalk supporting a mass of spore cells. The stalk cell dies and the spore cells enter a state of dormancy (Kelly *et al.*, 1979).

Differentiation in *Dictyostelium* relies on metabolic processing of an endogenous carbon source that also supplies energy and redox equivalents. The endogenous carbon source is normally cellular protein, and the system appears to be closed (Kelly *et al.*, 1979). The low gluconeogenic flux sustained from labelled aspartate leading to the accumulation of trehalose and cellulose has been interpreted in the sense that catabolized protein is not directed towards glucose synthesis. The aggregating slime mould catabolizes proteins through the tricarboxylic acid (TCA) cycle to obtain energy rather than carbon intermediates. Moreover, the total amount of carbohydrates does not change. There is neither gluconeogenesis nor glycolysis during the period extending from aggregation until culmination in a fruiting body. As metabolic transitions associated with *Dictyostelium* development are rather simple, they have been extensively modelled to show their dynamics (Wright and Reimers, 1988; Wright and Albe, 1994a,b).

In fact, it appears that there are two 'islands' of metabolism: on the one hand, the TCA cycle is providing energy using amino acids derived from protein degradation as substrates; on the other hand, the carbohydrate and sugar phosphates pool remains more or less constant with an active exchange between them (Figure 10.1). This state resembles the pre-gastrulation stages in the development of fertilized *Xenopus* oocytes in which no pyruvate kinase activity was detected *in vivo* and where energy is also obtained from amino acids and fatty acids catabolism. This blockage at the level of phosphoenol pyruvate has been interpreted as a 'dam' which maintains a certain pool size of precursors for anabolic purposes (Dworkin and Dworkin-Rastl, 1991). Wright and coworkers showed that stalk cells on the one hand and presumptive spore and stalk cells on the other hand behave as two relatively independent compartments of metabolic processes. For instance, UDP-glucose exists only in presumptive stalk and spore cells whereas glycogen and other end-product saccharides, once synthesized, are distributed among spore and stalk cells, where they accumulate or degrade (Figure 10.1).

Summing up, both developing organisms behave as highly dissipative, utilizing their own material or that of maternal origin to obtain energy and performing very few if any anabolic processes, namely growth of biomass.

10.4.2 ENERGY METABOLISM DURING EMBRYONIC DEVELOPMENT IN *CAENORHABDITIS ELEGANS*

C. elegans is a worm in which four developmental larval stages may be distinguished after egg eclosion: L1, L2, L3 and L4. The characteristic energy metabolism during embryonic development has been studied through the use of the non-invasive technique NMR (Wadsworth and Riddle, 1989). Typical ^{31}P spectra of cellular extracts characterize each one of these stages, indicating differences in the metabolic status of the larva. In addition, under specific environmental conditions larval development will run through a different pathway characterized by a highly resistant form, the Dauer larva, preceded by an L2d stage metabolically different from the L2 stage. During metabolic transitions, ^{31}P NMR spectra of constitutive mutants for Dauer larva development showed a significant decrease in ATP and ADP content with respect to the L2 normal larva and a correlative increase in inorganic phosphate and AMP signals in the spectra of ^{31}P NMR (Wadsworth and Riddle, 1989).

The activities of isocitrate dehydrogenase and isocitrate lyase were measured in larvae extracts at different developmental stages. It was observed that the larvae of the constitutive Dauer mutant failed to exhibit the increase in activity of isocitrate dehydrogenase normally produced after the L1 stage (Figure 10.2). The latter was correlated with an increased oxygen consumption in L2, L3 and L4 larvae. The data indicated a shift from the use of the glyoxylate shunt to an increased

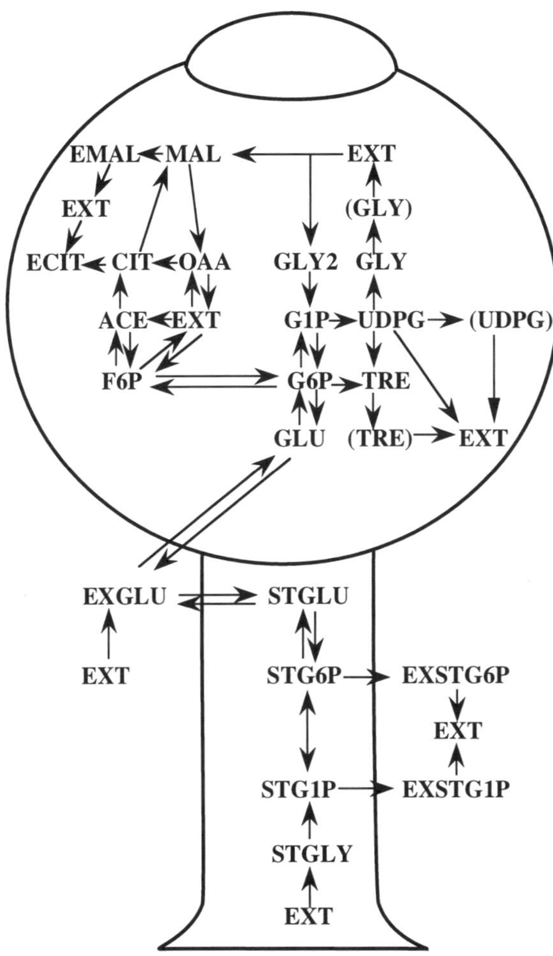

Figure 10.1 Metabolism of stalk and spore cells in *Dictyostelium discoideum*. ACE, acetate; CIT, citrate; EMAL and ECIT, extra-mitochondrial malate and citrate which is accumulating; EXGLU, exogenous glucose; EXSTG6P and EXSTG1P, stalk Glc-6-P and Glc-1-P excreted from the organism; EXT, external pool; F6P, Fru-6-P; G1P, spore Glc-1-P; G6P, spore Glc-6-P; GLU, spore glucose; GLY and (GLY), spore glycogen which is accumulating; GLY2, spore glycogen which is degraded; MAL, malate; OAA, oxalacetate; STG1P, stalk Glc-1-P; STG6P, stalk Glc-6-P; STGLU, stalk glucose; STGLY, stalk glycogen which is degraded; TRE and (TRE), trehalose which is accumulating; UDPG and (UDPG), UDP-glucose which is accumulating. (From Wright and Reimers, 1988, by permission of The American Society for Biochemistry and Molecular Biology.)

use of the TCA cycle (Wadsworth and Riddle, 1989). These studies clearly point to a protagonistic role of energy metabolism in the decision about the developmental pathway to be undertaken during a worm's embryonic life. Experimental evidence as applied to *Saccharomyces cerevisiae* is extensively treated in section 10.6 and in Chapter 11.

10.4.3 CHANGES IN pH ASSOCIATED WITH METABOLIC AND DEVELOPMENTAL TRANSITIONS

Microorganisms respond to changes in environmental conditions by varying patterns of gene expression. For example, the fungus *Aspergillus nidulans* grows over a wide range of temperature, osmolarity and ionic strength as well as pH (2.5–10) (Espeso *et al.*, 1993). The synthesis of extracellular enzymes and permeases in this fungus are not modulated by the intracellular pH (pH_i) homeostatic system, but by external pH (pH_o). It is thought that this system is widespread in nature but *A. nidulans* is the only microorganism in which the pH regulatory circuitry has been genetically dissected (Espeso *et al.*, 1993). In fact, mutations in the *pacC* regulatory gene, which mediates pH regulation, mimic the effect of growth at alkaline pH. The domain of action of the *pacC* gene, whose expression is regulated by pH_o, has been extended to small exported molecules such as penicillins. Penicillin production is higher at alkaline pH because the *ipnA* structural gene which encodes isopenicillin N synthetase is positively regulated in its transcription by the PacC regulatory protein (Espeso *et al.*, 1993).

Usually, developmental decisions are associated with changes in gene expression. One of the first and more established pieces of evidence for the involvement of pH_i in the regulation of development came from the activation of late events after fertilization of sea urchin eggs (Table 10.1) (Nuccitelli and Heiple, 1982; Busa, 1982; Busa and Nuccitelli, 1984). An increase in DNA and protein synthesis, RNA polyadenylation, nuclear membrane breakdown and chromosome condensation were elicited by treatment of the egg with NH_4^+. In the transition from the state of dormancy to vegetative life there is usually an alkalinization. Alkalinization has also been observed in the initiation of motility of spermatozoa of various species and after fertilization as well as at the entrance into the cell cycle after metabolic arrest in *S. cerevisiae*. All these data suggest a close link between pH_i and some cell cycle events, including DNA replication and mitosis (Chapter 9). An alkaline shift in pH_i of about 0.25 units per division cycle which peaks during S-phase

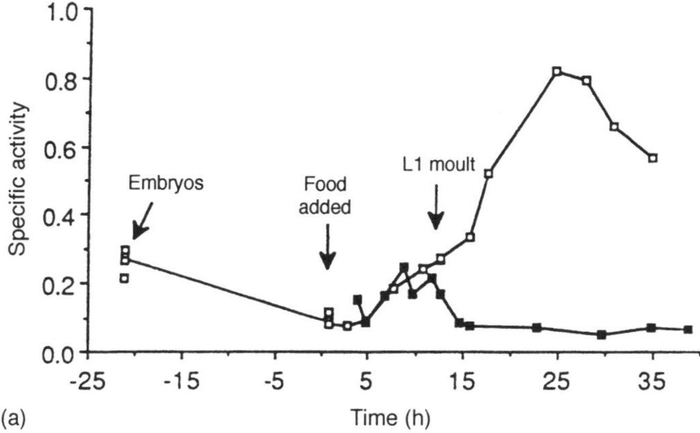

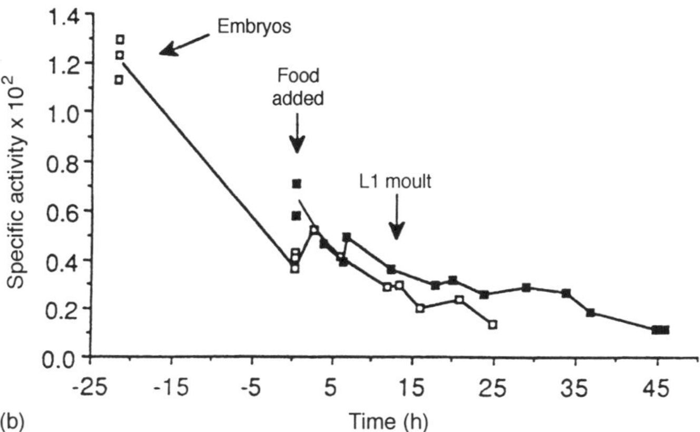

Figure 10.2 Developmental regulation of energy metabolism in *Caenorhabditis elegans*: specific activities of (a) isocitrate dehydrogenase and (b) isocitrate lyase during development. Eggs were isolated and allowed to hatch for 22 h in buffer at 25°C before the synchronized L1 larvae were placed on lawns of *E. coli* at 25°C to initiate development (0 h). Enzyme activities are from N2 embryos (–22 h) (open squares), larvae during the L1 (0–12 h), L2 (13–20 h), L3 (21–26 h) and L4 (27–35 h) stages, and from CB1372 (*daf-7*) (filled squares) larvae during the L1 (0–12 h), L2d (13–30 h) and dauer stages. (Reproduced from Wadsworth and Riddle, 1989, by permission of Academic Press.)

and mitosis has been reported in *D. discoideum* (Figure 10.3) (Aerts *et al.*, 1985). The rise in pH_i begins just prior to increased protein and DNA synthesis, whereas pH_i drops rapidly during or shortly after cell division. The cyclical pH_i changes are autonomous in that they continue when protein synthesis and DNA replication are fully blocked (Aerts *et al.*, 1985). Synchronous populations of vegetative cells with a lower pH_i than asynchronous populations of vegetative cells show a clear preference for the pre-spore differentiation pathway during development (Aerts *et al.*, 1987). Treatment of these cells with weak base ammonia results in an increase in pH_i and an abolition of the preference of the cells for the pre-spore pathway (Aerts *et al.*, 1987).

Table 10.1 Intracellular pH changes associated with cellular functions (reproduced from Nuccitelli and Heiple, 1982, by permission of Alan R. Liss, Inc.)

Cellular function	pH_i measurement technique	pH_i change	Net pH_i change	Nature of pH_i change
Fertilization: activation of development				
Eggs of sea urchin, *Lytechinus pictus*	Microelectrodes	6.84 to 7.26	+0.42	Permanent pH_i rise follows activation
Eggs of frog, *Xenopus laevis*	Microelectrodes and ^{31}P-NMR	7.4 to 7.7	+0.3	Permanent pH_i rise follows activation
Release from dormancy				
Spores of bacterium, *Bacillus cereus*	^{14}C-methylamine distribution	6.30 to 7.40	+1.1	Permanent pH_i rise accompanies transition from dormant to germinated spore
Bacillus megaterium	^{14}C-methylamine distribution	6.38 to 7.40	+1.02	Permanent pH_i rise accompanies transition from dormant to germinated spore
Spores of yeast, *Pichia pastoris*	^{31}P-NMR	6.0 to 7.0–7.4	+1.0–1.4	Permanent pH_i rise accompanies transition from dormant to germinated spore
Cysts of shrimp, *Artemia salina*	^{31}P-NMR	6.3–6.7 to ≥ 7.9	≥+1.2	Permanent pH_i rise accompanies transition from dormancy to developing embryo

Table 10.1 *continued*

pH_i influences rate of glycolysis				
Frog, *Rana pipiens*; muscle	^{14}C-DMO distribution	7.0 to 7.12	+0.12	Insulin stimulates a pH_i rise, which increases the rate of glycolysis by 43%
Human, red blood cells	^{14}C-DMO distribution	6.6 to 7.5	+0.9	Sucrose addition stimulates a pH_i rise, which increases rate of glycolysis by 200%
Cell cycle activation				
Lymphocytes	^{14}C-DMO distribution	7.2 to 7.6	±0.4	A transient pH_i rise occurs with the maximum at roughly the same time that DNA synthesis is maximal
Yeast, *Saccharomyces cerevisiae*	^{31}P-NMR	7.0 to 7.4	±0.4	A transient pH_i rise occurs 80 min before DNA synthesis is maximal
Ciliate, *Tetrahymena pyriformis*	^{14}C-DMO distribution	7.25 to 7.5	+0.25	Two transient pH_i rises occur: (1) prior to and (2) at the time of the maximal DNA synthesis
Slime mould, *Physarum polycephalum*	Microelectrode	7.0 to 7.4	±0.4	A transient pH_i rise occurs 30 min before mitosis
	Antimony microelectrode	6.0 to 6.6	±0.6	A transient pH_i rise occurs 30 min before mitosis
Sperm activation				
(1) Motility initiation: Sea urchin, *Lytechinus pictus*	^{14}C-methylamine distribution	~6.5 to ~7.0	+0.5	Na$^+$ addition triggers a permanent pH_i rise initiating sperm motility
Sea urchin, *Strongylocentrotus purpuratus*	^{14}C-methylamine distribution	~7.0 to ~7.4	+0.4	Na$^+$ addition triggers a permanent pH_i rise initiating sperm motility
(2) Acrosome reaction: Sea urchin *Lytechinus pictus*	^{14}C-methylamine distribution	~7.0 to ~7.1	+0.1	A transient pH_i rise accompanies the acrosome reaction
Sea urchin, *Strongylocentrotus purpuratus*	^{14}C-methylamine -diethylamine -DMO 9-Aminoacridine	~7.4 to ~7.6	+0.2	A transient pH_i rise accompanies the acrosome reaction

Table 10.1 *continued*

Stimulus-response coupling				
Human platelet, thrombin-stimulated activation	9-Aminoacridine and 6-carboxy-fluorescein fluorescence	7.0 to 7.3	+0.3	Thrombin stimulates a pH_i rise within 60–90 s
Rabbit neutrophil, f-Met-Leu-Phe chemotaxis stimulation	^{14}C-DMO distribution	7.0 to 7.1	+0.1	F-Met-Leu-Phe stimulates a small, transient, Ca^{2+}-dependent pH_i fall followed by a pH_i rise of 0.1 unit by 5 min after addition
Myocardial contractility				
Rabbit hearts	^{31}P-NMR	7.2 to 7.0	−0.2	A 0.2 pH_i unit fall is associated with a 50% depression in left ventricular pressure

Changes in pH_i have been found to accompany a variety of cellular functions, and in some of these the pH_i change is an essential part of the control mechanism. This table lists most of the specific cellular functions associated with the pH_i change. Two main trends are shown. 1. The higher pH_i values are associated with increased cellular activity or efficiency. Specifically, intracellular alkalinization accompanies increased rates of glycolysis, protein synthesis and (in some cases) DNA synthesis, and initiates sperm motility. 2. The magnitudes of the pH_i changes fall into two categories: 13 of the 19 examples listed exhibit a natural pH increase of 0.1–0.5 pH unit, whereas the four cases of release from dormancy exhibit a larger pH_i rise of 1.0–1.4 units. Such large pH_i increases will dramatically affect the kinetics of many intracellular enzymatic reactions.

However, an important exception to the regulation of the switch of a developmental process by pH_o instead of pH_i is given in the differentiation of the cellular slime mould *Dictyostelium discoideum*. Using 'sporogenous mutants' of *D. discoideum* (i.e. mutants which under monolayer conditions differentiate into a mixture of both stalk cells and spores), several authors have shown that incubation of cells in a low pH buffer favours stalk cell formation, while a high pH favours spores (Gross *et al.*, 1983; Town *et al.*, 1987). These effects of pH_o on phenotypic expression are accentuated by the addition of weak acids or weak bases at low or high pH, respectively. Evidence presented by Town *et al.* (1987) suggested that such a dramatic phenotypic regulation by pH_o occurred without an appreciable change in pH_i. Apparently, the ability of cells to maintain a constant pH_i in the face of acid or alkaline loads is probably due to the activation of ion transport systems under those conditions.

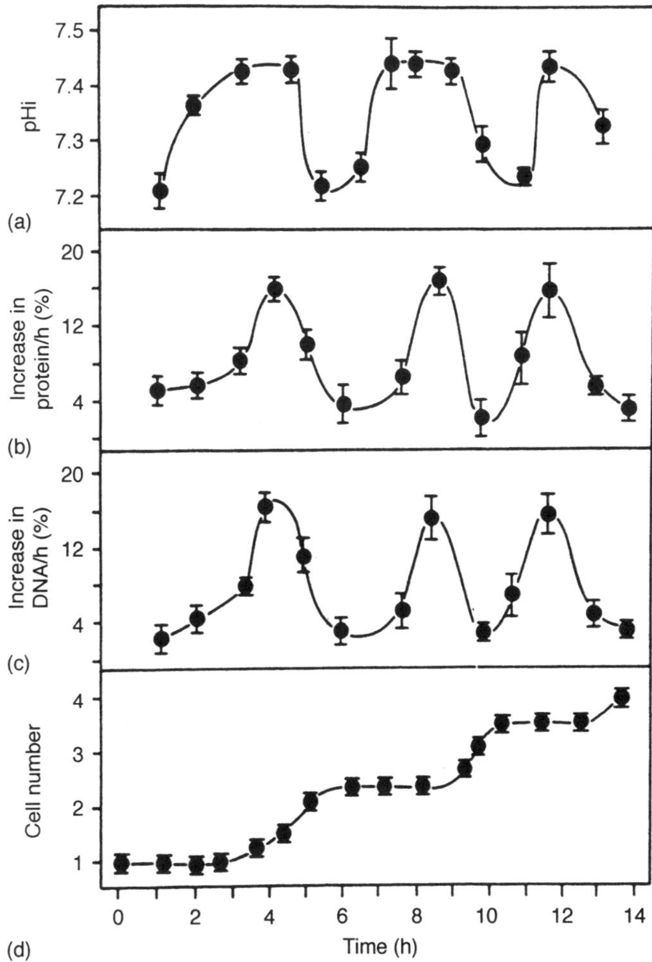

Figure 10.3 Cytoplasmic pH and the regulation of the *Dictyostelium* cell cycle. Changes in (a) pH$_i$, (b) protein synthesis, (c) DNA synthesis and (d) cell number during three successive cell cycles. At time zero, synchronous cell divisions of *D. discoideum* were induced by resuspending stationary-phase cells into fresh medium. (Reproduced from Aerts, Durston and Moolenaar, 1985, by permission of Cell Press.)

These changes in ionic fluxes and potentials could in turn feed back on the levels of second messengers, and hence modulate the relative activity of the stalk and spore pathways of differentiation (Town *et al.*, 1987). A clear connection with gene expression was shown for pH_o-regulated pre-stalk II and pre-spore II mRNAs (Town *et al.*, 1987).

10.5 DYNAMIC ORGANIZATION IN EARLY AMPHIBIAN EMBRYOS

It is well established that the unfertilized egg in nearly all species is a polarized cell. This polarity is usually most readily identified by the asymmetrical distribution of pigment granules, yolk, mitochondria or other organelles.

10.5.1 MITOCHONDRIAL CLOUD IN THE PROCESS OF MATURATION OF *XENOPUS* OOCYTES

The origins of egg asymmetry during oogenesis in *Xenopus* can be traced back to pre-vitellogenic stages (Wylie *et al.*, 1986; Elinson and Houliston, 1990). A characteristic aspect of the pre-vitellogenic oocyte is a dense cytoplasmic mass known as the mitochondrial cloud. The main function of this organelle was thought to be the source of oocyte mitochondria. Under Nomarski optics the cloud is seen to be a phase-dense mass connected to the developing filament system of the oocyte. At the beginning of vitellogenesis, the cloud breaks down and most of its fragments spread to the sub-cortical cytoplasm at one pole of the oocyte. Careful electron-microscopic observations of the cloud itself and its breakdown products reveal that they contain the germinal granules characteristic of the germ plasm of the egg (Heasman *et al.*, 1984, quoted in Wylie *et al.*, 1986).

The pole to which the mitochondrial cloud fragments spread becomes the vegetal pole later on during vitellogenesis. Clearly then, animal–vegetal polarity is already established at pre-vitellogenic stages, and at least one role of the mitochondrial cloud is to sequester the germinal granules and localize them to the correct region of the differentiating oocyte (Elinson and Houliston, 1990).

Staining of the pre-vitellogenic oocyte with antibodies against cytoskeleton components showed that the mitochondrial cloud and its breakdown products react strongly with anti-tubulin and anti-mammalian vimentin antibodies. The cloud was found to be surrounded by a shell of anti-cytokeratin staining material, which partially divides it into segments, thus presaging the breakdown of the cloud (Wylie *et al.*, 1986).

10.5.2 POLARITY OF THE SURFACE AND CORTEX OF THE AMPHIBIAN EGG AFTER FERTILIZATION AND UP TO THE FIRST CLEAVAGE

The *Xenopus* egg undergoes first cleavage in about 90 min after fertilization. The radially symmetrical egg becomes a bilaterally symmetrical embryo whose accomplishment is achieved by the formation of a second axis of polarity (Figure 4.6). Therefore, the future dorsal–ventral axis of the embryo is not predetermined in the egg, but becomes established about halfway between fertilization and first cleavage. The fertilized egg then divides synchronously 12 times in about six hours to form a 4000-cell blastula.

The dorsal–ventral cell polarity results from a massive cytoplasmic rearrangement which leads to the grey crescent formation, i.e. a lightly pigmented area, located equatorially, marking the dorsal side of the future embryo (Figure 10.4). The grey crescent forms by a rotation of the cortex relative to the cytoplasm. The structure that occupies the outer boundary of the egg plays a vital role in this rotation of the amphibian egg during early development. These structures, along with the properties of the cell surface (components associated directly with the external face of the plasma membrane), are part of the so-called **cortex**, i.e. the layer of cytoplasm that lies within about 1 μm depth of the plasma membrane (Stewart-Savage *et al.*, 1991). Apparently, the cortex rotates relative to the cytoplasm in an active way, i.e. it is energy-dependent (Vincent *et al.*, 1987), while the mechanism of rotation appears to involve an array of aligned microtubules in the cortex. The cortical microtubule array results from an asymmetrical microtubule growth outward from cytoplasm to cortex (Houliston and Elinson, 1991; Figure 4.6).

10.5.3 ENERGY METABOLISM IN *XENOPUS* EMBRYOS

It has been suggested that several similarities exist between tumour cells and cleaving amphibian embryos, despite the absence of glycolytic activity in the embryos (Figure 10.5) (Dworkin and Dworkin-Rastl, 1991). Cleaving embryos appear to be capable of glycolysis, but glycolysis is not sufficient to support cleavage, which is dependent upon oxidative phosphorylation. Exposure of frog embryos to iodoacetic acid did not inhibit cleavage, but affected gastrulation and completely blocked neurulation. Therefore, glycolysis did not appear to be required for cleavage to proceed but is required for subsequent morphogenesis (Dworkin and Dworkin-Rastl, 1989a,b, 1991).

When radiolabelled phosphoenolpyruvate (PEP) was injected into oocytes, eggs and blastomeres of cleaving embryos, two unexpected

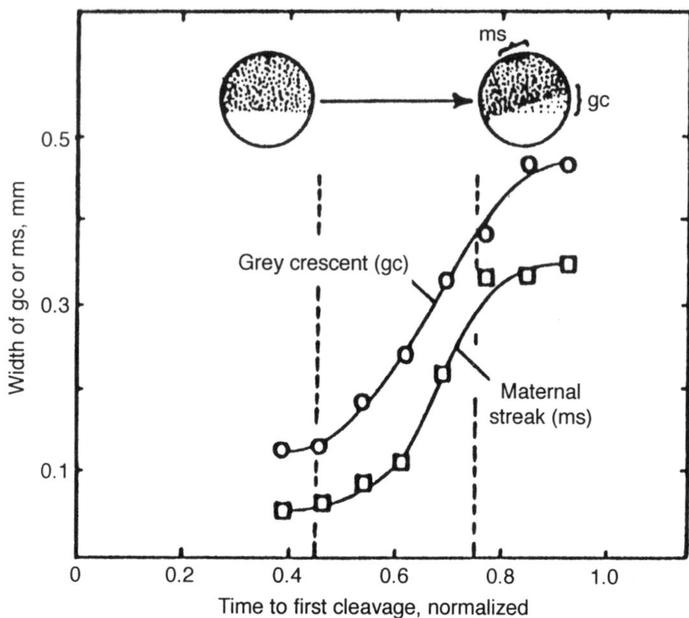

Figure 10.4 Kinetics of grey crescent formation in *Rana pipiens*. The width of the grey crescent was measured in approximately 20 eggs at various times (normalized decimal scale). Time 0.0 corresponds to fertilization, with first cleavage at 1.0. Movement of the polar body spot at the animal pole was also recorded; the spot elongates into a streak: the 'maternal streak' (ms). The equatorial movement of the pigment at the grey crescent is approximately equal in magnitude to the movement in the maternal streak, as if the whole cortex moves relative to deeper materials. (Reproduced from Gerhart, Black and Scharf, 1983, with permission.)

results arose: first, the Embden–Meyerhof pathway (EMP) appeared to operate in reverse; second, there was a lack of fructose-1,6-biphosphatase (F1,6Pase) (Figure 10.5) (Dworkin and Dworkin-Rastl, 1991). Additional data suggested that, in normal embryos, the EMP is blocked at the pyruvate kinase step. *Xenopus* embryos appear to express an allosteric isozyme of PK which has a high $K_{0.5}$ for PEP, is stimulated by FBP and is strongly inhibited at the ATP concentration found in embryos ($\sim$3 mM) (Dworkin and Dworkin-Rastl, 1989a,b, 1991).

Glutamate oxidation appears to be the main energy source up to gastrulation and the major carbon source for catabolic processes (and possibly anabolic processes as well) (Dworkin and Dworkin-Rastl, 1991).

The source of glutamate appears to be mitochondrial. The latter also happens in tumour cells (Cortassa and Aon, 1994b; Chapter 7).

Glycolysis is activated before gastrulation given by an increased ability during cleavage to oxidize exogenously supplied glucose both glycolytically and through the oxidative pentose phosphate shunt that levels off during gastrulation. An increase in the respiratory quotient which indicates increased utilization of carbohydrate, a decrease in cellular glycogen and an increase in lactic acid levels also correlate with this metabolic shift to glycolysis (Dworkin and Dworkin-Rastl, 1991, and references therein). The concomitant onset of glycolysis and gastrulation before morphological differentiation suggests that some aspect of the metabolism of the cleaving embryo might not be permissive for expression of differentiated phenotypes, while glycolytic metabolism is permissive. Therefore, the significance of pyruvate formation from mitochondrial amino acids in cleaving embryos may be the ability of this metabolic mode to support high rates of cell division without mass growth while simultaneously repressing differentiation (Figure 10.5) (Dworkin and Dworkin-Rastl, 1991).

10.6 METABOLISM, BIOENERGETICS AND GENE EXPRESSION DURING SPORULATION IN *SACCHAROMYCES CEREVISIAE*

Sporulation in *S. cerevisiae* is a dynamic process comprising a series of well ordered events initiated in response to nutrient starvation. The transition toward the quiescent final state involves growth arrest, premeiotic DNA synthesis (not accompanied by bud formation as occurs for mitotic DNA synthesis), mitochondrial DNA synthesis, recombination, first and second meiotic divisions, appearance of tetranucleated cells and asci. The occurrence of these events depends on the cellular genotype (e.g. presence of the two alleles of the mating type locus *MATα* and *MATa, IME1, RME1*), and a dynamic metabolic status not completely understood at present (Esposito and Klapholz, 1981; Miller, 1989)

From a metabolic viewpoint, sporulation is characterized by the absolute requirement of respiratory activity, gluconeogenic metabolism, breakdown and turnover of RNA, extensive proteolysis, accumulation of storage carbohydrates and degradation of glycogen (reviewed by Miller, 1989).

The dynamics of the sporulation process exhibits a number of identifiable steps, according to the ability of the cells to sporulate or resume growth upon transfer to water or rich medium, respectively (Simchem *et al.*, 1972). Readiness is defined from the competence to sporulate upon transference to water after exposure to sporulation

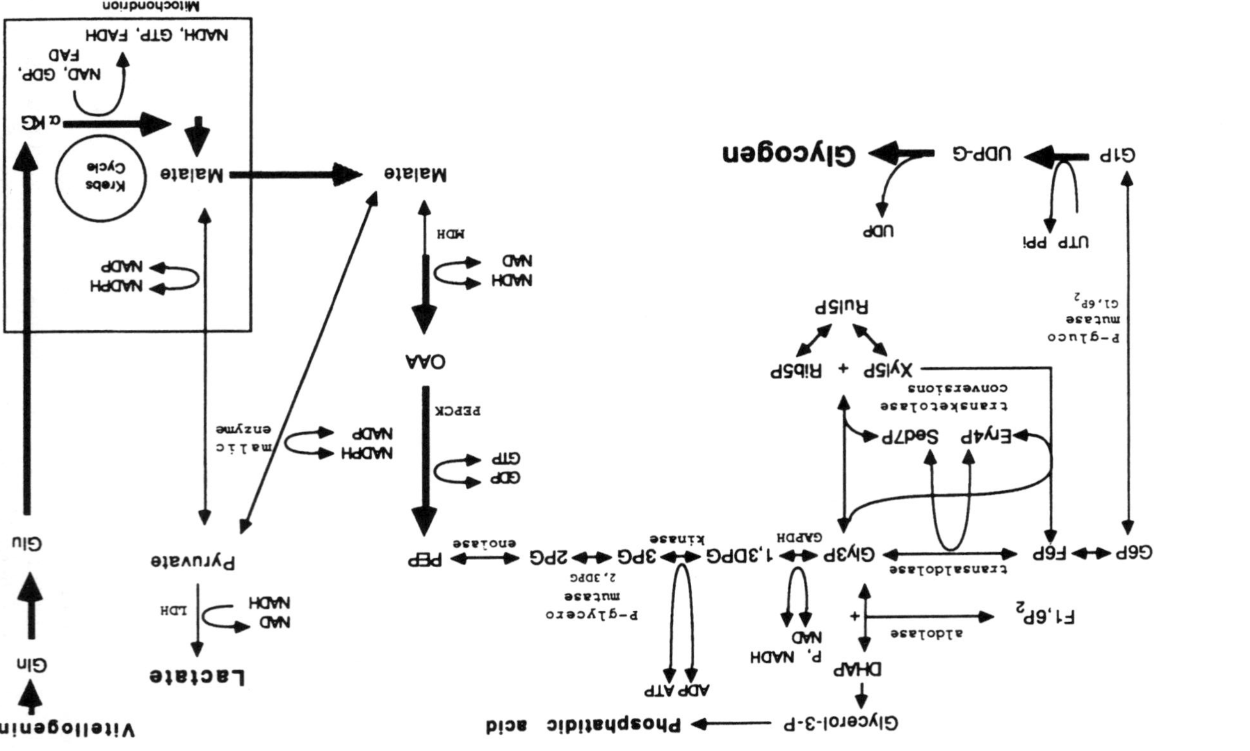

Vitellogenin
Gln
Glu
Lactate
LDH
NAD
NADH
Pyruvate
Krebs Cycle
α KG
NADPH
NADP
NAD, GDP, FAD
NADH, GTP, FADH
Malate
Mitochondrion
malic enzyme
NADPH
NADP
Malate
PEP
enolase
PEPCK
GDP
GTP
OAA
MDH
NADH
NAD
Phosphatidic acid
Glycerol-3-P
P-glycero mutase
2,3DPG
2PG
3PG
ADP ATP
kinase
1,3DPG
DHAP
P, NADH
NAD
GAPDH
Gly3P
Xyl5P + Rib5P
Ru5P
Glycogen
UDP
F1,6P2
aldolase
+
transaldolase
F6P
Ery4P
Sed7P
transketolase conversions
UDP-G
UTP PPi
G6P
P-gluco mutase
G1,6P2
G1P

medium. Readiness is achieved after 3–4 h of continuous exposure to sporulation medium even before the onset of premeiotic DNA synthesis. After 7–8 h in sporulation medium, cells exhibit commitment: they sporulate even if transferred to growth medium. There seems to be a temporary loss of plating efficiency after 6–7 h in sporulation medium during which the cells are neither committed nor resume growth upon transference to growth medium; such stage is called partial commitment. Simchem *et al.* (1972) attempted to correlate the transitions among these stages with the onset of the different steps of meiosis. From a dynamic view transitions between readiness and commitment may be identified as the achievement of a threshold value in the parametric space of the process(es) which trigger(s) reversible or irreversible transitions (Chapters 1 and 6). Honigberg and Esposito (1994) proposed that the reversal of commitment occurs when the normal progression of the meiotic cycle is somehow blocked. They hypothesized that if the time required to complete the meiotic steps driving to asci formation is shorter than the time needed to return to mitotic growth, the cells will follow the sporulation programme. If the time to complete the meiotic programme is longer, yeast cells will return to the proliferative cycle. From a bioenergetic point of view, as time elapses in the sporulation medium, cells slow down anabolic and catabolic rates (Aon, J.C. and Cortassa, 1995; Chapter 11). When yeast cells are challenged with

Figure 10.5 (see facing page) Inferred major pathways of carbon metabolism in *Xenopus* oocytes and early embryos. All reactions upstream of PEP are speculative. The interconversions between three-carbon and six-carbon compounds, indicated here by non-oxidative pentose phosphate shunt reactions, may occur through any of a large number of reactions that interconvert these compounds. DHAP, dihydroxyacetone; 1,3DPG, 1,3-diphosphoglycerate; 2,3DPG, 2,3-diphosphoglycerate; Ery4P, erythrose 4-phosphate; $F1,6P_2$, fructose 1,6-bisphosphate; F6P, fructose 6-phosphate; $G1,6P_2$, glucose 1,6-bisphosphate; G6P, glucose 6-phosphate; Gln, glutamine; Glu, glutamate; Gly3P, glyceraldehyde 3-phosphate; KG, α-ketoglutarate; LDH, lactate dehydrogenase; MDH, malate dehydrogenase; OAA, oxalacetate; P, phosphate, PEP, phosphoenolpyruvate; 2PG, 2-phosphoglycerate; 3PG, 3-phosphoglycerate; PPi, pyrophosphate; Rib5P, ribose 5-phosphate; Ru5P, ribulose 5-phosphate; Sed7P, sedoheptulose 7-phosphate; UDP-G, UDP-glucose; Xyl5P, xylulose 5-phosphate. The heavy lines indicate the direction of flux. (Reproduced from Dworkin and Dworkin-Rastle, 1989, by permission of Academic Press.)

growth medium, they do not have metabolic machinery ready to react to the environmental stimulus and if they are already engaged in the sporulation process they will continue it up to the end, i.e. spore formation. It appears that, once engaged in the sporulation process, the cell is energetically compromised so that it is compelled to dissipate energy and carbon until metabolic rates achieve the almost zero level depicted by spores.

An additional aspect of the relationship between energy metabolism, gene expression and sporulation concerns the link between the various events, all of them required for a successful sporulation. Are they synchronous or sequential? Do they belong to a common chain or are there several parallel control paths of sporulation?

It is commonly accepted that meiosis and sporulation events are controlled by two parallel pathways: starvation is the initial signal eliciting the process of sporulation; the other signal is provided by the heterozygocity at mating type locus (*MATa* and *MATα*). The latter signalling pathway comprises other genes, including *RME1* and *IME4* (Mitchell and Herskowitz, 1986; Simchen and Yassir, 1989). The genic product of *IME1* is believed to integrate signals from genetic and metabolic pathways. However, the nature of the signalling at the metabolic level remains unknown.

Since the simultaneous presence of the two loci of the mating type is a *sine qua non* condition for sporulation, it was considered that the events at any level of organization occurring during exposure to sporulation conditions in **a**/α strains, but not in genotypically related yet homozygous strains (*a/a* or *α/α*), were 'specific' to sporulation. Thus, the doubling of the DNA content and the onset of glycogen degradation that precedes spore wall formation were considered under the latter category (Roth and Lusnak, 1970).

An alternative perspective was to consider that those events were under genetic control – namely, under the control of the mating type locus. There are a number of processes occurring in *a/α*, *a/a* or *α/α* that are also essential for sporulation to occur, such as polysaccharide accumulation. In fact, mutants unable to accumulate polysaccharides are asporogenic. Therefore, it appears that there are several control paths for the sporulation process. One of those pathways is of a genetic nature: it is controlled by the mating type alleles, and others are parallel but equally relevant for the process. The importance of these control paths is that instability in the dynamics of processes belonging to those paths may determine at a higher level of organization the differentiation fate in yeast.

10.6.1 SUBSTRATE UPTAKE AND SPORULATION EFFICIENCY

A major factor controlling sporulation is the availability of nutrients. Starvation encourages it; nutrients discourage it. Nonetheless, the presence of a carbon source is required for an efficient sporulation to occur. The readiness concept is a clear indication that the presence of the carbon source is dispensable after 3–4 h of exposure to sporulation conditions. The justification of the requirement of a carbon source might be attributed to the fact that most yeast strains appear to require a minimal biomass growth during the first hours in sporulation medium (spm) for a successful sporulation. The initial increase in biomass, of course, requires carbon substrate consumption. We found that the achievement of a high efficiency of sporulation by *S. cerevisiae* cells strain CH1211 in the presence of acetate depends linearly on the average rate of acetate consumption. Figure 10.6a displays the linear correlation found between the percentage of sporulation versus the average rate of acetate consumption during the first 12 h in sporulation medium (Aon, J.C. and Cortassa, 1995; Chapter 11). When the sporulation efficiency was plotted against the average oxygen consumption rate, we could not find any correlation under the experimental conditions reported in that work, i.e. for cells derived from different phases of growth or challenged with sporulation medium at various cell densities (Figure 10.6a). A positive correlation between the rate of substrate consumption and the efficiency of sporulation coincided with a proportionality between the sporulation percentage attained after 72 h in spm and the increase in biomass and correspondingly the gluconeogenic and glyoxylate fluxes (Figure 10.6c). In agreement with these correlations, there are extensive data in the literature which indicate the importance of the gluconeogenic pathway: mutants defective in gluconeogenic enzymes are asporogenic (Dickinson *et al.*, 1983; Dickinson and Williams, 1986).

In the 1980s, Freese and co-workers postulated that a low rate of carbon source uptake was required for sporulation to occur (Chu *et al.*, 1981; Bautz-Freese *et al.*, 1982). These authors interpreted their hypothesis on the pH effect on sporulation as an effect on the uptake rate of acetate, but this hypothesis has never been substantiated by quantitative data on fluxes of carbon consumption. Aon, J.C. and Cortassa (1995) have provided evidence indicating that the decrease in the rate of acetate consumption observed in the course of the sporulation process does not follow the increase in pH. Moreover, in experiments in which the external pH was maintained by the use of a buffer, the rate of acetate consumption increased, as well as the growth observed, without affecting the success of sporulation. Apparently, only the time of appearance of

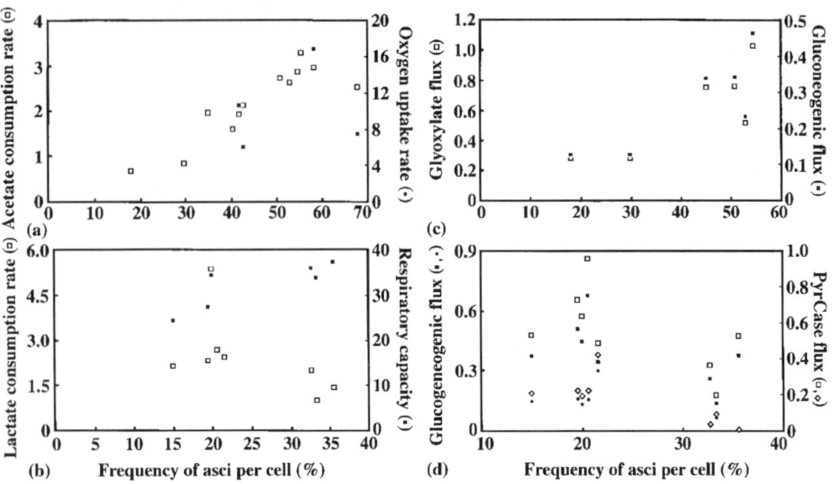

Figure 10.6 Quantitation of catabolic and anabolic fluxes as a function of sporulation efficiency in acetate and lactate sporulation medium. The average oxygen and substrate consumption rates were plotted as a function of sporulation frequency at 48 h in (a) acetate sporulation medium (ASM) and (b) lactate sporulation medium (LSM). The metabolic rates were averaged over the first 12 h in sporulation medium. Data were taken from experiments in which the cells were either harvested at different growth phases or transferred to sporulation medium at different cell densities. (c) Values of the estimated glyoxylate and gluconeogenic fluxes were averaged over the first 12 h in ASM. (d) Values indicate the calculated PyrCase and gluconeogenic fluxes corresponding to 0–2 h (diamonds) and 6–8 h (squares) periods in LSM. (Panels b, c and d are reproduced from Aon, J.C. and Cortassa, 1996, by permission of Academic Press.)

asci was delayed at lower pH compared with unbuffered sporulation medium. This experimental fact does not contradict the possibility that very high rates of acetate consumption, such as those displayed under growth conditions, might interfere with the sporulation process.

Unlike the experimental results obtained in acetate spm, in the presence of lactate the correlation between the rate of substrate consumption and the sporulation effciency was negative, i.e. when lactate consumption rate increases, sporulation efficiency at 72 h decreases (Figure 10.6b). This decrease in sporulation efficiency was accompanied by an increase in the rate of biomass synthesis and in the fluxes through the pyruvate carboxylase and gluconeogenic pathway. On the other hand, the respiratory capacity of the cells correlated positively with the sporulation efficiency (Figure 10.6b).

Trying to figure out the grounds of such conflicting behaviour exhibited by lactate and acetate with respect to the relations between metabolic fluxes and sporulation, we investigated the kinetics and energetics of each substrate consumption. Acetate is completely consumed during the sporulation process by the cells, while lactate is not. One possibility is that in lactate sporulation medium there is another limiting substrate, but this is unlikely since only a carbon source is required for sporulation to occur, while nitrogen, phosphate or sulphur sources should be absent. The other possibility is that there is a kinetic limitation to assimilation of the carbon source; this is supported by the fact that consumption rates in minimal growth medium are lower for lactate than for all the other substrates assayed (Table 8.1; Cortassa *et al.*, 1995). According to the correlations described above, the larger the carbon consumption flux, the higher are the anabolic fluxes. In fact, we found parallel tendencies between rates of substrate consumption and biomass increase. As lactate is incompletely consumed, it may be indicating some likely 'kinetic' limitation on use of this substrate as an efficient carbon source for anabolism under sporulation conditions. Acetate, on the contrary, was completely exhausted from sporulation medium.

The rates of oxygen consumption displayed in lactate sporulation medium are far larger than those displayed in acetate. The use of lactate requires the activity of a ferricytochrome c-dependent lactate dehydrogenase of mitochondrial localization (de Vries and Marres, 1987). The correlation of the sporulation efficiency with the respiratory capacity (Figure 10.6b) may be attributed to the amount of ferricytochrome c-dependent lactate dehydrogenase which belongs to an electron transport chain requiring oxygen as a final electron acceptor. Thus, a likely explanation of the positive correlation with the sporulation efficiency and respiratory capacity may be that there is an energy limitation due to the limiting amounts of ferricytochrome c-dependent lactate dehydrogenase, which was not the case in acetate (Aon, J.C. *et al.*, 1996).

10.6.2 FATE OF THE CARBON CONSUMED IN SPORULATION MEDIUM

The carbon consumed seems to be directed to both anabolism and catabolism. The rate of substrate consumption increased parallel to biomass growth and thereby to fluxes through anabolic, glyoxylate and gluconeogenic paths (Figure 10.6). The latter are mainly directed to polysaccharide synthesis. In acetate spm, the increased redirection of anabolism toward polysaccharide synthesis occurred together with a predominantly oxidative fate of the acetate consumed. Apparently, the total amount of carbon consumed during the first 12 h, when cells are supposed to synthesize all the precursors for the spore wall, determines the percentage of them that will be able to complete the sporulation process. The acetate consumed by growing cells was, however, insufficient to account for the carbon incorporated into biomass plus the calculated CO_2 released from catabolic activity. This fact was in agreement with a degradation of proteic material to fulfil catabolic requirements reported by other authors, as well as in other systems, such as *D. discoideum* (Wright and Reimers, 1988).

The fate of the acetate consumed was apparently more catabolic under sporulation with respect to growth conditions. This was reflected by a considerably lower ratio between the rate of oxygen and acetate consumption in growing cells (Figure 10.6) (Aon, J.C. and Cortassa, 1995), suggesting that a larger fraction of the carbon substrate was being oxidized during sporulation with respect to growth in minimal medium. The amount of acetate oxidized and the percentage of carbon diverted through glyoxylate and gluconeogenic pathways suggest that a more oxidative metabolism occurred under sporulation compared with growth conditions (Figure 10.6). Overall, acetate utilization is to a larger extent diverted to oxidative rather than to anabolic purposes as the sporulation programme develops. The results reported by Cortassa *et al.* (1995) suggested that the increased dissipation of the carbon source triggered an imbalance between catabolic and anabolic fluxes which might be involved in initiation of yeast differentiation.

10.6.3 QUANTITATIVE STUDIES OF CELLULAR ENERGETICS AND METABOLIC FLUXES DURING SPORULATION

The experimental evidence supports the proposition that both anabolic and catabolic fluxes should attain a given imbalance to sustain the sporulation process. With acetate, lactate or different cellular densities which induce distinct physiological conditions, we have been able to perturb the flux relationship between catabolism and anabolism. In lactate, it is quite likely that cells are energy-limited to accomplish sporulation; this would explain the correlation of the respiratory capacity and the sporulation efficiency. On the contrary, cells sporulating on acetate are mainly carbon-limited, as revealed by the correlation of sporulation frequency with anabolic fluxes.

Analysis of carbon and energy-coupling indices also points to the importance of a flux imbalance having to be met in order to accomplish the sporulation process. The **energy-coupling index** is a measure of how much of the ATP generated during catabolism is used for anabolism; it is, in fact, a ratio of the flux of ATP needed to support the observed rate of biomass increase to the rate of ATP generation by respiratory activity. The **carbon-coupling index** is the flux ratio of the carbon used for biomass synthesis with respect to the total flux of carbon consumed (Aon *et al.*, 1995). In the presence of lactate, the tendency is clear and confirms the previous correlations noted thus far: namely, that both carbon and

ATP coupling indices decrease with the sporulation efficiency supporting the fact that the larger the catabolic fluxes compared with the anabolic fluxes, the more successful will the sporulation process be. By contrast, the tendency is not so conspicuous with acetate since both carbon and ATP are required to support macromolecular synthesis in cells that will sporulate.

The energetic pattern of cellular events characterizing sporulation appears to be an imbalance between anabolic and catabolic fluxes. Both substrates perturb yeast bioenergetic status, provoking the flux imbalance through different mechanisms: lactate brings about an energy limitation, while acetate appears to trigger an anabolic carbon limitation.

10.6.4 SPORULATION AND CATABOLITE REPRESSION

The main result indicating a possible link between the commitment of yeast cells to sporulation and catabolite repression was obtained from studies where sporulation success was analysed as a function of increasing concentrations of the glucose analogue 2-deoxyglucose (2DG) (Figure 10.7). This catabolite repressor elicited a dramatic inhibition of sporulation efficiency along with a decrease in the rate of appearance of asci (Aon, J.C. and Cortassa, submitted). Global anabolic and catabolic fluxes were measured in order to discern the elements participating in the metabolic control mechanism of the observed inhibition of sporulation. The results indicated that repression of the acetate consumption flux increased with 2DG concentration while oxygen uptake exhibited a threshold inhibitory concentration of 2DG (~ 0.06 g l^{-1}) (Figure 10.7a). Additionally, oxygen uptake rate was repressed in only a small range, i.e. up to 40%.

A semilogarithmic plot of sporulation efficiency versus 2DG concentration rendered a straight line with a single slope, as did the acetate uptake, suggesting that the decrease of the sporulation success is related to the inhibition of the acetate consumption rate rather than to oxidative phosphorylation. In support of this interpretation, the sporulation efficiency displayed a linear correlation at a constant slope with the acetate consumption flux, whereas a two-phase behaviour was evident when oxygen consumption rate was plotted against sporulation efficiency (Figure 10.7a).

The anabolic fate of part of the acetate consumed when yeast is exposed to sporulation conditions was mentioned in the previous section. To confirm this, we calculated the anabolic fluxes through glyoxylate and gluconeogenic pathways. It could be verified that both fluxes were correlated proportionally with sporulation success. This is shown in Figure 10.7b; sporulation inhibition is probably due to

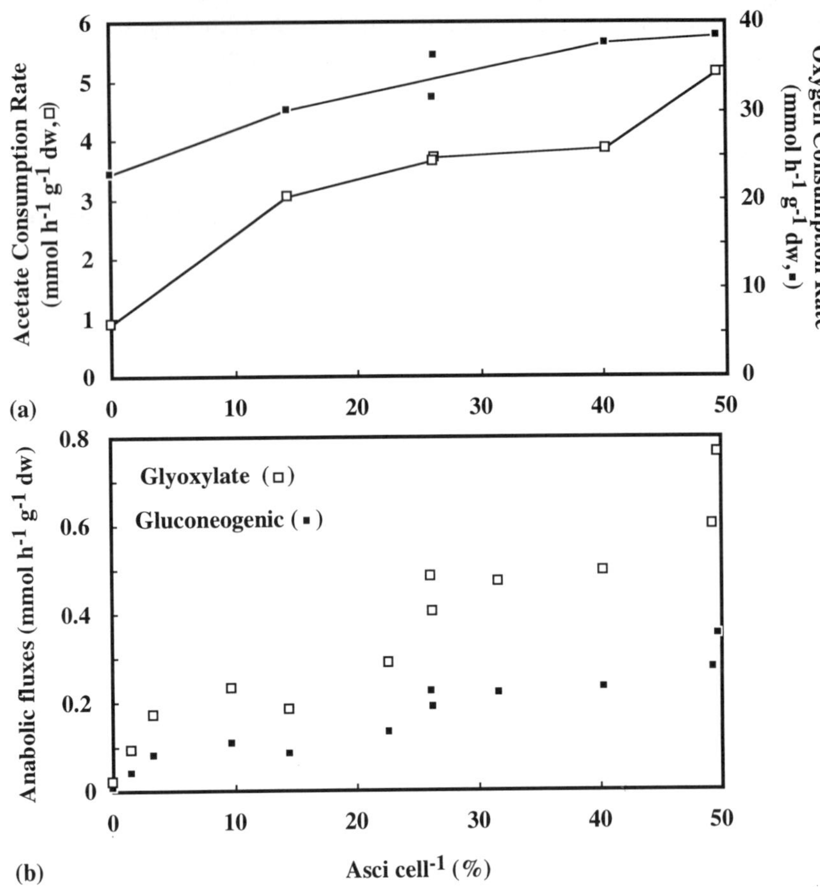

Figure 10.7 Catabolite repression of sporulation and metabolic fluxes in acetate sporulation medium in the presence of 2-deoxyglucose (2DG). The process of sporulation of *S. cerevisiae* strain CH1211 at 72 h was partially repressed by 2DG in acetate sporulation medium. (a) Average oxygen (filled squares) and acetate (open squares) consumption rates were plotted as a function of sporulation frequency at 72 h. The metabolic rates were averaged over the first 12 h in sporulation media. (b) Corresponding values of the estimated glyoxylate (open squares) and gluconeogenic (filled squares) fluxes were averaged over the first 12 h in acetate sporulation medium. Data were taken from experiments in which the cells were exposed to different 2DG concentrations in sporulation medium. (From Aon, J.C. and Cortassa, unpublished data.)

repression of anabolic pathways and consequently growth restriction. This is a remarkable result since the expression of *IME1* is sensitive to glucose-induced repression providing a link between metabolic and genetic control pathways of sporulation (Kassir *et al.*, 1988).

Overall, catabolite repression experiments further support the hypothesis that a flux imbalance may elicit the differentiation pathway in yeast. The acetate consumption rate modulated by catabolite repression appears as a major determinant of sporulation success. Flux imbalance may occur because either catabolic or anabolic fluxes are perturbed. In acetate sporulation medium, catabolite repression by 2DG mainly disturbs anabolic fluxes. Under these conditions, catabolism does not appear to control differentiation to a large extent. It could be inferred that in lactate sporulation medium, 2DG-induced repression of catabolism might limit sporulation. Thus, catabolite repression agrees with the notion that the imbalance between anabolism and catabolism is involved in the control of yeast cell differentiation.

10.7 THINKING ABOUT DEVELOPMENT IN THE LIGHT OF THE DYNAMIC ORGANIZATION CONCEPT

Development may be visualized as a process in which living systems travel through successive states in a more or less irreversible way; these are not stationary states of minimal entropy production but successive transient states in a changing parametric landscape (Figure 1.6). Such a view may be illustrated by the 'epigenetic landscape' of Waddington (1947) in which the system undergoing development evolves through valleys. All those transient states should be more or less unstable to allow the developing system to reach the adult or differentiated organism where many of their variables will attain a state of minimal dissipation, characterized by stability to perturbations in environmental and/or parametric conditions. The minimal dissipation is achieved after overcoming states of flux imbalance.

A major question is: to what extent do changes in metabolic status control the progression to sporulation? We propose the flux coordination hypothesis to interpret existing experimental evidence and discuss possible implications. Flux imbalance is a particular case of the flux coordination hypothesis that will be developed in Chapter 11. In particular, gluconeogenesis might act as a sensing device of the flux imbalance between catabolism and anabolism in the sense that the level of intermediates that serve as substrates for PEPCK or pyruvate carboxylase/ PEPCK varies according to the flux imbalance. When some anabolic substrates are missing (nitrogen, sulphur or phosphate), gluconeogenesis drives carbon towards carbohydrate storage even if, under those conditions, catabolism is the most conspicuous block of metabolism. The TCA cycle acts as a sink of products from all catabolic paths.

The data from yeast differentiation support the postulation that the gluconeogenic flux is permissive with respect to differentiation. On the other hand, glycolysis would favour growth (mass growth and/or proliferation). In *Xenopus laevis*, in the cleaving embryo before the mid-blastula transition, the main direction of flux is gluconeogenic, obtaining carbon from amino acid degradation (Dworkin and Dworkin-Rastl, 1991). Before the mid-blastula transition there is cleavage and specification of the main developmental axis, without increase in embryonic cell mass. After that step, when glycolysis is switched on, the embryo begins to grow and the map of the whole body appears to be already fixed.

Carbon catabolite repression demonstrates common control mechanisms of yeast cell proliferation and differentiation. Glucose-induced repression has been shown to be related to carbon and energy uncoupling and cell cycle arrest (Mónaco *et al.*, 1995; Aon *et al.*, 1995). Inhibition of sporulation by catabolite repression has been shown to depend on repression of anabolic fluxes. These results may be integrated into a common picture in which cell fate is determined by the coupling or balance between anabolism and catabolism, in turn regulated by catabolite repression (Chapter 11).

In spite of the difference in mechanisms, the imbalance between flows through anabolism and catabolism appears to have a direct and strong influence in driving cellular dynamics to a particular differentiation fate.

Cell growth and differentiation from the perspective of dynamics and thermodynamics of cellular and subcellular processes

11

This chapter develops the concept that spatio-temporal coordination between thousands of chemical reactions inside the cell concerns cellular metabolism and involves enzyme activity modulation by gene expression. It deals with the following questions. Might a balanced pattern of fluxes, genetically and environmentally determined, be a determinant of growth, proliferation or differentiation? Might that pattern of fluxes somehow mediate gene expression and, say, differentiation? In order to illustrate these concepts we will also be dealing specifically with cell commitment to transit through the cell cycle or to sporulation in *Saccharomyces cerevisiae* and will give experimental and theoretical support for the flux coordination hypothesis as a signal involved in proliferation or differentiation.

11.1 BIOTHERMOKINETICS OF GENE EXPRESSION AND CELL GROWTH AND PROLIFERATION

11.1.1 GROWTH AS A BALANCE OF FLUXES

Following a dynamic view of cell growth and proliferation, we consider growth as a dynamically coordinated, dissipative balance of fluxes, i.e. the result of fluxes of the different materials required, the interactions between those fluxes and the properties emerging from such interactions.

(a) Dynamic aspects

Growth as the result of the dynamic balance between synthesis and degradation of macromolecular components (e.g. DNA, RNA, proteins, polysaccharides) results in the typical growth curve shown by batch cultures of microorganisms with synthetic processes decreasing their rates with time and degradative ones increasing their rates achieving a pseudosteady state level which will persist for some time and then decline (Williams, 1990). Figure 11.1 shows the results of a numerical simulation of a typical growth curve in batch cultures of microorganisms at decreasing contents of available redox equivalents and high free-energy transfer bonds (NAD(P)H, ATP). Growth measured as the net accumulation of macromolecules, M, results from the dynamic balance of synthetic, J_s, and degradative, J_d, fluxes from catabolized substrate, S. As shown in Figure 11.1, J_s decreases with time, while J_d increases. A pseudo-steady state level of M is achieved for some time and then declines (Figure 11.1a,c); the latter corresponding to lower J_s than J_d fluxes. Figure 11.1 shows that the pseudo-steady state level of M also depends on the amount of fluxes which in turn are influenced by ATP and NAD(P)H contents (see next section).

The rates of accumulation of protein, RNA and DNA (or of any other extensive property, X, of the system) can be calculated using the first-order rate equation:

$$\frac{dX}{dt} = \mu X \tag{11.1}$$

where μ is in doublings per unit time (Bremer and Dennis, 1987). For DNA, ribosomes and protein, the rates of synthesis during periods of balanced growth are essentially equal to the rates of accumulation, since their turnover is negligible. For total RNA, however, the instantaneous synthesis rate is substantially higher than the accumulation rate because of the instability of mRNA and of spacer sequences in the primary rRNA and tRNA transcripts (Bremer and Dennis, 1987).

That metabolic variables attain balanced growth depends on how fast different parts of metabolism reach constancy, since different processes exhibit distinct relaxation times (Chapters 2 and 8). For example, in batch cultures of *S. cerevisiae* when considered separately, the specific oxygen (q_{O2}) and carbon dioxide (q_{CO_2}) uptake and production rates did not approach a balanced condition. However, the ratio between these two quantities rapidly approached the balanced condition (Beck and von Meyenburg, 1968; Barford *et al.*, 1982) which may indicate that the organism has more rapidly apportioned the total energy-producing capacity of the cell into a particular ratio of fermentation and respiration long before the full capacity of either system has been attained. In other words, the balance between the relative amounts of the energy-producing pathways of fermentation and respiration is favoured rather than the increase in fermentation or respiratory capacity *per se* (Barford *et al.*, 1982).

From this section we are led to suggest that the dynamic balance between fluxes is a property that determines cellular activities such as growth (i.e. the balance between macromolecule synthesis and degradation) and cell proliferation or division.

(b) Thermodynamic aspects

Thermodynamically, cellular processes such as metabolic reactions or cell growth may be characterized through systemic properties such as fluxes and the Gibbs free energy of the substrates being consumed by the cell (section 11.2). Biochemically and thermodynamically, cellular metabolism may be represented as a set of catabolic and anabolic fluxes coupled to each other through energy transducing events (Westerhoff and van Dam, 1987). The dynamic balance of anabolic and catabolic fluxes is associated with a dissipative pattern of matter and energy.

Analysis of the dynamics of cellular processes in the light of the concepts of dynamics and thermodynamics reveals that flux redirection at the cellular level may be accomplished by taking advantage of the intrinsic dynamic properties of cellular metabolism. Large dissipation rates through metabolic pathways or branch flux redirection may take place because of kinetic limitations. Kinetic limitation occurs when a metabolic pathway has attained the maximum possible rate of dissipation. The rate of dissipation (equation 11.2) suggests that at fixed ΔG_i and then fixed stoichiometry, the maximum in flux, J_i, may be set by the maximum velocity a catalyst is able to sustain through that path. Since the maximum velocity depends on enzyme concentration of rate-controlling steps, then the higher the enzyme concentration, the higher is the flux.

Flux redirection was found associated with carbon and energetic uncoupling (Aon *et al.*, 1995; Mónaco *et al.*, 1995). This redirection may be viewed as a bifurcation in the phase portrait dynamics of metabolic flux behaviour (Abraham and Shaw, 1987; Bailey *et al.*, 1987; Kauffman, 1989; Aon *et al.*, 1991; Cortassa and Aon, 1994b). What could be the nonlinear cellular mechanisms giving rise to such bifurcations in the dynamic behaviour of metabolic fluxes? We have recently investigated the *in vivo* regulation and control of glycolysis dynamics in *S. cerevisiae* experimentally. The stoichiometry of ATP production by glycolysis is itself autocatalytic (Cortassa *et al.*, 1990a). This autocatalytic feedback loop gives rise to sophisticated dynamics in addition to stable, asymptotic

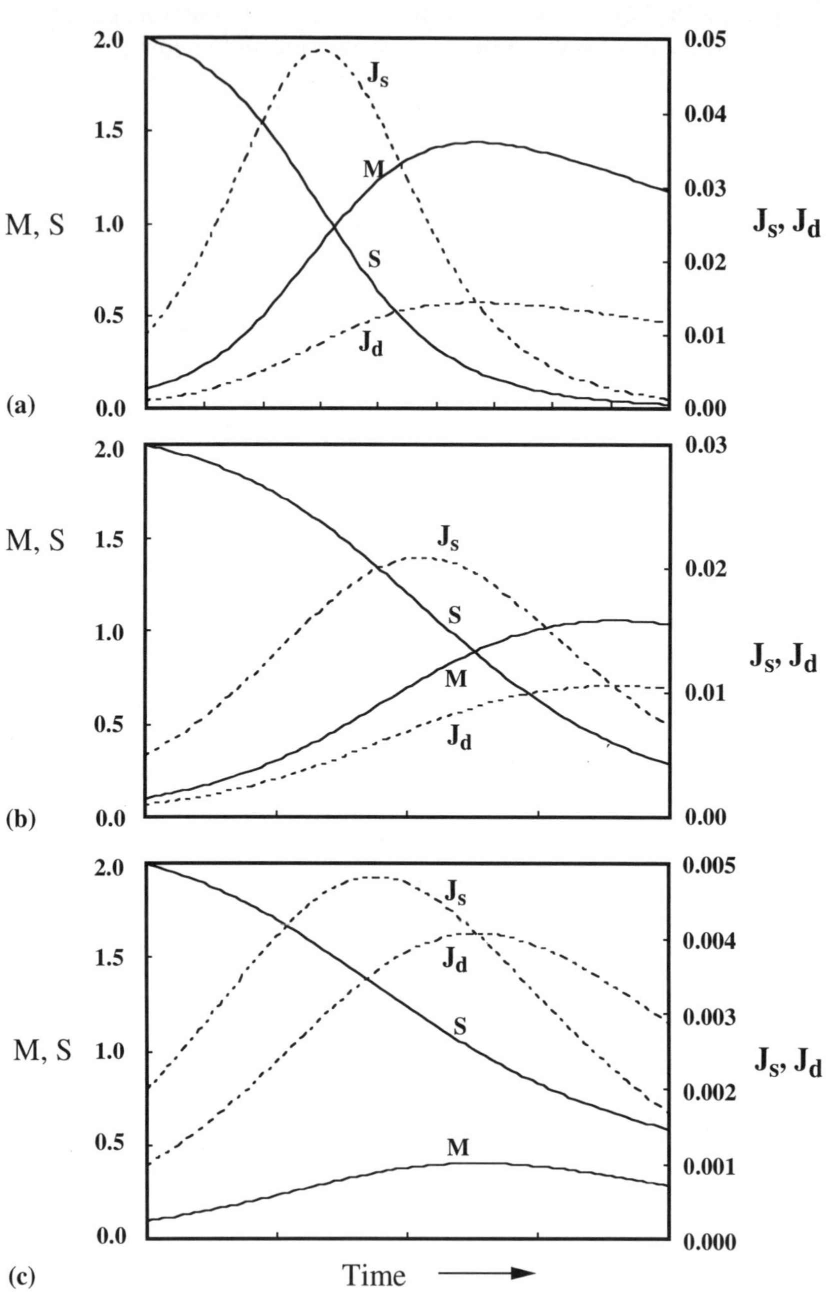

2.0
1.5
M, S 1.0
0.5
0.0
J_s
M
S
J_d
0.05
0.04
0.03
0.02
0.01
0.00
J_s, J_d
(a)
2.0
M, S 1.5
1.0
0.5
0.0
J_s
S
M
J_d
0.03
0.02
0.01
0.00
J_s, J_d
(b)
2.0
1.5
M, S 1.0
0.5
0.0
J_s
J_d
S
M
0.005
0.004
0.003
0.002
0.001
0.000
J_s, J_d
(c)
Time

steady states, and oscillatory ones (Chapter 7). On the basis of the results obtained in glycolysis, it is clear that pathway stoichiometry constitutes a built-in autocatalytic source of nonlinear behaviour. Changes in stoichiometry would then introduce the potential ability of the system to exhibit different dynamics in metabolic fluxes.

11.1.2 THE FLUX COORDINATION HYPOTHESIS

We emphasize the regulation of the degree of coupling between catabolic and anabolic fluxes as a regulatory mechanism of the growth rate. An imbalance between catabolic and anabolic fluxes may be associated with the onset of cell cycle arrest through growth limitation (Figure 11.2). In section 11.4 we apply the flux coordination hypothesis to cell division or

Figure 11.1 (see facing page) Cell growth as a result of a dynamic balance of synthetic and degradative fluxes. A simple model consisting of the autocatalytic synthesis of a macromolecular component (M) from a substrate (S). M is degraded according to a first-order kinetic law:

$$S + ATP + NADPH \rightarrow M \rightarrow$$

The model is mathematically expressed by two ODEs:

$$\frac{dM}{dt} = J_s - J_d$$

$$\frac{dS}{dt} = -J_s$$

with synthetic ($J_s = k_1$ S MATP NADPH) and degradative fluxes ($J_d = k_2$ M); k_1 and k_2 are the rate constants of synthesis and degradation of macromolecules (M), respectively. ATP and NADPH contents are parameters, i.e. they are assumed to be constant for simplicity. The parameter values were: $k_1 = 0.1$ mM^{-2} h^{-1}; $k_2 = 0.01$ h^{-1}; S_0 (initial concentration of substrate) $= 2.0$; ATP $= 1.0$. According to the units employed in the model, S, ATP, NADPH and M are in mM concentration units. From (a) to (c) the varying parameter is NADPH: (a) 0.5; (b) 0.25; (c) 0.1; indicating decreasing redox equivalent contents. The model was numerically integrated on a PC-60 III Commodore using the program SCoP with an Adams method (Duke University, 1989). Notice that when the flux of synthesis (J_s) intercepts the flux of degradation (J_d) the accumulation of M is maximal; thus, when $J_s > J_d$ the flux balance favours the synthesis of M and consequently it accumulates. On the contrary, when $J_s < J_d$ the flux balance favours the degradation of M, which begins to decline. The time, as well as the amount, reached by M depends on the metabolic status, e.g. available redox equivalents. (Reprinted from Aon and Cortassa, 1995, with kind permission from Elsevier Science Ltd, The Boulevard, Langford Lane, Kidlington OX5 1GB, UK.)

sporulation in *S. cerevisiae*. According to this hypothesis, some cellular mechanism of metabolic flux redirection toward product (either catabolic, organic acids, or anabolic, polysaccharides) or biomass (macromolecules) formation should be associated with the onset of cell division arrest. For instance, cell cycle control would correspond to a combination of interdependent, as well as autonomous, but synchronized cellular processes (Schlegel and Craig, 1991).

We take into account the following facts to formulate the hypothesis that flux coordination may be a signal for proliferation or differentiation:

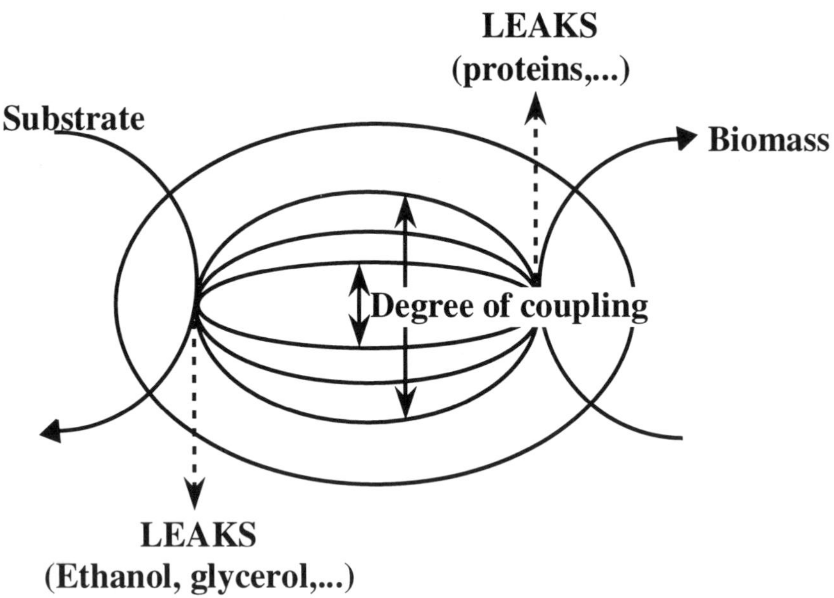

Figure 11.2 The flux coordination hypothesis: the degree of coupling between catabolic and anabolic fluxes as a phenomenon involved in the regulation of microbial proliferation and metabolic flux redirection. The hypothesis postulates that for higher degrees of coupling, i.e. longest arrow, the more coupled anabolic and catabolic fluxes will be, and that would correspond to lower leaks (dashed arrows). Under those conditions, cells will continue their mitotic cycling. Cells will leave the mitotic cycle whenever they are challenged by an unfavourable environment, or when environmental changes (e.g. oxygen, carbon source) induce a particular metabolic and energetic status that leads to a differential gene expression, which in turn induces metabolic flux redirection and lower growth rates. (Reprinted from Aon and Cortassa, 1995, with kind permission from Elsevier Science Ltd, The Boulevard, Langford Lane, Kidlington OX5 1GB, UK.)

- Growth results from the flux balance of synthetic and degradative processes (section 11.1.1).
- The flux is a target of control through gene expression (section 11.3).
- The length of the cell cycle is a potential (supra)regulatory mechanism of expression of genes possessing too large transcriptional units; these large transcriptional units would need a time too long with respect to the length of the cell cycle to be transcribed by RNA polymerase II (section 11.5).
- Flux imbalance between catabolic and anabolic processes, also called metabolic uncoupling, is related to the blocking of the yeast cell progression through the cell cycle (section 11.6).

A consequence of these four considerations is that the regulation of gene expression by the cell cycle would in turn regulate the coordination of fluxes as a signal for cell proliferation or differentiation.

11.2 TO WHAT EXTENT IS THE RATE OF ENERGY DISSIPATION BY AN ORGANISM GENETICALLY OR ENVIRONMENTALLY DETERMINED?

The answer to this question may begin with the analysis of the components of the free-energy dissipation – namely, fluxes and their conjugated force:

$$\sigma = \sum J_i \Delta G_i \tag{11.2}$$

Let us start with ΔG_i. The free-energy difference in every i-th enzyme-catalysed reaction of either catabolism or anabolism is fixed by the nature of the participating reactants and products. If ΔG_i is negative or positive, the reaction will occur either spontaneously or coupled to another process contributing the energy. The metabolic pathways stoichiometry (also called mechanistic stoichiometry) may change in different alternative catabolic pathways – for example, the glucose breakdown by the Embden–Meyerhoff or Entner–Doudoroff pathways – thus being genetically determined. Stoichiometry may also change due to environmental pressures such as oxygen levels (Verdoni *et al.*, 1990) and the change in P/O ratios of oxidative phosphorylation (Verdoni *et al.*, 1992). Additionally, the mechanistic stoichiometry may be influenced by differential expression of environmentally responsive genes (Chapter 9).

The next section shows how fluxes may be directly modulated by gene expression. Thus, the rate of processes and their effective stoichiometry may be genetically determined but environmental conditions will regulate the availability of effectors of the fluxes or modify the effective stoichiometry of fluxes, e.g. through flux redirection.

11.3 HOW CAN THE RECIPROCAL INFLUENCE OF GENETIC BACKGROUND AND ENVIRONMENT ON METABOLIC FLUXES BE EXPRESSED QUANTITATIVELY?

In the previous sections we analysed dynamic and thermodynamic aspects of cell free-energy dissipation. We shall now specify cellular mechanisms that could mediate both genetic and environmental determination of metabolic fluxes.

This section gives arguments to support the view that the target of cell energy dissipation could be either fluxes or free-energy differences. Specific cellular mechanisms that could mediate both genetic and environmental determination of metabolic fluxes are mentioned.

Metabolic fluxes may be directly influenced by gene expression through enzyme concentration. A mutational event anywhere in the genome can alter the kinetic parameters of catalytic activity essentially by:

- changes in the turnover number or the Michaelis constant, or both;
- changes in enzyme concentration resulting from changes in repression or induction or changes in gene dose (Kacser and Burns, 1981).

The maximal velocity (V_{max}) of an enzyme is the product of its concentration and the catalytic constant (k_{cat}) being the target of gene expression. Thus, by affecting the enzyme concentration, the flux may be modulated. In terms of MCA the flux is controlled by enzyme dosage of the rate-controlling steps of that pathway (Kacser and Burns, 1973; Heinrich and Rappoport, 1974; Fell, 1992). Therefore, enzyme concentration is a component of a rate law susceptible to genetic regulation, whereas substrates and products are both environmentally and genetically determined. This rationale assumes that enzyme velocity is proportional to enzyme concentration. It has been stated that the latter assumption fails when either there are different enzyme forms or channelling occurs (Liao and Delgado, 1993).

The matrix method of MCA has been applied to the Embden–Meyerhoff pathway and the branches to alcoholic fermentation and the citric acid cycle in *S. cerevisiae* growing in batch or chemostat cultures (Galazzo and Bailey, 1990; Aon *et al.*, 1996b; Cortassa and Aon, 1994a; Appendix 8A). According to the model used to interpret the results of chemostat cultures, four steps shared the control of the flux: sugar transport, an ATPase activity, the TCA cycle and the ADH branch to ethanolic fermentation (Chapter 7). When the rate of sugar uptake kinetics became independent of any intermediary metabolite, the sugar transport step turned out to be the only rate-controlling step of the flux (Cortassa and Aon, 1994a).

11.3.1 MODULATION OF METABOLIC FLUXES BY GENE EXPRESSION

(a) The role of gene expression in the modulation of fluxes through enzyme dosage: the control of the flux

Metabolic flux modulation may also be achieved through changes in the control of the flux, i.e. changing the nature of the rate-controlling steps. Having described which kinetic parameters may be changed by gene expression and in turn alter the values of the flux control coefficients, we then asked how the nature of the rate-controlling steps of the flux depends on gene expression. In order to answer this question we took into account the fact that the elasticity coefficients of MCA, which express the local dependence of the flux (or rate) of an individual enzyme, are the first derivative of the rate law of that enzyme. These elasticity coefficients are the prime data for constructing matrix E (Appendix 8) which, when inverted, gives the flux control coefficients in its first column and the concentration control coefficients in the other columns. The analytical expression of these flux control coefficients should contain information about the dependence of the flux on the enzyme concentration and other kinetic parameters, as well as reactants and products (Appendix 11).

We did a systematic survey of the analytical expressions of control coefficients for metabolic pathways with different topologies (linear, branched and circular) for enzyme-catalysed steps with various kinetics, i.e. linear, Michaelian, product inhibited and reversible (Appendix 11). We found that V_{max} values (i.e. the target of gene expression) appeared explicitly in the control coefficient expressions only when there is a reversible step. For linear pathways such reversibility should be in a step that exerts some control on the flux. In linear pathway topologies, control tends to be concentrated at the beginning of the path unless there is a strong dependence of the first step on a metabolite. For circular pathways, the V_{max} values of reversible steps (forward and backward reactions) appear in the analytical expression of control coefficients even if the reversible step does not exert control at all. It is worth remarking that only the V_{max} of forward and backward reactions of reversible steps appears in the expressions of all control coefficients of the pathway in which such steps are involved. For branched pathways, in all cases analysed, the V_{max} of the first enzymes after the branch appeared in the expression of the flux ratio control coefficient (Cortassa and Aon, unpublished data).

It then appears that flux control by gene expression depends mainly on pathway topology (configuration), i.e. linear, branched and cyclic, and less on the form of the rate law, i.e. Michaelian-type or allosteric. This is a surprising result since it would imply that the work of selection on phenotype through changes in the rate-controlling steps of the flux depends on an epigenetic character, i.e. pathway topology.

(b) Flux modulation may also be exerted by structural polymers

The intracellular amount and polymeric status of structural polymers such as actin and tubulin may modulate enzymatic fluxes (Luther and Lee, 1986; Roberts and Somero, 1987, Cortassa *et al.*, 1994a; Cortassa and Aon, 1994b). Regulation of enzymatic fluxes was shown to be dependent on microtubular protein concentration and its polymeric status (Chapters 6 and 7). For allosteric enzymes related to carbon metabolism such as PK, flux modulation by polymers appears to be due to regulation of the degree of self-association of the enzyme. The latter may be accomplished through stabilization of oligomeric forms with a higher degree of activity (Minton and Wilf, 1981; Cortassa *et al.*, 1994a). Thus, with gene dosage of structural polymers (gene-expression dependent) or changes in its polymeric status (epigenetic, environmentally dependent), enzymatic fluxes might in turn be modulated.

11.4 COORDINATION OF CATABOLIC AND ANABOLIC FLUXES AS A SIGNAL FOR PROLIFERATION AND DIFFERENTIATION

11.4.1 THE FLUX COORDINATION HYPOTHESIS AND METABOLIC COUPLING: AN EMPHASIS ON ITS DYNAMIC ASPECTS

Our view of metabolic coupling emphasizes the dynamics of cellular bioenergetics – namely, instantaneous carbon and energy flux ratio balances. Globally, the rate of energy dissipation by a cell is given by the product between fluxes, J_i, and their conjugated forces, ΔG_i (equation 11.2). ΔG_is are related to pathway stoichiometry (i.e. methylglyoxal bypass), P/O ratios (Verdoni *et al.*, 1992) or H^+/organic acids stoichiometry of symport mechanisms (Ten Brink and Konings, 1982; Verdoni *et al.*, 1990). In equation 11.2, at fixed ΔG_i, gene expression may change the control of the flux through changes in enzyme dosage.

We developed a method of quantifying the dynamics of cellular metabolism and energetics (Chapter 8; Cortassa *et al.*, 1995) and this was applied to growing yeast on different carbon sources and sporulating yeast on different gluconeogenic substrates (Cortassa *et al.*, 1995; Aon, J.C. and Cortassa, 1995; Aon, J.C. *et al.*, 1996). Carbon and energy balances are expected to vary together with fluxes through the central amphibolic pathways (see Chapter 8). This method provided us with an analytical tool to evaluate the uncoupling in carbon and energy undergone by *S.*

cerevisiae cells progressing through the cell cycle and during the process of sporulation. For that purpose, we quantified metabolic coupling or the degree of coupling between catabolic and anabolic fluxes through carbon and energetic coupling indexes (Aon *et al.*, 1995; Cortassa *et al.*, 1995). Metabolic coupling may be considered from two main points of view:

- From the carbon side, coupling is defined as the amount of carbon assimilated as cell material (biomass) related to the total amount of carbon consumed for anabolic and catabolic purposes. A carbon coupling index (J_r^C) was defined as the ratio of fluxes of carbon incorporated into biomass over the total carbon consumption flux (equation 11.7).
- The energetic view takes into account how much of phosphorylation and redox potentials produced in catabolism are used to fulfil synthetic requirements. An energetic coupling index (J_r^{ATP}) was defined as the ratio of the ATP flux required to sustain the synthesis of biomass over the ATP flux produced in catabolism (either oxidative or fermentative) (equation 11.8).

The following formalism summarizes the approach described above:

$$J_c^C = 2J_{EtOH} + J_{CO_2} + 3J_{gly} \tag{11.3}$$

$$J_c^{ATP} = J_{EtOH} + 1.4J_{O_2} - J_{gly} \tag{11.4}$$

Equations 11.3 and 11.4 are the carbon and energy fluxes obtained by catabolism of glucose. J_{EtOH}, J_{CO_2}, J_{O_2} and J_{gly} are the ethanol, carbon dioxide, oxygen and glycerol fluxes, respectively. The ATP flux by catabolism was obtained considering a P/O ratio of 1.0.

$$J_a^C = \text{Total amount of carbon necessary to synthesize 1 g of yeast} \times \mu \tag{11.5}$$

$$J_a^{ATP} = \text{Total amount of ATP necessary to synthesize 1 g of yeast} \times \mu \tag{11.6}$$

Equations 11.5 and 11.6 represent the carbon and energy fluxes anabolically required to synthesize one gram of yeast biomass multiplied by the growth rate, μ (Cortassa *et al.*, 1995).

The carbon, J_r^C, and energy, J_r^{ATP}, coupling indices were calculated as the ratio of either carbon or ATP anabolic fluxes over the corresponding catabolic ones, as previously described (Cortassa *et al.*, 1995; Aon *et al.*, 1995):

$$J_r^C = \frac{J_a^C}{J_c^C} \tag{11.7}$$

$$J_r^{ATP} = \frac{J_a^{ATP}}{J_c^{ATP}} \tag{11.8}$$

This type of analysis gives a more thorough insight into the nature of metabolic coupling than biomass growth yields, since it takes into account instantaneous fluxes of ATP and biomass and not only a global biomass accumulation and substrate consumption.

Metabolic uncoupling occurs when a large proportion of the carbon and/or energy is leaked through microbial products or energy-spilling processes (Figure 11.2) (see also Aon *et al.*, 1995, and references therein). Microorganisms displaying low growth rates and metabolic uncoupling apparently show the highest capacity to dissipate metabolic free energy through high yield of metabolite production (Tempest and Neijssel, 1984; Linton, 1990). In *S. cerevisiae* growing in chemostat cultures under conditions of oxidative metabolism at low glucose concentrations, most of the ATP generated by catabolism is utilized for biomass synthesis; high biomass yields are then obtained (high degree of coupling) (Rieger *et al.*, 1983; Cortassa and Aon, 1994b; Aon *et al.*, 1995). In the absence of oxygen or when yeast cells display respiro-fermentative metabolism, since a great deal of the ATP generated by catabolism is not used for growth, the microorganism is said to be uncoupled (Aon *et al.*, 1996b). Under these conditions ethanol is the main catabolic product.

11.4.2 PROLIFERATION OF *S. CEREVISIAE* AND THE FLUX COORDINATION HYPOTHESIS

The eukaryotic cell cycle comprises two main stages: mitosis (M) and DNA synthesis (S), separated by two gap periods, G_1 and G_2 (Hartwell, 1974). The view of dependent ordered events emerged from studies with temperature-sensitive cdc mutants of *S. cerevisiae*, namely that the execution of late events in the cell cycle depends on the prior completion of early events (Hartwell, 1991, and references therein; section 5.1; Chapter 3).

(a) Growth rate and the length of cell cycle phases in *S. cerevisiae*

We applied the flux coordination hypothesis to the proliferation of *S. cerevisiae* in either batch or chemostat cultures. It appears that the length of G_1 is dependent on both carbon and nitrogen sources; although the ensemble of S + G_2 + M phases also varies, G_1 showed wider differences (Table 11.1). In chemostat cultures, the length of the budding phase was independent of the growth rate, i.e. the length of the cell cycle. Because budding occurred in a constant period, the increase in the mean generation time at lower growth rates is due to the lengthening of the single cell phase.

A relationship appears to exist between the growth rate of *S. cerevisiae* and the lapse for which cells rest at different stages of their cell cycle.

Table 11.1 Length (in minutes) of cell cycle phases and cell growth rate in *Saccharomyces cerevisiae* (reproduced from Aon and Cortassa, 1995, with permission from Elsevier Science Ltd, The Boulevard, Langford Lane, Kidlington OX5 1GB, UK)

Carbon and nitrogen sources	Doubling time	G1 phase length	S+G2+M	Reference[e]
Glucose (2%), glutamate	144	70	74	1
Glucose (2%), ammonia	156	94	62	1
Glucose (2%), leucine	240	214	26	1
YPD[a]	84.6	16.8	67.8	2
SF[b]	130	39	91	2
SEY[c]	195	101	94	2
SE[d]	258	146	112	2
Glucose (2%), ammonia	125	35	90	3
Glucose (2%), glutamine	110	30	80	3
Glucose (2%), proline	400	95	305	3
Glucose (2%), ammonia	92	5	87	4
Galactose (2%), ammonia	181	30	151	4
Ethanol (2%), ammonia	435	283	152	4

(a) 1% yeast extract, 2% peptone and 2% dextrose (Slater *et al.*, 1977).
(b) 2% fructose, 0.7% YNB (Slater *et al.*, 1977).
(c) 0.3% yeast extract, 2% ethanol and 0.7% YNB (Slater *et al.*, 1977).
(d) 2% ethanol, 0.7% YNB. (Slater *et al.*, 1977).
(e) 1, Johnston *et al.* (1980); 2, Slater *et al.* (1977); 3, Rivin and Fangman (1980); 4, Barford and Hall (1976).

Table 11.1 summarizes several reported data which show the dependence of the cell cycle phase length of *S. cerevisiae* as a function of the nature of carbon or nitrogen sources. Beck and von Meyenburg (1968) were the first to relate the cell cycle to the growth rate of yeast, by measuring the frequency of non-budded and budded cells as a function of growth conditions. In these studies, the percentage of budding cells was shown to increase almost linearly with the growth rate; the length of the budding phase was almost constant. Because budding occurred in a constant period, the increase in the mean generation time at lower growth rates was interpreted as being due to the lengthening of the single cell phase. Variable growth rates, i.e. doubling times, with different sources of carbon (Slater *et al.*, 1977) or nitrogen (Johnston *et al.*, 1980) mainly depend on changes in the length of the G_1 phase of the cell cycle (however, see Rivin and Fangman, 1980).

The lengthening of the single cell phase (G_1) was found to be associated with:

- low substrate (glucose) fluxes;
- derepressed (high levels) of respiratory enzymes;
- repressed levels of glycolytic enzymes such as aldolase.

In contrast, when the mean generation time decreased (and therefore the length of G_1 was shorter) correlatively with substrate flux, respiratory enzyme levels decreased and aldolase increased (Beck and von Meyenburg, 1968; Munch *et al.*, 1992b; Mónaco *et al.*, 1995).

(b) The length of the cell cycle as a temporal determinant of gene expression

Within the context of a dynamic view of cellular organization implying that a dissipative pattern of fluxes is mediating between gene expression and cellular growth, proliferation or differentiation, in the next step we ask about the way in which the dynamics of gene expression is regulated.

Different paths have been taken to answer this question. For eukaryotic cells, one of those views assumes a collection of operationally independent genes, each controlled through interactions with a number of *cis*-acting promoter elements and *trans*-acting transcription factors which diffuse at random throughout the nucleoplasmic space. Another view proposes that, in eukaryotic cells, the nucleus is organized in such a way that specific genes occupy distinct sites and that spatially ordered RNA synthesis, processing and transport deliver mature RNAs to particular sites in the cytoplasm (Cook, 1989; Jackson, 1991). Genome organization would be influenced by nuclear architecture through establishment of domains and gene location, and by driving the formation of complexes with the cytoskeleton (Cook, 1989; Jackson, 1991; Georgiev *et al.*, 1991).

An alternative mechanism of regulation of gene expression determined by the length of the cell cycle has been proposed through the intron-delay hypothesis (Gubb, 1993). The essence of this hypothesis is that the length of a transcription unit is important in modulating gene expression in a cell cycle-dependent manner. Taylor (1960) and Prescott and Bender (1962) demonstrated, through measurement of incorporation of RNA precursors, that there is little or no RNA synthesis during mitosis. By *in situ* hybridization experiments (Shermoen and O'Farrell, 1991) it was detected that nascent Ubx transcripts of *Drosophila* are aborted as a cell progresses through mitosis. Shermoen and O'Farrell suggested that the abortion of nascent transcripts is a general phenomenon and affects all RNA polymerase II transcripts since the transcription of all genes is interrupted during mitosis (Taylor, 1960; Prescott and Bender, 1962).

The time to complete a cell cycle depends on the organism (8 min in *Drosophila*, 30 min in frog, 2 h in the yeast *Saccharomyces* and 3 h in mouse) and on the physiological condition (Table 11.1). RNA polymerase traverses a gene from about 30 s to 11 h, the variation depending on the size of the transcription unit (from 1 kb to 2000 kb). Chain elongation rates of 30 nucleotides per second (1.8 kb min^{-1}) have been described for all RNAs, but different RNA polymerase II start sites function with very different efficiencies (De Robertis and De Robertis, 1988; Alberts *et al.*, 1989). Polymerase II transcripts (destined to leave the nucleus as mRNA molecules) comprise more than half of the RNA synthesized by the cell. However, most of the RNA in these transcripts is unstable, thus short-lived (only 3–5% of cytoplasmic RNA is mRNA) (Alberts *et al.*, 1989).

Taking into account that there is no gene transcription during mitosis and that genes have transcription units of different length, the temporal lengths of the cell cycle in general and of each phase of the cycle (e.g. G_1, S) in particular are potentially (supra)regulatory of gene expression. This is so because mitosis (and probably DNA synthesis) disrupts transcription, thus precluding the expression of genes whose required transcription time exceeds the length of the cell cycle. Abortion of transcripts occurs for genes possessing too large transcription units that need too long a time with respect to the length of the cell cycle to be transcribed by RNA polymerase II. The latter mechanism has been demonstrated as a candidate in regulating the expression of large genes (Shermoen and O'Farrell, 1991). As a consequence of abortion of nascent transcripts at each mitosis, transcription of each gene will have to begin anew in the next cell cycle.

If the regulation of gene expression by cell cycle length is a likely mechanism operating in the cell, the factors that influence the length of the cell cycle stages are significant and need to be known. This is explored further in the next section.

(c) Transcription rates and the cellular growth rate

It is known that transcription serves as a primary level of control in gene expression and that subsequent post-transcriptional events can play important roles in establishing the final levels of a gene product (Cleveland and Yen, 1989).

The rates of accumulation of protein, RNA and DNA or the rate of cell division (or of any other extensive property, X, of the system) can be calculated using first-order kinetics for rates of variation of any extensive property.

The mean generation time τ_d is defined as:

$$\tau_d = \frac{ln2}{\mu} \qquad (11.9)$$

By analogy with equation 11.9 we may also define a mean polymerization time of RNA, τ_p:

$$\tau_p = \frac{ln2}{\tau_{1/2}} \qquad (11.10)$$

where $\tau_{1/2}$ is:

$$\tau_{1/2} = \frac{L}{S} \qquad (11.11)$$

L being the length of the transcription unit (in kb) and S the speed of polymerization (in kb min^{-1}) (Shermoen and O'Farrell, 1991). Considering that transcription mainly occurs in the G_1 phase of the cell cycle, the mean time of transcription τ_{tr} may be written as follows:

$$\tau_{tr} = \frac{ln2}{G_1 l} \qquad (11.12)$$

with $G_1 l$ as the length of the G_1 phase of the cell cycle.

The two main assumptions underlying expressions 11.11 and 11.12 are that RNA polymerase progresses through a gene at a uniform rate, and that gene transcription is negligible during the S and G_2 phases of the cell cycle. By comparing $\tau_{1/2}$ and τ_{tr}, we may be able to realize if a gene will be transcribed or its transcript aborted in the nucleus. We assume that gene transcription or abortion of transcripts will occur as a function of the lapse during which a cell stays in G_1 and the length of the gene transcription unit.

We analysed whether the reported length of G_1 ($G_1 l$ in equation 11.12) in *S. cerevisiae* growing in the presence of glucose and ammonia (~ 10 min for a doubling time of ~ 90 min: Table 11.1) could be time enough to allow transcription of genes coding for the enzymes involved in the main catabolic pathways of yeast (Table 11.2). According to the flux coordination hypothesis, lower growth rates and therefore longer cell cycles and G_1 phases would fit with larger transcription units for respiratory enzymes and longer transcription times. Chain elongation rates (S in equation 11.11) of 30 nucleotides per second, or 1.8 kb min^{-1} (De Robertis and De Robertis, 1988; Alberts *et al.*, 1989) were used; we then calculated the mean polymerization time, τ_p (equation 11.10) and the mean time of transcription, τ_{tr} (equation 11.12). For a length of G_1 ($G_1 l$, equation 11.12) of 10 min, a τ_{tr} of 6.93 min was calculated, i.e. each transcription would use 6.93 min in G_1. The mean value of τ_p obtained from the data of Table 11.2 was 1.073 ± 0.43 min for glycolytic enzymes and 0.82 ± 0.49 min for characterized TCA cycle enzymes and electron carriers from oxidative phosphorylation. According to these results, the length of G_1 would not regulate the expression of glycolytic enzymes through the transcription rate (Table 11.2). The transcription of glycolytic genes would fit with the shorter times for cell cycle progression needed by the cells at higher growth rates. This is in agreement with experimental data, i.e. during fermentative growth of *S. cerevisiae* the glycolytic genes are among the most efficiently expressed, the glycolytic enzymes comprising over 30% of soluble cell protein (Moore *et al.*, 1991). Results obtained in chemostat cultures of *S. cerevisiae* indicate that the glycolytic flux is not regulated at either transcriptional or translational levels (Sierkstra *et al.*, 1992). This conclusion was reached by measuring mRNA levels and enzyme activities at different dilution rates in continuous cultures. These results are in agreement with the absence of regulation of the glycolytic flux by different lengths of the G_1 phase. It has indeed been shown that the levels of several mRNAs corresponding to different glycolytic enzymes, such as enolase, phosphoglycerate kinase, pyruvate kinase, pyruvate decarboxylase and alcohol dehydrogenase, are regulated in response to carbon source (Denis *et al.*, 1983; Moore *et al.*, 1991). Such regulation of glycolytic mRNA levels would run through a mechanism independent of the length of the cell cycle. A comparative study of the steady-state levels of all the glycolytic mRNAs during exponential growth of *S. cerevisiae* on glucose or lactate showed that those stationary states were induced to a different extent by glucose. Two peaks of maximal induction by glucose were observed for phosphofructokinase and pyruvate kinase, both encoded by mRNAs that were shown to be co-regulated at translational levels in *S. cerevisiae* (Moore *et al.*, 1991).

However, the transcript length of important enzymes and enzymatic complexes of the TCA cycle and oxidative phosphorylation are missing in Table 11.2 – for example, pyruvate dehydrogenase ($\sim 3 \times 10^6$ kDa and 60 polypeptide chains) and α-oxoglutarate dehydrogenase (380 000 kDa and 8 polypeptide chains). These enzyme complexes are likely to be controlled by the length of G_1. On this basis, our working hypothesis would suggest that non-fermentable or gluconeogenic substrates such as acetate, ethanol, glycerol, lactate or pyruvate, which need TCA cycle enzymes and electron carriers to be catabolized by *S. cerevisiae*, would induce longer cell cycles or lower growth rates. This corresponds to what has been observed experimentally (Cortassa *et al.*, 1995).

11.4.3 DIFFERENTIATION OF *S. CEREVISIAE* AND THE FLUX COORDINATION HYPOTHESIS

Diploid *S. cerevisiae* cells undergo the developmental process of sporulation in response to nitrogen, phosphate and sulphur deprivation. Sporulation involves transition from mitotic to meiotic cycles; its occurrence depends on genetic background (two alleles of the mating type locus *Mat*α and *Mata*, *IME1*, *RME1*) and a certain metabolic status

Table 11.2 Transcript lengths and MW of glycolytic, TCA cycle, respiratory enzymes and electron carriers of *Saccharomyces cerevisiae* (reprinted from Aon and Cortassa, 1995, with kind permission from Elsevier Science Ltd, The Boulevard, Langford Lane, Kidlington OX5 1GB, UK)

Glycolytic enzymes	MW	No. of polypeptide chains	Transcript length (kb)	Reference
Phosphoglucomutase (PGM 1 yeast)	112 000	4 (28 000 each)	1.3	Rodicio and Heinisch, 1987; Heinisch, 1986
Aldolase	150 000	4	4.5	Rodicio and Heinisch, 1987
Triosephosphate isomerase	50 000	2	1.5	Rodicio and Heinisch, 1987
GAPDH	145 000	4	4.35	Rodicio and Heinisch, 1987
Phosphoglycerate kinase (PGK1)	43 500	1	2.3	Rodicio and Heinisch, 1987; Van der Aar, 1990
Phosphoglycerate mutase	57 000	2	1.7	Rodicio and Heinisch, 1987
Enolase	85 000	2	2.5	Rodicio and Heinisch, 1987
Pyruvate kinase	230 000	6.9	4	Rodicio and Heinisch, 1987
Lactate dehydrogenase	150 000	4	4.5	Rodicio and Heinisch, 1987
Alcohol dehydrogenase (ADH1, ADH2)[a]	37 000	4	2.4(ADH1) 2.55(ADH2)	Bennetzen and Hall, 1982
Phosphoglucose isomerase (PGI1)	60 000	2	1.8 (2.2)	Aguilera and Zimmerman, 1986
PFK1, PFK2 (fructose-6-phosphate 2 kinase)	800 000	8 (4α and 4β subunits)	3.6 (mRNA)	Heinisch, 1986; Kretschner *et al.*, 1991
Pyruvate carboxylase (PYC1, chromosome VII; PYC2, chromosome II)			2.4	Walker *et al.*, 1991
Hexokinase (HXKII)			1.65	Frolich *et al.*, 1984

Table 11.2, *continued*

Tricarboxylic acid cycle (TCA) enzymes	MW	No. of polypeptide chains	Transcript length (kb)	Reference
Pyruvate dehydrogenase	3.1×10^6	60 (MW$_{av}$=50 000)		Lehninger, 1977
Citrate synthase	100 000			Lehninger, 1977
Isocitrate dehydrogenase	380 000	8		Lehninger, 1977
α-Oxoglutarate dehydrogenase	2.1×10^6			Srivastava and Bernhard, 1986; Lehninger, 1977
α-Ketoglutarate dehydrogenase (KE1 of the KGDC)	2.5×10^6		3.7	Repetto and Tzagoloff, 1989; Srivastava and Bernhard, 1986; Lehninger, 1977
Succinate dehydrogenase	100 000	2	1.3	Lombardo and Scheffler, 1989
Fumarase (FUM1)	200 000	4	1.5	Wu and Tzagoloff, 1987
Malate dehydrogenase			1.7 (mRNA)	McAlister-Henm and Thompson, 1987
Aldehyde dehydrogenase (ALDH, mitochondrial)	60 000	4	2.0	Saigal et al., 1991

Table 11.2, *continued*

Electron transport enzymes and carriers	MW	No. of polypeptide chains	Transcript length (kb)	Reference
NADH dehydrogenase	78 000			Saigal *et al.*, 1991
Succinate dehydrogenase (SDH)	100 000		1.3	Lombardo and Scheffler, 1989; Saigal *et al.*, 1991
Cytochrome c oxidase[d,b] (Cox5, Cox6 subunits 5, 6)		b	1.17, 0.85	Koerner *et al.*, 1985; Wright *et al.*, 1984
Cytochrome b (Cob1, Cob2)[d]		c	6.2	Nobrega and Tzagoloff, 1985
Iso-2-cytochrome c (CYC1)[d]			2.2	Montgomery *et al.*, 1980
F1-ATPase α and β subunits (ATP$_1$, ATP$_2$)	54 575		1.7, 1.92	Takeda *et al.*, 1985, 1986
CYTOCHROME C:				
14 kDa Subunit VII	14 000		1.9	Schoppink *et al.*, 1989
11 kDa Subunit VIII	11 000		1.447	Schoppink *et al.*, 1989
17 kDa Subunit VI			1.177	Schoppink *et al.*, 1988
QH2-CYCC reductase (44 kDa core protein)	44 000		2.2	Tzagoloff *et al.* 1986

(a) In the liver, cytosolic alcohol dehydrogenase catalyses the following reaction:

ethanol Ý acetaldehyde

and mitochondrial acetaldehyde (ALDH) catalyses the conversion:

acetaldehyde Ý acetate

In *S. cerevisiae* it appears that a mitochondrial alcohol dehydrogenase (ADH3) is responsible for the initial oxidation (Saigal *et al.*, 1991).
(b) Seven non-identical subunits: four nuclear and three mitochondrial.
(c) Six exons and five introns.
(d) All these regulated by catabolite repression.

not completely defined at present (Esposito and Klapholz, 1981; Miller, 1989) (Chapter 10). The emergent view with respect to the control of sporulation by the genetic background includes a series of genes exerting activation or inhibition on each other. *MATα* and *MATa* genes are required to inactivate *RME1*, which acts as a sporulation inhibitor in *MATα/MATα* or *MATa*/MAT.a or in *mata2/matα1* cells (Mitchell and Herskowitz, 1986). In turn *RME1* product inhibits the expression of *IME1*, which is apparently essential for sporulation (Kassir *et al.*, 1988). It has been hypothesized that the expression of *IME1* integrates signals derived from the genetic and nutritional control paths of sporulation (Kassir *et al.*, 1988). Other members of the *IME* family have been identified more recently, such as *IME2* and *IME4* which interact with *IME1* (Shah and Clancy, 1992). The previously mentioned genes are apparently essential for normal regulation of sporulation. However, the function of gene products and the detailed biochemical mechanisms through which they exert their action are ignored. On the other hand, there is a series of genes differentially expressed exclusively during sporulation in *S. cerevisiae* (namely, *SGA* and the cluster containing *SPS1*, *SPS2* and *SPS3* coding sequences) which are dispensable for meiosis and spore formation (Yamashita and Fukui, 1985; Smith and Segall, 1986).

Less attention has been paid to the physiological changes occurring concomitantly with the sporulation process. It is known that sporulation is characterized by an absolute requirement of respiratory activity, gluconeogenic metabolism, a typical pattern for synthesis and degradation of polysaccharides and extensive proteolysis (reviewed by Miller, 1989). On the other hand, our experimental data point out a large energy dissipation which occurs during sporulation as compared with growth conditions. During sporulation, anabolic fluxes are lower and catabolic ones much larger than in growing yeast cells (Cortassa *et al.*, 1995; Aon, J.C. and Cortassa, 1995).

Taking the above phenomenology into account, the following question arises within the framework of the flux coordination hypothesis: is the *IME1* gene product involved in the flux imbalance associated with the triggering of yeast sporulation and, if so, how?

From an experimental series, we have postulated that sporulation of *S. cerevisiae* is triggered by a transient increase in energy dissipation through a decrease in the rate of biomass production followed by a decrease in the rate of catabolic processes. Such uncoupling would activate the process leading to the formation of a cell (asci) containing four haploid spores (Aon, J.C. and Cortassa, 1995). Two facts support the proposal that the uncoupling between catabolic and anabolic fluxes may be a signal for differentiation:

- The environmental signal is a nutrient deficiency in N, P and S which results in a lack of anabolic substrates leading to a decrease in the growth rate.
- Such a decrease in anabolic processes is followed by a decrease in catabolic fluxes during exposure to sporulation medium, which would tend to decrease the dissipation of energy and carbon.

(a) The imbalance between catabolic and anabolic fluxes as a regulatory phenomenon of sporulation efficiency in *S. cerevisiae*

We have quantified the influence of metabolic fluxes on the initiation of sporulation after studies carried out in the presence of different carbon sources (Chapter 10; Figure 10.6).

Our experimental evidence is in agreement with the postulation that a certain threshold imbalance between anabolic and catabolic fluxes should be attained to trigger the sporulation process. The flux balance of metabolism was perturbed with acetate, lactate or different cellular densities. In lactate, it is quite likely that cells are energy-limited to complete sporulation, which explains the correlation observed between respiratory capacity and sporulation efficiency (Figure 10.6b). In contrast, cells sporulating on acetate are mainly carbon-limited, as revealed by the correlation of sporulation frequency versus anabolic fluxes (Figure 10.6c).

Carbon and energy coupling indices (equations 11.7 and 11.8) also point to the importance of a flux imbalance to be met in order to accomplish the sporulation process. In lactate sporulation medium, the tendency is clear and confirms the previous correlations noted thus far: namely, that both carbon and energy coupling indices decrease with the sporulation efficiency. Accordingly, the larger the catabolic with respect to anabolic fluxes, the better will be the sporulation achieved. In contrast, the tendency is not so conspicuous in the presence of acetate since carbon and ATP are both required to sustain sporulation efficiently.

11.4.4 FLUX IMBALANCE AS A TRIGGER OF CELLULAR MORPHOGENESIS: GENE OVEREXPRESSION AND FLUX IMBALANCE

Another prediction of the flux coordination hypothesis is that flux imbalance would mediate subcellular remodelling and cellular transformation. One way to produce flux imbalance or alteration of coordination between catabolic and anabolic fluxes is to over- or under-produce a given protein or a metabolite (for example, by expression of specific foreign proteins or by defective mutants, respectively). Section 11.5 shows that temperature-sensitive cell cycle mutants of *S. cerevisiae* bearing mutations defective in cdc gene products have either catabolic or anabolic fluxes and their degree of coupling impaired concomitantly with the arrest of cell proliferation (Aon *et al.*, 1995).

Foreign gene overexpression is another experimental test of our working hypothesis. If the latter is correct, cellular flux imbalances introduced by foreign gene expression should have consequences at the level of metabolic coupling which is somehow reflected at the cellular level. In fact, if the unbalanced flux is not directed to biomass synthesis it results in morphological alterations with respect to a standard cell shape. In the following we present various reported data that illustrate the plausibility of the flux coordination hypothesis, i.e. that prokaryotic and eukaryotic cells are known to react to unbalanced overproduction of protein, mediated by either vectors, malignant transformation or drugs, with changes in their morphology. According to our hypothesis, every cellular process (e.g. protein synthesis) requires a temporal and spatial coordination for appropriate proliferation of organelles in order to accommodate the produced protein.

The effect of overexpression of genes coding for enzymes involved in PEP, pyruvate and oxalacetate metabolism, such as PEPCK or PEPcarboxylase, has been studied in *E. coli* cells. This is a timely tool for achieving rational manipulation in metabolic engineering (Liao *et al.*, 1994). Overexpression of PEPCK or PEPcarboxylase changed the level of expression of many other unlinked genes and also led to futile cycling and/or toxic effects (Liao *et al.*, 1994). This result is in agreement with flux imbalance due to changes in metabolic coupling.

Stationary phase *E. coli* cells in which fumarate reductase was amplified 30- to 40-fold responded by an increase in membrane phospholipid biosynthesis such that the lipid/protein ratio remained nearly constant at 0.4 (Weiner *et al.*, 1984). The cells accommodate the extra membrane lipid and protein by synthesizing a tubular structure composed of a lipid–protein core 16 nm in diameter, surrounded by a 9 nm electron-dense outer ring composed of an array of the catalytic subunits of fumarate reductase. The tubules are attached to the cytoplasmic membrane and appear to branch and grow to lengths of several microns, indicating that they must traverse the bacterial cell several times (Weiner *et al.*, 1984).

Large electron-dense inclusions are formed in *S. cerevisiae* cells which overexpress catalase A (Figure 11.3) (Binder *et al.*, 1991). Yeast cells transformed with the *CTA1* gene coding for catalase A, cloned into the multicopy vector YEp352, exhibit electron-dense inclusions of morphologically amorphous protein aggregates containing catalase A, clearly separated from the rest of the cytoplasm (Figure 11.3b,c). The

PMA2 gene, presumed to be the isogene of the *PMA1* gene, encodes the major yeast plasma membrane H^+-ATPase (Figure 11.4a). Immunoelectron microscopy of *S. cerevisiae* cells with overexpressed PMA2 protein revealed a dramatic increase in proliferation of PMA2 gold-labelled structures, probably derived from the endoplasmic reticulum (Figure 11.4b) (Supply *et al.*, 1993). The proportion of sections exhibiting the phenotype of PMA2-labelled structures with a multicopy vector of *PMA2* increased to about 40% compared with 10% of the 'normal' phenotype (Figure 11.4c).

It has been shown that diploid *S. cerevisiae* strains undergo a dimorphic transition that involves changes in cell shape and the pattern of cell division and results in invasive filamentous growth in response to starvation for nitrogen (Gimeno *et al.*, 1992). Several genes related to nitrogen starvation either directly or indirectly have been implied in such morphological transition. Impaired ammonia uptake, like amino acid uptake, related to *shr3* mutations induces more extensive and exaggerated pseudohyphal growth than wild type on low ammonia medium. Constitutive activation of the RAS signal transduction pathway and consequent elevated intracellular cAMP, as in strains carrying the dominant *RAS2^{val19}* mutation, show greatly enhanced pseudohyphal growth (Gimeno *et al.*, 1992). These strains with activated RAS/cAMP pathway are very sensitive to nitrogen starvation. Apparently, the mating type locus controls the dimorphic transition since only *MATa/MATα* diploids and neither *MATa* nor *MATα* haploids nor *MATa/a* or *MATα/α* diploids show pseudohyphal growth (Gimeno *et al.*, 1992).

The ability of certain yeast strains to grow on methanol as the sole source of carbon and energy is accompanied by the induction of numerous peroxisomes (Figure 11.5) (Roggenkamp *et al.*, 1989). Under fully induced conditions, more than 10 peroxisomes per yeast cell can be found. Large cuboid, completely crystalline organelles were only observed in cells grown in methanol-limited chemostat cultures at low dilution rates and are due to the synthesis (and crystallization) of excessive amounts of alcohol oxidase protein. In cells from the exponential phase in batch cultures, rounded organelles are predominant (Veenhuis *et al.*, 1992). The yeast *Hansenula polymorpha*, when transformed with a high copy-number vector harbouring the cloned methanol oxidase (*MOX*) gene, showed no detectable increase in the expression of *MOX*. However, two structural *MOX* mutants allowed overproduction of a fully active enzyme upon transformation at quantities of about two thirds of

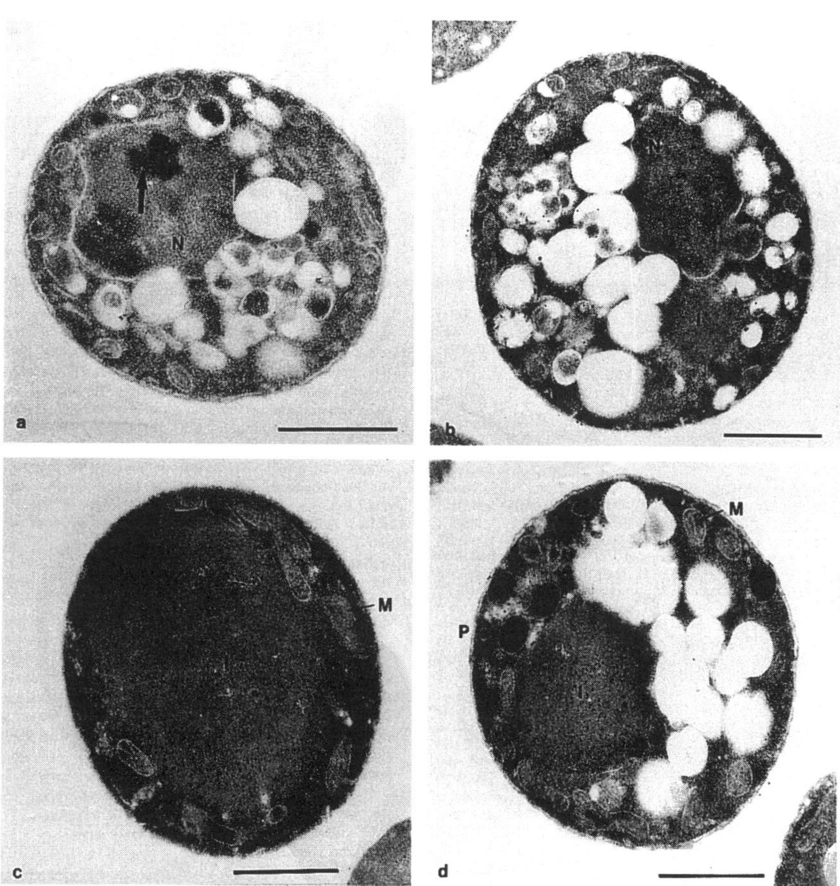

Figure 11.3 Overexpression of catalase A in the yeast *Saccharomyces cerevisiae* induces inclusion body formation: electron micrographs of inclusion bodies and subcellular localization of catalase A. Inclusion bodies and peroxisomes are labelled when reacted with anti-catalase antibody and protein A-gold (a–d) and represent amorphous, homologous or heterologous protein aggregates which are morphologically identifiable in the cytoplasm of the host cell. Note labelled inclusion body within a nucleus (arrow, a). I, inclusion body; N, nuclei; M, mitochondria; P, peroxisome. Bars = 1 μm. (Reproduced from Binder, Schanz and Hartig, 1991, by permission of Wissenschaftliche Verlagsgesellschaft mbH, Stuttgart.)

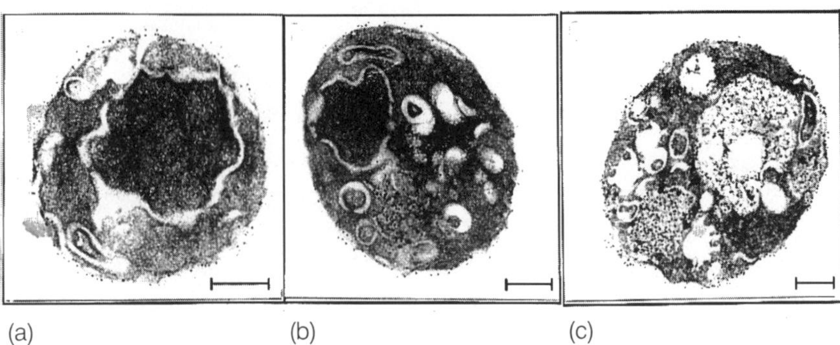

(a) (b) (c)

Figure 11.4 Proliferation of intracellular structures upon overexpression of the PMA2 ATPase in *S. cerevisiae*: electron micrographs of yeast cells expressing (a) PMA1 or (b) PMA2 ATPase. PMA ATPases tagged with antibodies raised against the cytoplasmic domains of the PMA1 protein were gold-labelled. Cells expressing the PMA1 ATPase are located, as expected, in the plasma membrane (a), whereas about 10% of the sections of those cells expressing PMA2 ATPase display a proliferation of intracellular membranes which are extensively gold-labelled (b). The proportion of sections exhibiting the labelling pattern of (b) increases dramatically to about 40% when PMA2 ATPase is overexpressed in a multicopy vector (c). (From Supply *et al.*, 1993, by permission of The American Society for Biochemistry and Molecular Biology.)

the total cellular protein. The overproduced protein was imported into peroxisomes, altering their morphology and stability in cell lysates (Roggenkamp *et al.*, 1989). Peroxisomes showed a drastic increase in size and a rectangular appearance instead of round or elliptic forms (Figure 11.5a,b). A large number of mutants in peroxisomal assembly (pas-mutants) and/or peroxisomal functions (fox-mutants) have been isolated (for a review, see Kunau and Hartig, 1992). Fox-mutants are defective in structural genes of β-oxidation enzymes and possess morphologically normal peroxisomes. The pas-mutants isolated are distributed among 13 complementation groups and represent three different classes. Peroxisomes are either morphologically not detectable (type I) or present but non-proliferating (type II), both types with mislocalized peroxisomal proteins. The third class of pas-mutants contains peroxisomes that are normal in size and number along with distinct mislocalized peroxisomal proteins (Kunau and Hartig, 1992). Transcriptional regulation of four genes encoding peroxisomal proteins (*PAS1, CTA1, FOX2, FOX3*) is regulated similarly as the morphology of peroxisomes: repressed by glucose and induced by oleic acid (Kunau and Hartig, 1992).

Growth of *S. cerevisiae* by budding involves a number of rearrangements of the cytoskeleton and the secretory apparatus of the

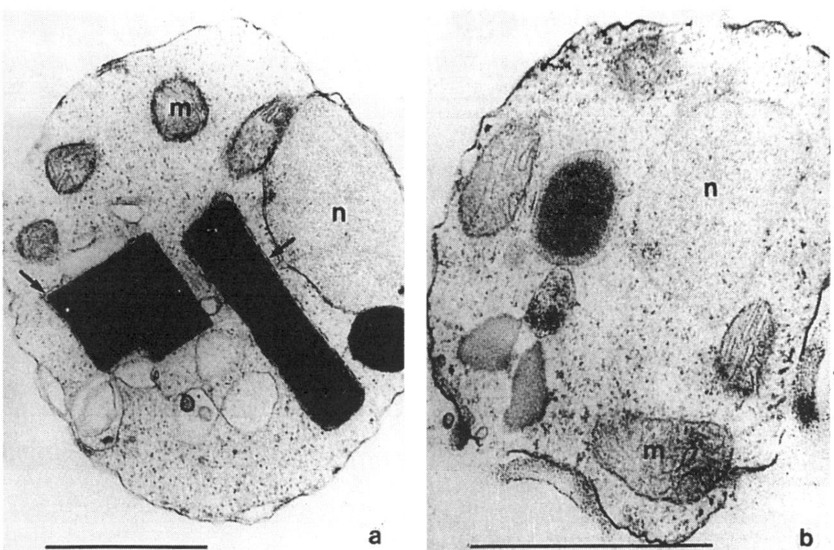

Figure 11.5 Formation of giant peroxisomes by overproduction of the crystalloid core protein methanol oxidase (MOX) in *Hansenula polymorpha*: electron micrograph of (b) a regular peroxisome in non-transformed cells (strain MCT-75) and (a) the methylotrophic yeast *H. polymorpha* transformant Q3N overproducing MOX. Yeast cells were grown under derepressed conditions with 3% glycerol as the carbon source to stationary growth phase (45 h). Arrows point to the peroxisomal membrane in transformed cells. Bars = 1 μm; m, mitochondria; n, nucleus; p, peroxisome; v, vacuole. (Reproduced from Roggenkamp, Didion and Kowallik, 1989, by permission of the American Society for Microbiology, ASM Press.)

cell during the proliferative cycle. The shape of the bud is determined by the balance between apical growth (directed towards the tip of the bud) and isotropic bud growth during this period (Lew and Reed, 1993). A protein kinase, cdc28, undergoes changes in activity through the cell cycle by associating with distinct groups of cyclins. These various cyclin/cdc28 complexes participate in the regulation of cell cycle progression, including the commitment step known as 'Start' (Hartwell *et al.*, 1974; Lew and Reed, 1993). Inactivation of cyclin/cdc28 kinase by three different methods, involving different temperature or nutritional shifts, all hyperpolarized the cortical actin distribution to the tip of the bud (Lew and Reed, 1993). Overexpression of cyclins 1 or 2 accelerated the transition to depolarized growth, with a high proportion of cells displaying cortical actin patches spread evenly over the entire bud surface and with spherical buds rather than oval. Cyclin 3 overexpression had an opposite effect: the cells retained an apical secretion pattern for longer than the wild type cells and showed elongated buds and cortical actin patches at the tip of the bud.

11.5 ADDITIONAL EXPERIMENTAL SUPPORT OF THE FLUX COORDINATION HYPOTHESIS

11.5.1 METABOLIC UNCOUPLING AND FLUX REDIRECTION ASSOCIATED WITH CELL CYCLE ARREST OF TEMPERATURE-SENSITIVE CELL DIVISION CYCLE (CDC) MUTANTS OF *S. CEREVISIAE*

The hypothesis of a coordinated pattern of fluxes involved in a cell's commitment to divide has been experimentally tested in temperature-sensitive cell division cycle (cdc) mutants of *S. cerevisiae*. The bioenergetic status associated with cell proliferation arrest in cdc mutants after the shift to the restrictive temperature was analysed. We hypothesized that in *S. cerevisiae* the uncoupling between anabolic and catabolic fluxes may be associated with the impairment of growth (macromolecular synthesis) and cell cycle arrest.

The main experimental results have been obtained with several cdc mutants of *S. cerevisiae* in comparison with the wild type strain (WT) A364A for cdc mutations. Cdc mutants arrest division at a unique stage of the cell cycle regardless of the stage they were at when the temperature shifted (Hartwell, 1991). We used several cdc mutants which blocked at different stages of the cell cycle, i.e. G_1 (*cdc28*, *cdc35*, *cdc19*), S (*cdc21*) and M (*cdc17*). These cell cycle mutants of *S. cerevisiae* were obtained by Hartwell (1967) from the parental haploid A364A strain by mutagenesis with N-methyl-N'-nitro-N-nitrosoguanidine (Hartwell *et al.*, 1970).

We comparatively measured catabolic and anabolic fluxes and carbon recovery in strains WT A364A and several cdc mutants of *S. cerevisiae* in asynchronous batch cultures subjected to transitions to the restrictive temperature. The results in Table 11.3 reveal metabolic uncoupling in cdc mutants, i.e. most of the substrate consumed was not directed to biomass synthesis. Y_{glc} measures the amount of biomass synthesized with respect to the substrate consumed over a time lapse of 6–8 h after the shift to the restrictive temperature. Table 11.3 shows the Y_{glc} measured at the restrictive temperature (37°C) for the WT A364A strain and the cdc mutants. The results obtained indicate that cdc mutants synthesized half or less of the biomass produced by the WT strain. Only 12% and 8% of the carbon consumed was recuperated as cells by cdc17 and cdc21, respectively (Table 11.3).

Figure 11.6 is intended to clarify further the meaning of carbon and energy coupling indices according to results obtained with cdc28 and the WT strain in the presence or absence of 8-hydroxyquinoline (8-HQ), an

Table 11.3 Bioenergetic characterization of temperature-sensitive cell division cycle mutants of *Saccharomyces cerevisiae* (reproduced from Aon, Mónaco and Cortassa, 1995, by permission of Academic Press)

Strain	Terminal phenotype	cdc gene product	References	Y_{glc} (g mol^{-1})	RQ	Temperature (°C)	Coupling indices in glucose	
							J_r^c	J_r^{ATP}
WT A364A	Willd type strain[g] with respect to cdc mutations		Hartwell, 1967, 1973; Hartwell *et al.*, 1973	72 ± 5[a]	1.02	25 37	0.5 0.48	0.28 0.34
cdc28-1 (G1)	Uninucleate, unbudded cells that grow into shmoos (EX.P. = 0.45)	Protein kinase	Pringle and Hartwell, 1981; Reed *et al.*, 1985	40 ± 3[b]	7.4	25 37	0.11 0.032	0.14 0.083
cdc35-1 (G1)	Uninucleate, unbudded cells which lack SPB satellite (EX.P. = 0.29)	Adenylate cyclase	Casperon *et al.*, 1985; Harold, 1990	26 ± 2[c]	3.4	25 37	0.2 0.036	0.34 0.21
cdc19-1 (G1)	Unbudded, uninucleated cells (EX.P. = 0.44)	Pyruvate kinase	Pringle and Hartwell, 1981; Gillies and Benoit, 1983	15 ± 2[d]	3.1	25 37	0.206 0.028	0.4 0.18
cdc21-1 (S)	Singly budded, uninucleated cells (EX.P = 0.67)	Thymidylate synthase	Pringle and Hartwell, 1981; Harold, 1990	8 ± 1.5[e]	27[h]	25 37	0.24 0.034	0.48 0.24
cdc17-1 (M)	Budded, binucleated cells (EX.P = 0.87)	DNA polymerase I	Burke and Church, 1991; Carlson, 1987	12 ± 3[f]	2.8	25 37	0.15 0.032	0.28 0.17

The execution points (EX.P.) biomass growth yield in glucose (Y_{glc}) and respiratory quotient (RQ) correspond to those determined at 37°C.
J_r^c and J_r^{ATP} are the carbon and energy coupling indices which are defined by equations 11.7 and 11.8.
Standard deviation (SD): (a) n=11; (b) n=6; (c) n=3; (d) n=3; (e) n=2; (f) n=2.
(g) All cdc mutants are derived from strain A364A (Hartwell, 1967, 1973; Hartwell *et al.*, 1973). However, strain A364A is not the parental of several cdc mutants. For instance, cdc28-1 (185-3-4) and cdc21-1 (146-2-3) were derived from mutants 23019 and 17026 by outcrosses, respectively (Hartwell, 1973). These strains are not isogenic with their parental A364A since 23019 and 17026 were found to contain more than one mutation (Hartwell, 1973). According to tetrad analysis cdc17 (4028) harbours two independently segregating temperature mutations from which only one produced the terminal phenotype (Hartwell *et al.*, 1973).
(h) The abnormally high RQ of cdc21 is due to the rho⁻, petite, character of this mutant.

inhibitor of RNA synthesis (Aon *et al.*, 1995). Cdc28 exhibited both high energy and carbon uncoupling (J_r^{ATP} = 0.083 compared with 0.34 of the WT strain, and J_r^C = 0.032 compared with 0.48 exhibited by the WT strain), i.e. cdc28 utilized only 8.3% of the ATP flux (compared with 34% of WT) produced by catabolism for biosynthetic needs and directed only 3% of the carbon flux (compared with 48% of WT) to biomass synthesis (Figure 11.6). The growth arrest elicited by 8-HQ is consistent with carbon (J_r^C = 0.05) rather than energy (J_r^{ATP} = 0.27) uncoupling. In all cdc mutants tested, the carbon coupling index was more severely affected than the energy coupling index, which is a bioenergetic indication of a more pronounced carbon rather than energy uncoupling (Table 11.3).

Apparently, a shift to ethanolic fermentation occurred after the transition to 37°C which was revealed by high RQs (Table 11.3). Over the whole period at 37°C, cdc mutants diverted 50–60% of the carbon recovered to ethanol production and only 5–12% to cell biomass synthesis (Aon *et al.*, 1995). The abnormally large value of RQ observed in cdc21 was due to the rho⁻ character of this cdc strain. The WT strain exhibited purely respiratory glucose catabolism (RQ ≈ 1.0). It should be pointed out that the respiratory type of glucose breakdown shown by WT A364A is not the usual one for wild type strains of *S. cerevisiae* whose glucose catabolism, at high concentrations of this substrate, is mainly fermentative (Fraenkel, 1982).

According to genetic and molecular biology studies performed in the past 15 years, the cell cycles of yeast and *Xenopus* are thought to be driven by a 'clock' whose biochemical basis appears to be the periodic synthesis and degradation of cyclin coupled with activation and inactivation of the *CDC2pombe/CDC28cerevisiae* 34-kd serine/threonine protein kinase (p34[cdc2]) by dephosphorylation and phosphorylation mechanisms, respectively (Forsburg and Nurse, 1991; Hartwell, 1991; Kirschner, 1992). Since the *cdc28* gene product appears to act at the interface between signal transduction and G_1/S transition in budding yeast (Forsburg and Nurse, 1991), it could be that some connection exists between p34[cdc2] phosphorylation–dephosphorylation mechanism and the observed carbon and energy uncoupling, though at present we are unable to suggest its nature.

11.5.2 GLUCONEOGENIC SUBSTRATES INDUCE SUPPRESSION OF THE TERMINAL PHENOTYPE OF CDC28 AT THE RESTRICTIVE TEMPERATURE

In the previous section we showed that several cdc mutants exhibited carbon and energy uncoupling together with activation of the fermentative pathway associated with cell proliferation arrest independently of the stage of the cell cycle at which they were when

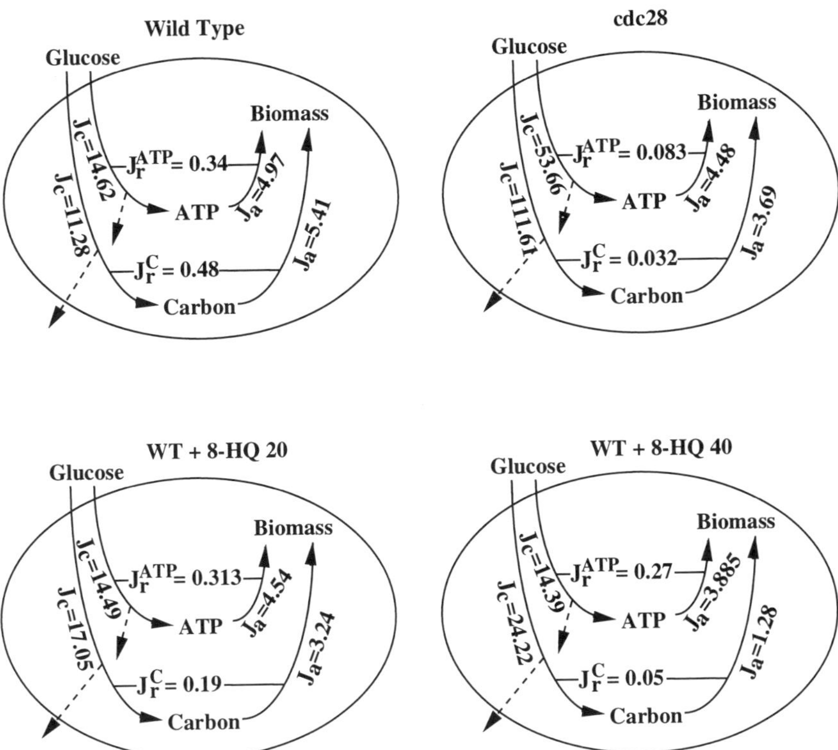

Figure 11.6 Energy and carbon uncoupling of the WT strain in the absence or presence of 20 or 40 μg ml^{-1} of 8-HQ, and cdc28 mutant. The anabolic (J_a), ATP and carbon fluxes were calculated from the ATP or carbon required to synthesize 1 g of biomass multiplied by the growth rate (μ) as described in Aon *et al.* (1995) and Cortassa *et al.* (1995). The catabolic (J_c) ATP was calculated as twice the q_{glc} plus 1.3-fold the q_{O_2} averaged over the time lapse from the temperature shift until 6 h later. J_r is the ratio of the indicated fluxes, providing a measure of the carbon and energy coupling index (see text). (Reprinted from Aon and Cortassa, 1995, with kind permission from Elsevier Science Ltd, The Boulevard, Langford Lane, Kidlington, OX5 1GB, UK.)

blocked by temperature (Table 11.3). Further investigation on this topic revealed that all cdc mutants utilized, in contrast with WT A364A, were repressed by glucose (Mónaco *et al.*, 1995). We then formulated the following question: can the terminal phenotype of cdc28 at the restrictive temperature be suppressed by effecting the temperature transition under derepressing conditions, i.e. in the presence of gluconeogenic substrates? We chose cdc28 because its terminal phenotype shows a clear and aberrant change in shape called 'shmoo' (Hartwell, 1974). We performed the shift to the restrictive temperature in the presence of 1% acetate (Figure 11.7, I and II) or 1% glycerol (Figure 11.7, III and IV). The 'shmoo' phenotype did not appear in either of the gluconeogenic substrates (Figure 11.7, II and IV). The terminal phenotype suppression by glycerol or acetate was reversible since cdc28 yeast cells grown in either glycerol or acetate did show the 'shmoo' at the restrictive temperature when readapted to glucose (Figure 11.7, V).

This result clearly shows an environmentally induced change in cell shape. By directly acting on metabolic pathways through induction of the cell to utilize different paths to those associated with cell proliferation arrest, i.e. fermentative, we were able to block the change in cell shape (shmoo).

Defective cdc gene products of the cell cycle mutants of *S. cerevisiae* are related to either sugar catabolism or anabolic fluxes. In fact, *cdc35* gene product (adenylate cyclase) has been involved in the general mechanism of glucose sensing by the yeast cell regulating the influx of sugar to the cell and the triggering of glucose-induced signalling pathways such as the RAS-adenylate cyclase pathway (Casperon *et al.*, 1985; Harold, 1990; Thevelein, 1992). In addition, it appears that the glucose uptake by the cell is the rate-controlling step of the glycolytic flux of *S. cerevisiae* growing aerobically in chemostat cultures subjected to carbon, phosphate or nitrogen limitation (Cortassa and Aon, 1993a, 1994a; Chapter 7). On the other hand, cdc mutants exhibiting primary defects which affect anabolism (blocking of nucleic acid synthesis) are cdc17 and cdc21, i.e. in DNA polymerase I (Carlson, 1987b; Burke and Church, 1991) and thymidylate synthase (Pringle and Hartwell, 1981; Harold, 1990), respectively.

These results indicate that glucose-induced catabolite repression may be associated with carbon and energy uncoupling along with the cell cycle arrest (Mónaco *et al.*, 1995). Carbon catabolite repression regulates several genes that code for enzymes of the Embden–Meyerhoff, citric acid cycle, gluconeogenesis and oxidative phosphorylation pathways, whose expression is controlled by several *cis*- or *trans*-acting regulatory genes (Gancedo and Gancedo, 1986; Entian, 1986; Gancedo, 1992). The results obtained indicate that invertase, respiratory enzymes and FBPase are catabolite repressed in cdc mutants while they are not in the wild type (for cdc mutations) A364A (Mónaco *et al.*, 1995). This is a remarkable result. We chose cdc mutants of *S. cerevisiae* whose defective genes at the restrictive temperature correspond to different enzymes and their characteristic execution points distributed at different stages of the cell cycle (Aon *et al.*, 1995). However, the cdc mutants used and WT A364A are not isogenic (Hartwell, 1973; Aon *et al.*, 1995). A simple interpretation might be that catabolite repression is associated with inactivation of the cdc gene products at 37°C. In fact, we observed that all cdc mutants with the exception of cdc21(rho⁻) and cdc19 (defective in PK at 37°C), were able to grow at 37°C when plated in the presence of glycerol. In a catabolite-derepressed yeast cell it appears that, even at the restrictive temperature, cdc gene products might still be active since cell division ·was not blocked under those conditions; yet a lower requirement of the cdc gene product in the presence of glycerol, which supports slower proliferation rates, cannot be ruled out (Mónaco *et al.*, 1995).

Taken together, the data presented suggest that those defective gene products, by impairing catabolic or anabolic fluxes, are involved in the carbon and energetic uncoupling associated with the yeast cell division arrest. The fact that, under catabolite derepressed conditions, yeast cells are carbon and energetically coupled may be taken as another indication of the association of impairment of yeast growth and proliferation with uncoupled catabolic and anabolic fluxes.

11.6 CONCLUDING REMARKS

The main aim of this chapter has been to show that flux coordination as an expression of the cell's dynamic organization is likely to be involved in the regulation of cell growth, division and sporulation in *S. cerevisiae* through the degree of coupling between carbon and energy fluxes (Aon and Cortassa, 1995). Morphogenetic events such as polar or isotropic growth of the bud in *S. cerevisiae* appear to be determined by the balance between macromolecular synthesis of the cell wall and its polarized or isotropic secretion to the bud surface (Lew and Reed, 1993, 1995). The polarization of cytoskeleton components (namely, actin and microtubules) along the mother–bud axis plays an evident role during yeast morphogenesis (Chapter 4). The fact that GTP or GDP-binding proteins able to interact with actin change their subcellular distribution according to the bound nucleotide (Chant, 1994) indicates that cellular morphogenesis may result from coordination between cellular energetics, metabolic fluxes and rearrangements of cytoskeleton components. Interestingly, metabolic fluxes appear to depend on the polymeric status, concentration and topological arrangement of actin and microtubular protein (Aon *et al.*, 1996a; Chapters 6 and 7). It is likely that other eukaryotic cells might be subjected to the same type of regulation. It has been shown that the Cdc28 protein kinase complex involved in yeast cell cycle regulation can exhibit different degrees of association and activity (Wittenberg *et al.*, 1987). On the other hand, the CDC28 kinase appears to

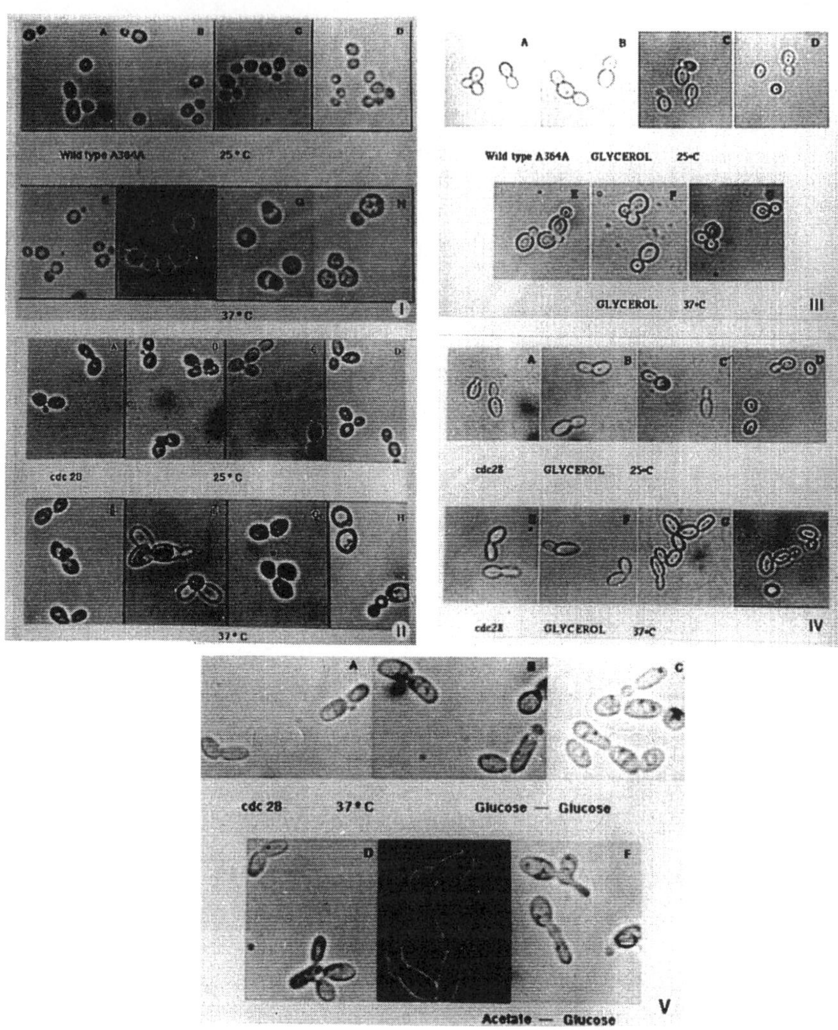

be, at least in part, tightly associated with the cytoplasmic matrix (Wittenberg *et al.*, 1987). When all these observations are taken together, it can be suggested that the activity of the Cdc28 protein kinase complex could be regulated by different cytoskeleton arrangements through the distinct extent of association of the complex, as was shown for pyruvate kinase (Chapters 6 and 7).

Figure 11.7 (see facing page) Metabolically induced suppression of the change in cell shape of the cdc28 temperature-sensitive mutant of *S. cerevisiae* with gluconeogenic substrates. Yeast cells were grown in YNB medium containing 1% acetate (I–II), glycerol (III–IV) or glucose (V) as carbon sources, as described before (Aon *et al.*, 1995; Mónaco *et al.*, 1995). Briefly, batch cultures of cells, previously adapted for 48 h to either substrate, were started at ~ 0.3 OD_{540} and grown at the permissive temperature (25°C) through two mass doublings before shifting to the restrictive temperature (37°C). I(A–D) correspond to cells kept at 25°C at time zero of the shift, 2 h, 6 h and 26 h later, respectively; I(E–H) 2 h, 6 h, 10 h and 26 h after the shift to 37°C, respectively. II(A–D) correspond to time zero of the shift, 2 h, 6 h and 10 h after the shift, respectively; II(E–H) 2 h, 6 h, 10 h and 26 h after the shift to 37°C, respectively. III(A–D) correspond to time zero of the shift, 2 h, 4 h and 6 h later, respectively; III(E–G) 2 h, 4 h and 6 h after the shift to 37°C, respectively. IV(A–D) correspond to time zero of the shift, 2 h, 4 h and 6 h after the shift, respectively; IV(E–H) 2 h, 4 h, 6 h and 26 h after the shift to 37°C, respectively. Whenever acetate or glycerol were used as carbon sources, experiments were performed in parallel in the presence of glucose with the strains under study. When cdc28 was grown in the presence of gluconeogenic carbon sources, after the experiment had finished, the mutant growing at the permissive temperature was readapted to glucose during 48 h. After readaptation, the cells were grown through two mass doublings at 25°C and shifted to 37°C again. V(A–C) correspond to 2 h, 6 h and 10 h after the shift, respectively, of cells always grown in the presence of glucose; V(D–F) 2 h, 6 h and 10 h after the shift to 37°C, respectively, of cells grown in the presence of acetate and readapted afterwards to glucose. The identity of the strain was checked through cell morphology, i.e. the appearance of shmoos, and physiological variables such as biomass growth yields, substrate consumption and products excreted. The terminal phenotype of cdc mutants, budding index, nuclear morphology, amount of biomass, cell growth and cell volume were quantitated as previously described (Aon *et al.*, 1995; Mónaco *et al.*, 1995). In all experiments, the WT A364A strain was cultured in parallel to cdc28. All the experiments were performed three times for independent cultures. (Reprinted from Aon and Cortassa, 1995, with kind permission from Elsevier Science Ltd, The Boulevard, Langford Lane, Kidlington OX5 1BG, UK.)

Through bypassing the fermentative pathway with gluconeogenic substrates (glycerol or acetate), we have been able to suppress in a reversible way the change in cell shape shown by the temperature-sensitive cdc28 mutant of *S. cerevisiae* at the restrictive temperature. The fermentative pathway was shown to be associated with carbon and energy uncoupling and cell cycle arrest in cdc28 as well as in cdc35, cdc19, cdc17 and cdc21 mutants.

Goodwin (1986) suggested that the explanation of cell form and transformation is not in the genetic programme but in the dynamic order of the whole process. We have shown that this may be a fair presumption for prokaryotic or eukaryotic microorganisms.

APPENDIX 11 THE CONTROL OF THE FLUX

As a way of characterizing the interaction between genotype and environment, we attempted to find out how the control structure depends on genetic or environmental features. We did so through an analysis of the control coefficients for pathways of known structure (e.g. linear, branched or circular) and whose enzymes have defined kinetic properties (e.g. Michaelian-type, product-inhibited, allosteric). Based on the matrix method with the simplest pathway possible for each topology, we derived expressions for the flux control coefficients.

Given the linear topology:
the matrix of elasticity coefficients has the form:

$$S \xrightarrow{v_1} A \xrightarrow{v_2} P \tag{11.13}$$

$$\begin{bmatrix} 1 & 1 \\ \varepsilon_A^{v_1} & \varepsilon_A^{v_2} \end{bmatrix} \tag{11.14}$$

where both rate expressions depend proportionally on their substrate concentrations:

$$v_1 = k_1 S; \quad v_2 = k_2 A \tag{11.15}$$

Both steps being irreversible and without product feedback, the expressions of the flux and concentration control coefficients are:

$$C_{v_1}^J = 1; \quad C_{v_2}^J = 0; \quad C_{v_1}^A = -1; \quad C_{v_2}^A = 1 \tag{11.16}$$

If both enzymes exhibit Michaelian kinetic behaviour instead of linear kinetics, the expressions for the flux and metabolite control coefficients are as follows:

$$C_{v_1}^J = 1; \quad C_{v_2}^J = 0; \quad C_{v_1}^A = \frac{-K_A + A}{K_A}; \quad C_{v_2}^A = \frac{K_A - A}{K_A} \tag{11.17}$$

If both enzymes exhibit Michaelian-like kinetics with an inhibition of v_1 due to an excess of A and of v_2 by an excess of P, the expressions for the flux control coeffients are:

$$C_{v_1}^J = \cfrac{1}{\left(1 + \cfrac{K_S A}{(K_S K_A + K_A S + K_S A)}\right)} \tag{11.18}$$

$$C_{v_2}^J = \cfrac{1}{\left(1 + \cfrac{(K_A K_S + K_A S + K_S A)}{K_S A}\left(1 - \cfrac{A}{K_P K_A + K_P A + K_A P}\right)\right)} \tag{11.19}$$

On the other hand, when we consider that the first step is reversible, ruled by:

$$v_1 = \frac{k_1\dfrac{S}{K_S} - k_3\dfrac{A}{K_A}}{K_A K_S + K_A S + K_S A}; \quad v_2 = \frac{k_2 A}{K_A + A} \tag{11.20}$$

the flux control coefficients are as follows:

$$C^J_{v_1} = \frac{K_A}{\dfrac{K_S A(K_A + A)}{K_A K_S + K_A S + K_S A} + \dfrac{K_A(k_1 S K_A) + A(k_3 A K_S)}{k_1 S K_A - k_3 A K_S}} \tag{11.21}$$

$$C^J_{v_2} = \frac{k_3 K_S + k_3 S + k_1 S}{\left(k_1 S K_A - k_3 A K_S + \dfrac{(K_A K_S + K_A S + K_S A)(K_A^2 k_1 S + K_S k_3 A^2)}{K_S A(K_A + A)} \right)} \tag{11.22}$$

The next metabolic topology modelled was a branched pathway of the form:

$$\begin{array}{c}
 \overset{\textstyle v_2}{\nearrow} \; P \\[-2pt]
S \xrightarrow{\;\;v_1\;\;} A \\[-2pt]
 \underset{\textstyle v_3}{\searrow} \; Q
\end{array} \tag{11.23}$$

ruled by the following rate laws:

$$v_1 = \frac{k_1 S}{K_S + S}; \quad v_2 = \frac{k_2 A}{K_2^A + A}; \quad v_3 = \frac{k_3 A}{K_3^A + A} \tag{11.24}$$

The matrix of elasticity coefficients has the following form:

$$\begin{bmatrix} 1 & 1 & 1 \\ \varepsilon_A^{v_1} & \varepsilon_A^{v_2} & \varepsilon_A^{v_3} \\ 0 & -v_3^r & v_2^r \end{bmatrix} \tag{11.25}$$

where v_3^r and v_2^r stand for the relative rate of v_2 and v_3 with respect to the total flux v_1, respectively.

$$C^J_{v_1} = 1 \qquad C^J_{v_2} = 0 \qquad C^J_{v_3} = 0 \tag{11.26}$$

On the other hand, the flux ratio control coefficients depend explicitly on rate constants, namely k_2 and k_3:

$$C^{J_r}_{v_1} = \frac{-A(K_2^A - K_3^A)(k_2 K_3^A + k_2 A + k_3 K_2^A + k_3 A)}{\{k_2 K_2^A(K_3^A)^2 + 2k_2 K_2^A K_3^A A + k_2 K_2^A A^2 + k_3 K_3^A(K_2^A)^2 + \ldots} \tag{11.27}$$

$$\ldots + 2k_3 K_2^A K_3^A A + k_3 K_3^A A^2\}$$

$$C^{Jr}_{v2} = \frac{-K^A_3(K^A_2 + A)(k_2 K^A_3 + k_2 A + k_3 K^A_2 + k_3 A)}{\{k_2 K^A_2 (K^A_3)^2 + 2k_2 K^A_2 K^A_3 A + k_2 K^A_2 A^2 + k_3 K^A_3 (K^A_2)^2 + \dots}$$
$$\overline{\qquad\qquad\qquad \dots + 2k_3 K^A_2 K^A_3 A + k_3 K^A_3 A^2\}} \qquad (11.28)$$

$$C^{Jr}_{v3} = \frac{K^A_2(K^A_3 + A)(k_2 K^A_3 + k_2 A + k_3 K^A_2 + k_3 A)}{\{k_2 K^A_2 (K^A_3)^2 + 2k_2 K^A_2 K^A_3 A + k_2 K^A_2 A^2 + k_3 K^A_3 (K^A_2)^2 + \dots}$$
$$\overline{\qquad\qquad\qquad \dots + 2k_3 K^A_2 K^A_3 A + k_3 K^A_3 A^2\}} \qquad (11.29)$$

Finally we considered a cyclic pathway of the form:

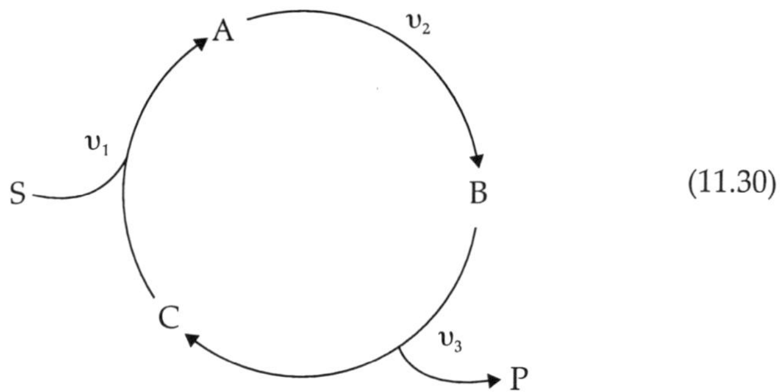

$$(11.30)$$

ruled by the following rate equations:

$$v_1 = \frac{k_1 \dfrac{S}{K_S}\dfrac{C}{K_C}}{K_S K_C + K_S C + K_C S}; \quad v_2 = \frac{k_2 A}{K_A + A}; \quad v_3 = \frac{k_3 B}{K_B + B} \qquad (11.31)$$

The matrix of elasticity coefficients exhibits the following form:

$$\begin{bmatrix} 1 & 1 & 1 \\ \varepsilon^{v1}_C - \varepsilon^{v1}_A \dfrac{C}{A} & \varepsilon^{v2}_C & \varepsilon^{v3}_C - \varepsilon^{v3}_B \dfrac{C}{B} \\ -\varepsilon^{v1}_A & \varepsilon^{v2}_A - \varepsilon^{v2}_B \dfrac{A}{B} & \varepsilon^{v3}_A - \varepsilon^{v3}_C \dfrac{A}{C} \end{bmatrix} \qquad (11.32)$$

The flux control coefficients resulting from the inversion of matrix 11.32 do not explicitly depend on the rate constants of any step:

$$C^J_{v1} = \frac{K_B C(K_S K_C + K_S C + K_C S)}{\{K_B K_C K_S C + K_B K_S C^2 + K_B K_C CS + K_B K_C K_S B + \dots}$$
$$\overline{\qquad\qquad \dots + K_C K_S B^2 + K_B K_C BS + K_C B^2 S\}} \qquad (11.33)$$

$$C_{v_2}^J = 0 \tag{11.34}$$

$$C_{v_3}^J = \frac{(K_B + B)(K_S + S)K_C B}{\{K_B K_C K_S C + K_B K_S C^2 + K_B K_C CS + K_B K_C K_S B + \dots \atop \dots + K_C K_S B^2 + K_B K_C BS + K_C B^2 S\}} \tag{11.35}$$

Dynamic coupling and spatio-temporal coherence in cellular systems

12

The basic idea of the concept of dynamic organization developed in Chapter 2 is that function, understood as the spatio-temporal coherence of events, results from the intrinsic dynamics of the processes taking place in living systems. This intrinsic dynamics may be either autonomous or forced by environmental (extra- or intracellular) perturbations. Dynamic organization results from the evolution of the spatio-temporal organization of biological processes between successive levels of organization. The temporal evolution of a biological process into successive dynamic regimes is achieved through instabilities (bifurcations) which give rise to the emergence of collective behaviour of, for example, supramolecular structures, enzyme activity, cells and spatio-temporal coherence.

We stress that cell function might arise from self-organized transitions between levels of organization (see Chapters 2, 6 and 11). This idea was substantiated in an allometric interpretation of the relationship existing between the characteristic space dimension, E_c, and relaxation time, T_r. An important feature of the concept of dynamic organization is the fact that time is not explicitly taken as an independent variable but through the dynamic information at a given level of organization, contained implicitly in the E_c and T_r of cellular processes of different natures: electrical (e.g. electronic transport, ionic fluxes, build up of ionic gradients, potential-gated ionic channels), mechanical (e.g. cell elongation, cell adhesion, cell wall deformation, turgor pressure), chemical (e.g. macromolecules synthesis, enzyme concentration changes, enzymatic catalysis) or mechano-electrical (e.g. stretch-potential activated channels) (Aon and Cortassa, 1989, 1993).

Mechanisms describe how and where the transitions between levels of organization may arise in the parametric space. For the latter it becomes crucial to know how biological processes of different natures are coupled to each other.

The present chapter attempts to show some examples of the enormous diversity of mechanisms through which the main physiological processes of cells may achieve coupling.

12.1 DYNAMIC COUPLING OF (SUB)CELLULAR PROCESSES

The coupling between cytoplasmic and membrane energetics as well as the mutual interconversions of different free-energy intermediates (e.g. phosphorylation potential, electrochemical gradients, electrical fields) may be viewed as an important and unifying principle in cellular physiology valid for either plant, animal or bacterial cells (Hochachka, 1988; Aon and Cortassa, 1989).

A coupling between two processes is dynamic in the sense that one (or both) influences the steady state or the time-dependence of the other process coupled to it (Chapter 7). At supramolecular–subcellular levels we may have processes spatially localized, with similar or large differences in time relaxation (e.g. membrane potential and cellulose synthesis), or processes spatially delocalized with similar temporal relaxation (ATP synthesis, NADH reoxidation in mitochondria and glycolysis). It has been shown that the steady state value of a fast-relaxing process had dynamic consequences for the transient and steady state behaviour of slowly relaxing ones (Figure 13.3).

The nature of the common intermediate that couples two or more processes varies according to the processes considered. For instance, in microbial metabolism the energy supplied by catabolic substrates is translated into phosphorylation, redox or transmembrane (electro)chemical potentials that in turn drive energetically uphill reactions such as macromolecule synthesis or active or facilitated transmembrane movement of substrates and ions.

12.1.1 ENERGY COUPLING IN CHEMIOSMOTIC SYSTEMS

The general principle of energy coupling in chemiosmotic systems is based on the transmembrane circulation of protons (Figure 12.1) (Mitchell, 1961; Racker, 1976; Maloney, 1987; Skulachev, 1991; Harold, 1991) or sodium (for a review, see Skulachev, 1991, and references therein). In a chemiosmotic cycle, some energy-dissipating reaction drives an energy-consuming step (Figure 12.1). Depending on the way in which ion extrusion occurs, bacteria posses two main forms of transport:

primary and secondary (Konings, 1985; Maloney, 1987). In the former, ion movement is associated with chemical transformation (ATP hydrolysis) while in the latter no chemical transformation takes place and only spatial arrangements change, as reactants (products) distribute over the membrane (Figure 12.1) (Maloney, 1987; Harold, 1991). Examples of primary ion pumps in bacteria include ion-motive ATPases, electron transport chains and light-driven cation or anion pumps; instances of secondary ion-coupled carriers include symport (cotransport) and antiport (counter-transport or exchange diffusion) (Maloney, 1987).

To summarize Mitchell's initial hypothesis (Skulachev, 1991): in the membrane of a topologically closed vesicle there is an H^+ pump which

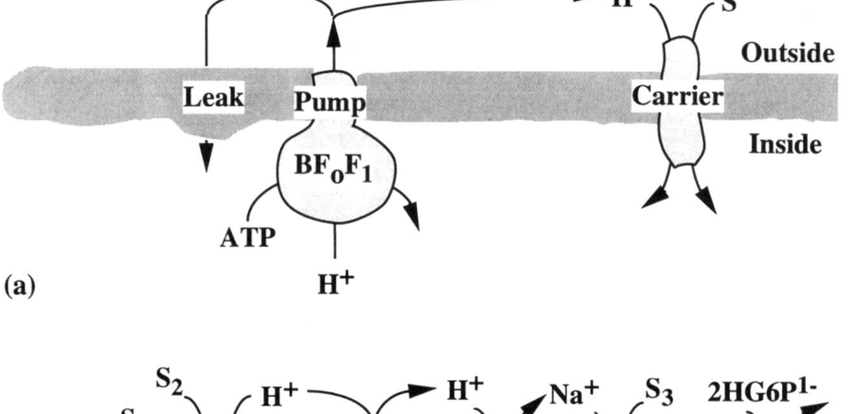

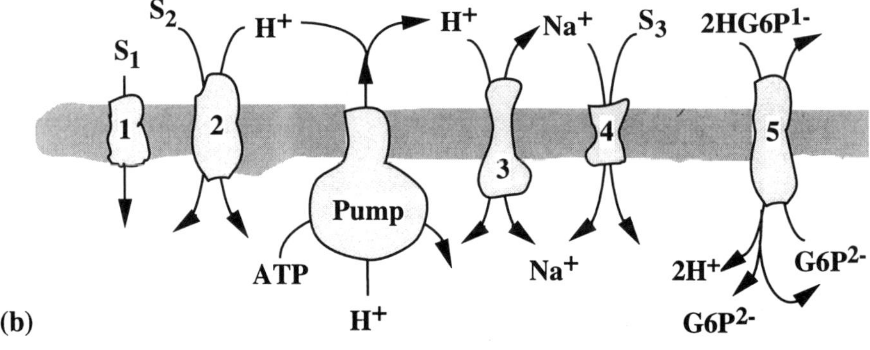

Figure 12.1 Chemiosmotic cycles. (a) H^+ extrusion by the bacterial H^+-translocating ATPase (BF_0F_1) establishes an electrochemical proton gradient. H^+ re-enters by either a non-specific (leak) pathway or a specific reaction in which re-entry is coupled to transport of solute, S. (b) An expanded chemiosmotic circuit: scheme (a) has been enlarged to indicate how subsidiary circulations might be linked indirectly to that of H^+. The chemiosmotic transporters are (1) uniport, (2) H^+/S_2 symport, (3) H^+:Na^+ antiport, (4) Na^+/S3 symport and (5) neutral anion exchange. G6P, glucose 6-phosphate. (Redrawn from Maloney, 1987, by permission of the American Society for Microbiology, ASM Press.)

uses an energy source to carry out the uphill translocation of H^+, and the $\Delta\mu_{H^+}$ formed is utilized by $\Delta\mu_{H^+}$ consumers to perform useful work.

The osmotic work of transporting solutes across membranes is driven by $\Delta\mu_{H^+}$, $\Delta\mu_{Na^+}$, ATP or phosphoenolpyruvate (PEP) energy as in the phosphotransferase system (PTS) (see below).

For a solute S, accumulated by a $\Delta\mu_{H^+}$ driven system in a negatively charged compartment, the accumulation ratio $[S]_{in}/[S]_0$ is described (Skulachev, 1991) by:

$$RT ln \frac{[S]_{in}}{[S]_0} = -(n + z)F\Delta\psi - nRT ln \frac{[H^+]_{in}}{[H^+]_0} \qquad (12.1)$$

where n is the number of H^+ ions symported with one S molecule, and z is the number of positive charges on S.

For a monovalent cation C^+ which is translocated electrophoretically by a uniporter, the accumulation ratio is described by:

$$RT ln \frac{[C^+]_{in}}{[C^+]_0} = -F\Delta\psi \qquad (12.2)$$

If ΔpH is the only driving force, equation 12.1 is simplified as follows:

$$\frac{[S]_{in}}{[S]_0} = \frac{[H^+]_{in}}{[H^+]_0} \qquad (12.3)$$

In acidophilic, neutrophilic and alkalophilic bacteria the different constituents of $\Delta\mu_{H^+}$ are (Skulachev, 1991):

$$\Delta\mu_{H^+} \approx RT ln \frac{[H^+]_0}{[H^+]_i} \qquad \text{(acidophiles)} \qquad (12.4)$$

$$\Delta\mu_{H^+} = F\Delta\psi + RT ln \frac{[H^+]_0}{[H^+]_i} \qquad \text{(neutrophiles)} \qquad (12.5)$$

$$\Delta\mu_{H^+} \approx 0 \qquad \text{(alkalophiles)} \qquad (12.6)$$

The alkaline pH of the medium creates kinetic and thermodynamic difficulties in the use of the H^+-cycle (Skulachev, 1991) which may be solved by substitution of Na^+ for H^+. Alkalophilic and alkalotolerant species require Na^+ to survive, as in marine bacteria (Skulachev, 1991). Na^+ extrusion forms $\Delta\psi$ in the right direction (that was a problem in microorganisms facing alkaline media because of the inversion of the pH gradient, i.e. higher inside) while downhill Na^+ influx uses $\Delta\mu_{Na^+}$, allowing chemical, osmotic or mechanical work in spite of the direction of the pH gradient.

It is known that the electron transfer systems and the ATPase complex are coupled via the proton motive force. When a bacterial cell has high concentrations of reducing equivalents and a low phosphate potential, electron transfer can generate a $\Delta\mu_{H^+}$ which subsequently can drive the synthesis of ATP. In this context, oxidation of an electron donor via the respiratory chain supplies the energy for the transport of solutes (Michels *et al.*, 1979; Konings, 1985). In the reversed situation, ATP hydrolysis can generate a $\Delta\mu_{H^+}$ which subsequently can drive the synthesis of reducing equivalents.

12.1.2 COUPLING BETWEEN CYTOPLASMIC-BASED ENERGETICS AND MEMBRANE-BASED TRANSPORT PROCESSES

The coupling of cytoplasmic ATP-synthetic processes (respiration or fermentation) to those requiring membrane-based ATP (e.g. ion and solute transport) are manifest by cells undergoing metabolic transitions. Physiologically, during metabolic aerobic–anaerobic transitions several energy-transducing events take place which implicate alternation between cytoplasmic energy intermediates (e.g. phosphorylation and redox potentials, DNA supercoiling) and membrane electrochemical, solute and osmotic gradients (Mitchell, 1961; Atkinson, 1977; Verdoni *et al.*, 1990; Drlica, 1990; Cortassa and Aon, 1993b).

The resistance of mammalian cells to hypoxia is greatly dependent on the extent of depression of ATP synthetic rates and of processes requiring membrane-based ATP. The usual metabolic arrest observed during hypoxic conditions is achieved by means of a reversed or negative Pasteur effect, i.e. reduced or unchanging glycolytic flux at reduced O_2 availability. Coupling of metabolic and membrane function is achievable in spite of the lower energy turnover rates by maintaining membranes of low permeability (Hochachka, 1988). When metabolic and membrane functions are uncoupled, the cells (tissues or organisms) necessarily become sensitive to hypoxia. Ion and electrical potentials typically cannot be sustained because of energy insufficiency and high membrane permeabilities; therefore, metabolic arrest and stabilized membrane functions are the most effective strategies for extending tolerance to hypoxia (Hochachka, 1988).

The activity of energy-consuming membrane processes will cause a dissipation of $\Delta\mu_{H^+}$; to counterbalance such a dissipation the proton pump should continuously extrude protons. Upon an aerobic–anaerobic transition of *Streptococcus lactis* cells, incubated at external pH ranging from 5 to 8, the most prominent effect was on the separate electrical and chemical gradients rather than on the electrochemical H^+ gradient itself

(Maloney, 1983) (Figure 12.2). A decrease of the pH gradient from 110 to 11 mV and a concomitant increase of the membrane potential (from 52 to 114 mV) were measured. These interconversions of the electrochemical gradients, provoked by the external pH alkalinization from around pH 5.5 to 7, were followed by a linear increase of the glycolytic rate (Figure 12.3). Correlatively, an increase in the phosphorylation potential occurred, at least in the intracellular pH range of 5.8 to 6.8 (Figure 12.4). When *S. lactis* cells were incubated at pH 7, addition of 1 to 10 μM carbonylcyanide-p-trifluoromethoxyphenyl-hydrazone (FCCP, a protonophore) gave a two- to three-fold stimulation of glycolytic rate as expected if pump currents were greater than leak (non-ATPase) currents (Maloney, 1983). The stimulation of glycolysis by FCCP is expected of cells possessing functional BF_0F_1. Because ATP levels did not change after FCCP treatment (Maloney, 1977), increased synthesis of ATP from glycolysis would be required to balance increased hydrolysis by BF_0F_1 as membrane permeability to protons becomes elevated in the presence of the H^+ conductor (Maloney, 1977). The proton extrusion activity in *S. faecalis* is dependent on cytoplasmic rather than medium pH and the activity declines as the cytoplasmic pH is raised beyond 7.7. It has been also observed that a decrease in the pH below 7.6 stimulates the H^+-ATPase biosynthesis (Kobayashi, 1985). When the pH of the medium was raised from 5.9 to 8.7, a degradation of the enzyme was verified.

These results clearly illustrate the interconvertibility between different free-energy intermediates associated with the coupling between membrane-based transport and free-energy conversion by cytoplasmic processes.

12.1.3 ENERGY RECYCLING AS A COUPLING MECHANISM BETWEEN METABOLIC END-PRODUCT EXCRETION AND MEMBRANE TRANSPORT

Metabolite gradients represent a form of energy that can be used in certain cases as metabolic energy. The efflux of end products of metabolism (organic acids) can occur via specific transport proteins in symport with protons, and this process can lead to the generation of a $\Delta\mu_{H^+}$ which subsequently can drive energy-requiring processes. The latter mechanism has been called energy recycling (Michels *et al.*, 1979; Konings, 1985). The production of metabolic energy by substrate level phosphorylation during the breakdown of the energy source and by the excretion of metabolic end-products can be calculated if the metabolic pathways and the H^+/end-product stoichiometries of the efflux process are known.

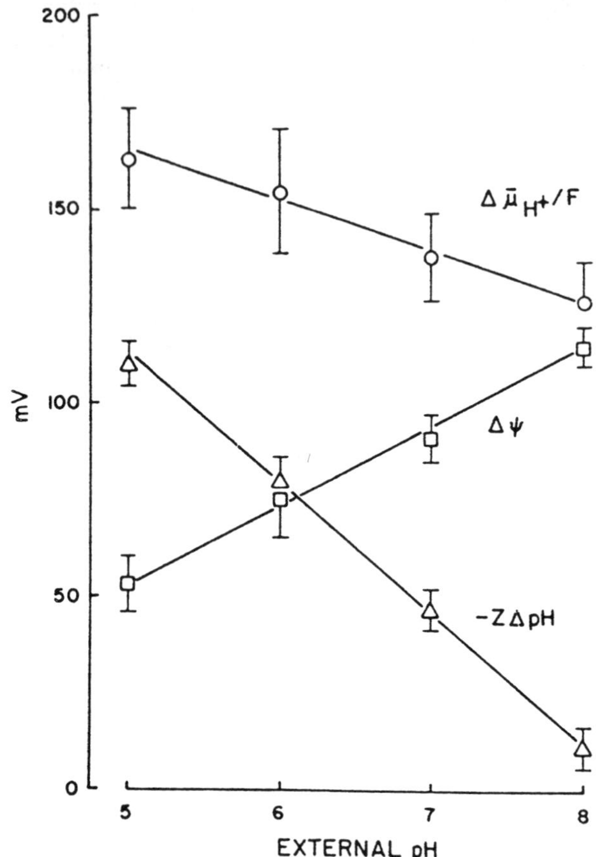

Figure 12.2 The electrochemical gradient composition of *Streptococcus lactis* during glycolysis. Transmembrane electrical, $\Delta\psi$, pH, ΔpH and the proton motive force, $\Delta\mu_{H^+}$, were estimated after 20–40 min of glycolysis at the indicated external pH. (Reproduced from Maloney, 1983, by permission of the American Society for Microbiology, ASM Press.)

The energy recycling model (ERM) initially postulated by Konings and colleagues proposed that the efflux or translocation of metabolites according to the electrochemical solute gradient ($\Delta\mu_{OA}$) can result in the generation of $\Delta\mu_{H^+}$. Based on the fact that proteins of secondary transport systems are symmetrical and can translocate solutes in both directions across the membrane, ERM describes the reverse process of solute uptake via secondary transport systems (Konings, 1985). The direction of transport will be determined by the direction of the driving force. Experimental evidence showed that energy recycling by metabolic

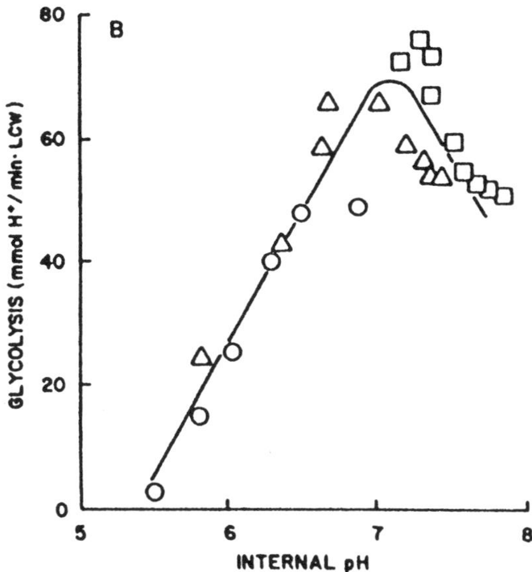

Figure 12.3 Glycolytic rate as a function of the intracellular pH. Electrical and chemical gradients were exchanged by treatment of S. *lactis* cells with 5 μM valinomycin after glycolysis for 20 min at external pH 5.0 (circles), pH 6.0 (triangles) or pH 7.0 (squares). LCW, litre of cell water. (Reproduced from Maloney, 1983, by permission of the American Society for Microbiology, ASM Press.)

end products could occur anaerobically (Verdoni *et al.*, 1990). An energy gain of at least 20–45% by lactate excretion was shown to occur in *Streptococcus cremoris* growing in batch culture (Ten Brink and Konings, 1982a,b). The main suggestion in energy recycling is that excretion of metabolic end-products (e.g. organic acids) in facultative anaerobes (*S. cremoris*) appears as an alternative mechanism, other than ATP hydrolysis, for creating an electrochemical gradient (Michels *et al.*, 1979; Ten Brink and Konings, 1982b). We have found energy recycling in *Pseudomonas mendocina* growing in complex media in the absence of exogenous electron acceptors (Verdoni *et al.*, 1990). The energy produced by glucose fermentation and organic acid excretion was calculated according to the general formula proposed in Ten Brink and Konings (1982) but adapted to *P. mendocina*, an organism utilizing the Entner–Doudoroff pathway and exhibiting a mixed fermentation under these extreme conditions:

$$\frac{\text{ATP equivalents produced}}{\text{moles of glucose consumed}} = \frac{2 + 3(n-1)}{p} \qquad (12.7)$$

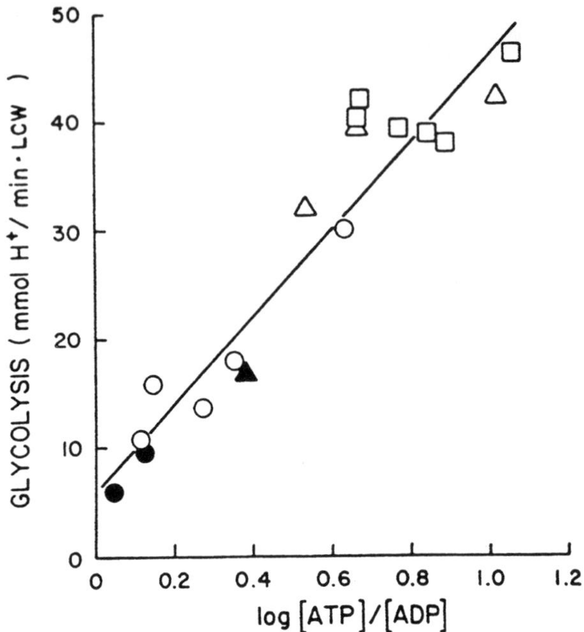

Figure 12.4 Correlation between the energetic status of S. *lactis* and the glycolytic rate. Nucleotide ratios and steady state glycolytic fluxes were determined for different phosphorylation potentials obtained by manipulation of the membrane electrochemical gradients. (Reproduced from Maloney, 1983, by permission of the American Society for Microbiology, ASM Press.)

where n represents the number of protons translocated with one molecule of either lactate, formate or acetate, and p represents the number of protons translocated per molecule of ATP synthesized. The organic acids are assumed to be independently transported since their individual gradients were added (Verdoni *et al.*, 1990):

$$n = \frac{(\Delta\psi - \Delta\mu_{OA})}{\Delta\mu_{H^+}} \tag{12.8}$$

When the Entner–Doudoroff pathway functions together with acetate production, two molecules of ATP per molecule of glucose are synthesized. Three molecules of organic acids comprising acetate, lactate and formate are assumed to be produced. Since one proton per molecule of lactate, formate or acetate is produced intracellularly and the three organic acids are monovalent and negatively charged, 3(n – 1) represents the number of translocated protons which contribute to the generation of the proton motive force.

According to Figure 12.5, there seems to have been an average energy gain of 1.5 (p = 2) to 1 (p = 3) mol of ATP equivalents per mole of glucose, reaching maximum values in the stationary phase of growth (Verdoni *et al.*, 1990). Since $\Delta\mu_{OA}$ and $\Delta\mu_{H^+}$ are opposing forces and there appears to be energy recycling, it may be stated that $\Delta\mu_{OA}$ is always more important than $\Delta\mu_{H^+}$ after the first day of culture according to equation 12.8. The very low Y_{glc} (4.26 g mol^{-1}) was in contrast with the Y_{ATP} calculated, as the ATP needed to produce 1 g of biomass was 469 mmol under anaerobic conditions. Considering Y_{ATP} values of 14 for anaerobic bacterial cultures growing in rich medium with a single carbon-substrate energy source, 71 mmol of ATP are required for the synthesis of 1 g of biomass. These observations suggest an uncoupling between anabolic and catabolic processes, i.e. the ATP produced in catabolism is not used for biomass production. Most of the ATP produced (85%) and the glucose consumed were not associated with growth.

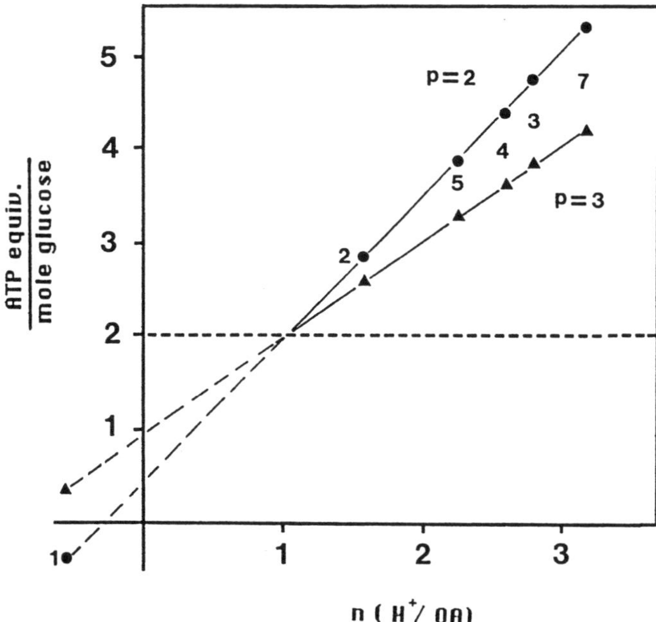

Figure 12.5 Energy recycling in anaerobic batch cultures of *Pseudomonas mendocina*. The stoichiometry n was calculated after experimental determinations of ΔpH, $\Delta\psi$ and $\Delta\mu_{H^+}$. The ATP equivalents produced per mole of glucose consumed were obtained as described in Verdoni *et al.* (1990). The number next to each point indicates the day of culture; p is the H$^+$/ATP stoichiometry. (Reproduced from Verdoni *et al.*, 1990, by permission of the American Society for Microbiology, ASM Press.)

We investigated whether the observed uncoupled metabolism was due to the activity of the membrane energy-consuming processes which, through the uncoupling effect of organic acids, will cause a $\Delta\mu_{H^+}$ dissipation. A coupling between electrochemical, $\Delta\mu_{H^+}$, and solute, $\Delta\mu_{OA}$, gradients should exist for continuous OA excretion to occur. The $\Delta\mu_H^+$ must be lower than the $\Delta\mu_{OA}$, or the $\Delta\mu_{H^+}$ must dissipate faster than the $\Delta\mu_{H^+}$ generated by H^+-OA efflux (Michels *et al.*, 1979). When the membrane ATPase is inhibited with DCCD a collapse of ΔpH occurs (Baronofsky *et al.*, 1984); then $\Delta\mu_{H^+}$ is lower, which favours a continuous efflux of end-products. A continuous ATP-fuelled H^+ extrusion by the proton pump will be operating to regulate and maintain $\Delta\mu_{H^+}$ and pH_i compatible with growth. Indirect evidence of the presence and the putative role of a proton pump in *P. mendocina* was provided by experiments with DCCD (Verdoni *et al.*, 1990). Control cultures of *P. mendocina* with respect to DCCD-treated ones (1 mM) consumed the same amount of glucose, though 30% more ATP was needed to synthesize 1 g of biomass. More direct evidence that the effect of DCCD on *P. mendocina* growth was ATPase-mediated came from a partial characterization of the enzyme activity. A membrane-bound ATPase activity was detected in particles obtained from *P. mendocina* sedimenting at 80 000 g and 150 000 g. The enzyme showed an apparent K_m of 0.08 mM for ATP and optimal activity at pH 8.0 (Verdoni *et al.*, 1990). According to the sensitivity to inhibitors, it was partially characterized as a E1–E2 type (P-ATPases) (Nelson and Taiz, 1989).

Evidence in favour of pH_i regulation not only being a function of H^+-ATPase inhibition, but also depending on passive membrane permeation or organic acid excretion, came from experiments with cultures run anaerobically in the presence of exogenous neutralized formic acid and DCCD (Verdoni *et al.*, 1990; Figure 12.6). Figure 12.7 summarizes the experimental data.

These results led us to suggest that energy recycling contributes to greater yields of at least 30% in anaerobiosis even under conditions of metabolic uncoupling. Therefore, according to the results obtained in batch cultures, the energy-recycling mechanism appears to be associated at least in part with growth.

The high cytoplasmic concentration of organic acids appeared as incompatible with the relative alkaline intracellular pH observed in *P. mendocina* (Verdoni *et al.*, 1990). An active (ATP-fuelled) proton extrusion by the H^+-ATPase or proton-driven active transport systems as for lactic acid export in streptococci to alkalinize the cell's interior (Otto *et al.*, 1980;

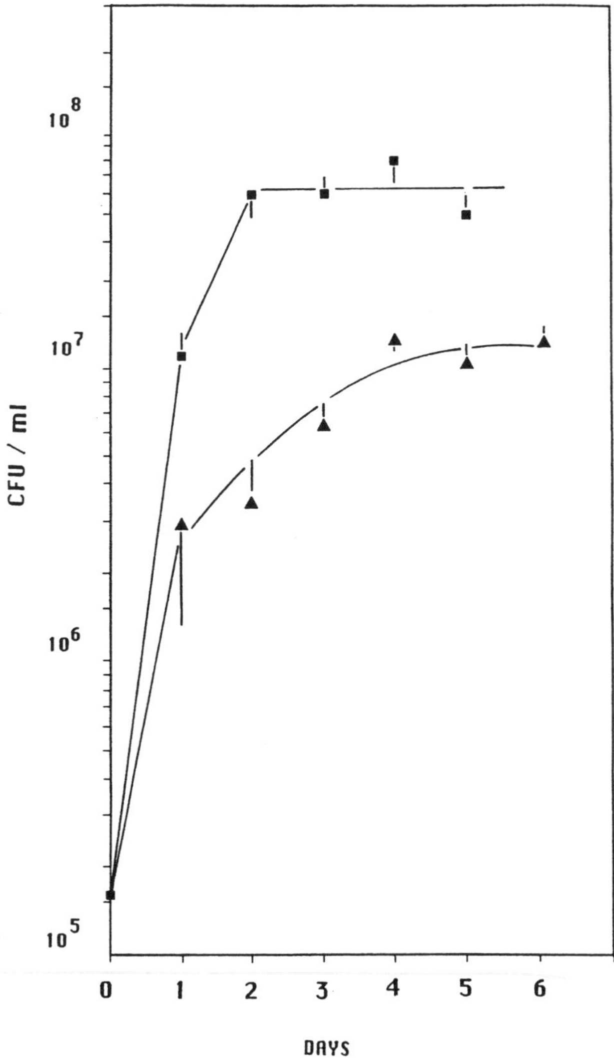

Figure 12.6 Growth uncoupling of *P. mendocina* by passive leak of excreted organic acids during anaerobiosis. *P. mendocina* was cultured at pH 7.1 in the presence (triangles) or absence (squares) of 50 mM formate – 5mM DCCD. (Reproduced from Verdoni *et al.*, 1990, by permission of the American Society for Microbiology, ASM Press.)

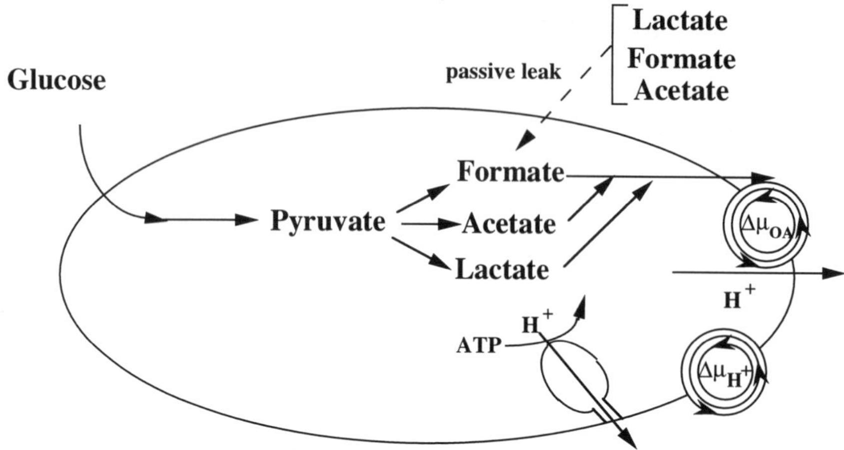

Figure 12.7 A schematic view of energy recycling in *P. mendocina*. The membrane ATPase (probably a E1–E2 type ATPase) uncouples bacterial growth from energy metabolism. The scheme proposes that the passive leak of organic acids (OA) (lactate, formate, acetate) expends the electrochemical gradient previously generated by the symport mechanism of OA/H$^+$. Experimental evidence presented in Figures 12.5 and 12.6 support this scheme.

Baronofsky *et al.*, 1984) may accommodate these observations. Anaerobic fermenters are known to tolerate fairly wide variations in internal proton concentration (Kashket, 1981). Another factor involved in pH control is the pH-dependence of the H$^+$-ATPase whose range of optimum pH is 6–8; an increase in cytoplasmic pH above 8 decreases proton pumping and less alkalinization of the cytoplasm takes place (Kashket, 1981; Kobayashi, 1985).

12.1.4 COUPLING BETWEEN GLYCOLYSIS AND THE PHOSPHOENOLPYRUVATE:CARBOHYDRATE PHOSPHOTRANSFERASE SYSTEM

In eubacteria the phosphotransferase system (PTS) imports extracellular glucose into the bacterium and phosphorylates it into glucose-6-phosphate (G6P) using the high-energy phosphate bond of phosphoenolpyruvate (PEP) (Postma, 1986; Postma *et al.*, 1993). In this way, two energetic problems are solved simultaneously: the uphill transport of sugar from the medium to the cytoplasm, and the initiation of glycolysis.

Regardless of the microorganism or carbohydrate, all PTSs that have been characterized catalyse the following overall process:

$$\text{P-enolpyruvate+carbohydrate}_{out} \xrightarrow{\text{PTS}} \text{pyruvate}_{in} + \text{carbohydrate} - \text{P}_{in}$$

The phosphoryl transfer potential of PEP is higher than that of ATP. PEP is nonetheless equivalent to ATP since, during glycolysis, one ATP molecule is derived from one PEP molecule in the pyruvate kinase reaction. The fact that more than one ATP equivalent must be expended per monosaccharide unit for both transport and subsequent ATP-dependent phosphorylation is an important reason why the PTS is found mainly in obligate and facultative anaerobic bacteria, which synthesize ATP by substrate-level phosphorylation under anaerobic conditions (Postma *et al.*, 1993).

The bacterial PTS system comprises two set of events (Postma, 1986):

- generation of phosphorylated phosphocarrier proteins, which takes place essentially in the cytoplasm;
- phosphorylation and translocation of the sugar substrates which utilize the membrane-bound enzyme II.

Enzyme I is a unique protein kinase that uses PEP to phosphorylate HPr and the fructose-induced HPr-like protein FPr. P-HPr will phosphorylate any constitutive or inducible IIIsugar proteins present in a cell extract. P-HPr interacts with a number of constitutive and inducible sugar-specific Enzyme IIs. The Enzyme IIs form phosphorylated intermediates, and the specific sugar is phosphorylated and translocated (Postma, 1986; Waygood, 1987).

We formulated a mathematical model of the PTS system coupled to glycolysis as shown in the scheme of Figure 12.8. A brief model description is given in Appendix 12. One relevant feature of this model is given by the branch point in which PEP goes further into glycolysis or triggers the phosphorylation–dephosphorylation cascade of the PTS proteins which lastly phosphorylate glucose while it is transported into the cell. In the model, the phosphorylation cascade is captured by the interconvertible cycle A↔A $\sim$P catalysed by enzymes E1 and E2 (Figure 12.8 and Appendix 12B).

An intriguing question addressed by the model concerned the regulatory properties of the branch point at PEP: how much of the PEP is derived downstream to glycolysis or to the PTS system? In other words, what is the optimum flux-partition regime physiologically allowed for these two competing pathways?

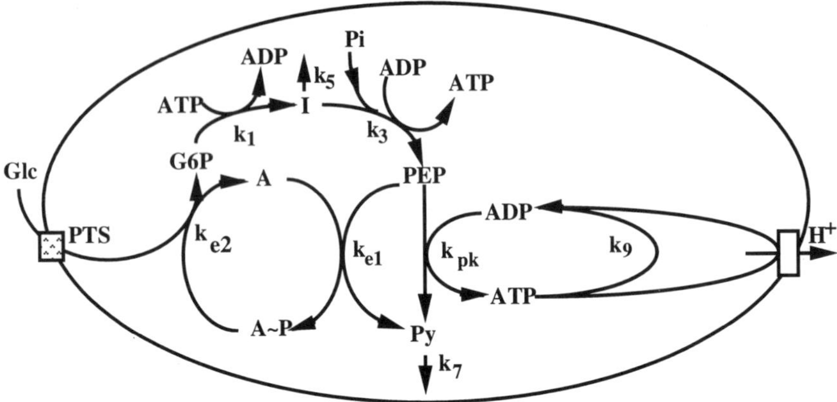

Figure 12.8 A scheme for the phosphotransferase (PTS) system of sugar transport. The circular topology of the glycolytic pathway with the PTS systems is put in evidence by the scheme. The phosphate group of PEP may flow in the branch to either downstream glycolysis or through the cascade of phosphorylation covalent modification (in the model the cascade is lumped in A–A~P) toward glucose (Glc) phosphorylation and uptake.

Figure 12.9 shows a systematic study of the steady state concentration values reached by the main state variables of the model, i.e. A, ATP and G6P, as a function of the logarithm of the ratio of the maximum velocities of the enzymes PK and E1 catalysing the branch (Appendix 12B). In the ratio of the rates of the enzymes catalysing the branch, one of the enzymes is allowed to vary whilst the other remains constant; for example, for the ratio k_{pk}/k_{e1}, k_{pk} varies for a fixed k_{e1}. Figure 12.9a,b shows the values for which the system behaves within a reasonable physiological range. In order to answer the question formulated, we have calculated partition coefficients which depict the ratio between the flux of PK (for γ_{pk}) or E1 (for γ_{e1}) over the sum of fluxes of both enzymes. This ratio informs a measure of the percentage of the flux processed by one or the other enzyme. It may be clearly seen that for the range in which the model behaves with acceptable ATP concentrations, the partition coefficient equals 0.5. In other words, an optimum flux partition of 50% at the PEP branch point is necessary to avoid an energetic collapse of the system (Figure 12.9a–c).

12.1.5 ENERGY-COUPLING MECHANISMS THROUGH MEMBRANE PROCESS DYNAMICS

During metabolic compromise elicited by, say, energy load, it has been shown that the genesis of arrhythmias may be produced by energy-driven oscillations in potassium currents. The oscillations in potassium currents produced cyclical changes in the cardiac action potential (O'Rourke *et al.*, 1994). The periodic changes in membrane ionic current observed in guinea-pig cardiomyocytes were linked to intrinsic oscillations of energy metabolism. By altering the rates of glycolysis or oxidative phosphorylation with exogenous substrates or metabolic inhibitors, a change in energy metabolism initiated the change in membrane conductance. Provision of glucose to cardiomyocytes displaying the oscillatory response interrupted the cycle of oscillations as well as the addition of the glucose analogue 2-deoxyglucose. As observed in yeast, an oscillatory pattern can be sustained in cardiac cells; an increase in flux above or a decrease below a crucial range of glycolytic rates 'pushes' the system out of the oscillatory domain (Aon *et al.*, 1991; O'Rourke *et al.*, 1994).

12.2 ELECTROCONFORMATIONAL COUPLING

Electromagnetism is one of the fundamental forces present in nature. The generation of electromagnetic fields occurs when moving charged particles (e.g. molecules) or structures. This is the conceptual basis of the electroconformational coupling which has been proposed as a means for a membrane-bound ATPase to transduce energy from dynamic electrical fields (Tsong and Astumian, 1988). Proteins of cell membranes are subjected to the influence of electrical fields which have magnitudes of $100–500\,kV\ cm^{-1}$. The rationale underlying the electroconformational coupling is that the energy of an electrical field can be transduced to perform chemical work, such as active transport or ATP synthesis. Coupling of an electrical field and a membrane protein may be a conformational one, i.e. electroconformational coupling (Tsong and Astumian, 1988).

Conformational equilibria of most biological molecules have different electrical properties and are thus capable of interacting with an electrical field. The shift in the chemical equilibrium constant (K) by an electrical field of strength E is given by the generalized thermodynamic relation expressed by the van't Hoff equation (Tsong and Astumian, 1988):

$$\left[\frac{\delta(ln\mathrm{K})}{\delta\mathrm{E}}\right]_{p\mathrm{VT}} = \frac{\Delta\mathrm{M}}{\mathrm{RT}} \tag{12.9}$$

where $\Delta\mathrm{M}$ is the difference of macroscopic electrical moment of two chemical species. Charged amino acids and prosthetic groups of proteins are targets of electrical field interaction. Each amino acid contributes with an electrical dipole of 3.5 debye (D) that do not cancel each other in a common conformational structure of the protein backbone; for example,

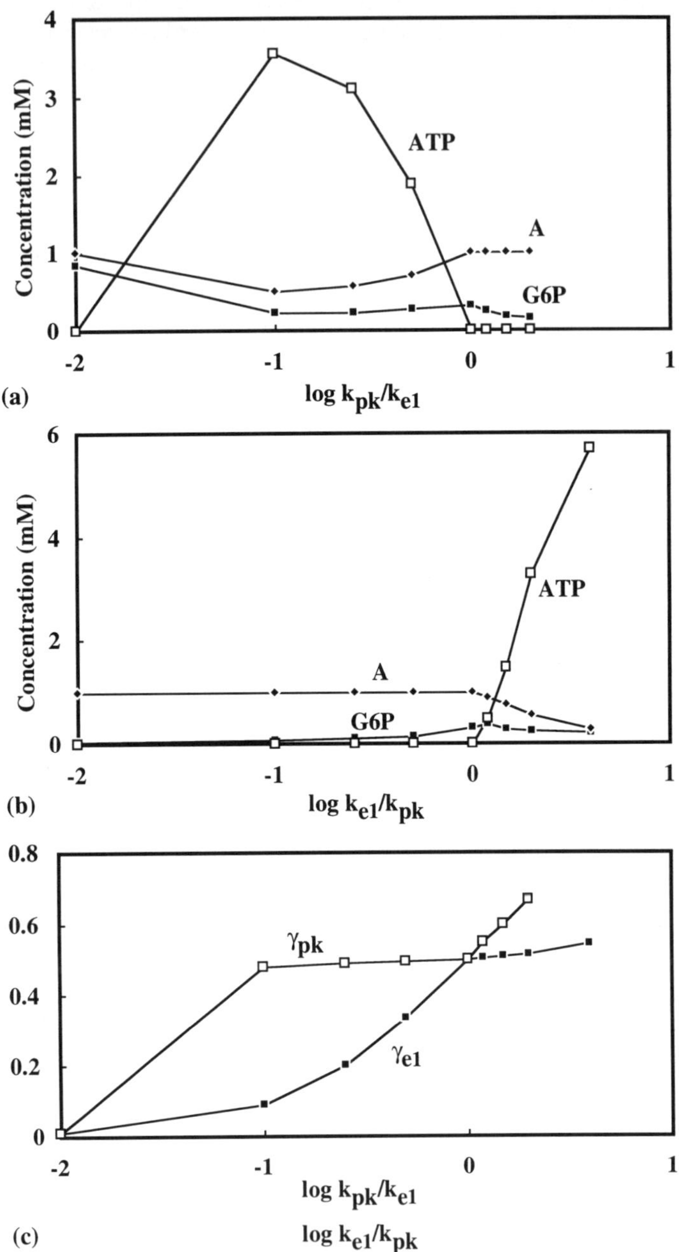
4
Concentration (mM)
3
2
1
0
ATP
A
G6P
-2
-1
0
1
log k_pk/k_el
(a)

6
Concentration (mM)
4
2
0
ATP
A
G6P
-2
-1
0
1
log k_el/k_pk
(b)

0.8
0.6
0.4
0.2
0
γ_pk
γ_el
-2
-1
0
1
log k_pk/k_el
log k_el/k_pk
(c)

a transmembrane helix 5 nm long is an electrical dipole equivalent to 120 D (Tsong and Astumian, 1988). It has been shown that membrane-bound enzymes may absorb energy from electrical fields in that way, transducing electrical energy into chemical work (Tsong and Astumian, 1988; Tsong, 1989). ATPases of chloroplasts, thermophilic bacteria and mitochondria have been induced to synthesize ATP from ADP and P_i with pulsed electrical fields. In case of ATP synthases, the free energy contained in the electrical field is absorbed and coupled directly to drive an endergonic reaction like ATP synthesis. In the case of the mitochondrial ATPase, the maximum energy transferred is ΔM times E, ΔM being the difference in the molar electrical moment of the two conformers and E the field strength. For instance, the interaction energy for a conformational change involving an electrical moment, ΔM, of 200 D under an electrical field of 400 kV cm^{-1} ($\Delta\psi$ of 200 mV) would be ~ 4 kcal mol^{-1} (16.8 kJ mol^{-1}) (Tsong, 1989). The free energy transmitted through an oscillating electrical field has been shown to be utilized by an Na$^+$,K$^+$-ATPase to pump Na$^+$ and K$^+$ against their respective concentration gradients (Tsong and Astumian, 1986; Tsong, 1989). An excess influx of Rb$^+$ (K$^+$) only, against its electrochemical gradient, was

Figure 12.9 (see facing page) Parametric domains of effective functioning of the PEP branch to the PTS or further glycolysis in a PTS model. The rate constants related to the covalent cascade of phosphate transfer from PEP to glucose (Glc) or the rate constant of the pyruvate kinase reaction at the branch (k_{e1} and k_{pk}, respectively) were varied one at a time. In (a) and (b) the logarithms (log) of the ratios were varied through the numerator k_{pk}, (a) at constant $k_{e1} = 0.5$ s^{-1}, or k_{e1} (b) at constant $k_{pk} = 0.5$ s^{-1}, keeping the denominator constant. In the y-axis the steady state values of metabolites ATP, A and G6P are represented. The simulations of (a) and (b) are summarized in (c) in order to answer the question of how much flux should be partitioned into both competing pathways in order to be still physiologically compatible, i.e. with appreciable steady state values of ATP. The flux partition coefficient at the PEP branch was calculated either as the relative flux of PK (γ_{PK}) or Enzyme I (γ_{e1}) over the sum of fluxes through the steps catalysed by both enzymes. Equations 12.40–12.46 were numerically integrated on a PC-60 III Commodore using the program SCoP with an Adams method (Duke University, 1989) with the following parameters: $k_{e1} = k_{e2}$ (s^{-1}) $= 0.5$; k_1 (mM^{-1} s^{-1}) $= 0.3$; k_3 (mM^{-2} s^{-1}) $= 0.1$; k_5 (s^{-1}) $= 0.1$; k_7 (s^{-1}) $= 0.1$; k_9 (s^{-1}) $= 0.05$; $\tau = 0.83$; dp $= 0.998$; K_g (mM) $= 1.0$; K_{pe} (mM) $= 0.1$; K_{pk} (mM) $= 0.5$; $K_p = 1.0$. The kinetic parameters of the proton pump were: $K_M = 2$ mM; $V_M = 0.5$ mM s^{-1}. The total nucleotide (C_A) and phosphate pool (P_T) were both 10 mM, whereas the total sum of substrate A involved in the covalent modification cycle (TA) was 1 mM. The steady states of the plotted variables were attained by simulation from the following initial values (mM): Glc $= 100$; G6P $= 0.7$; ATP $= 0.001$; I $= 0.001$; Pyr $= 0.001$; PEP $= 0.001$; A $= 0.9$.

induced by the a.c. field (Tsong and Astumian, 1986). Optimal values of both field strength and a.c. frequency were $20\,V\,cm^{-1}$ and $1\,kHz$, respectively. For voltage-induced ATP synthesis in submitochondrial particles, a time threshold of about $10\,\mu s$ has been described. Since the RC time constant was expected to be a few orders of magnitude smaller than $10\,\mu s$, it was interpreted that the time threshold was related to the relaxation time of an electrical field-induced conformational change of the enzyme (Tsong and Astumian, 1986).

12.3 CHANGES IN GENE EXPRESSION COUPLED TO TRANSPORT PROCESSES, ENERGY METABOLISM AND HORMONAL EFFECTS

Chapters 9 and 10 developed the idea that morphology and organization are the outcome of interactions among multiple genes and the environment. Chapter 11 shows how the reciprocal interaction between the environment (intra- or extracellular) and the genetic background can be quantitatively expressed through the flux coordination hypothesis. Due to phenotypic complexity, genes and cell form as well as enzyme activity and fitness (reproductive success) of a microorganism to the environment are not unambiguously related (Dykhuizen *et al.*, 1987; Harold, 1991). A genic product participates in an intricate, braided web of causes and effects; for example, an enzyme takes part in a vastly complex metabolic network (Harold, 1990, 1991). Chapters 9–11 present numerous examples in either growing, proliferating or differentiating organisms in which outside signals change the pattern of gene expression and switch the phenotype of a cell.

Hormones are main extracellular signals that mediate the spatio-temporal coordination of cell responses in the face of environmental changes. Auxin is involved in the control of polar cell differentiation in higher plants (Schnepf, 1986). In plant morphogenesis, it has been shown that the effect of auxin on membrane and wall properties, cytoplasmic and nuclear events reflects the close and interdependent relationship of all these cellular events (Taiz, 1984; Schnepf, 1986; Aon and Cortassa, 1989). If this is the case, what is the nature of the signal(s) sent by the plasmalemma to the cytoplasm and nuclei, and how are they transduced into different metabolic responses and nuclear activity? We have suggested elsewhere (Aon and Cortassa, 1989) that the signal could be carried out by ions (De Loof, 1986) or proton currents ('proticity') (Berry, 1981) generated by redox or photoredox processes and ATP cleavage reactions (Welch, 1984). A hypothetical ionic signal could be more rapidly sent and conducted. This would explain the short time lag between auxin application and mRNA-induced accumulation (10 min later) (Theologis,

1986). A family of auxin-regulated RNAs from soybean begins to accumulate within 2.5 min after application of auxin, whose distribution is altered in response to gravistimulation (McClure and Guilfoyle, 1989). A sequence of events for explaining rapid gene regulation by auxin has been proposed (Theologis, 1986). The auxin stimulus would be transduced by the membrane into different electrical signals, i.e. ionic fluxes or H^+ currents (proticity), which might induce a different nuclear activity and hence cellular differentiation. In the following we analyse the plausibility of this hypothesis in different cellular systems; namely, for insulin and effectors such as glucose or calcium.

The production and secretion of insulin rise during food consumption and fall during fasting. Insulin multiple effects on cell metabolism and stimulation of protein synthesis appear to be due to multilevel controls: transcriptional and post-transcriptional events such as those modifying mRNA stability or translatability. The beta cells of the pancreatic islets of Langerhans respond to changes in glucose concentration by varying the rates of insulin synthesis and secretion. Studies show that gene regulation can proceed via a metabolic or energetic signal from glucose flux transmitted through the insulin enhancer (German *et al.*, 1990; German, 1993). Studies of insulin secretion by the beta cells in the pancreatic islets of Langerhans show that preproinsulin mRNA levels rise four- to 10-fold in response to glucose stimulation. This rise in mRNA levels apparently results from both an increase in transcription rate and an increase in messenger half-life (German *et al.*, 1990). Experimental evidence shows that insulin promoter activity increases, reaching a maximum at 1 mM glucose in beta cells expressing hexokinase I (German, 1993). It was also found that pyruvate can stimulate the insulin promoter in cells expressing the bacterial gluconeogenic enzyme phosphoenolpyruvate carboxykinase. The author concluded that the intermediates of anaerobic glycolysis between fructose 1,6-diphosphate and phosphoenolpyruvate are essential for beta cell glucose sensing (German, 1993).

Insulin gene expression was studied in primary cultures of rat islet cells transfected with portions of the rat insulin I gene 5′-flanking sequence linked to the reporter gene chloramphenicol acetyltransferase (*CAT*). The transcriptional response to glucose was tested by growing the transfected cells in media containing glucose for 48 h prior to harvest (German *et al.*, 1990). At between 1 and 16 mM glucose, CAT enzyme activity rose 10-fold, with a half-maximal response at 6 mM glucose. Since the response to glucose was sensitive to the calcium channel blocker verapamil, it appeared that calcium played a role in the transcriptional response of insulin expression to glucose (German *et al.*, 1990).

Insulin action on gene expression in cultured hepatocytes can be glucose-dependent or independent (Decaux *et al.*, 1989, 1991). Aldolase B and L-type pyruvate kinase mRNAs accumulate in response to the presence of both glucose and insulin (Decaux *et al.*, 1989). Inhibition of the L-type pyruvate kinase gene expression that occurs when either glucose or insulin (or both) is removed from the medium is not associated with accumulation of phosphoenol pyruvate carboxykinase mRNA, which argues against the involvement of intracellular cyclic AMP.

The endocrine beta cells of the pancreatic islets of Langerhans have the crucial physiological role of responding to hyperglycaemia through insulin secretion. Glucose-induced pleiotropic effects in the beta cell include regulation of gene expression. It has been suggested that L-type pyruvate kinase gene expression is stimulated by a transcriptional activator from carbohydrate metabolism that accumulates in the presence of insulin (Decaux *et al.*, 1991). The two isoforms L and L′ of pyruvate kinase are encoded by the L-type pyruvate kinase gene. The upstream promoter and adjacent first exon give rise to the mRNA of the L′ isoform, whereas the downstream promoter and associated leader exon are responsible for the synthesis of the mRNA for the L isoenzyme (Marie *et al.*, 1993; for a review, see Vaulont and Kahn, 1994). The transcription initiated at the downstream promoter is regulated by nutritional factors, as it occurs with the mRNA of the L-type pyruvate kinase in rat liver when fasted animals are re-fed with carbohydrates. This effect appears to be a direct consequence of hyperglycaemia since it can be elicited in rat hepatocytes, maintained in culture in the presence of insulin, by simply raising the glucose concentration of the culture medium (Decaux *et al.*, 1989). Insulinoma beta cells responded by strong induction of L-type pyruvate kinase (15-fold above basal level) following a rise in extracellular glucose from 3.5 to 33 mM (Marie *et al.*, 1993). The effect was specific to the pyruvate kinase gene since neither proinsulin I mRNA nor glucokinase mRNA were increased in glucose-stimulated insulinoma beta cells. The effect on L-type pyruvate kinase gene expression was detectable within 2 h of glucose supplementation, reaching a maximum after 8–12 h.

The glucose signalling mechanisms of pyruvate kinase gene induction or the insulin secretion may be clearly dissociated. The insulin secretory response is known to require the complete metabolic conversion of glucose through glycolytic and mitochondrial oxidative pathways, whereas only glucose phosphorylation appears necessary for induction of the pyruvate kinase gene (Marie *et al.*, 1993, and references therein).

Calcium is an important intracellular signalling ion. Rise of intracellular levels of calcium is an apparent transduction mechanism of extracellular stimuli, including hormones and neurotransmitters. It has been shown that Ca^{2+} influx into sympathetic neurons causes a significantly greater and faster increase of [Ca^{2+}] in the nucleus than in the cytosol. Similar patterns were observed when Ca^{2+} entered the neuron after voltage-dependent channel   activation (K$^+$-induced depolarization or field stimulation), agonist-induced Ca^{2+} release (acetylcholine) or ionophore application (ionomycin) (Przywara *et al.*, 1991). Experiments performed with acridine orange, a DNA and RNA vital stain, completely prevented the differential rise in nuclear [Ca^{2+}]$_i$ but did not abolish the stimulated rise in the cytosolic regions of sympathetic neurons (Przywara *et al.*, 1991). These results allowed the authors of this work to argue that the chromatin material may contain the releasable Ca^{2+} pool and that acridine orange displaces Ca^{2+} from its normal binding sites in the nucleus, leaving no releasable pool of Ca^{2+} to respond to subsequent stimulation.

12.3.1 DNA CONFORMATIONAL-INDUCED CHANGES IN GENE EXPRESSION

Modifications at the level of intermediary metabolism and gene expression have been shown by prokaryotic cells when subjected to changes in environmental conditions such as oxygen level, osmolarity or nutrient availability (Neidhardt, 1987). The induction of several genes elicited by anaerobiosis or osmotic signals depends on DNA topology – namely, the degree of DNA supercoiling (Higgins *et al.*, 1987; Ni Bhriain *et al.*, 1989; Pruss and Drlica, 1989). In prokaryotic cells, the DNA is under configurational stress and the maintenance of that stress is apparently determined by the relative levels of activity of two enzymes, topoisomerase I (coded by *topA* gene) and DNA gyrase (coded by *gyrA*, *gyrB* genes) (Gellert *et al.*, 1976, 1983; Wang, 1985; Drlica, 1990). The level of transcription of the genes coding for topoisomerase subunits is in turn regulated by the levels of DNA supercoiling (Menzel and Gellert, 1983).

In bacterial cells, DNA is under negative superhelical tension which is introduced enzymatically by DNA gyrase and relaxed by topoisomerase I (Drlica, 1990). In the absence of ATP, gyrase removes supercoils. Gyrase is more active on a relaxed DNA substrate and topo I on a more negatively supercoiled one. Negative supercoiling is considered an energetically activated state favouring three-dimensional configurations which allow DNA to wrap around proteins and facilitates loop formation (Drlica, 1990). Negative supercoiling compacts DNA, enhances binding of intercalating agents, and facilitates formation of structures such as cruciforms and Z-form DNA (Drlica, 1990). Thus, intracellular factors that control the negative supercoiling of DNA are likely to mediate changes in gene expression triggered by environmental stimuli.

Changes in gene expression through DNA supercoiling may be affected by the ability of promoters to be transcribed (Higgins *et al.*, 1988). DNA bending and looping in promoter regions may be involved in the response of these regions to DNA supercoiling (Higgins *et al.*, 1988). The activated energetic state of negative supercoiling may provide energy for strand separation and assist the formation of an open complex presumed to play a role in recognition by RNA polymerase. Extrusion of cruciforms is strongly influenced by DNA supercoiling, ionic strength and temperature. Although only a small percentage of promoters are sensitive to medium osmolarity, a large number of genes are affected two- to three-fold in its expression (Drlica, 1984, 1990; Higgins *et al.*, 1988).

The *E. coli* chromosome is divided into different domains (about 40), each of them independently supercoiled. Environmental factors such as osmotic stress or oxygen levels may have different effects on different domains, adding some specificity (Higgins *et al.*, 1988).

(a) Changes in DNA supercoiling accompanying transcription

Cook and co-workers (1992) found strong evidence supporting the fact that RNA polymerase introduced negative superturns in DNA during transcription. Both wild type *E. coli* cells and $\Delta topA$ mutants harbouring plasmids with highly transcribed genes such as *phoA*, coding for alkaline phosphatase, and *tet*, coding for a membrane-bound tetracycline-resistant protein, cloned in opposite orientation, accumulate negative supercoils at a rate of 5.7 turns per second after 1 min induction of transcription. The accumulation of helical turns when both genes are cloned in parallel orientations is much slower, due to relaxation by diffusion of superturns (Cook *et al.*, 1992). Much higher accumulation of supercoiled DNA was obtained in $\Delta topA$ mutants. The experiments also showed that both gyrase and topoisomerase I can remove superturns at a rate equal to their introduction by RNA polymerase, supporting the view that the topoisomerases are the primary determinants of the level of DNA supercoiling in *E. coli* cells.

12.3.2 CHANGES IN GENE EXPRESSION COUPLED TO METABOLIC TRANSITIONS

Aerobic–anaerobic transitions in facultative anaerobes involve shifts in energy metabolism – for example, from respiratory to fermentative (Clark, 1989) – occurring concomitantly with changes in the pattern of expression of at least 18 proteins, among them some glycolytic enzymes (Smith and Neidhardt, 1983) (Figure 9.7). This fact raises the question of whether coupled changes in gene expression and energetic metabolism occur mediated by free-energy intermediates. *In vitro* experiments indicated that the activity of gyrase depends on the ATP/ADP ratio rather than on the ATP concentration itself (Westerhoff *et al.*, 1988). Thus, the possibility exists that the DNA supercoiling senses the phosphorylation potential of the cell and acts as an 'energy intermediate' to transduce signals coming from energy metabolism to regulate gene expression.

As changes in osmolarity and oxygen tension correlate with changes in the ionic distribution across the plasma membrane, the PMF components (namely, the transmembrane ΔpH and the electrical potential, $\Delta\psi$, may act as a free-energy intermediate influencing DNA supercoiling. Aerobically, *E. coli* showed a $\Delta\mu_{H^+}$ which was mainly accounted for by $\Delta\psi$ (40–60 mV) with a negligible ΔpH (< 10 mV) (Figure 12.10). The magnitude and composition of the $\Delta\mu_{H^+}$ were not significantly modified in the first few minutes after the shift to anaerobic growth (Figure 12.10). Twelve minutes after nitrogen began to be flushed into the culture flask, the $\Delta\mu_{H^+}$ collapsed. A pH gradient (alkaline inside) of up to 0.8 pH units was built up, probably mediated by K^+ or Na^+ influx to compensate charge movements, since the $\Delta\psi$ remained nil in anaerobiosis (Figure 12.10). These results indicated that the changes in membrane electrochemical gradients did not occur simultaneously with the drop in ATP and ADP levels (see below). The temporal scale in which changes in $\Delta\mu_{H^+}$ occurred is closer to that of topoisomerase activities than to the ATP/ADP ratio.

The ATP/ADP ratio determined (Cortassa and Aon, 1993b) agreed well with those observed in other strains of *E. coli* cells growing in minimal medium, which fluctuated around 8 (Kashket, 1982). In spite of a 35% decrease in the intracellular concentration of ATP, the total adenine nucleotide pool was apparently depleted, keeping the constancy of the ATP/ADP ratio (Figure 12.11). This result is in agreement with those reported by Dietzler *et al.* (1979b), who also found that a 35% depletion of the adenylate pool did not affect the energy charge when nitrogen-starved cells were treated with sodium azide. With respect to the relationship between the phosphorylation potential and $\Delta\mu_{H^+}$, assuming an intracellular phosphate concentration of 10 mM, an H^+/ATP ratio of 4 may be calculated for an internal pH of 7.5. This H^+/ATP ratio was higher than reported data (Kashket, 1982; Maloney, 1987). We could not rule out the possibility that the membrane potential of the *E. coli* strain HW103 was particularly low.

A close temporal correlation was observed between a rapid increase in DNA supercoiling and the decrease in the ratio of activities of topoisomerase I with respect to DNA gyrase, following aerobic to

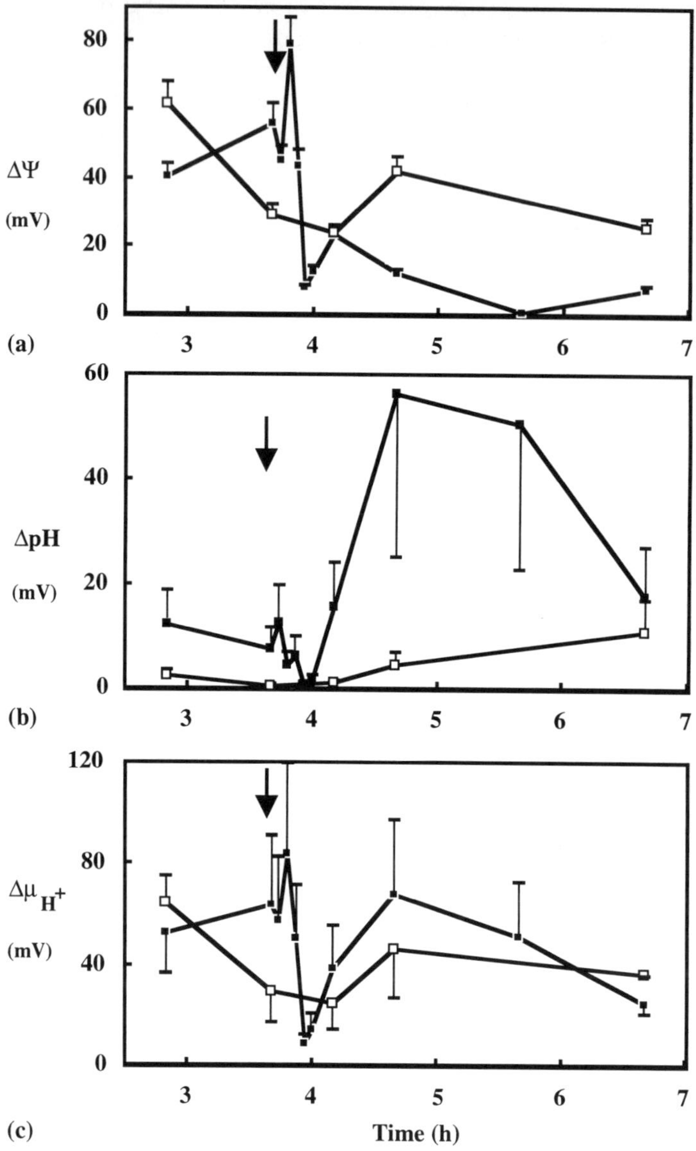

Figure 12.10 Change in *E. coli* proton motive force upon aerobic to anaerobic transitions. The components of the Δμ$_{H^+}$ were determined through the distribution across the plasma membrane of radioactive dimethyldiphenylphosphonium and DMO, for Δψ and ΔpH, respectively. (a) Evolution of Δψ; (b) evolution of ΔpH; (c) Δμ$_{H^+}$ calculated from the addition of the Δψ and ΔpH components. Labels correspond to cells in aerobiosis (open squares) and those undergoing an anaerobic shift where indicated by an arrow (filled squares). Each point represents the mean of two experiments run in duplicate ± SEM. (From Cortassa and Aon, unpublished data.)

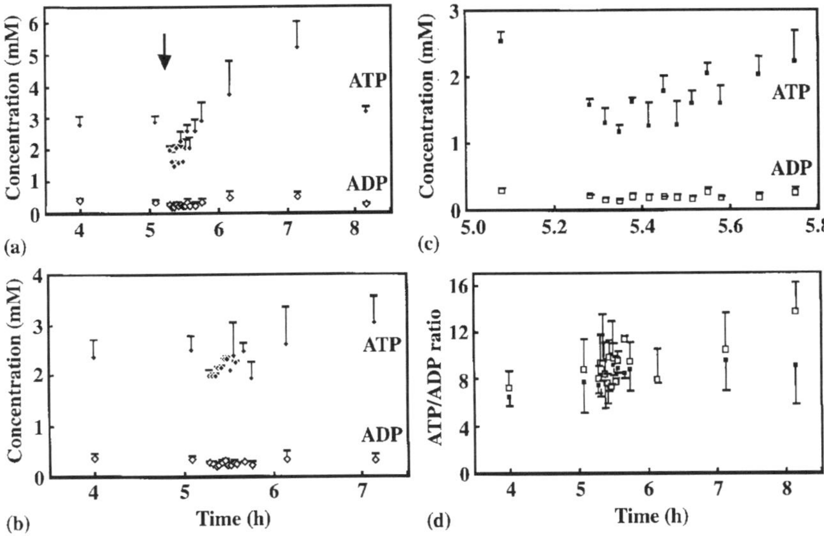

Figure 12.11 Evolution of adenine nucleotides after the transition to anaerobic conditions. Samples (1 ml) of *E. coli* batch cultures, grown as described in Cortassa and Aon (1993b), were taken at the indicated times. The ordinates in (a)–(c) indicate the intracellular concentrations of ATP (filled diamonds and squares) and ADP (open diamonds and squares) determined in cell extracts from the culture, (a) subjected to an anaerobic transition where indicated by the arrow, or (b) kept in aerobiosis, respectively. (c) Zoom of (a) around the anaerobic shift. (d) Ratio ATP/ADP for the culture shifted to anaerobiosis (open squares) or kept in aerobic conditions (filled squares). Each point represents the mean of two experiments run in quadruplicate ± SEM. (From Cortassa and Aon, 1993b, reprinted by permission of Kluwer Academic Publishers.)

anaerobic transitions in batch cultures of exponentially growing *E. coli* (Figure 12.12). The decrease in the ratio of topoisomerase activities was mainly determined by the decrease in the specific activity of topoisomerase I (Figure 12.13). DNA gyrase only slightly decreased its activity, as could be measured in cell-free extracts prepared after the onset of anaerobiosis (Figure 12.13). These results, together with the constancy of the ATP/ADP ratio, suggest that the latter is not the main free-energy intermediate effector of DNA supercoiling under the experimental conditions reported in Cortassa and Aon (1993b).

Similar experiments, although performed under conditions far different from those used in Cortassa and Aon (1993b), were reported by Hsieh and colleagues (Hsieh *et al.*, 1991a). Apparently an opposite conclusion was reached from experiments with *E. coli* cells growing in rich medium at high concentrations of glucose (1%) and phosphate (100

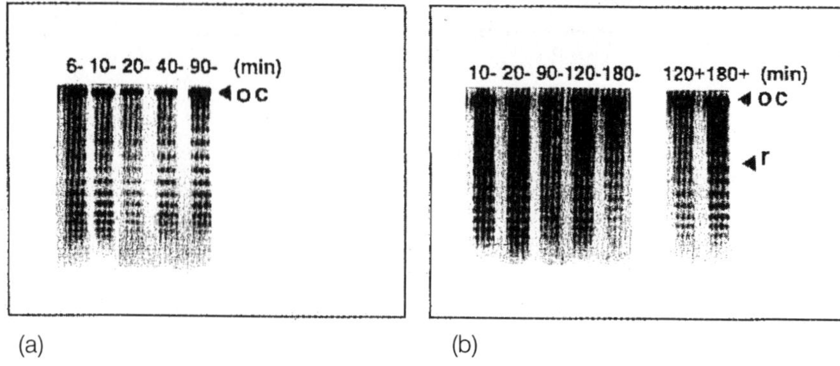

Figure 12.12 Change in pBR322 supercoiling upon aerobic to anaerobic transitions. Autoradiogram of pBR322 topoisomers isolated from *E. coli* HW103 cell culture at the indicated times on top of each lane after the onset of anaerobiosis, and electrophoresed in an agarose gel containing chloroquine 10 µg ml⁻¹. At the right of the times indicated on top of each lane, + or − signs mark that the culture was under aerobic (control culture) or anaerobic conditions, respectively. The more supercoiled the DNA, the faster it runs under the electrophoretic conditions used to separate pBR322 topoisomers. OC, open circle; r, relaxed DNA. (From Cortassa and Aon, 1993b, reprinted by permission of Kluwer Academic Publishers.)

mM). A transient decrease in the DNA supercoiling was reported, together with a reduction of the ATP/ADP ratio (Hsieh *et al.*, 1991a).

A plausible explanation of the different results obtained by both groups of researchers may be attempted. The *E. coli* cells of Hsieh and colleagues were grown in rich medium and therefore glucose catabolism may not be providing the precursors for biomass synthesis; additionally, the osmotic pressure in the culture medium, which is known to alter DNA supercoiling and ATP/ADP ratio (Hsieh *et al.*, 1991b), is much higher in rich medium supplemented with glucose and phosphate: less than 200 mOs under the conditions used in Cortassa and Aon (1993), compared with 500 mOs in rich medium. We have noticed that the state of DNA supercoiling is in fact slightly higher when isolated from cells growing in LB medium and the distribution of DNA topoisomers is remarkably less spread (Cortassa, unpublished data). In addition, the ATP/ADP ratio we determined is higher than that reported in LB medium (between 0.7 and 2) (Hsieh *et al.*, 1991a); therefore this comparison strongly suggests that, in addition to the ATP/ADP ratio, another factor should play a major role in the *in vivo* control of DNA supercoiling. A similar conclusion was reached by Meury and Kohiyama (1992) with *E. coli* mutants affected in K⁺ transport systems. In that work

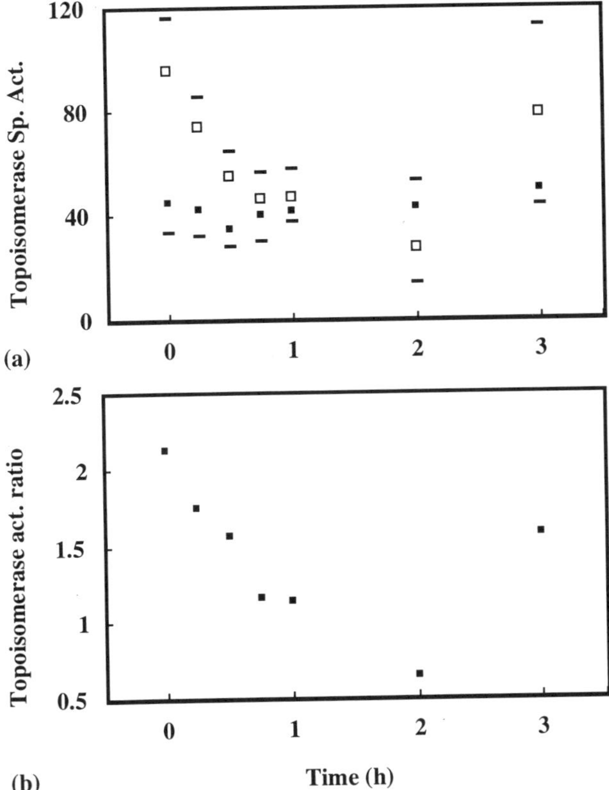

Figure 12.13 Topoisomerase activities in cell-free extracts of *E. coli*. Topoisomerases were assayed and their activities quantified as described in Cortassa and Aon (1993b). (a) Topoisomerase activities are reported in arbitrary units as a function of time after the transition to anaerobiosis: topoisomerase I, open squares; DNA gyrase, filled squares. (b) Relative enzyme activities as calculated from the data shown in (a). Each point represents the mean of three experiments run in duplicate ± SEM. (From Cortassa and Aon, 1993b, reprinted by permission of Kluwer Academic Publishers.)

there was no clear correlation between variations in the ATP/ADP ratio and changes in DNA supercoiling.

Jensen and co-workers have demonstrated a close correlation between the ATP/ADP ratio and the linking number of pBR322 in *E. coli* cells under starvation conditions (Jensen *et al.*, 1995). The cells' ATP and ADP levels were changed by IPTG induction of the H⁺-ATPase. When the level of expression varied from 400% to 15% with respect to the wild-type level, the ATP/ADP ratio varied from 0.3 to 12 (8 being the ratio found in growing cells) whereas the linking number changed from 6 to zero. These authors were unable to modulate either the energy charge of the cells or the degree of pBR322 supercoiling in growing cells even in the presence of a poor energy source such as succinate (Jensen *et al.*, 1995). These results suggest that regulation of DNA supercoiling by the ATP/ADP ratio would not occur under *in vivo* growth conditions, since the latter appears to be highly buffered.

The results shown in Figure 12.13 offer an explanation for the increase in the level of DNA supercoiling by the decrease in the topoisomerase I/DNA gyrase ratio of activities. Already 15 min after the onset of anaerobiosis, both topoisomerase I and DNA gyrase activities had decreased (Figure 12.13). However, a significantly lower activity level of topoisomerase I than that of gyrase was measured which resulted in a two-fold net decrease of the topoisomerase I/gyrase ratio, one hour after anaerobiosis (Figure 12.13). This decrease may be due either to changes in the relative amount of enzyme protein or to the presence of effectors.

At present, little is known about the regulation of topoisomerase I activity. This was very low in cells of *Bradyrhizobium* sp. strain 32H1, when grown under low oxygen and high potassium, which are required to express the bacteroid-associated functions in this nitrogen-fixing symbiont (Gober and Kashket, 1989). A decrease in the activity of topoisomerase I associated with an increase in the intracellular K^+ and low membrane potential has been reported (Gober and Kashket, 1989). Our results show the same tendency, since the decrease of topoisomerase I activity in the aerobic to anaerobic transition was associated with a drastic drop in $\Delta\psi$ after 12 min (Figure 12.10). The observed drop in $\Delta\psi$ could well be due to an influx of K^+ or Na^+ to the cell to compensate charge movements upon H^+ extrusion, as revealed by the slight alkalinization shown in Figure 12.10. In bacteria, K^+ uptake is electrogenic and is associated with the regulation of pH_i, mediating cellular responses to changes in oxygen tension as well as osmotic pressure (Higgins *et al.*, 1987). A direct link may exist between the $\Delta\mu_{H^+}$ and topoisomerase I activity. The link might be achieved through either pH_i alkalinization or higher K^+ or Na^+ levels due to interconversion of ΔpH and $\Delta\psi$ upon the onset of anaerobiosis.

12.3.3 CHANGES IN GENE EXPRESSION INDUCED BY OSMOLARITY CHANGES

Intracellular ionic strength has been postulated as a 'second message' transmitting the effects of extracellular osmolarity to modulate a variety of intracellular functions (Higgins *et al.*, 1987). Potassium is a primary osmolyte in many species providing the principal means of balancing external osmolarity and maintaining turgor. Potassium traffic in *E. coli* is regulated by two uptake systems: a low-affinity, constitutive system (TrK), and a high-affinity, osmotically induced system (Kdp). Osmotic upshock causes an initial flow of water from the cell, reducing turgor, which immediately induces *kdp* expression, and K^+ is accumulated until turgor is restored, whereupon *kdp* is repressed (Higgins *et al.*, 1987). Betaine is a compatible solute (i.e. it can be accumulated intracellularly with minimal perturbation of normal metabolism or enzyme function) of *E. coli* and *S. typhimurium* that at high osmolarity accumulates to molar concentrations by either uptake or synthesis. Betaine uptake is mediated by two transport systems, *ProP* and *ProU* (Higgins *et al.*, 1987). *ProU* is induced in response to the turgor-induced accumulation of intracellular K^+ whereas *kdp* is induced directly by changes in cell turgor. These two genetic loci (*kdp* and *proU*) are osmoregulated at the transcriptional level. Ionic strength could exert its regulatory effects on gene expression by changing the conformation of a regulatory protein, and hence its binding to the *proU* promoter. Another plausible model suggested is that ionic strength alters the RNA polymerase–promoter interactions themselves through perturbation of the normal conformation of B-DNA (Higgins *et al.*, 1987).

Osmostress-induced changes in gene expression are not limited to bacteria. In *Saccharomyces cerevisiae*, when cells are exposed to a severe salt shock (up to 1.4 M NaCl) or 0.8 M sucrose, nine [³H]-leucine labelled proteins show elevated rates of synthesis. Among the proteins characterized, heat shock proteins as well as glycerol-3-phosphate dehydrogenase have been identified (Varela *et al.*, 1992). Whether these changes are associated with DNA-conformational changes remains to be seen.

12.4 COMMUNITY DYNAMICS AND COUPLING BETWEEN CELLS: SYNCHRONIZATION OR MUTUAL STIMULATION MECHANISMS AT THE CELL POPULATION LEVEL

Chapter 8 described cell population dynamics in at least two competing microbial communities. Two types of microbe–microbe interactions were described: commensalism and mutualism. Cell populations exhibiting coherent behaviour are considered as mutually synchronous when the frequency and phases of individual oscillating cells are the same (Winfree, 1980; Aon et al., 1992). When they remain the same over many oscillatory periods, they must be synchronized by some mechanism (Chapter 3), unless the differences in magnitude of the kinetic parameters between individual cells are very slight. Although previous studies of yeast cell synchronization during glycolytic oscillations suggested that cell interaction would involve a diffusible glycolytic intermediate (Ghosh *et al.*, 1971; Chance *et al.*, 1973; Aldridge and Pye, 1976), the metabolic basis of that interaction remained elusive.

Before macroscopic damping of NAD(P)H oscillations, single yeast cells oscillated in phase with the cell population (Aon *et al.*, 1992). At high cell densities the cell population oscillated in a sustained way or with drastically reduced damping. After macroscopic damping, single cells still oscillated ($\sim$50%), but individual oscillators were out of phase. They also progressively lost oscillatory dynamics.

Experimental evidence and model calculations supported the notion that the flux balance between NAD reduction (ethanol catabolism) and NADH oxidation, against the backdrop of redox neutral glycolysis, constituted the basic biochemical mechanism that determined the dynamic behaviour of the cell population (Aon *et al.*, 1992; Aon and Cortassa, unpublished data).

Highly dense yeast cell suspensions were shown to excrete ethanol and the latter compound appeared to play an essential role in the oscillations. Oscillatory transients triggered by glucose addition alone could be observed in washed, resuspended yeast cells preincubated with ethanol. The latter experimental evidence and the fact that starved yeast cells were able to metabolize glucose and ethanol simultaneously (see Aon *et al.*, 1992, for additional experimental evidence; Beauvoit *et al.*, 1993) are in agreement with the possibility that extracellular ethanol affects the oscillations of individual cells.

Ethanol production is the main pathway of NAD^+ regeneration in anaerobic yeast (Gancedo and Serrano, 1989). Utilization of ethanol in yeast implicates a first oxidation to acetaldehyde by alcohol dehydrogenase II and a second oxidation to acetic acid catalysed by an aldehyde dehydrogenase. Acetate metabolism proceeds via acetyl-CoA synthetase followed by the operation of TCA and glyoxylate cycles (Gancedo and Serrano, 1989).

Our observations that the amplitude of oscillations varied less than proportionally with cell density, and that the damping of the oscillations was more rapid at lower cell densities, cannot be explained without invoking mutual sensing between yeast cells (Aon *et al.*, 1992). Therefore, we considered the possibility of interaction between the cells. It was likely that a secreted metabolic product of the cells could be taken up again, be further metabolized and then influence the oscillatory behaviour. Experimental evidence showed that the likely candidate was the ethanol formed by glycolysis. The role of ethanol and other glycolytic intermediates (acetaldehyde and pyruvate) was further assessed by studying the kinetics of damping of oscillations at low or high cell densities in the presence of different concentrations of metabolites. The results obtained at high cell densities are shown in Figure 12.14. Both at low and high cell concentrations, acetaldehyde increased the damping factor. Ethanol and pyruvate at low cell densities did not induce changes in damping, and at high densities the lower concentrations of metabolites especially appeared to increase the damping. The role of ethanol in promoting oscillations at high cell densities was further confirmed by treatment of yeast cell suspensions with pyrazole, an inhibitor of alcohol dehydrogenase II (Aon *et al.*, 1992). The role of acetaldehyde as an inducer of damping of oscillations was confirmed later (Richard *et al.*, 1994). These authors demonstrated that cyanide promoted the conversion of acetaldehyde to lactonitril, keeping the acetaldehyde concentration at levels lower than 0.1 mM. An analysis of the dependence of oscillation damping on cyanide concentration revealed that at high cyanide concentration this respiratory inhibitor was competing with ethanol formation by trapping acetaldehyde. Ethanol levels decreased at cyanide concentrations higher than 9 mM, resulting in damping of the oscillations. This is consistent with the hypothesis that ethanol would act as a synchronizer of glycolytic oscillations (Aon *et al.*, 1992; Richard *et al.*, 1994).

We have formulated an explanation for much of the experimental observations in the form of a mathematical model (Figure 12.15; Appendix 12A). The oscillations observed in highly dense populations of yeast cells after addition of cyanide can be fully explained on the basis of a metabolic model where there is a variable degree of balance between two catabolic pathways. These are the pathways reducing NAD(P), such as ethanol catabolism, and the pathways oxidizing NAD(P)H, such as dihydroxyacetone phosphate reduction. The (im)balance between these two types of pathway functions against the background of glycolysis, which, in principle, operates at a much higher throughput (in the presence of cyanide). Because glycolysis coupled to alcoholic fermentation is redox neutral in terms of causing no net NAD(P) reduction, yet depends dynamically on the redox state of NAD, the dynamic redox balance is an important determinant of glycolytic oscillations. The cells affect each other's oscillations merely by affecting their dynamic redox balance, because of an increasing concentration of extracellular ethanol with time and cell density.

The 'co-substrate model', leading to this explanation postulates that:

- ethanol is utilized in an oxidative pathway;
- the flux balance between NAD reduction and NADH oxidation at the single cell level is the biochemical determinant of the autonomous cellular glycolytic oscillations;

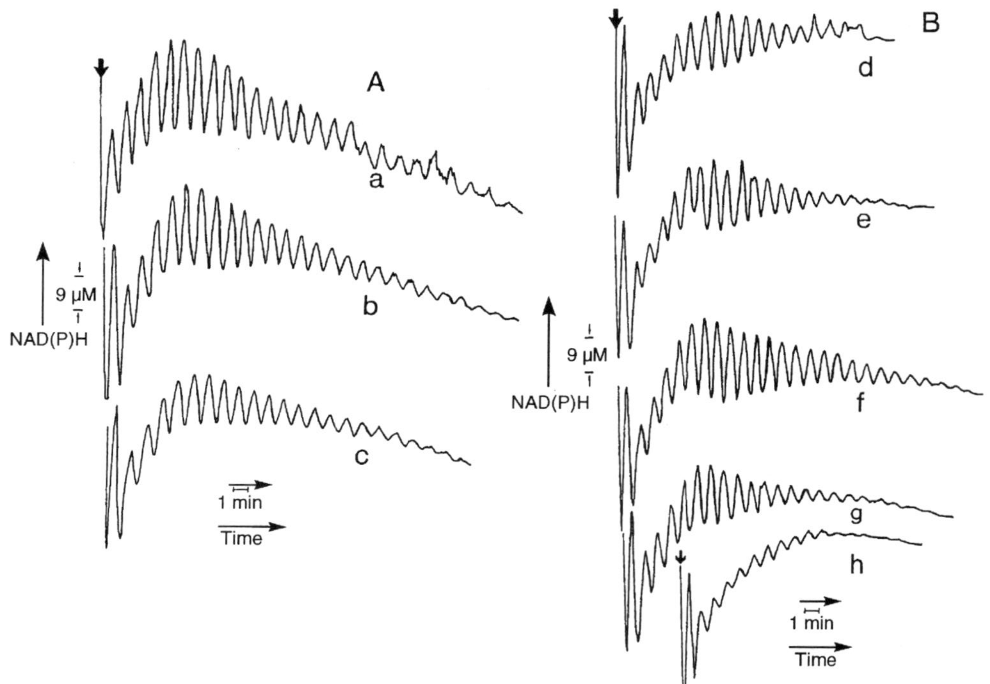

Figure 12.14 Yeast glycolytic oscillations at high cell concentrations in the presence of (A) ethanol and (B) pyruvate and acetaldehyde. Starved yeast cells (3×10^8) were preincubated for 2 min at 25°C in the cuvette of the fluorimeter with stirring, in the absence (a,d) or in the presence of two different concentrations of ethanol (b,c), pyruvate (e,f) or acetaldehyde (g,h): 0.1 mM (b,e,g), 1 mM (c,f,h), before addition of 20 mM glucose followed by 15 mM KCN (indicated by the arrow). The oscillations in NAD(P)H were monitored as described in Aon *et al.* (1991, 1992).

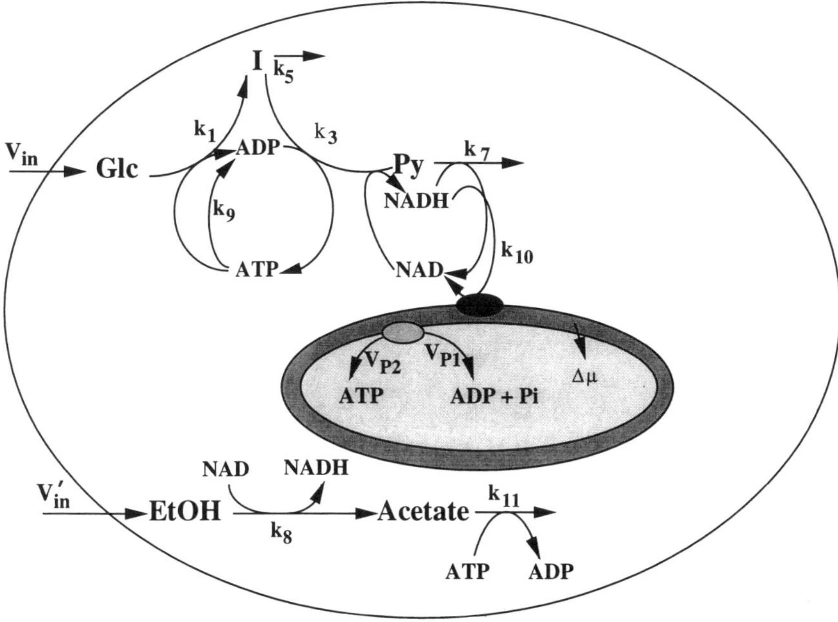

Figure 12.15 The co-substrate model represents a yeast cell fermenting glucose (Glc), irreversibly, by the Embden–Meyerhoff pathway and ethanol (EtOH) through the ethanol oxidative pathway. V_{in} and V_{in}' are the cell uptake rates of glucose or ethanol, respectively.

- ethanol produced by glycolysis and diffusing into the medium subsequently affects the oscillatory dynamics of the cells by affecting the redox balance.

12.4.1 HETEROGENEITY IN OSCILLATORY DYNAMICS CONTRIBUTES TO MACROSCOPIC DAMPING OF THE OSCILLATIONS

Heterogeneity introduced by dispersion of cells over the cell cycle does not explain the damping of the oscillations (Aon *et al.*, 1992). We therefore evaluated the contribution of the loss of mutual synchrony to the macroscopic damping of oscillations by simulating the co-substrate model under conditions of cell–cell interaction. A number of cells were simulated individually under precisely the same constant external circumstances. Cell heterogeneities in membrane permeability coefficients to ethanol and different rate constants were introduced. Their calculated NADH concentrations were averaged. In Figure 12.16 'homogeneous' implies that no differences in kinetic parameters exist between cells. Simulation studies showed that if cells in a synchronized (in phase) population exhibit, individually, autonomous sustained oscillations even if they are heterogeneous in their permeability to ethanol, they will not show damping when averaged (Figure 12.16c,d). These results are in agreement with the notion that heterogeneity in permeability does not induce changes in dynamics. On the other hand, a cell population heterogeneous in other kinetic constants of glycolysis, and consequently showing a mixture of autonomous damped and sustained oscillations, dampens the macroscopic oscillation (Figure 12.16a,b). The results obtained support the notion that differences in oscillatory dynamics at the single cell level contribute to the macroscopic damping by loss of synchrony: individual oscillators go out of phase.

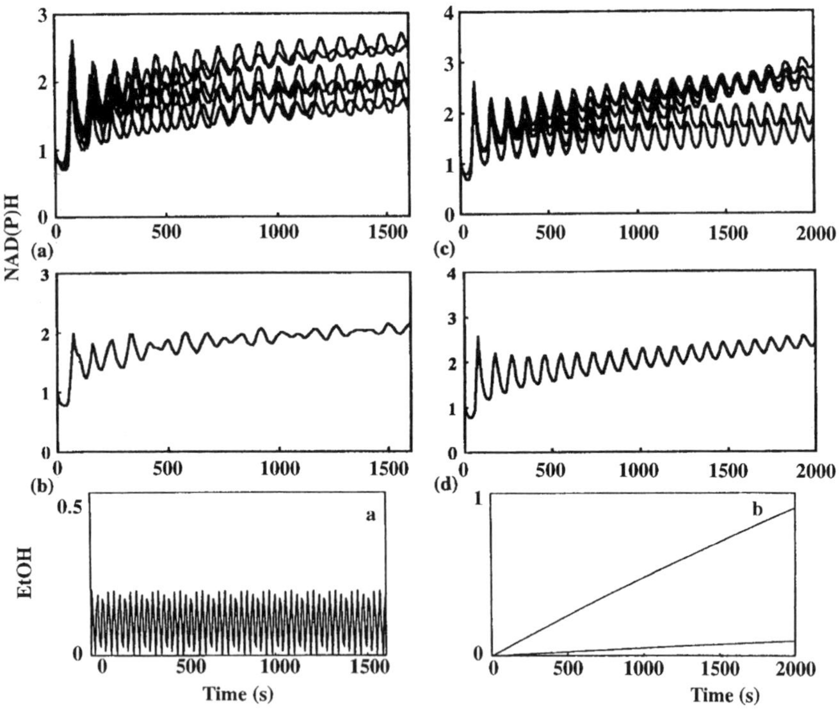

12.4.2 ETHANOL CONSUMPTION PROVIDES THE BASIS FOR SYNCHRONOUS BEHAVIOUR

We calculated whether limited ethanol catabolism could affect glycolytic oscillations. For this purpose we considered the ethanol degradation pathway together with anaerobic glycolysis in a stoichiometric model of yeast metabolism. We calculated the effect of the co-metabolism of glucose and ethanol on the dynamic behaviour of individual cells (Figure 12.17). The calculations showed that influx of ethanol into its oxidative pathway (Figure 12.16) could lead to enhanced damping of the oscillations of individual cells (Figure 12.17a,b). This provided an explanation for our experimental observation that added ethanol accelerated damping of the oscillations (Aon *et al.*, 1992). In these calculations, the simulated NADH oscillations could accompany a sustained increase in reduction of the NAD(P) pool; in essence, oscillations seemed to come to an end because the NAD(P) became too reduced.

Figure 12.16 (see facing page) Simulation of the co-substrate model for a population of interacting oscillating cells with or without mutual synchronization. The ODEs system (equations 12.10–12.16) with the changes specified in equations 12.29 and 12.30 was solved with the following parameter values: (a) V_{in} (mM s^{-1}) = 0.12 (for the three cells showing sustained oscillations), 0.1 (for the three cells showing damped oscillations); k_1 (mM^{-1} s^{-1}) = 0.03, 0.04; k_3 (mM^{-3} s^{-1}) = 0.06, 0.08; k_5 (s^{-1}) = k_7 (mM^{-1} s^{-1}) = 0.1; k_8 (mM^{-1} s^{-1}) = 6.25 × 10^{-4}; k_9 (s^{-1}) = 0.05; k_{11} (mM^{-1} s^{-1}) = 0.01; k_{12} (s^{-1}) = 0.37, 3.7 × 10^{-3}, 3.7 x 10^{-4}; K_p (= $\tau V_M[1 - \Delta\mu/\Delta p°]^6$) (mM s^{-1}) = 0.3097; τ = 0.83; $\Delta p°(V)$ = –0.3; $\Delta\mu/\Delta p°$ = 0.01. The kinetic parameters of the proton pump were: K_M = 2 mM; V_M = 0.5 mM s^{-1}. The total nucleotide (C_A) and phosphate pool (P_t) were both 10 mM, while the pyridine nucleotide pool (C_N) was 4 mM. The initial values of the state variables (mM) were the following: Glc = 0.157; ATP = 1; I = 0.0031; NADH = 1; Pyr = 0.394; EtOH$_i$ = 1; Acetate = 0.01. The forcing function was a harmonic simulated with a sinusoidal wave generator with amplitude = 0.1 mM and period 30 s. Inset (a): the y-axis corresponds to ethanol concentration (mM). (b) Here the average of the six oscillating cells of (a) is represented. Notice that the loss of synchrony between oscillating cells in (a) is reflected as a damped oscillation at the cell population level. It is expected that by summing up more non-synchronized oscillating cells, the damping will be more severe. (c,d) V_{in} (mM s^{-1}) = 0.12; k_1 (mM^{-1} s^{-1}) = 0.03; k_3 (mM^{-3} s^{-1}) = 0.06, 0.08; k_5 (s^{-1}) = k_7 (mM^{-1} s^{-1}) = 0.1; k_8 (mM^{-1} s^{-1}) = 6.25 × 10^{-4}; k_9 (s^{-1}) = 0.05; k_{11} (mM^{-1} s^{-1}) = 0.01; k_{12} (s^{-1}) = 0.37, 3.7 × 10^{-3}, 3.7 × 10^{-4}; K_p (= $\tau V_M[1 - \Delta\mu/\Delta p°]^6$) (mM s^{-1}) = 0.3097; τ = 0.83; $\Delta p°(V)$ = –0.3; $\Delta\mu/\Delta p°$ = 0.01. The kinetic parameters of the proton pump, the total nucleotide (C_A) and the phosphate pool (P_t), the nicotinamide nucleotide pool (C_N) and the initial values of the state variables (mM) were as described in (a) and (b). The forcing function was a linear increase in ethanol concentration. Inset (b): the y-axis corresponds to ethanol concentration (mM). (d) Average of the six oscillating cells of (c). When the individual oscillators are in phase (c), the oscillatory behaviour of the cell population (d) does not damp.

The decrease in the damping rate as well as the appearance of regions of undamped oscillations at high cell densities could be simulated by increasing the rate of ethanol input or the extracellular ethanol concentration up to an optimum concentration (Figure 12.17c). A further increase of the ethanol then again inhibited the oscillations (Figure 12.17b,c). Thus, a prediction of this model is that an increase in extracellular ethanol concentration at high cell densities is likely to increase the ethanol input to the cells and dampen the oscillations. This corresponds with our experimental data. By feeding the ethanol produced by glycolysis back into the ethanol oxidative pathway of the model (accounting for the fact that this ethanol is diluted by the extracellular volume) (Figure 12.16), it was calculated that oscillations may indeed be enhanced by increased cell densities (compare lower and middle traces in Figure 12.17d). Notably, in the latter case, extra added ethanol tended to dampen the oscillations (the upper trace in Figure 12.17d).

That oscillations do depend on the cell concentration is because yeast cells excrete ethanol; they are also able to catabolize external ethanol, and this catabolism affects their tendency to oscillate. Essentially, the reason is that the flux balance between NAD reduction and NADH oxidation sustained by glycolysis and ethanol oxidation is affected by extracellular ethanol. At high cell densities, the ethanol produced by the cells is sufficient to bring the extracellular concentrations of this compound to a level that enhances the oscillations. The oscillations primarily result from autonomous oscillations of the individual cells. Because of slight kinetic heterogeneity, the oscillation of the population dampens more quickly than that of the individual cells (Aon *et al.*, 1992).

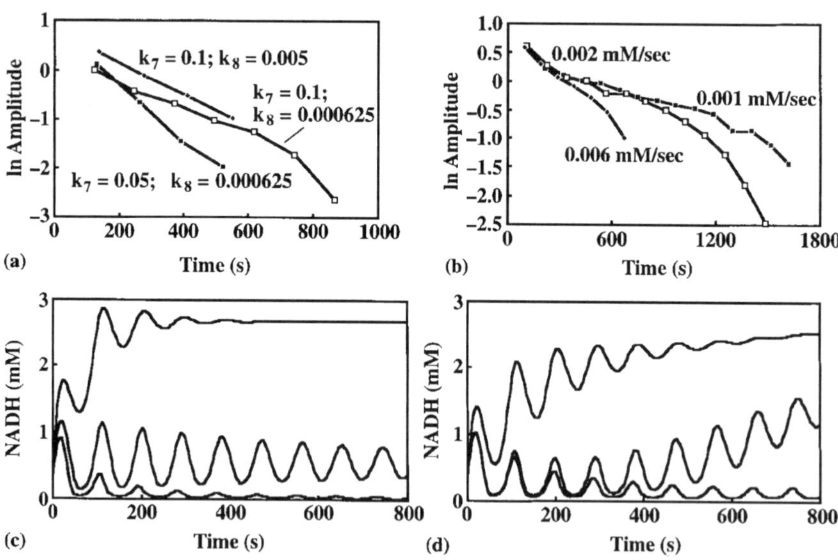

12.5 CHEMICAL, MECHANICAL AND ELECTRICAL COUPLING BETWEEN PROCESSES AT DIFFERENT LEVELS OF ORGANIZATION IN PLANT CELL GROWTH

The dynamics of the electromechanochemical changes undergone by a plant cell which is, concomitantly, elongating and changing its surface by cell wall synthesis was formulated through the Bemchem model (Aon and Cortassa, 1989). The model used to describe from a unified and dynamic view the biological, electrical, mechanical and chemical aspects

Figure 12.17 (see facing page) Dynamic analysis of the co-substrate model. Effect of different rates of (a) ethanol oxidation, (b) ethanol input, (c) constant extracellular ethanol concentrations and (d) increasing cell density and added ethanol on the dynamics of glycolytic oscillations. Numerical simulation of the experimentally observed kinetics of oscillation decay were performed. The ODEs system, equations 12.10–12.19 (Appendix 12A), was integrated with the following parameter values: (a) V_{in} (mM s^{-1}: glucose uptake) = 0.12; V'_{in} (mM s^{-1}: ethanol uptake) = 0.002; k_1 (mM^{-1} s^{-1}) = 0.025; k_3 (mM^{-3} s^{-1}) = 0.05; k_5 (s^{-1}) = 0.1; k_9 (s^{-1}) = 0.05; k_{11} (mM^{-1} s^{-1}) = 0.01; τ = 0.83; $\Delta\mu/\Delta p°$ = 0.01. The kinetic parameters of the proton pump were: K_M = 2 mM; V_M = 0.5 mM s^{-1}. The total nucleotide (C_A) and phosphate pool (P_t) were both 10 mM, while the pyridine nucleotide pool (C_N) was 4 mM. The initial values of the state variables (mM) were as follows: Glc = 0.157; ATP = 1; I = 0.0031; NADH = 3 (a,b), 1 (c); Pyr = 0.394; EtOH = 1; Acetate = 0.01. The three curves shown correspond to different rates of ethanol oxidation (k_8) and NADH reoxidation (k_7). (b) The three curves shown correspond to different rates of ethanol input (V'_{in}) as indicated, and were simulated with the same parameter values as in (a) except that: k_7 (mM^{-1} s^{-1}) = 0.05; k_8 (mM^{-1} s^{-1}) = 6.25 × 10^{-4}. (c) The co-substrate model was modified as follows. (i) Reaction 5 (equation 12.22) was assumed also to oxidize NADH in a saturable fashion: V_5 = k_5[I][NADH]/(K_{h5} + [NADH]); (ii) an extra reaction (V_{10}) oxidizing NADH was added: V_{10} = k_{10}[NADH]/(K_{h10} + [NADH]); (iii) a saturable function for reaction 7 (equation 12.23) was considered: V_7 = k_7[Pyr][NADH]/(K_{h7} + [NADH]). It was assumed that the intracellular and extracellular ethanol concentrations had not equilibrated. The following parameter values were changed with respect to (a): k_{10} = 0.02; k_5 = 0.2; k_8 = 1 × 10^{-6}; k_9 = 0.15; k_P = 0.25; K_{h5} = 0.01; k_{h7} = 0.1; k_{h10} = 0.1; C_A = 8. The initial values of the state variables (mM) were the following: Glc = 5.72; ATP = 1.15; I = 0.0146; NADH = 0.37; Pyr = 2.62; EtOH = 0, 2660 and 9310 (for the lower, middle and upper traces, respectively); Acetate = 0.0000192. (d) It was assumed that the intracellular and extracellular ethanol concentrations have equilibrated. As a consequence, the ethanol concentration is diluted by a factor of $1/V_r$, where V_r is the ratio of the intracellular to the extracellular volume. V'_{in} is now absent from equation 12.15, which becomes:

$$\frac{d[EtOH]}{dt} = V_r V_7 - V_8$$

V_7 is defined as in (c). The effect of increasing cell density (lowest versus middle line) and added ethanol (top versus middle line) were simulated with the equations, parameters and initial values of the state variables as in (c), except that: EtOH = 0.012 and V_r = 0.0004 (lowest curve); EtOH = 0.12 and V_r = 0.004 (middle curve); EtOH = 0.52 and V_r = 0.004 (upper curve).

of plant cell growth, and how it could regulate cell wall synthesis at the plasmalemma, takes into account the supramolecular–cellular levels of organization for its description. This is reflected by the state variables that have been chosen. The interdependent relations between mechanical, electrical and biochemical events at supramolecular–cellular levels were formulated by a system of 14 simultaneous, coupled, nonlinear ordinary differential equations relating membrane potential, cell water uptake, turgor pressure, cell elongation, membrane strain, ATP and nucleotide–sugar precursor concentrations, and protein and polysaccharide synthesis (Figure 12.18). Further various algebraic equations which consider conservation relations, functions of state variables and an exogenous stimulating function were also taken into account. For a complete mathematical formulation of the Bemchem model, see Aon and Cortassa (1989).

Plant cell expansion is intimately associated with cell wall extension. The cell wall is an envelope which changes in time and space, as the cell progresses along the growth gradient at the meristema (Roland *et al.*, 1982). A dynamic spatio-temporal relationship has been inferred for cortical plant cells during their differentiation along the growth gradient of the mung bean hypocotyl. The observations made in mung bean hypocotyls reveal the coherent and self-organized nature of the various processes implied in plant cell differentiation in the apice, i.e. meristema, and even for isolated algal or fungi cells. One of the impediments in the elucidation of the dynamics of plant cell growth and the expansion of the cell wall has been its complexity and the loss, by current biochemical analysis, of cellular integrity and, with it, some functional aspects (Bacic and Delmer, 1981; Roland *et al.*, 1982; Delmer, 1983; Taiz, 1984).

12.5.1 ROLE OF IONIC CURRENTS

Currents and growth are closely related. While ionic currents generally enter tip-growing cells at the apex and leave them laterally behind the tip, their directions are supposed to be driven by polarly distributed ion pumps and channels in the plasma membrane (Schnepf, 1986, and references therein; Aon and Cortassa, 1989, and references therein). The ATP-fuelled H^+-pumps may also account for endogenous electric fields which, through electrophoretic mechanisms, modulate membrane dynamics (e.g. transport, enzyme activity, water uptake) and membrane topography. Self- or lateral-electrophoretic mechanisms may account for cytoplasmic and membrane electrical fields, which constitute another way of transducing transmembrane and transcellular electrochemical gradients into cellular functions. Unequal distribution of ion pumps and

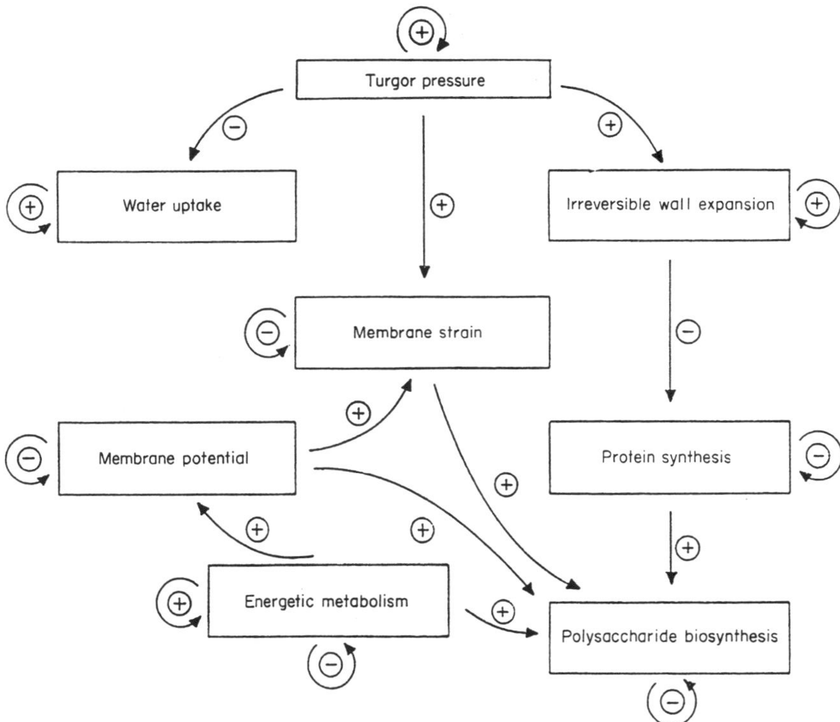

Figure 12.18 State variables of the Bemchem model and their interrelationships. Arrows indicate feedback by the state variable (e.g. turgor pressure) on its own behaviour. Arrow direction indicates the influence exerted by each state variable on another state variable (e.g. turgor pressure on membrane strain); + or − signs indicate positive or negative feedback. (Reproduced from Aon and Cortassa, 1989, by permission of Academic Press.)

channels within the plasma membrane are a clear manifestation of cell polarization. Alternatively, pumps and leaks need only be functionally segregated, and not necessarily asymmetrically distributed. A transcellular current may result from local regulation of transporters such that they are active at only one end of, say, a zygote (Kropf, 1992).

After fertilization in embryos of the marine brown algae *Fucus* and *Pelvetia* spp., the young zygotes establish a primary developmental axis in accordance with vectorial cues (Kropf, 1992; Chapter 4). Rhizoid growth is the first obvious morphological expression of the inherent polarity (Figure 12.19). Growth is localized to one end of the developmental axis, creating a pear-shaped zygote.

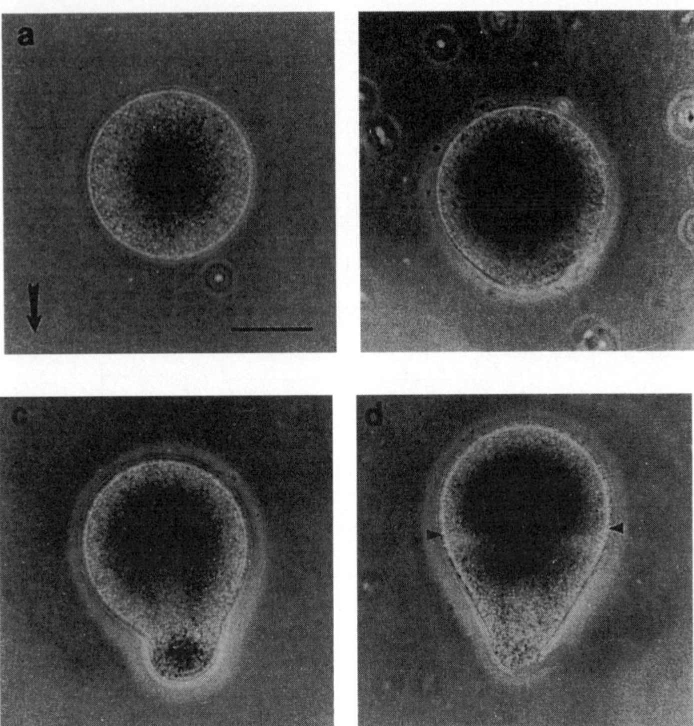

Figure 12.19 Establishment of polarity during the first cell cycle of *Pelvetia* zygotes: photomicrographs after fertilization. (a) 3 h; (b) 12 h, just before germination; (c) 16 h, in mitosis with a well defined rhizoid; (d) 22 h, first cytokinesis forming a larger thallus cell and a smaller rhizoid cell. Arrowheads mark the division plane, and the arrow indicates the direction of unilateral light. Bar = 50 μm. (Reproduced from Kropf, 1992, by permission of the American Society for Microbiology, ASM Press.)

The axis is established by localization of cytoplasmic components; in this way the egg cytoplasm is restructured into a polar cell. A first step in establishing polarity is the selection of an axis, which is made in fucoid zygotes by sensing environmental gradients. Zygotes can perceive an impressive array of gradients – for example, unilateral light of high irradiance, heat gradients, osmotic and ionic gradients, electrical fields or a neighbour (Kropf, 1992). The polar zygote results from localization of macromolecular determinants to the different ends of the developing embryonic axis, causing regional cytoplasmic differentiation. Manifestations of the latter are localizations of polysaccharides in the cell wall, organelles in the cytoplasm, the secretory apparatus, ionic fluxes and the cytoskeleton. Some of these localizations are temporally associated with formation of a labile axis, whereas others occur as the axis is irreversibly fixed in space (Harold, 1990; Kropf, 1992).

Several findings reported in the literature (reviewed by Kropf, 1992) have established a number of facts associated with axis formation:

- Polarized secretion begins within 4 hours of fertilization.
- An ionic current (partially carried by Ca^{2+}) accompanies axis establishment.
- Disruption of F-actin prevents photopolarization.
- The medium need contain only K^+ salts.

It is believed that environmental cues favour the development of an initial asymmetry probably involving the accumulation of calcium channels in the region of the prospective rhizoid. This localized calcium flux would be amplified progressively, generating a transcellular ionic current with an associated electric field and gradient of calcium concentration across the cytoplasm (Harold, 1990; for a review, see Jackson and Heath, 1993). A major assumption of this model is that ionic currents precede and cause localized secretion. It may well be that localized secretion causes the current flow. Localized secretion is observed at least as early as ionic currents, and perhaps localized secretion of vesicles containing ion channels, in particular conductive K^+ channels, sets up the current flow (Kropf, 1992). According to this hypothesis, the essence of axis formation is polarized secretion, not rearrangements of transport proteins in the plasmalemma.

Cytoplasmic gradients may act as macroscopic cues of, say, secretory processes. Tip growth in rhizoid cells of *Pelvetia* embryos was associated with a cytoplasmic longitudinal pH gradient in which the cytosol at the tip was 0.3 to 0.5 units more acidic than the cytosol at the base of the cell (Gibbon and Kropf, 1994). Two major observations suggested that the cytoplasmic pH gradient and tip growth are causally related:

- dissipation of the cytoplasmic gradient with propionic acid (a membrane-permeant weak acid) inhibited growth;
- gradients of up to 0.3 units of pH linearly correlated with growth rates of up to $\sim 3.0\ \mu m\ h^{-1}$ (Gibbon and Kropf, 1994).

12.5.2 BIOELECTROMECHANOCHEMICAL (BEMCHEM) MODELLING OF PLANT CELL GROWTH

The dynamic behaviour of the model actually depends on the structure of the system: the state variables and their interactions are represented by a set of simultaneous ODEs (Figures 12.18 and 13.2). From a sensitivity analysis by computer simulation experiments, some insights were obtained about the relative influence of the different state variables and parameters on the dynamics of plant cell expansion and cellulose synthesis.

The results obtained could be synthetically expressed in the following way.

1. Changes associated with the transmembrane potential difference ($\Delta\psi$) directly influence, qualitatively and quantitatively, cell wall synthesis (Figure 13.3). The observed effect of $\Delta\psi$ on polysaccharide synthesis could be obtained by changing parameters either directly associated to the dynamics of $\Delta\psi$, i.e. H^+ conductance, or indirectly related ones such as glucose input flux and the kinetic constants of energetic metabolism (Aon and Cortassa, 1989).

   A plausible mechanism which could mediate the observed effect of $\Delta\psi$ on cellulose synthesis could be the polarization of the cellulose synthase complexes by a self-electrophoretic mechanism (Jaffe, 1977). The electrophoretic mechanism of charged macromolecules along the plasmalemma would polarize *in vivo* the synthesis of cellulose to the tip of the cell. In a hypothetical manner, a similar mechanism has recently been suggested for explaining the possible significance of the self-maintained endogenous electrical currents in fungal hyphae. However, it must also be taken into account that not only tip growth occurs but diffusive growth as well, i.e. incorporation of cell wall material along the cell surface (Roland, 1973; Taiz, 1984). Transmembrane electric potentials could also exert their influence by directly affecting membrane enzymatic activity associated with cellulose synthesis (Bacic and Delmer, 1981; Delmer *et al.*, 1982), membrane lipid composition (Clementz *et al.*, 1986) or energy-transducing mechanisms through membrane enzymes (Westerhoff *et al.*, 1986; Tsong and Astumian, 1988; Tsong, 1989). Triggering of phospholipase A2 activity by an electrical field in the range of 25–175 mV, usually found in cells, has been demonstrated (Thuren *et al.*, 1987).

2. Not only quantitative changes in $\Delta\psi$ but also qualitative changes, i.e. oscillatory hyperpolarized $\Delta\psi$ and oscillatory hyperpolarizing–depolarizing of the membrane, influenced the dynamics of cell wall synthesis.

   The lower rate and different dynamics (linear or sigmoidal) of cellulose synthesis in the presence of a constant or an oscillatory $\Delta\psi$ could be explained as the difference existing for the polarization mechanism of the cellulose synthase complexes in an analogous manner to that observed in Con A receptors. The electromigration of cell surface Con A receptors in *Xenopus* myoblasts in the case of pulsed electrical fields is interrupted by periods of diffusional randomization during interpulse intervals (Poo, 1981; Lin-Liu *et al.*, 1984). The self-sustained oscillatory behaviour may simulate a pulsed endogenous electrical field.

It has been experimentally and theoretically demonstrated that transmembrane enzymes (e.g. Na^+, K^+-ATPase) can be expected to absorb free energy from an oscillating electric field and transduce that to chemical or transport work (Westerhoff *et al.*, 1986; Tsong, 1989). The addition of indole-3-acetic acid (IAA) to corn coleoptiles resulted in membrane hyperpolarization followed by oscillations of $\Delta\psi$ in the hyperpolarized phase (Felle *et al.*, 1986). The latter result is strongly reminiscent of those obtained in our simulations. In *Nitella*, oscillations of $\Delta\psi$ with periods of 45 s, 15 min and 1 h have been described and would correspond, at least those with periods of 15 min and 1 h, to oscillations of an electrogenic H^+ pump and a K^+ channel (Fishan *et al.*, 1986).

3. The dynamics of $\Delta\psi$ in the Bemchem model showed a rapid hyperpolarization which could well be simulating the hyperpolarization response of plant cells to the auxin treatment (Cleland *et al.*, 1977; Brummer *et al.*, 1985).

   It is known that auxins also cause a rapid acidification of the cytoplasm and the periplasmic space (Cleland, 1971, 1976; Brummer and Parish, 1983; Taiz, 1984; Brummer *et al.*, 1985; Marre and Denti-Ballarin, 1985; Clementz *et al.*, 1986), an effect thought to be related to the short-term elongation response of plant cells to the hormone (Ray, 1985). Other authors argue that continuous elongation growth would depend on changes in $\Delta\psi$ through changes in transmembrane ion gradients (Brummer *et al.*, 1985).

4. Different steady values for $\Delta\psi$ were obtained corresponding to different steady values of ATP (Aon and Cortassa, 1989). The latter correlates well with experimental results obtained in *Characean* perfused cells since it had been shown that $\Delta\psi$ hyperpolarization was dependent upon the adenine nucleotide levels (Mimura and Tazawa, 1986).

5. Certain state variables such as protein–polysaccharide synthesis and cell elongation, although mathematically related, behaved as temporally unrelated. In Figure 12.20 it can be seen that for a significant amount of protein synthesized there was no cellulose produced. However, after a critical value (i.e. threshold behaviour), both state variables become directly and proportionally related. For both coleoptiles and internodes, it has been repeatedly shown that

indole-3-acetic acid (IAA) increases the rate of cell wall synthesis (Abdul-Baki and Ray, 1971; Kutschera and Briggs, 1987). Electron microscopy studies of maize coleoptile cell walls have revealed that IAA may rapidly stimulate cell wall synthesis in the epidermal cells, which comprise only a small percentage of the cells of the whole organ (Kutschera and Briggs, 1987). In pea internodes, the IAA-induced change in the mechanical property of the tissue (higher plastic extensibility) was detected preferentially in the outer cell layer. Using the precursor of new matrix non-cellulosic polysaccharides (mostly uronic acid and pentose residues) of growing cell walls (myo-[2-^{3}H(N)]inositol ([^{3}H]Ins), Kutschera and Briggs (1987) showed that its incorporation was stimulated in epidermal cell walls within 15 min by IAA without a detectable effect in the cortical cylinder during the first hour after hormone treatment.

Substantial evidence supports the notion that continued RNA and protein synthesis is required for auxin to induce cell elongation for the continued production of extracellular substances and the maintenance of the secretory apparatus (Nooden and Thimann, 1963;

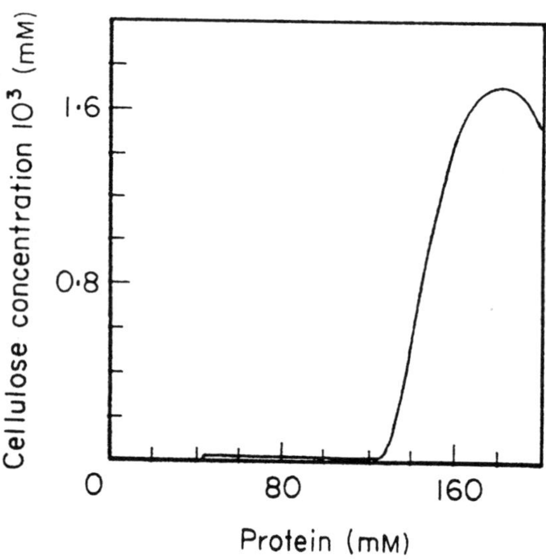

Figure 12.20 Phase–space analysis of polysaccharide and protein synthesis in the Benchem model. The initial values of the state variables and parameters are specified as in Aon and Cortassa (1989), except for $K_{qd} = 9 \times 10^{-3}$; $\phi_g = 1 \times 10^{-3}$ (see Table 1 in Aon and Cortassa, 1989, for definition of other model parameters K_{qd} and ϕ_g). (Reproduced from Aon and Cortassa, 1989, by permission of Academic Press.)

Roland, 1973; Chrispeels, 1976; Bates and Cleland, 1979; Zurfluh and Guilfoyle, 1980; Theologis and Ray, 1982). After reaching a maximum the protein–polysaccharide synthesis relationship became inversely proportional (Figure 12.20).

We have also shown that there was no close relationship between cell elongation and polysaccharide synthesis for the first part of the curve, since a significant amount of carbohydrate was synthesized for still no appreciable cell elongation (Figure 1.4). In other words, cell elongation and polysaccharide synthesis correspond to two stages of the same phenomenon. In fact, first the cell synthesizes its wall and then it elongates. Thus, the model predicts that cell wall synthesis could be elicited prior to an exponential elongation of the cell. Roland *et al.* (1982) had stated that, for the growth gradient existing in mung bean hypocotyls, cells at first acquired the capacity for a massive elaboration of wall subunits and after that they reached the exponential phase of extension in which both thickness and order were maximal (Figure 1.4a–c). In the epidermal cells of pea internodes, evidence was presented which suggested that wall synthesis and wall loosening may be causally linked processes (Kutschera and Briggs, 1987). Yet the equations describing cell extension are for single isolated cells (Cosgrove, 1986); no straightforward conclusions can be drawn for multicellular tissues where parameters such as water diffusivity (Cosgrove, 1981) and geometrical constraints could be important.

Several of the effects described by the Bemchem model reflected systemic behaviour. Systemic properties would imply variables with different relaxation times which optimize each other succesively and hierarchically. The time scale for a significant variation of $\Delta\psi$ is milliseconds–seconds, while for cellulose synthesis it is minutes–hour (Figure 13.3). The membrane strain behaves as a relatively fast process since it takes milliseconds to attain the steady state. Under those conditions, turgor pressure and $\Delta\psi$ have varied little for a relatively great change in the membrane strain. These results suggested that electromechanical events are earlier processes affecting and conditioning the future behaviour of plasma membrane, i.e. self-electrophoresis of plasmalemma-proteins, and H^+ pumps. A proposed rationale for understanding these apparently temporally unrelated phenomena ($\Delta\psi$ and cellulose synthesis dynamics) was described in Chapters 2, 6 and 7. Briefly, subcellular processes occurring in widely different temporal scales but showing spatio-temporal scaling may exhibit self-organized transitions. These transitions, from microscopic (local) to macroscopic (global) organization, may be affected by polymerization–depolymerization of cytoskeleton components at the cellular level (Chapters 6 and 7). In a series of sequential events moving in different time scales, the fast variables will reach their stationary state before any significant change in the slower processes is observed (Reich and Sel'kov, 1981). This has been rationalized through the concept of dynamic coupling (see also Chapters 1 and 7).

The conditioning of plasma membrane behaviour by electromechanical events could also be related to the high degree of organization, both spatially and temporally, observed in the plasmalemma of growing fungi, algae and plant cells (Herth, 1985; Ryser, 1985; Wessels, 1986). This particular membrane topography would suggest a self-organized phenomenon (Nicolis and Prigogine, 1977; Roland et al., 1982). It is known that reacting–diffusing, nonlinear systems far from thermodynamic equilibrium, as is the case with enzymatic reactions taking place in the plane of the membrane, would behave as pattern-generating (Schiffmann, 1982) and signal-propagating mechanisms (Ricard, 1987) which exhibit several dynamic behaviours, i.e. hysteresis and oscillations (Katchalsky and Spangler, 1968; Hervagault and Thomas, 1987) (see, for example, Chapter 5).

6. According to a current hypothesis, the H^+ pump would be also a turgor detector. Reported data suggest that high turgor inhibits H^+ extrusion but that H^+ extrusion can be activated in turgid tissue by cytoplasmic acidification (Kinraide and Wisse, 1986). Thus, the H^+ pumping activity raises the possibility of unifying mechanical and electrical events at the plasmalemma. In algae, low turgor stimulates the active transport of ions into the vacuole, while excessive turgor elicits leakage of ions outward (for a review, see Harold, 1982).

12.6 CARBON AND ENERGY COUPLING FOR METABOLITE AND BIOMASS PRODUCTION

Assimilatory and dissimilatory flows are needed to incorporate carbon substrates into biomass. The **assimilatory flow** is defined as the substrate flux directed towards biomass synthesis (**anabolism**) whereas the **dissimilatory flow** is the amount of substrate needed to release energy for other purposes such as transport, futile cycles and synthesis of biomass (**catabolism**) (Gommers et al., 1988). The oxidation of an auxiliary energy source results in a redirection of the carbon flow from dissimilation into assimilation. A benefit, then, of dual substrate use might be the improvement of the proportion of the primary substrate being assimilated. A general relationship between the energy content of a given growth substrate (kJ g^{-1} carbon substrate) and the growth yield (g dw g^{-1} substrate carbon) has been derived by Linton (1989). The growth yield is directly proportional to the heat of combustion when the latter is less than 46 kJ g^{-1}. Thus, the yield is energy-limited during growth on a single C-compound that has a heat of combustion less than 46 kJ g^{-1} carbon. The addition of a second substrate that can be oxidized to provide additional energy, but not carbon, should overcome this limitation and result in an increase in yield. However, this is true only when the auxiliary substrate has a heat of combustion greater than the first substrate (Gommers *et al.*, 1988; Linton, 1989) (Figure 8.7a).

We have modelled the catabolic behaviour of a microorganism such as *E. coli* fermenting glucose (first substrate) by the Embden–Meyerhoff (EM) pathway and consuming formate as a second substrate whose oxidation through anaerobic electron transport to form CO_2 gives additional energy (Aon and Cortassa, 1991). The model is described by a set of six nonlinear differential equations and two conservation relations which were solved numerically (Appendix 12).

12.6.1 THERMODYNAMIC BEHAVIOUR OF MIXED SUBSTRATE CONSUMPTION

By feeding the model with different input ratios of both substrates, formate/glucose (S_2/S_1), we were able to vary the steady state behaviour of thermodynamic efficiency at low ratios, i.e. higher rates of uptake of glucose than formate (Figure 12.21). **Thermodynamic efficiency**, defined as the ratio between the output power of ATP synthesis and the input catabolic power, was coincident with maximum in ATP yields for optimum molar ratios of the two substrates. This allows the prediction that higher biomass yields may be obtained through a variable, high, thermodynamic efficiency by directing fluxes through catabolic pathways with different ATP yields.

By comparing single- and mixed-substrate models (Cortassa *et al.*, 1990a; Aon *et al.*, 1990a, 1991; Aon and Cortassa, 1991) significantly greater yields in ATP were shown to be obtained in the two-substrate case. The maximum in ATP yield was obtained for an optimum input ratio of the two substrates. A useful prediction of this analysis is that higher biomass yields may be obtained by directing fluxes through catabolic pathways with different stoichiometry of ATP production.

We have been able to show that mixed-substrate consumption was involved in cell–cell interaction in yeast cell suspensions at concentrations higher than 10^7 cells (Aon *et al.*, 1992). The dependence of the damping of glycolytic oscillations and the amplitude per cell upon

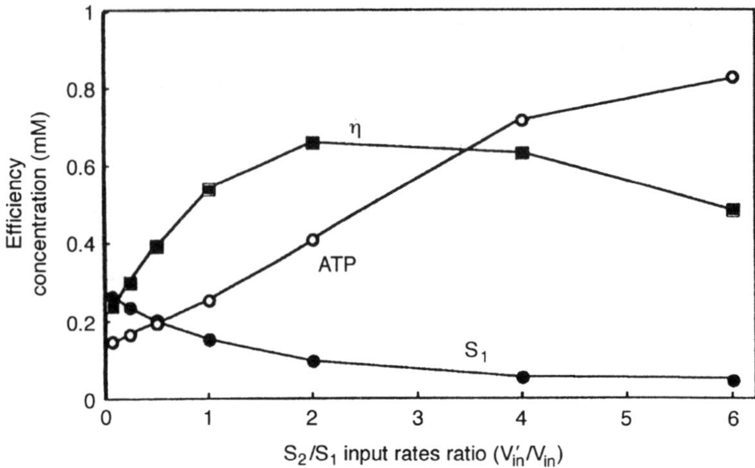

Figure 12.21 Steady state behaviour of thermodynamic efficiency (η), S_1 and ATP as a function of substrate input ratios. Model simulations were carried out by varying the input rates of S_2 (V'_{in}: equation 12.61) keeping that of S_1 constant (V_{in} = 0.12 mM s^{-1}: equation 12.56). The following parameter values were used: k_1 = 0.3 mM^{-1} s^{-1}; k_3 (mM^{-2} s^{-1}) = k_5 (s^{-1}) = k_7 (s^{-1}) = 0.1; k_9 = 0.05 s^{-1}; k_{10} = 0.2 mM^{-2} s^{-1}; γ_1 = 2; γ_3 = 4; σ = 1.0; $\Delta G°_{ATP}$ = 7.3 kcal mol^{-1}; $\Delta G°_{SL}$ = –47 kcal mol^{-1}; $\Delta G°_{FC}$ - –20.3 kcal mol^{-1}; RT = 0.596. The kinetic parameters of the H$^+$ pump were: K_M = 2 mM; k_P = 0.391 mM s^{-1}. The total nucleotide (ΣC_A) and phosphate pool (P_t) were both 10 mM. The initial values of the state variables (mM) were: S_1 = 0.157; ATP = 0.552; I = 0.0031; L = 0.394; S_2 = 0.157 and CO_2 = 0.01. In all cases the ATP concentration plotted is one tenth of the actual value. (From Aon and Cortassa, *Biotechnology and Bioengineering*, copyright 1991 John Wiley & Sons, Inc.)

cell density, in the absence and presence of exogenously added ethanol, pyruvate and acetaldehyde, could be explained on the basis of a co-substrate model (glucose and ethanol) feeding glycolysis and reductive ethanol catabolism, respectively (Figures 12.14 and 12.15).

APPENDIX 12A MODEL FORMULATION

12A.1 ANAEROBIC GLYCOLYSIS

A model of glycolysis and its interactions with mitochondrial functions was developed by Aon *et al.* (1991). Briefly, the model represents a yeast cell fermenting glucose (Glc), irreversibly, by the Embden–Meyerhoff pathway (Figure 12.15). Only stoichiometric relationships are included, i.e. no allosteric mechanisms are taken into account. The utilization of energy-rich intermediates (I: glyceraldehyde 3-phosphate) for synthesis and the non-glycolytic ATP-consuming processes were modelled as described by Cortassa *et al.* (1990a). The reactions describing either ATP synthesis (V_{P_2}) or hydrolysis (V_{P_1}) by the mitochondrial H^+ pump were assumed to follow Michaelis–Menten kinetics as a function of ADP or ATP concentrations, respectively (equation 12.28). The flux (V_P) of ATP synthesis or hydrolysis was dependent upon the proton motive force $\Delta\mu_{H^+}$. The mitochondrial ATPase was simulated only in the sense of ATP hydrolysis that was previously demonstrated to trigger glycolytic oscillations through a decrease in $\Delta\mu_{H^+}$ by, for example, respiratory inhibitors (Aon *et al.*, 1991).

12A.2 ETHANOL OXIDATION PATHWAY

The additional ethanol oxidative pathway, comprising four steps leading to acetyl CoA, was modelled as two lumped steps. The first step, leading to acetate, consumes two molecules of NAD; the second step transforms acetate into acetyl phosphate with ATP consumption, first, and into AcCoA afterwards (Figure 12.15).

12A.3 MATHEMATICAL FORMULATION OF THE MODEL

The model represented in Figure 12.15 is described by a set of seven coupled nonlinear differential equations (12.10–12.16) and three conservation equations (12.17–12.19):

$$\frac{d[\text{Glc}]}{dt} = V_{in} - V_1 \tag{12.10}$$

$$\frac{d[\text{ATP}]}{dt} = -2V_1 + 4V_3 - V_9 - V_p - V_{11} \tag{12.11}$$

$$\frac{d[\text{I}]}{dt} = V_1 - V_3 - V_5 \tag{12.12}$$

$$\frac{d[\text{NADH}]}{dt} = V_3 - V_7 + 2V_8 - V_{10} - V_5 \tag{12.13}$$

$$\frac{d[\text{Pyr}]}{dt} = V_3 - V_7 \tag{12.14}$$

$$\frac{d[\text{EtOH}]}{dt} = V'_{in} - V_8 \tag{12.15}$$

$$\frac{d[\text{Acetate}]}{dt} = V_8 - V_{11} \tag{12.16}$$

$$C_A = [ATP] + [ADP] \qquad (12.17)$$

$$C_N = [NAD] + [NADH]; \qquad (12.18)$$

$$P_t = P_i + 2[I] + [ATP] \qquad (12.19)$$

with the following fluxes:

$$V_1 = k_1[ATP][Glc] \qquad (12.20)$$

$$V_3 = k_3[ADP][I][P_i][NAD] \qquad (12.21)$$

$$V_5 = k_5[I] \qquad (12.22)$$

$$V_7 = k_7[Pyr][NADH] \qquad (12.23)$$

$$V_8 = k_8[EtOH][NAD] \qquad (12.24)$$

$$V_9 = k_9[ATP] \qquad (12.25)$$

$$V_{10} = k_{10}[NADH] \qquad (12.26)$$

$$V_{11} = k_{11}[Acetate][ATP] \qquad (12.27)$$

$$V_P = k_P \frac{[ATP]}{K_M + [ATP]}; \qquad k_P = \tau V_{max}\left(1 - \frac{\Delta\mu_{H^+}}{\Delta p^0}\right)^6 \qquad (12.28)$$

The co-substrate model was simulated under the following conditions: 'homogeneous' or 'heterogeneous' cell populations interacting through different types of ethanol signals, e.g. sinusoidal, linear, square pulses (Figure 12.16). By conditions of cell-to-cell interaction is meant that the ethanol pool is split in two. In equation 12.29, $[EtOH]_e$ and $[EtOH]_i$ represent the extracellular and intracellular pools of ethanol, respectively. $[EtOH]_e$ will be forced from 'outside the cell' with different exogenous, time-dependent forcing functions or 'signals'(SCoP, Duke University, 1989):

- harmonic (sine wave generator);
- linear (simulated with a right-triangular sawtooth with positive slope);
- squarewave (square wave generator);
- perpulse (periodic pulses following a lag duration, delay).

Although only two of these functions, i.e. harmonic and linear, are shown (Figure 12.16), extensive calculations were carried out with all the others.

This strategy is intended to represent a cell that is receiving a signal from the medium, which in this case is ethanol, the substrate of one of the two catabolic pathways (Figure 12.15). The simulations were carried out for 'homogeneous' or 'heterogeneous' cell populations. Essentially, the changes introduced in the ODEs system (12.10–12.16) are as follows. The rate of input of ethanol (V'_{in}), instead of being constant, is now defined by:

$$V_{12} = k_{12}([EtOH]_e - [EtOH]_i) \qquad (12.29)$$

where k_{12} with dimension s^{-1} results from the following relationship:

$$k_{12} = \frac{S}{V} P_{EtOH} \qquad (12.30)$$

S and V are the cell surface and volume, respectively, and P is the permeability coefficient. The S/V ratio of 1.5 μm^{-1} was obtained considering a yeast cell of 4 μm diameter and 30 nl volume. Permeability coefficients for ethanol of 10^{-5} to 10^{-9} cm s^{-1} were used (Jones, 1988).

12A.4 INFLUENCE OF CELL DENSITY ON GLYCOLYTIC CELLS' AUTONOMOUS OSCILLATIONS THROUGH EXTRACELLULAR ETHANOL

The ethanol produced in reaction 7 (represented by k_7) was assumed to be rapidly equilibrating across the plasma membrane, hence diluting out into the bulk medium. At this lower concentration, it was then again allowed to react into the bottom pathway of Figure 12.15. A saturable function for reaction 7 was considered (Figure 12.17).

APPENDIX 12B PTS-GLYCOLYSIS COUPLED MODEL

Glycolysis was modelled in two lumped steps. The first five steps (upper part) or preparatory phase (Lehninger, 1977) are represented by G6PI, with rate constant k_1, and I representing the pool of glyceraldehyde 3-phosphate. The second phase of glycolysis (lower part), is lumped in IPEPPyr, with rate constants k_3 and k_{pk}. Pyruvate is considered to be further processed with a first-order rate constant k_7. From PEP the model considers a branch point towards Pyr or the PTS system.

The phosphorylation–dephosphorylation cascade of the PTS system depicted in the model of Figure 12.8 has been simplified to one loop of a moeity conserved cycle in which substrate A is phosphorylated by enzyme E_1 with rate constant k_{e1} and PEP as phosphate donor, and dephosphorylated by enzyme E_2 with rate constant k_{e2} to phosphorylate glucose to G6P.

Membrane-based as well as non-glycolytic ATP-consuming processes were also taken into account. These processes are captured by k_9 and the membrane H^+-ATPase.

$$V_{e1} = k_{e1} \frac{[PEP][A]}{K_{PE} + [PEP]} \tag{12.31}$$

$$V_{e2} = k_{e2} \frac{[Glc]([TA] - [A])}{K_G + [Glc]} \tag{12.32}$$

$$V_1 = k_1[ATP][G6P] \tag{12.33}$$

$$V_2 = k_3[ADP][I]([P_t] - [I] - [ATP] - [PEP] - [G6P] - [TA] + [A]) \tag{12.34}$$

$$V_5 = k_5[I] \tag{12.35}$$

$$V_{PK} = k_{PK} \frac{[PEP][ADP]}{K_P + [PEP]} \tag{12.36}$$

$$V_7 = k_7[Py] \tag{12.37}$$

$$V_9 = k_9[ATP] \tag{12.38}$$

$$V_P = k_P \frac{[ATP]}{K_M + [ATP]} \tag{12.39}$$

$$\frac{d[Glc]}{dt} = -V_{e2} \tag{12.40}$$

$$\frac{d[G6P]}{dt} = V_{e2} - V_1 \tag{12.41}$$

$$\frac{d[ATP]}{dt} = -V_1 + V_2 - V_9 - V_P + V_{PK} \tag{12.42}$$

$$\frac{d[I]}{dt} = 2V_1 - V_5 - V_2 \tag{12.43}$$

$$\frac{d[Py]}{dt} = V_{e1} - V_7 + V_{PK} \tag{12.44}$$

$$\frac{d[PEP]}{dt} = V_2 - V_{e1} - V_{PK} \tag{12.45}$$

$$\frac{d[A]}{dt} = V_{e2} - V_{e1} \tag{12.46}$$

$$\Sigma C_A = [ATP] + [ADP] \tag{12.47}$$

APPENDIX 12C MODEL OF MIXED SUBSTRATE CATABOLISM

The model is described by the following equations (Aon and Cortassa, 1991):

$$V_1 = k_1[ATP][Glc] \tag{12.48}$$

$$V_3 = k_3[ADP][I]([P_t] - 2[I] - [ATP]) \tag{12.49}$$

$$V_5 = k_5[I] \tag{12.50}$$

$$V_7 = k_7[L] \tag{12.51}$$

$$V_9 = k_9[ATP] \tag{12.52}$$

$$V_{10} = k_{10}[ADP][Fo]([P_t] - 2[I] - [ATP]) \tag{12.53}$$

$$V_{11} = k_{11}[CO_2] \tag{12.54}$$

$$V_P = k_P \frac{[ATP]}{K_M + [ATP]} \tag{12.55}$$

where k_1, k_3, k_5, k_7, k_9, k_{10}, k_{11} and k_P are rate constants.

The reactions describing the ATP consumption by an H^+ pump, V_P, were assumed to follow Michaelis–Menten behaviour as a function of ATP (the parameter k_P is defined in equation 12.28). The system is described by the following set of ODEs, including a conservation relationship for adenine nucleotides (Aon and Cortassa, 1991):

$$\frac{d[Glc]}{dt} = V_{in} - V_1 \tag{12.56}$$

$$\frac{d[ATP]}{dt} = -2V_1 + 4V_3 + V_{10} - V_9 - V_P \tag{12.57}$$

$$\frac{d[I]}{dt} = V_1 - V_3 - V_5 \tag{12.58}$$

$$\frac{d[L]}{dt} = V_3 - V_7 \tag{12.59}$$

$$\frac{d[CO_2]}{dt} = V_{10} - V_{11} \tag{12.60}$$

$$\frac{d[F_0]}{dt} = V'_{in} - V_1 \tag{12.61}$$

$$\Sigma C_A = [ATP] + [ADP] \tag{12.62}$$

The thermodynamic functions, fluxes (J) and forces (ΔG) were evaluated through the instantaneous level of metabolites by the following equations:

$$J_a = -2V_1 + 4V_3 + V_{10} \tag{12.63}$$

$$J_{S1} = \alpha V_3 \tag{12.64}$$

$$J_{FC} = \beta V_{10} \tag{12.65}$$

$$\Delta G_{ATP} = \Delta G^{\circ}_{ATP} + RT ln \frac{[ATP]}{(P_t - 2I - [ATP])(\Sigma Ca - [ATP])} \qquad (12.66)$$

$$\Delta G_s = \alpha \left[\Delta G^{\circ}_s + RT ln \frac{[L]^2}{[Glc]} \right] + \beta 0.5 \left[\Delta G^{\circ}_{FC} + RT ln \frac{[CO_2]}{[F_0]} \right] \qquad (12.67)$$

$$\alpha = \frac{\phi L}{\phi L + [CO_2]} ; \quad \beta = \frac{[CO_2]}{\phi L + [CO_2]} ; \quad \phi = \frac{k_7}{k_{11}} \qquad (12.68)$$

where α and β are variable weighing coefficients depending on the products of glucose (Glc) and formate (Fo) pathways; ϕ is the ratio between the apparent first-order rate constants of disappearance, k_7 and k_{11}, of both products, L and CO_2, from the Glc and Fo pathways, respectively.

Thermodynamic efficiency is defined as follows (Aon and Cortassa, 1991):

$$\eta = \frac{\Delta G_{ATP} J_a}{\Delta G_s (J_{SL} + J_{FC})} \qquad (12.69)$$

where η varies between 0 and 1, being always positive since J_a and ΔG_{ATP} are positive, and ΔG_s and $J_{SL} + J_{FC}$ are negative

Conclusions and outlook: models, facts and biocomplexity

13

If, then, the unitary chemical ground-plan of animal growth exists, we must think of it as deformable in space and time. Just as D'Arcy Thompson was able to transform one morphological shape into another... so the chemical ground-plan is transformed in time and space, by systematic deformation of Cartesian coordinates.

(J. Needham, 1934, quoted by Rosen, 1967)

This final chapter seeks to summarize in a direct form our experience with different biological or biochemical systems and their modelling. It is our attempt to distil the main aspects of our learning concerning problems and limitations we realized during this intellectual adventure. One main purpose is to give some clues about the living complexity. We will refer continuously to the whole book, intending to allow the reader to 'iterate' on the perspectives we have adopted throughout.

13.1 APPROACHES TO COMPLEXITY

As observers of an object, i.e. a biological system, researchers are restrained both in the particular hypothetical scheme, model or paradigm they choose for looking at the object, and in the perturbation method of study (Aon, 1995) The latter always gives information from a certain level of organization, depending on whether it is, for example, a fluorescent molecule or particle, a laser beam, a radioactive precursor or a magnetic probe. On the other hand, biological systems are complex (manifold) objects, whose spatio-temporal complexity is characterized by two main features: counter-intuitivity and non-reducibility (Rosen, 1967; Stein, 1989). The first feature encourages the use mathematical models in order to deal with biological systems, whereas the second stimulates the study of whole systems, taking into consideration that many of the possible behaviours a system is able to exhibit are likely to evaporate whenever it is reduced to some of its components. A fairly general and important result is that complexity usually results in instability rather than stability (Murray, 1989).

We are always confronted with the problems of ignorance and error which arise whenever we analyse multidimensional systems (Abraham, 1987). It is common to find contradictory experimental evidence in the literature with respect to the same phenomenon in either the same or different biological systems. When these differences are not due to obvious experimental procedures, it may be thought that researchers are exploring different phases of the system's dynamic behaviour. Strictly speaking, different parametric regions of the phase space dynamic behaviour of the system may correspond to different attractors (Chapter 1). Some of these attractors could have large or small basins, which in the former case implies that in departing from widely differing initial parameters the system will still evolve to the same attractor. Yet for attractors with small basins even slightly different initial parametric conditions will drive the system to drift toward another attractor, thus explaining different experimental results for apparently similar conditions (Chapter 1). When experimental conditions poise the behaviour of the system around bifurcation points (where the dynamics is suddenly and drastically altered), the change in the system's motion may be even more dramatic. Qualitative changes may be envisaged, for example, from asymptotic steady states to limit cycle (oscillations) or even chaotic behaviour (Chapters 3 and 7). Quantitatively, the performance behaviour of phenomena may also exhibit divergent results.

Let us synthesize with a paradigmatic biological example the main features of the considerations above: the coupling between microtubular protein (MTP) and enzyme dynamics. Figures 1.7 and 6.11 emphasize the multidimensional character of the spatio-temporal regulation of cell function, i.e. the changes in steady state values of PEP or polymerized MTP versus the simultaneous variation of the total amount of MTP, C_o, and its rate of polymerization, K_{Pol} (Figure 1.8). The multidimensional character of regulation results, for this particular case, in a 'line of possible bifurcation points', i.e. joining the points that limit the stable and unstable branches of steady states in the individual bifurcation diagrams. The shaded parts of Figure 1.8 correspond to a 'parametric surface' of stable dynamic behaviour which results from the concomitant balance of the total amount of MTP and its rate of polymerization. However, one may think of a 'hypersurface' instead of a surface since the studied parameters 'lump' other ones. In fact, the variations of C_o and K_{Pol} may result from the 'lumping' of several physiological processes which, in turn, entrain different functional properties. For instance, actin or tubulin polymerization is sensitive to cytoplasmic Ca^{2+}, phosphoinositide levels (Janmey, 1994) and pH changes (Suprenant, 1991). These physiological variables will affect K_{Pol}. In the same vein, C_0 may be modulated by changes in gene expression (Ben-Ze'ev, 1986; Cleveland, 1988). Figure 1.8 shows that as the amount of protein increases either the irreversible transitions in PK activity or the level of polymerized protein, C_P, occurs at lower rates of polymerization, K_{Pol}. The transitions in MTP polymeric status entrain irreversible changes in the catalytic properties of PK as well as in its conformational state – namely, aggregation as a pentamer (Figure 6.11).

Figure 13.1 shows the temporal evolution of several processes of chemical or physical (e.g. mechanical) nature that have been described as associated with plant cell growth. At first sight the occurrence of some of these processes happens at similar or widely different time scales. Following auxin treatment, cell elongation, pH and glucose incorporation closely correlate in time, i.e. the three events appreciably change in the first hour after hormone treatment. However, the three processes relax more or less rapidly after the perturbation (i.e. auxin presence). Therefore, a plant cell which is elongating and changing its cell wall by polysaccharide synthesis is a complex biological system whose several implicated processes scale not only in time but also in space (Table 2.1). A main problem arises when we attempt to model such a system mathematically since several of these processes must be somewhat functionally coupled to give the emergent functional behaviour which is cell elongation.

We have formulated a mathematical model of this system (Figure 13.2). An important unsolved problem is how those several different processes are coupled and how the spatio-temporal functional scaling is effected by the cell. This is probably one of the main questions that we have addressed in the whole book. Let us be more precise. We asked how fast processes, such as changes in electrochemical gradients at the plasmalemma or Ca^{2+} waves or polarization, relate to processes that occur much more slowly, such as cell elongation and cell wall synthesis (Figure 13.3). By fast, we mean those processes that relax quickly toward a perturbation, i.e. membrane strain or electrical potential reaction perturbed by auxin.

The basis of our efforts in approaching this problem are given in Chapters 2, 6 and 9. Essentially, we have verified that main functions of complex cellular systems (for example, growth, proliferation, energy transduction; Table 2.1) are the result of several coupled processes functioning simultaneously at distinct levels of organization. The nature of the coupling may be manifold, i.e. electrochemical, mechanochemical, electromechanical, being driven by diverse mechanisms as well (Chapters

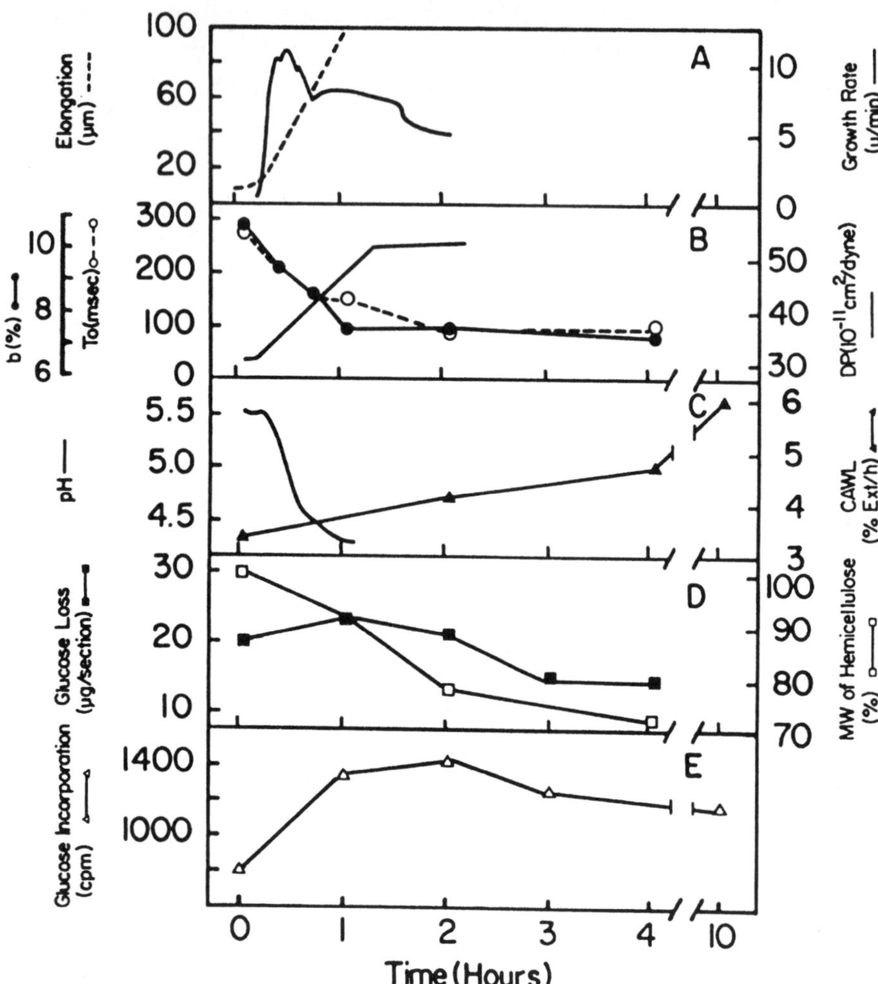

Figure 13.1 Summary time-course of the biomechanochemical events during auxin-induced elongation of *Avena* coleoptiles. (Reproduced from Taiz, *Annual Review of Plant Physiology*, Vol. 35, 1984, with permission by Annual Reviews Inc.)

7 and 12) that interconvert among each other following the particular fate of the cellular physiological status. But, as stated above, although the various levels of organization function simultaneously, they do so in different spatio-temporal scales. Accordingly, we have attempted a new characterization of a level of organization based on functional properties. The occurrence of a biological process at distinguished dimensions of space (E_c) and time (T_r) differentiates a level of organization (Tables 2.1

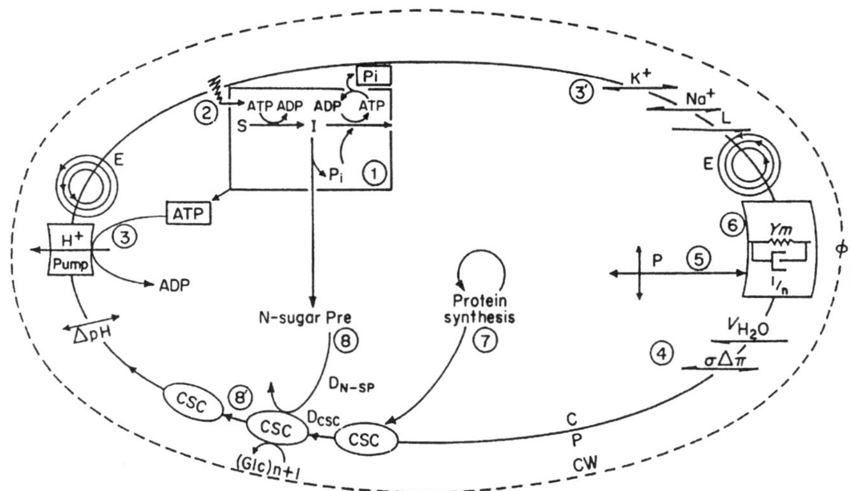

Figure 13.2 The bioelectromechanochemical (Bemchem) model of plant cell growth. The circled numbers correspond to the biomechanical, bioelectrical and biochemical events according to: (1) the set of reactions representing the energetic metabolism which gives the ATP, ADP and inorganic phosphate (P_i) cellular levels; (2) exogenous input stimulating function of glucose (S); (3) the ATP-driven plasmalemma H^+ pump in the presence of a difference of pH between the extra- and the intracellular medium (ΔpH) which contributes to the electrogenic component of membrane potential ($\Delta\psi$); (3') the K^+ outward flux and the Na^+ inward flux driven by the concentration differences across the plasmalemma and the leakage current (L) which contributes to the non-electrogenic component of $\Delta\psi$; (4) the osmotic potential difference which above the turgor pressure (P) constitutes the driving force for water uptake (V_{H_2O}) which also depends on the hydraulic conductance (L_c) of the cell; (5) the turgor pressure (P) which above the yield turgor threshold (minimum turgor pressure necessary for growth) (p_c) constitutes the driving force for cell elongation (cell wall irreversible expansion); the yield turgor threshold depends on the extensibility (or cell wall yielding coefficient) of the cell wall; (6) the membrane strain modelled as a Voigt–Kelvin body subjected to turgor pressure (P) and $\Delta\psi$ (E) compression (electromechanical coupling); (7) protein synthesis; (8) nucleotide–sugar precursor (N-sugar Pre) of cellulose synthesis at the plasmalemma and its diffusion coefficient ($D_{n\text{-sp}}$); (8') cellulose–synthetase complexes at the plasmalemma (CSC) and its diffusion coefficient in the plane of the membrane (D_{csc}). C, cell; P, plasmalemma; CW, cell wall. (Reproduced from Aon and Cortassa, 1989, by permission of Academic Press.)

and 2.2; Figures 2.2 and 2.4). It could be found that there is a spatio-temporal 'window' given by E_c and T_r, at which (sub)cellular processes may show transitions from microscopic (local) to macroscopic (global) behaviour. Those transitions happen at unstable points in the dynamic behaviour of assembly–disassembly of cytoskeleton components that

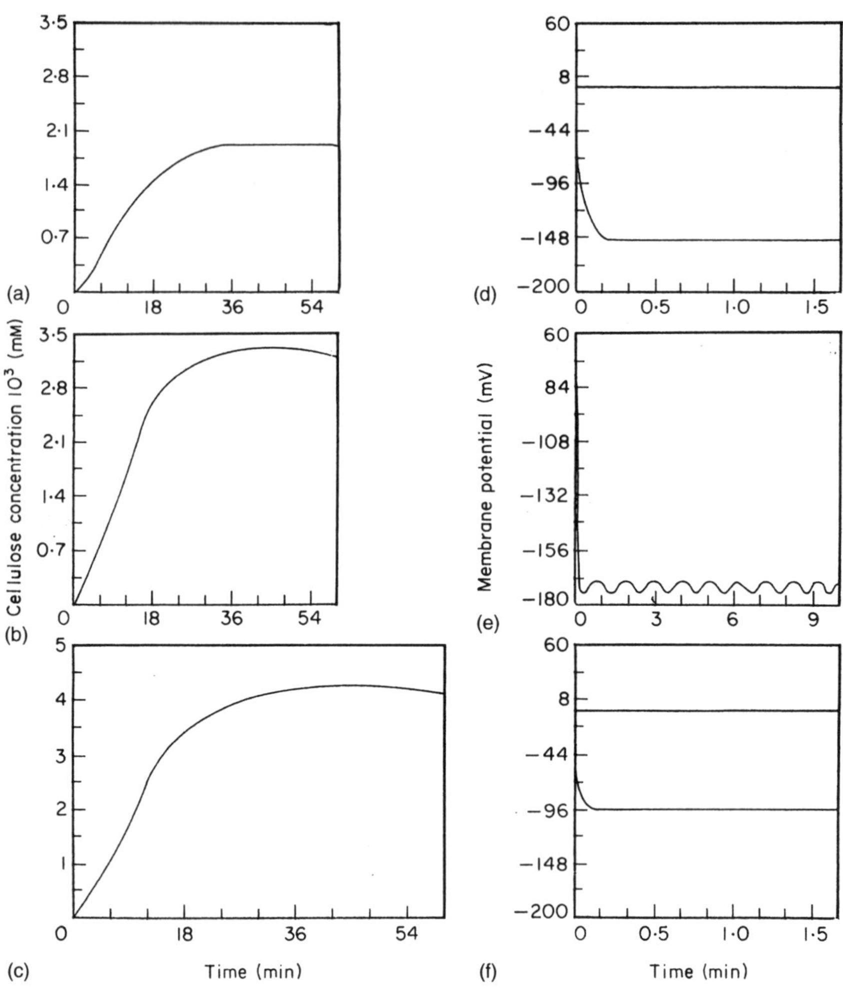

Figure 13.3 An example of how fast-relaxing processes can affect slow-relaxing ones according to the Bemchem model. Different dynamic behaviours of the membrane potential, $\Delta\psi$ (d–f) may affect the long-range time-course evolution of cell wall synthesis (a–c). The differential dynamics of the membrane potential were obtained by changing parameters of the model related to the electrochemical properties of the plasmalemma such as ion conductance (proton, potassium and sodium) and transmembrane pH, as well as parameters related to the energetic metabolism (substrate input, and rate constants). (Reproduced from Aon and Cortassa, 1989, by permission of Academic Press.)

have been shown to be candidate processes to affect at the cellular level functional global transitions (Chapters 6 and 7). We have been able to show that one of the simplest self-organized phenomena, i.e. bistability or multiple stationary states (Chapter 5), may occur in the coupled dynamics of MTP and a biochemical reaction as a function of two relevant cellular parameters, the total amount of MTP and its rate of polymerization (see the above paragraphs and Chapters 1 and 6). These two parameters in turn regulate the geometry of the lattice of MTP and, through it, the percolation threshold, P_c, and the fractal dimension, D (Feder, 1988; Rabouille *et al.*, 1992). In Chapter 6 we verified that the subsystem which describes the dynamics of MTP assembling/disassembling displays bistability when isolated, whereas the enzyme dynamics was only monostable. We concluded that it is the dynamics of MTP assembly which entrains the enzyme kinetics. Subtle alterations in the topology of supramolecular structures could also have regulatory properties on enzyme-catalysed metabolic fluxes (Chapter 7). Structural microscopic features such as molecular interactions, subtly modulated by relevant physiological parameters such as pH, Ca^{2+} or second messenger levels, determine macroscopic geometrical characters of the lattice that change the P_c (Figures 2.5 and 2.6). Therefore, the dynamics of chemical reactions can be either entrained by the architectural dynamic arrangements of cytoskeleton components, or at unstable parametric regions can give rise to self-organized behaviour whose emergence adds new functional properties to cells, i.e. adaptive conduct following changes of the intra- or extra-cellular environment. Taken together, these results reinforce the idea that polymerization–depolymerization of cytoskeleton components may be involved in the passage from microscopic to macroscopic order at the cellular level. Besides, the existence of abrupt transitions at limit points in the dynamic behaviour of MTP polymeric status, together with PK activity, further suggests a link between P_c and self-organization of cytoskeleton components, and their involvement in the synchronization of cellular activities farther apart in the cytoplasm of living cells.

We come back to the original question of how processes functionally related, that scale in space and time, yield the main functions of living cells. We propose that cell function is essentially given by the spatio-temporal coherence exhibited transiently by cells such as, say, waves of ions, secretory processes, gene expression or rearrangements of the cytoskeleton. According to this view, cellular function is the result of the dynamic integration of simultaneously occurring processes that, at the scale of micrometres and minutes, may synchronize each other by self-organized passage from local to global behaviour. An important aspect of this proposal is that spatio-temporal scaling of physiological processes happens in the putative fractal percolation cytostructure of the cytoplasm. The supramolecular organization of the cytoskeleton as a percolation lattice adds a feasible physicochemical mechanism of coherence, through the passage from local to global behaviour in the cytoplasm, at the percolation threshold, P_c. Below P_c the cytoplasm would behave as loosely coupled, independent dynamic subsystems (Chapter 6). These fractal structures clearly show spatial scaling, i.e. they are self-similar at different spatial magnifications (Figures 2.2 and 2.4).

What about temporal scaling? For this aspect we have suggested that the percolation threshold, a main feature of percolation clusters, may be related to the 'transition point' described in the (allometric) spatio-temporal scaling shown by several biological processes of different natures (Chapters 2 and 6). According to the spatio-temporal scaling of biological processes, microtubular protein as well as other cytoskeleton components (e.g. actin) which organize as fractal percolation lattices exhibit self-organization in the spatio-temporal 'window' of the 'transition point' (Figures 2.2 and 2.4; see also Chapter 6), after which new functional properties emerge (Chapters 6 and 7). Coincidently, percolation lattices exhibit transitions between local to global connectedness at their threshold level of occupancy, i.e. the percolation threshold. These pieces of evidence are in agreement with the percolation threshold of fractal cytostructures as being related to the 'transition point' at which polymerization–depolymerization of cytoskeleton components takes place.

13.2 LEVELS OF ORGANIZATION AND REDUCTIONISM

> ... when we come to discuss complexity and reduction of biological complexity, there are many different levels in a biological organism that might be chosen as the reference level of distinguishing 'parts' (... molecule, organelle, cell, organ, organism, population of organisms, etc). ... Although most biochemists and molecular biologists would have a predilection to choose the molecular level there is really no 'fundamental' level of reality that has an obvious priority ...
>
> Peacocke, 1983

In the quotation above, we agree with the fact that no single level of organization has a 'priority' over the others (Figures 2.1–2.3; Tables 2.1 and 2.2). Essentially, this is because we are dealing with functions which emerge from the whole, i.e. either cell, organ or organism. We have emphasized that most of the main functional properties of cells, such as energy transduction, solute transport, action potentials, macromolecule polymerization, cell growth and division, are placed in T_r ranges of 1 to 3 (seconds to several minutes), and E_c of -7 to -6 (around micrometres). Molecular properties involved in those processes span T_r values of -6 to -10 (μs to ps) and E_c values of -8 to -10 (nanometers to angstroms) (Table 2.1). There is a wide spatio-temporal span between cellular functions and molecular properties. For molecular properties to extend their range of action to higher spatio-temporal dimensions, some organizing principles implying their coherent spatio-temporal behaviour should be invoked. We have proposed, based on self-organization principles, that bifurcations in the behaviour of living systems may help us to rationalize transitions between levels of organization. Macroscopic patterns arising beyond bifurcations express the emergent functional properties of living systems (Chapters 1, 2 and 5–8). It is likely that those new macroscopic properties would correspond to new functions when integrated to a 'wilful machine' such as a cell.

In the analysis of biological complexity we should distinguish:

- the level of observation or information given by the particular method of perturbation;
- the phenomenon we search to explain, which takes us to the third point:
- the level of explanation chosen for the particular phenomenon.

From the definition of a level of organization according to spatial and temporal characteristic dimensions (Chapter 2), our level of explanation depends on the spatio-temporal dimensions of the process under observation. One appropriate question should be: does our perturbation method provide information about the level of organization corresponding to the phenomenon we want to explain? The existence of structural as well as functional levels of organization in biological systems is provided by the length and temporal relaxation scales sensed by different probes as applied in perturbation of physiological processes (Table 2.2). A similar spatio-temporal scaling to that shown by biological processes may be demonstrated for probes such as fluorescent or magnetic types, used to monitor the spatio-temporal behaviour of biological phenomena (Figures 2.3 and 2.4). Furthermore, inhibition of processes operating at lower spatio-temporal levels of organization in physiological phenomena, such as energy transduction, macromolecular organization and cell-to-cell interaction, also hampers the appearance of spatio-temporal coherence at higher levels of organization (Chapter 2).

For instance, if we are trying to explain the appearance of macroscopic spatio-temporal patterns such as those observed in a photobiochemical system (Chapter 5), the question is: what shall we look at? If we inspect Table 2.1 we see that several spatio-temporal levels of organization may be distinguished in this particular system which range from pigment excitation to the spatial patterns. We know that macroscopic spatio-temporal patterns do not arise either if we inhibit the water-splitting system or if a gradient of the electron acceptor is absent even with an active water-splitting system. Thus, a macroscopic cue is needed, i.e. the gradient, and also the intrinsic nonlinear kinetics of the photobiochemical system which gives rise to a bistable phenomenon that may explain the appearance of the alternating bands of blue (oxidized DCIP) and green (reduced DCIP) (Chapter 5).

In a more biological context, if we want to explain how the cell membrane invaginates to endocyte, for example, a virus, there are again many levels of organization implied in endocytosis (it may well be exocytosis). Let us suppose that we utilize a fluorescent molecular probe that integrates to the membrane and changes its emission properties following the endocytotic process. Those changes in emission properties will reveal an average result of changes in membrane fluidity or the thermotropic mesophormism of lipids around the probe. Are those changes related to the endocytosis process? Definitely, yes. Are those changes an adequate level of explanation, supposing our question is how endocytosis happens? We conjecture not, since endocytosis is a supramolecular phenomenon and we are probing the membrane at the molecular level. This is true even though microenvironments sensed by the probe depend on the membrane's macroscopic properties in which it is integrated. The membrane's curvature radii or the interfacial tension are properties of the whole ensemble of molecules as the trans-gauche isomerization of the hydrophobic chains of phospholipids is a molecular property. Therefore, according to our question about how endocytosis happens, what we seek to understand is how global properties such as the curvature radii, interfacial tension or the asymmetry of the bilayer, that occur in the spatio-temporal 'window' of endocytosis, are influenced by, for example, molecular composition of the bilayer, ionic concentration or catalytic properties of the surface.

Overall, these observations and caveats lead us into how to look and what to look for. Seemingly, there is no fundamental level of observation to explain the appearance of spatio-temporal patterns in living systems.

13.3 DEVELOPMENT IN THE FRAMEWORK OF DYNAMIC ORGANIZATION

A main problem of development concerns the mechanism(s) through which an environmental cue (e.g. a gradient of a morphogen), of either chemical, mechanical or electrical nature, induces a cell's commitment to a specific developmental path (Chapters 4, 9 and 10). It is becoming increasingly apparent that the understanding of how a cell senses a developmental field depends crucially on the spatio-temporal organization of its cytostructure (Chapters 4, 6, 7 and 10). This emergent view is accompanied by the recognition that prominent steps of cellular metabolism related to main metabolic pathways and gene expression, such as transcription and translation, occur associated with the cell's dynamic scaffolds, either nuclear or cytoplasmic (Chapters 6, 7, 9 and 10). The nuclear matrix is the structural framework which organizes the DNA in the nucleus and has a direct role in the regulation of gene expression. The cytoskeleton is part of a tissue matrix system which forms a structural and functional bridge from the DNA to the cell periphery and beyond to the extracellular matrix as well as other cells. Overall, the cytoarchitecture consists of linkages and interactions that interlock the nuclear matrix, the cytoskeleton and the cell periphery.

The cell's position in a tissue could be sensed by cell cytoarchitecture, since the genome itself appears to be responsive to major controlling signals such as cell shape and surface contact from the intra- and extra-cellular macromolecular networks. Thus, conceivably, genome expression may be modulated by cell shape and surface signals. Several studies have shown that much of the macromolecular metabolism of the cell, including DNA, RNA and protein synthesis, responds to changes in cell shape. According to the involvement of the nuclear matrix components in the DNA rearrangements, i.e. sister chromatid exchanges at the site of DNA replication, it has been proposed that alterations in the nuclear matrix structure could contribute to genetic instability. Genetic instability might place errors within the genetic apparatus that may be the basis of progression and the formation of tumour cell heterogeneity (Chapters 9 and 10). The models that have been proposed concerning the organization of the ground plan of living cells as a 'percolation cluster' or as a 'tensegrity structure' (Chapter 6) provide a conceptual framework from which may be built up an appropriate field theory that correctly and quantitatively answers the question of 'what consequences of (known or unknown) natural laws are responsible for the characteristic spatial order that is observed in organisms' (Goodwin, 1986). It has been postulated that threshold responses occur in percolation clusters as a consequence of the threshold property exhibited by these 'random fractals' (Chapter 6). Candidate morphogens such as retinoic acid and the fibroblast growth factor may show a genuine threshold response (Chapter 4) or the concomitant action of alkaloids on mammalian cell division and cytoskeleton morphology (Chapter 6). Auxin-induced plant cell elongation shows a threshold response as well, when the initial growth rates of coleoptile sections are evaluated as a function of indoleacetic acid concentration (Cleland, 1972).

13.4 FINAL REMARKS

It is suggested that the integration in time and space of cellular processes at different levels of organization is achieved at parametric sets where the inherent dynamics of biological processes becomes unstable. Relevant biological processes, such as cellular signalling by ions or second messenger waves as well as metabolic fluxes, were analysed to search for instabilities in their dynamics. Above a threshold, the dynamics of bistable amplifiers driven by nonlinear mechanisms, such as autocatalysis and cooperation, may become unstable in their local cytoplasmic domains. When that instability occurs beyond the percolation threshold of a percolation cluster at a site belonging to the spanning cluster, the local dynamics of a process implying an ion, metabolite or second messenger may extend to the whole cellular field.

The consequences of dynamic coupling on the spatio-temporal regulation *in vivo* of subcellular processes, i.e. subcellular dynamics, were analysed (Chapters 7, 11 and 12). The concept of dynamic coupling may shed light on 'dynamic diseases', a concept introduced by Glass and Mackey (1979, 1988). A dynamic disease is an abnormal, pathological pattern of temporal or spatial organization in a physiological control system. For instance, various sleep disorders are malfunctions of our circadian temporal organization; cardiac arrhythmias and ventricular fibrillation are maladies in the temporal and spatial organization of the pacemakers and muscular activity of the heart (Glass and Mackey, 1979; Winfree, 1987; Tyson and Kagan, 1988). Neurological disorders and cardiac arrhythmia may be viewed as originating from chaotic dynamics of physiological parameters such as electrical activity. A recent hypothesis, based on chaos theory and the geometrical concept of fractals, proposes that physiological ageing is associated with a generalized loss of an age-related deprivation of complex variability in multiple physiological processes including cardiovascular control, pulsatile hormone release and electroencephalographic potentials (Lipsitz and Goldberger, 1992). Perturbations in the dynamic organization of physiological processes in living systems are one of the keystones of disease and ageing (Olsen and Degn, 1985; Glass and Mackey, 1988; Lloyd and Rossi, 1993).

Is there causality in complex biological systems which comprise processes, functioning at many simultaneous levels of organization? First of all, one should always be sceptical when faced with causality in complex systems. We have shown that, at a certain period of time in the evolution of a system, two related variables may appear as unrelated while two unrelated or only indirectly related ones appear as related (Chapters 1 and 12). In the next period of time the situation may be reversed. What is more appropriate is to look at biological processes (e.g. macromolecule synthesis, cell division, vesicle secretion) as emergent properties depending on the simultaneous or contemporary operation of processes at different levels of organization. At instability points of their dynamics, macroscopic, self-organized behaviour may appear. The passage from local to global connectedness in the fractal cytoarchitecture implies that the dynamic organization of metabolic reactions may become functionally synchronized at different cytoplasmic locations through local instability in the dynamics of biochemical devices.

We and other researchers (Churchland and Sejnowski, 1988; Churchland, 1989; Paton, 1993, and references therein) have the impression that there is no unique, 'royal' way to model systems whose functional properties are manifest at several simultaneous levels of organization, such as cellular systems. Metaphors, with the appropriate biological constraints, will have to be used such as 'cell-as-machine', 'cell-as-society', 'cell-as-text' or 'cell-as-field', exhibiting parallel distributed processing (Paton, 1993). Cognitive neuroscience is one of the fields that more strikingly supports this view. Instead of a single model spanning all levels of organization to give a unified account of the whole complexity, one approach is to think of the coexistence of a 'chain of models' linking adjacent levels of organization, which in turn implies the coexistence of explanations at several levels.

References

Abdul-Baki, A.A. and Ray, P.M. (1971) Regulation by auxin of carbohydrate metabolism involved in cell wall synthesis by pea stem tissue. *Plant Physiol.*, **47**, 537–544.

Abraham, R.H. (1987) Dynamics and self-organization, in *Self-Organizing Systems. The Emergence of Order* (eds F.E. Yates, A. Garfinkel, D.O. Walter and G.B. Yates), Plenum Press, New York, pp. 599–613.

Abraham, R.H. and Shaw, Ch.D. (1984a) *Dynamics – The Geometry of Behavior. Part II: Chaotic Behavior*, Aerial Press, Inc., Santa Cruz, California.

Abraham, R.H., and Shaw, Ch.D. (1984b) *Dynamics – The Geometry of Behavior. Part I: Periodic Behavior*, Aerial Press, Inc., Santa Cruz, California.

Abraham, R.H., and Shaw, Ch.D. (1987) Dynamics: A visual introduction, in *Self-Organizing Systems. The Emergence of Order* (eds F.E. Yates, A. Garfinkel, D.O. Walter and G.B. Yates), Plenum Press, New York, pp. 543–597.

Adams, A.E. and Pringle, J.R. (1984) Relationship of actin and tubulin distribution to bud growth in wild type and morphogenetic mutant *Saccharomyces cerevisiae. J. Cell Biol.*, **98**, 934–945.

Adelman, W.J. and Fitzhugh, R. (1975) Solutions of a Hodgkin–Huxley equation modified for potassium accumulation in a peri-axonal space. *Fed. Proceed.*, **34**, 1322–1329.

Aerts, R.J., Durston, A.J. and Moolenaar, W.H. (1985) Cytoplasmic pH and the regulation of the *Dictyostelium* cell cycle. *Cell*, **43**, 653–657.

Aerts, R.J., Durston, A.J. and Moolenaar, W.H. (1986) Cytoplasmic pH and glycolysis in the *Dictyostelium discoideum* cell cycle. *FEBS Lett.*, **196**, 167–170.

Aerts, R.J.A., Durston, A.J. and Konijn, T.M. (1987) Cytoplasmic pH at the onset of development in *Dictyostelium discoideum. J. Cell Sci.*, **87**, 423–430.

Agladze, K.I. and Krinsky, V.I. (1982) Multiarmed vortices in an active chemical medium. *Nature*, **296**, 424–426.

Aguilera, A. and Zimmermann, F.K. (1986) Isolation and molecular analysis of the phosphoglucose isomerase structural gene of *Saccharomyces cerevisiae. Mol. Gen. Genet.*, **202**, 83.

Aihara, K. and Matsumoto, G. (1982) Temporally coherent organization and instabilities in squid giant axons. *J. Theor. Biol.*, **95**, 697–720.

Aihara, K. and Matsumoto, G. (1983) Two stable steady states in the Hodgkin–Huxley axons. *Biophys. J.*, **41**, 87–89.

Aihara, K., Matsumoto, G. and Ikegaya, Y. (1984) Periodic and non-periodic responses of a periodically forced Hodgkin–Huxley oscillator. *J. Theor. Biol.*, **109**, 249–270.

Aisenberg, A.C. (1961) *The Glycolysis and Respiration of Tumors*, Academic Press, London.

Alberts, B., Bray, D., Lewis, J. *et al.* (1989) *Molecular Biology of the Cell*, Garland Publishing, New York.

Albery, W.J. and Knowles, J.R. (1976) Evolution of enzyme function and the development of catalytic efficiency. *Biochemistry*, **15**, 5631–5638.

Aldridge, J. and Pye, E.K. (1976) Cell density dependence of oscillatory metabolism. *Nature*, **259**, 670–671.

Aleksander, I. and Morton, H. (1990) *An Introduction to Neural Computing*, Chapman & Hall, London.

Alexander, M. (1994) *Biodegradation and Bioremediation*, Academic Press, San Diego.

Alexander, M.A. and Jeffries, T.W. (1990) Respiratory efficiency and metabolite partitioning as regulatory phenomena in yeasts. *Enzyme Microb. Technol.*, **12**, 2–19.

Amchenkova, A.A., Bakeeva, L.E., Chentsov, Y.S. *et al.* (1988) Coupling membranes as energy-transmitting cables. I. Filamentous mitochondria in fibroblasts and mitochondrial clusters in cardiomyocytes. *J. Cell Biol.*, **107**, 481–495.

Amoroso, M.J. (1990) *Tésis doctoral*, Universidad Nacional de Tucumán, Argentina.

Ancel, P. and P. Vintemberger (1948) Recherches sur le determinisme de la symmetrie bilaterale dans l'oeuf des amphibiens. *Bull. Biol. Fr. Belg.*, Suppl., **31**, 1–182.

Andersen, K.B. and von Meyenburg, K. (1980) Are growth rates of *Escherichia coli* limited by respiration? *J. Bacteriol.*, **144**, 114–123.

Anderson, A.J. and Dawes, E.A. (1990) Occurrence, metabolism, metabolic role, and industrial uses of bacterial polyhydroxyalkanoates. *Microbiol. Rev.*, **54**, 450–472.

Aon, J.C. and Cortassa, S. (1996) Metabolic rates during sporulation of *Saccharomyces cerevisiae* on acetate. *Antonie van Leeuwenhoek*, **69**, 257–265

Aon, J.C., Rapisarda, V.A. and Cortassa, S. (1996) Metabolic rates regulate the success of sporulation in *Saccharomyces cerevisiae*. *Exp. Cell Res.*, **222**, 157–162.

Aon, M.A. (1995) *Los Límites de la Racionalidad*, Ed. Catálogos, Buenos Aires.

Aon, M.A. and Cortassa, S. (1989) The regulation of plant cell growth. A bioelectromechanochemical model. *J. Theor. Biol.*, **138**, 429–456.

Aon, M.A. and Cortassa, S. (1991) Thermodynamic evaluation of energy metabolism in mixed substrate catabolism. Modelling studies of stationary and oscillatory states. *Biotechnol. Bioeng.*, **37**, 197–204.

Aon, M.A. and Cortassa, S. (1993) An allometric interpretation of the spatio-temporal organization of molecular and cellular processes. *Mol. Cell. Biochem.*, **120**, 1–14.

Aon, M.A. and Cortassa, S. (1994) On the fractal nature of cytoplasm. *FEBS Lett.*, **344**, 1–4

Aon, M.A. and Cortassa, S. (1995) Cell growth and differentiation from the perspective of dynamical organization of cellular and subcellular processes. *Prog. Biophys. Mol. Biol.*, **64**, 55–79.

Aon, M.A., Cortassa, S., Hervagault, J.F. and Thomas, D. (1989a) pH-induced bistable dynamic behaviour in the reaction catalyzed by glucose-6-phosphate dehydrogenase and conformational hysteresis of the enzyme. *Biochem. J.*, **262**, 795–800.

Aon, M.A., Thomas, D. and Hervagault, J.F. (1989b) Spatial patterns in a photobiochemical system. *Proc. Natl. Acad. Sci. USA*, **85**, 516–519.

Aon, M.A., Cortassa, S. and Thomas, D. (1990a) Modèle du métabolisme énergétique. Etudes theoretiques de l'efficacité pour différentes conditions des bésoins énergétiques et d'entrée du substrat, in *Proceedings of the Colloque Societé de Microbiologie Francaise on Stratégie d'Utilisation des Substrats pour la Production de Métabolites Microbiens* (eds J.M. Lebeault and J.G. Pan), Société de Microbiologie Française, Paris, pp. 57–65.

Aon, M.A., Westerhoff, H.V., Cortassa, S. *et al.* (1990b) Influence of mitochondrial function and substrate uptake on glycolytic oscillations in yeast. *Biochim. Biophys. Acta* (European Bioenergetic Conference), **6**, 96.

Aon, M.A., Cortassa, S., Westerhoff, H.V. *et al.* (1991) Dynamic regulation of yeast glycolytic oscillations by mitochondrial functions. *J. Cell Sci.*, **99**, 325–334.

Aon, M.A., Cortassa, S., Westerhoff, H.V. and van Dam, K. (1992) Synchrony and mutual stimulation of yeast cells during fast glycolytic oscillations. *J. Gen. Microbiol.*, **138**, 2219–2227.

Aon, M.A., Mónaco, M.E. and Cortassa, S. (1995) Carbon and energetic uncoupling are associated with block of division at different stages of the cell cycle in several cdc mutants of *Saccharomyces cerevisiae*. *Exp. Cell Res.*, **217**, 42–51.

Aon, M.A., Cáceres, A. and Cortassa, S. (1996a) Heterogeneous distribution and organization of cytoskeletal proteins drive differential modulation of metabolic fluxes. *J. Cell. Biochem.*, **60**, 271–278.

Aon, M.A., Cortassa, S. and Verdoni, N. (1996b) Bioengineering of biocatalysts for biotransformation, in *Fifth Annual International Symposium on Gas, Oil and Environmental Biotechnology*, Chicago (in press).

Arora, K.K. and Pedersen, P.L. (1988) Functional significance of mitochondrial bound hexokinase in tumor cell metabolism. Evidence for preferential phosphorylation of glucose by intramitochondrially generated ATP. *J. Biol. Chem.*, **263**, 17422–17428.

Ashby, W.R. (1960) *Design for a Brain*, 2nd edn, Wiley, New York.

Ashizawa, K., Fukuda, T. and Cheng, S. (1992) Transcriptional stimulation by thyroid hormone of a cytosolic thyroid hormone binding protein which is homologous to a subunit of pyruvate kinase M1. *Biochemistry*, **31**, 2774–2778.

Atkinson, D.E. (1977) *Cellular Energy Metabolism and its Regulation*, Academic Press, London.

Atri, A., Amundson, J., Clapham, D. and Sneyd, J. (1993) A single-pool model for intracellular calcium oscillations and waves in the *Xenopus laevis* oocyte. *Biophys. J.*, **65**, 1727–1739.

Bacic, A. and Delmer, D.P. (1981) Stimulation of membrane-associated polysaccharide synthetases by a membrane potential in developing cotton fibers. *Planta*, **152**, 346–351.

Baggetto, L.G. (1992) Deviant energetic metabolism of glycolytic cancer cells. *Biochimie*, **74**, 959–974.

Bailey, J.E. (1991) Toward a science of metabolic engineering. *Science*, **252**, 1668–1675.

Bailey, J.E. and Ollis, D.F. (1977) *Biochemical Engineering Fundamentals*, McGraw-Hill, New York.

Bailey, J.E., Axe, D.D., Doran, P.M. *et al.* (1987) Redirection of cellular metabolism: analysis and synthesis. *Ann. New York Acad. Sci.*, **506**, 1–23.

Baker, D.A. and Hall, J.L. (1988) Introduction and general principles, in *Solute Transport in Plant Cells and Tissues* (eds D.A. Baker and J.L. Hall), Longman Scientific & Technical, London, pp. 1–27.

Barbier, E.B. (1987) The concept of sustainable development. *Environ. Conservation*, **14**, 101–110.

Barford, J.P. (1990) A general model for aerobic yeast growth: continuous culture. *Biotechnol. Bioeng.*, **35**, 921–927.

Barford, J.P. and Hall, R.J. (1976) Estimation of the length of cell cycle phases from asynchronous cultures of *Saccharomyces cerevisiae*. *Exp. Cell Res.*, **102**, 276–284.

Barford, J.P. and Hall, R.J. (1981) A mathematical model for the aerobic growth of *Saccharomyces cerevisiae* with a saturated respiratory capacity. *Biotechnol. Bioeng.*, **23**, 1735–1762.

Barford, J.P., Pamment, N.B. and Hall, R.J. (1982) Lag phases and transients, in *Microbial Population Dynamics* (ed. M.J. Bazin), CRC Press, Boca Raton, Florida, pp. 56–89.

Barnes, G., Louie, K.A. and Botstein, D. (1992) Yeast proteins associated with microtubules *in vitro* and *in vivo*. *Mol. Biol. Cell*, **3**, 29–47.

Baron, A.C., Tobin, T.H., Wallsgrove, R.M. and Tobin, A.K. (1994) A metabolic control analysis of the glutamine synthetase/glutamate synthase cycle in isolated barley (*Hordeum vulgare* L.) chloroplasts. *Plant Physiol.*, **105**, 415–424.

Baronofsky, J.J., Schreurs, W.J.A. and Kashket, E.R. (1984) Uncoupling by acetic acid limits growth of and acetogenesis by *Clostridium thermoaceticum*. *Appl. Environ. Microbiol.*, **48**, 1134–1139.

Bates, G.W. and Cleland, R.E. (1979) Protein synthesis and auxin-induced growth: inhibitor studies. *Planta*, **145**, 437–442.

Bautz-Freese, E., Chu, M.I. and Freese, E. (1982) Initiation of yeast sporulation by partial carbon, nitrogen and phosphate deprivation. *J. Bacteriol.*, **149**, 840–851.

Bayley, P.M. (1990) What makes microtubules dynamic? *J. Cell Sci.*, **95**, 329–334.

Baylies, M.K., Bargiello, T.A., Jackson, F.R. and Young, M.W. (1987) Changes in abundance or structure of the *per* gene product can alter periodicity of the *Drosophila* clock. *Nature*, **326**, 390–392.

Beauvoit, B., Rigoulet, M., Bunoust, O. *et al.* (1993) Interactions between glucose metabolism and oxidative phosphorylations on respiratory-competent *Saccharomyces cerevisiae* cells. *Eur. J. Biochem.*, **214**, 163–172.

Beck, C., and von Meyenburg, H.K. (1968) Enzyme patterns and aerobic growth of *Saccharomyces cerevisiae* under various degrees of glucose limitation. *J. Bacteriol.*, **96**, 479–486.

Belmont, L.D., Hyman, A.A., Sawin, K.E. and Mitchison, T.J. (1990) Real-time visualization of cell cycle-dependent changes in microtubule dynamics in cytoplasmic extracts. *Cell*, **62**, 579–589.

Ben-Jacob, E. and Garik, P. (1990) The formation of patterns in non-equilibrium growth. *Nature*, **343**, 523–530.

Bennetzen, J.L. and Hall, B.D. (1982) The primary structure of the *Saccharomyces cerevisiae* gene for alcohol dehydrogenase I. *J. Biol. Chem.*, **257**, 3018.

Ben-Ze'ev, A. (1986) The relationship between cytoplasmic organization, gene expression and morphogenesis.*TIBS*, **11**, 478–481.

Berden, J.A., Schoppink, P.J. and Grivell, L.A. (1988) A model for the assembly of the ubiquinol: cytochrome c oxidoreductase in yeast, in *Molecular Basis of Biomembrane Transport* (eds F. Palmieri and E. Quagliariello), Elsevier, Amsterdam, pp. 195–208.

Bereiter-Hahn, J. (1976) Dimethylaminostyryl-methylpyridinium iodide (DASPMI) as a fluorescent probe for mitochondria *in situ*. *Biochim. Biophys. Acta*, **423**, 1–14.

Bereiter-Hahn, J., Seipel, K.H. and Voth, M. (1983) Fluorimetry of mitochondria in cells vitally stained with DASPMI or rhodamine 6 GO. *Cell Biochem. Function*, **1**, 147–155.

Bereiter-Hahn, J., Stubig, C. and Heymann, V. (1995) Cell cycle-related changes in F-actin distribution are correlated with glycolytic activity. *Exp Cell Res.*, **218**, 551–560.

Berger, F. and Brownlee, C. (1993) Ratio confocal imaging of free cytoplasmic calcium gradients in polarising and polarised *Fucus* zygotes. *Zygote*, **1**, 9–15.

Berridge, M.J. (1990) Calcium oscillations. *J. Biol. Chem.*, **265**, 9583–9586.

Berridge, M.J. (1993) Inositol trisphosphate and calcium signalling. *Nature*, **361**, 315–325.

Berridge, M.J., Cobbold, P.H. and Cuthbertson, K.S.R. (1988) Spatial and temporal aspects of cell signalling. *Phil. Trans. R. Soc. London, Ser. B*, **320**, 325–343.

Berry, D.R., Franco, C.M.M. and Smith, J.E. (1981) *Current Developments in Yeast Research*, Pergamon Press, Toronto, pp. 393–398.

Berry, M.N. (1981) An electrochemical interpretation of metabolism. *FEBS Lett.*, **134**, 133–138.

Berry, M.N., Gregory, R.B., Grivell, A.R. *et al.* (1987) Linear relationships between mitochondrial forces and cytoplasmic flows argue for the organized energy-coupled nature of cellular metabolism. *FEBS Lett.*, **224**, 201–207.

Bershadsky, A.D. and Gelfand, V.I. (1981) ATP-dependent regulation of cytoplasmic microtubule disassembly. *Proc. Natl. Acad. Sci. USA*, **78**, 3610–3613.

Binder, M., Schanz, M. and Hartig, A. (1991) Vector-mediated overexpression of catalase A in the yeast *Saccharomyces cerevisiae* induces inclusion body formation. *Eur. J. Cell Biol.*, **54**, 305–312.

Bisson, L.F. and Fraenkel, D.G. (1983) Involvement of kinases on glucose and fructose uptake by *Saccharomyces cerevisiae*. *Proc. Natl. Acad. Sci. USA*, **80**, 1730–1734.

Bisson, L.F. and Fraenkel, D.G. (1984) Expression of kinase-dependent glucose utpake in *Saccharomyces cerevisiae*. *J. Bacteriol.*, **159**, 1013–1017.

Boiteux, A., Goldbeter, A. and Hess, B. (1975) Control of oscillating glycolysis of yeast by stochastic, periodic, and steady source of substrate: A model and experimental study. *Proc. Natl. Acad. Sci. USA*, **72**, 3829–3833.

Borckmans, P., Dewel, G., Walgraef, D. and Katayama, Y. (1987) The search for Turing structures. *J. Statist. Phys.*, **48**, 1031–1044.

Boucherie, H. (1985) Protein synthesis during transition and stationary phases under glucose limitation in *Saccharomyces cerevisiae*. *J. Bacteriol.*, **161**, 385–392.

Boyer, P.D. (1970) *The Enzymes*, Academic Press, London.

Brachet, J. (1934) Etude du métabolisme de l'oeuf de grenouille (*Rana fusca*) au cours du développement. *Arch. Biol.*, **45**, 611–727.

Breitenbach, M. and Lachkovics, E. (1983) The yeast genome in yeast differentiation, in *Secondary Metabolism and Differentiation in Fungi* (eds J.W. Bennett and A. Ciegler), Marcel Dekker, New York, pp. 307–374.

Bremer, H. and Dennis, P.P. (1987) Modulation of chemical composition and other parameters of the cell by growth rate, in Escherichia coli *and* Salmonella typhimurium: *Cellular and Molecular Biology* (eds F.C. Neidhardt, J.L. Ingraham, K. Brooks Low *et al.*), American Society for Microbiology, Washington, pp. 1527–1542.

Breton, J., Thomas, D. and Hervagault, J.F. (1985) Experimental evidence for multistability in a photobiochemical system. *Eur. J. Biochem.*, **152**, 509–514.

Bronstein, W.W. and Knull, H.R. (1981) Interaction of muscle glycolytic enzymes with thin filament proteins. *Can. J. Biochem.*, **59**, 494–499.

Brooks, S.P.J. and Storey, K.B. (1991) Where is the glycolytic complex? A critical evaluation of present data from muscle tissue. *FEBS Lett.*, **278**, 135–138.

Brown, G.C. (1992) Control of respiration and ATP synthesis in mammalian mitochondria and cells. *Biochem. J.*, **284**, 1–13.

Brown. R.M. and Montezinos, D. (1976) Cellulose microfibrils. Visualization of biosynthetic and orienting complexes in association with the plasma membrane. *Proc. Natl. Acad. Sci. USA*, **73**, 143–147.

Bruinenberg, P.M., de Bot, P.H.M., van Dijken, J.P. and Scheffers, W.A. (1983) The role of redox balances in the anaerobic fermentation of xylose by yeasts. *Eur. J. Appl. Microbiol. Biotechnol.*, **18**, 287–292.

Bruinenberg, P.M., van Dijken, J.P., Kuenen, J.G. and Scheffers, A. (1985) Oxidation of NADH and NADPH by mitochondria from the yeast *Candida utilis. J. Gen. Microbiol.*, **131**, 1043–1051.

Brummer, B. and Parish, R.W. (1983) Mechanism of auxin-induced plant cell elongation. *FEBS Lett.*, **161**, 9–13.

Brummer, B., Bertl, A., Potrykus, I. *et al.* (1985) Evidence that fusicoccin and indole acetic acid induce cytosolic acidification of *Zea mays* cells. *FEBS Lett.*, **189**, 109–113.

Budrene, E.O. and Berg, H.C. (1995) Dynamics of formation of symmetrical pattern by chemotactic bacteria, *Nature*, **376**, 49–52

Bunow, B., Kernevez, J.-P., Joly, G. and Thomas, D. (1980) Pattern formation by reaction-diffusion instabilities: Application to morphogenesis in Drosophila. *J. Theor. Biol.*, **84**, 629–649.

Burden, D.W. and Eveleigh, D.E. (1990) *Yeast Technology*, Springer Verlag, Berlin, Heidelberg, pp. 199–227.

Burke, D.J. and Church, D. (1991) Protein synthesis requirements for nuclear division, cytokinesis, and cell separation in *Saccharomyces cerevisiae. Mol. Cell. Biol.*, **11**, 3691–3698.

Busa, W.B. (1982) Cellular dormancy and the scope of pHi-mediated metabolic regulation, in *Intracellular pH. Its Measurement, Regulation and Utilization in Cellular Functions*, vol. 15 (eds R. Nucitelli and D.W. Deamer) Alan R. Liss, New York, pp. 417–426.

Busa, W.B. and Nuccitelli, R. (1984) Metabolic regulation via intracellular pH. *Am. J. Physiol.*, **246**, R409–R438.

Bustamante, E. and Pedersen, P.L. (1977) High aerobic glycolysis of rat hepatoma cells in culture: role of mitochondrial hexokinase. *Proc. Natl. Acad. Sci. USA*, **74**, 3735–3739.

Bustamante, J.O. (1994a) Nuclear electrophysiology. *J. Membrane Biol.*, **138**, 105–112.

Bustamante, J.O. (1994b) Open states of nuclear envelope ion channels in cardiac myocytes. *J. Membrane Biol.*, **138**, 77–89.

Cáceres, A., Banker, G., Steward, O. *et al.* (1984) MAP2 is localized to the dendrites of hippocampal neurons which develop in culture. *Dev. Brain Res.*, **13**, 314–318.

Cairns, J., Overbaugh, J. and Miller, S. (1988) The origin of mutants. *Nature*, **335**, 142–145.

Cameron, D.C. and Tong, I.-T. (1993) Cellular and metabolic engineering. An overview. *Appl. Biochem. Biotechnol.*, **38**, 105–140.

Campbell, D. (1989) An introduction to nonlinear dynamics, in *Lectures in the Sciences of Complexity*, vol. 1 (ed. D.L. Stein), Addison-Wesley, Santa Fe Institute, pp. 3–105.

Caplan, S.R. (1973) A thermodynamic and kinetic approach to the study of biological models, with particular reference to membranes containing immobilized enzyme. *Biochimie*, **55**, 967–973.

Caplan, S.R. and Essig, A. (1983) *Bioenergetics and Linear Nonequilibrium Thermodynamics. The Steady State.* Harvard University Press, Cambridge. MA.

Carlier, M.F. (1992) Nucleotide hydrolysis regulates the dynamics of actin filaments and microtubules. *Phil. Trans. R. Soc. Lond. B*, **336**, 93–97.

Carlier, M.F., Melki, R., Pantaloni, D. *et al.* (1987) Synchronous oscillations in microtubule polymerization. *Proc. Natl. Acad. Sci. USA*, **84**, 5257–5261.

Carlson, M. (1987) Regulation of sugar utilization in *Saccharomyces* species. *J. Bacteriol.*, **169**, 4873–4877.

Casperon, G.F., Walker, N. and Bourne, H.R. (1985) Isolation of a gene encoding adenylate cyclase in *Saccharomyces cerevisiae*. *Proc. Natl. Acad. Sci. USA*, **82**, 5060–5063.

Cassimeris, L. (1993) Regulation of microtubule dynamics instability. *Cell Motil. Cytoskel.*, **26**, 275–281.

Chabre, M. (1985) Visual phototransduction. *Annu. Rev. Biophys. Biophys. Chem.*, **14**, 331–360.

Chance, B., Estabrook, R.W. and Ghosh, A. (1964) Damped sinusoidal oscillations of cytoplasmic reduced pyridine nucleotides in yeast cells. *Proc. Natl. Acad. Sci. USA*, **51**, 1244–1251.

Chance, B., Williamson, G., Lee, I.Y. *et al.* (1973) Synchronization phenomena in oscillations of yeast cells and isolated mitochondria, in *Biological and Biochemical Oscillators* (eds B. Chance, K. Pye, A.K. Ghosh and B. Hess), Academic Press, New York, pp. 285–300.

Chant, J. (1994) Cell polarity in yeast. *TIG*, **10**, 328–333.

Chapman, R.L. and Staehelin, L.A. (1985) Plasma membrane 'rosettes' in carrot and sycamore suspension culture cells. *J. Ultrastruct. Res.*, **93**, 87–91

Chen, Y.D. and Hill, T.L. (1987) Theoretical studies on oscillations in microtubule polymerization. *Proc. Natl. Acad. Sci. USA*, **84**, 8419–8423.

Chesbro, W. (1988) The domains of bacterial growth. *Can. J. Microbiol.*, **34**, 427–435.

Cheung, Ch.-W., Cohen, N.S. and Raijman, L. (1989) Channeling of urea cycle intermediates *in situ* in permeabilized hepatocytes. *J. Biol. Chem.*, **264**, 4038–4044.

Chialvo, D.R. and Jalife, J. (1987) Non-linear dynamics of cardiac excitation and impulse propagation. *Nature*, **330**, 749–752.

Chrispeels, M.J. (1976) Biosynthesis, intracellular transport and secretion of extracellular macromolecules. *Annu. Rev. Plant Physiol.*, **27**, 19–38.

Chu, M.I., Hartig, A., Bautz-Freese, E. and Freese, E. (1981) Adaptation of glucose grown *Saccharomyces cerevisiae* to gluconeogenic growth and sporulation. *J. Gen. Microbiol.*, **125**, 421–430.

Churchland, P.S. (1989) *Neurophilosophy. Toward a Unified Science of the Mind/Brain*, MIT Press, Cambridge, MA.

Churchland, P.S. and Sejnowski, T.J. (1988) Perspectives on cognitive neuroscience. *Science*, **242**, 741–745.

Clark, A.J., Cotton, N.P.J. and Jackson, J.B. (1983) The relation between membrane ionic current and ATP synthesis in chromatophores from *Rhodopseudomonas capsulata*. *Biochim. Biophys. Acta*, **723**, 440.

Clark, D.P. (1989) The fermentation pathways in *Escherichia coli*. *FEMS Microbiol. Rev.*, **63**, 223–234.

Clarke, F., Stephan, P., Morton, D. and Weidemann, J. (1985) Glycolytic enzyme organization via the cytoskeleton and its role in metabolic regulation, in *Regulation of Carbohydrate Metabolism*, vol. 2 (ed. R. Beitner), CRC Press, Boca Raton, pp. 1–31.

Clegg, J.S. (1983) *Coherent Excitations in Biological Systems*, Springer Verlag, Berlin.

Clegg, J.S. (1984a) Properties and metabolism of the aqueous cytoplasm and its boundaries. *Am. J. Physiol.*, **246**, R133–R151.

Clegg, J.S. (1984b) Intracellular water and the cytomatrix: some methods of study and current views. *J. Cell Biol.*, **99**, 167–171.

Clegg, J.S. (1991) Metabolic organization and the ultrastructure of animal cells. *Biochem. Soc. Trans.*, **19**, 985–991.

Clegg, J.S. (1992) Cellular infrastructure and metabolic organization. *Curr. Top. Cell. Regul.*, **33**, 3–14.

Cleland, R.E. (1971) Cell wall extension. *Annu. Rev. Plant Physiol.*, **22**, 197–222.

Cleland, R.E. (1972) The dosage–response curve for auxin-induced cell elongation: A reevaluation. *Planta*, **104**, 1–9.

Cleland, R.E. (1976) Kinetics of hormone-induced H^+ excretion. *Plant Physiol.*, **58**, 210–213.

Cleland, R.E., Prins, H.B.A., Harper, J.R. and Higinbotham, N. (1977) Auxin-induced membrane hyperpolarization in avena coleoptiles. *Plant Physiol.*, **59**, 395–397.

Clementz, T., Christiansson, A. and Wieslander, A. (1986) Transmembrane electrical potential affects the lipid composition of *Acholeplasma laidlawii*. *Biochemistry*, **25**, 823–830.

Cleveland, D.W. (1988) Autoregulated instability of tubulin mRNAs: a novel eukaryotic regulatory mechanism. *TIBS*, **13**, 339–343.

Cleveland, D.W. and Mooseker, M.S. (1994) Cytoskeleton. Editorial Overview. *Curr. Opin. Cell Biol.*, **6**, 1–2.

Cleveland, D.W. and Yen, T.J. (1989) Multiple determinants of eukaryotic mRNA stability. *Nature New Biol.*, **1**, 121–126.

Cook, D.N., Ma, D., Pon, N.G. and Hearst, J.E. (1992) Dynamics of DNA supercoiling by transcription in *Escherichia coli*. *Proc. Natl. Acad. Sci. USA*, **89**, 10603–10607.

Cook, P.R. (1989) The nucleoskeleton and the topology of transcription. *Eur. J. Biochem.*, **185**, 487–501.
Cook, P.R. (1991) The nucleoskeleton and the topology of replication. *Cell*, **66**, 627–635.
Cooper, S. (1991) *Bacterial Growth and Division*, Academic Press, San Diego.
Cooper, T.G. (1982) Nitrogen metabolism, in *The Molecular Biology of the Yeast Saccharomyces. Metabolism and Gene Expression* (eds J.N. Strathern, E.W. Jones and J.R. Broach), Cold Spring Harbor Laboratory Press, Cold Spring Harbour NY, pp. 39–99.
Cornish, A., Greenwood, J.A. and Jones, C.W. (1988) The relationship between glucose transport and the production of succinoglucan exopolysaccharide by *Agrobacterium radiobacter. J. Gen. Microbiol.*, **134**, 3111–3122.
Cornish-Bowden, A. (1991) Failure of channelling to maintain low concentrations of metabolic intermediates. *Eur. J. Biochem.*, **195**, 103–108.
Cortassa, S. and Aon, M.A. (1993a) Sugar uptake controls catabolism in yeast. Metabolic control analysis of glycolysis and the branch to Krebs cycle and fermentative pathways in *Saccharomyces cerevisiae*, in *Metabolic Compartmentation in Yeasts. Sixteenth International Specialized Symposium on Yeasts* (eds W.A. Scheffers and J.P. van Dijken), Pasmans, La Haya, pp. 116–118.
Cortassa, S. and Aon, M.A. (1993b) Altered topoisomerase activities may be involved in the regulation of DNA supercoiling in aerobic to anaerobic transitions in *Escherichia coli. Mol. Cell. Biochem.*, **126**, 115–124.
Cortassa, S. and Aon, M.A. (1994a) Metabolic control analysis of glycolysis and branching to ethanol production in chemostat cultures of *Saccharomyces cerevisiae* under carbon, nitrogen, or phosphate limitations. *Enzyme Microb. Technol.*, **16**, 761–770.
Cortassa, S. and Aon, M.A. (1994b) Spatio-temporal regulation of glycolysis and oxidative phosphorylation *in vivo* in tumor and yeast cells. *Cell Biol. Int.*, **89**, 687–714.
Cortassa, S., Aon, M.A. and Thomas, D. (1990a) Thermodynamic and kinetic studies in a stoichiometric model of energetic metabolism under starvation conditions. *FEMS Microbiol. Lett.*, **66**, 249–256.
Cortassa, S., Sun, H., Kernevez, J.P. and Thomas, D. (1990b) Pattern formation in an immobilized bienzyme system: A morphogenetic model. *Biochem. J.*, **269**, 115–122.
Cortassa, S., Aon, M.A. and Westerhoff, H.V. (1991) Linear nonequilibrium thermodynamics describes the dynamics of an autocatalytic system. *Biophys. J.*, **60**, 794–803.
Cortassa, S., Cáceres, A. and Aon, M.A. (1994a) Microtubular protein in its polymerized or non-polymerized states differentially modulates *in vitro* and intracellular fluxes catalyzed by enzymes related to carbon metabolism. *J. Cell. Biochem.*, **55**, 120–132.
Cortassa, S., Aon, M.A. and Westerhoff, H.V. (1994b) Temporal self-organization and linear non-equilibrium thermodynamics are not mutually exclusive in autocatalytic systems, in *Biothermokinetics* (ed. H.V. Westerhoff), Intercept, Andover, pp. 197–205.

Cortassa, S., Aon, J.C. and Aon, M.A. (1995) Fluxes of carbon, phosphorylation and redox intermediates during growth of *Saccharomyces cerevisiae* on different carbon sources. *Biotechnol. Bioeng.*, **47**, 193–208.

Cortes, P.M. (1992) Analysis of the electrical coupling of root cells: implications for ion transport and the existence of an osmotic pump. *Plant, Cell Environ.*, **15**, 351–363.

Coschigano, P.W., Miller, S.M. and Magasanik, B. (1991) Physiological and genetic analysis of the carbon regulation of the NAD-dependent glutamate dehydrogenase of *Saccharomyces cerevisiae*. *Mol. Cell. Biol.*, **11**, 4455–4465.

Cosgrove, D.J. (1981) Analysis of the dynamic and steady-state responses of growth rate and turgor pressure to changes in cell parameters. *Plant Physiol.*, **68**, 1439–1446.

Cosgrove, D.J. (1986) Biophysical control of plant cell growth. *Annu. Rev. Plant Physiol.*, **37**, 377–405.

Crabtree, H.G. (1929) Observations on the carbohydrate metabolism of tumors. *Biochem. J.*, **23**, 536–545.

Crick, F. (1970) Diffusion in embryogenesis. *Nature*, **225**, 420–422.

Crick, F. (1989) The recent excitement about neural networks. *Nature*, **337**, 129–132.

Cross, M.C. and Hohenberg, P.C. (1994) Spatiotemporal chaos. *Science*, **263**, 1569–1570.

Cross, S.S. (1994) The application of fractal geometric analysis to microscopic images. *Micron*, **25**, 101–113.

Cuneo, P., Magri, E., Verzola, A. and Grazi, E. (1992) 'Macromolecular crowding' is a primary factor in the organization of the cytoskeleton. *Biochem. J.*, **281**, 507–512.

Curie, P. (1894) Symétrie dans les phénomènes physiques. *J. Phys. Theor. Appl.*, *Sér.* 3, **III**, 393–415.

Davidson, E.H. (1990) How embryos work: a comparative view of diverse modes of cell fate especification. *Development*, **108**, 365–389.

Davies, D.D. (1973) Control of and by pH. *Symp. Soc. Exp. Biol.*, **27**, 513–529.

Davis, A., Sage, C.R., Wilson, L. and Farrell, K.W. (1993) Purification and biochemical characterization of tubulin from the budding yeast *Saccharomyces cerevisiae*. *Biochemistry*, **32**, 8823–8835.

Dean, A.C.R. and Hinshelwood, C. (1966) *Growth, Function and Regulation in Bacterial Cells*, Oxford University Press, London.

Decaux, J.F., Antoine, B. and Kahn, A. (1989) Regulation of the expression of the L-type pyruvate kinase gene in adult rat hepatocytes in primary culture. *J. Biol. Chem.*, **264**, 11584–11590.

Decaux, J.F., Marcillat, O., Pichard, A.L. *et al.* (1991) Glucose-dependent and -independent effect of insulin on gene expression. *J. Biol. Chem.*, **266**, 3432–3438.

de Jong, L., van Driel, R., Stuurman, N. *et al.* (1990) Principles of nuclear organization. *Cell Biol. Int. Rep.*, **14**, 1051–1074.

Dekker, R.F.H. (1982) Ethanol production from D-xylose and other sugars by the yeast. *Biotechnol. Lett.*, **4**, 411–416.

Delbruck, M. (1949) in *Unités Biologiques Douées de Continuité Génetique* Editions du CNRS, Paris, pp. 33–34.

Delmer, D.P. (1983) Biosynthesis of cellulose. *Adv. Carbohydr. Chem. Biochem.*, **41**, 105–153.

Delmer, P.D., Benziman, M. and Paday, E. (1982) Requirement of a membrane potential for cellulose synthesis in intact cells of *Acetobacter xylinum*. *Proc. Natl. Acad. Sci. USA*, **79**, 5282–5286.

De Loof, A. (1986) The electrical dimension of cells: The cell as a miniature electrophoresis chamber. *Int. Rev. Cytol.*, **104**, 251–352.

De Loof, A., Callaerts, P. and Vanden Broeck, J. (1992) The pivotal role of the plasma membrane–cytoskeletal complex and of epithelium formation in differentiation in animals. *Comp. Biochem. Physiol.*, **101A**, 639–651.

Den Hollander, J.A., Ugurbil, K., Brown, T.R. and Shulman, R.G. (1981) Phosphorus-31 nuclear magnetic resonance studies of the effect of oxygen upon glycolysis in yeast. *Biochemistry*, **20**, 5871–5880.

Den Hollander, J.A., Ugurbil, K. and Shulman, R.G. (1986) ^{31}P and ^{13}C NMR studies of intermediates of aerobic and anaerobic glycolysis in *Saccharomyces cerevisiae*. *Biochemistry*, **25**, 212–219.

Denis, C.L., Ferguson, J. and Young, E.T. (1983) mRNA levels for the fermentative alcohol dehydrogenase of *Saccharomyces cerevisiae* decrease upon growth on a nonfermentable carbon source. *J. Biol. Chem.*, **258**, 1165–1171.

Deretic, V., Dikshit, R., Konyecsni, W.M. *et al.* (1989) The algR gene, which regulates mucoidy in *Pseudomonas aeruginosa*, belongs to a class of environmentally responsive genes. *J. Bacteriol.*, **171**, 1278–1283.

De Robertis, E.D.P. and De Robertis, E.M.F. (1988) *Biología Celular y Molecular*, El Ateneo, Buenos Aires.

Devreotes, P. (1989) Cell–cell interactions in *Dictyostelium* development. *TIG*, **5**, 239–245.

de Vries, S. and Marres, C.A.M. (1987) The mitochondrial respiratory chain of yeast. Structure and biosynthesis and the role in cellular metabolism. *Biochim. Biophys. Acta*, **895**, 205–239.

Dickinson, J.R. and Williams, A.S. (1986) A genetic and biochemical analysis of the role of gluconeogenesis in sporulation in *Saccharomyces cerevisiae*. *J. Gen. Microbiol.*, **132**, 2605–2610.

Dickinson, J.R., Dawes, I.W., Boyd, A.S.F. and Baxter, R.L. (1983) ^{13}C NMR studies of acetate metabolism during sporulation of *Saccharomyces cerevisiae*. *Proc. Natl. Acad. Sci. USA*, **80**, 5847–5851.

Dietzler, D.N., Leckie, M.P., Sternheim, W.L. *et al.* (1979a) Regulation of glycogen synthesis and glucose utilization in *Escherichia coli* during maintenance of the energy charge. *J. Biol. Chem.*, **254**, 8276–8287.

Dietzler, D.N., Leckie, M.P., Lewis J.W. *et al.* (1979b) Evidence for new factors in the coordinate regulation of energy metabolism in *Escherichia coli*. *J. Biol. Chem.*, **254**, 8295–8307.

Dills, W.L. Jr (1993) Nutritional and physiological consequences of tumour glycolysis. *Parasitology*, **107**, S177–S186.

Dirick, L. and Nasmyth, K. (1991) Positive feedback in the activation of G1 cyclins in yeast. *Nature*, **351**, 754–757.

DiTella, M., Feiguin, F., Morfini, G. and Cáceres, A. (1994) Microfilament-associated growth cone component depends upon tau for its intracellular localization. *Cell Motil. Cytoskel.*, **29**, 117–130.

Dix, J.A. and Verkman, A.S. (1990) Mapping of fluorescence anisotropy in living cells by ratio imaging. Application to cytoplasmic viscosity. *Biophys. J.*, **57**, 231–240.

Doedle, E. (1986) *AUTO Manual*, California Institute of Technology, Pasadena.

Does, A.L. and Bisson, L.F. (1989) Comparison of glucose uptake kinetics in different yeasts. *J. Bacteriol.*, **171**, 1303–1308.

Driever, W. and Nüsslein-Volhard, Ch. (1988a) A gradient of *bicoid* protein in *Drosophila* embryos. *Cell*, **54**, 83–93.

Driever, W. and Nüsslein-Volhard, Ch. (1988b) The *bicoid* protein determines position in the *Drosophila* embryo in a concentration-dependent manner. *Cell*, **54**, 95–104.

Driever, W. and Nüsslein-Volhard, Ch. (1989) The *bicoid* protein is a positive regulator of hunchback transcription in the early *Drosophila* embryo. *Nature*, **337**, 138–143.

Drlica, K. (1984) Biology of bacterial deoxyribonucleic acid topoisomerases. *Microbiol. Rev.*, **48**, 273–289.

Drlica, K. (1990) Bacterial topoisomerases and the control of DNA supercoiling. *TIG*, **6**, 433–437.

Dunlap, J.C. (1990) Closely watched clocks. *TIG*, **6**, 159–165.

Dupont, G. and Goldbeter, A. (1992) Oscillations and waves of cytosolic calcium: Insights from theoretical models. *Bioessays*, **14**, 485–493.

Dworkin, M.B. and Dworkin-Rastl, E. (1989a) Metabolic regulation during early frog development: Glycogenic flux in *Xenopus* oocytes, eggs, and embryos. *Dev. Biol.*, **132**, 512–523.

Dworkin, M.B. and Dworkin-Rastl, E. (1989b) Metabolic regulation during early frog development: Flow of glycolytic carbon into phospholipids in *Xenopus* oocytes and fertilized eggs. *Dev. Biol.*, **132**, 524–528.

Dworkin, M.B. and Dworkin-Rastl, E. (1991) Carbon metabolism in early amphibian embryos. *TIBS*, **16**, 229–234.

Dykhuizen, D.E., Dean, A.M. and Hartl, D.L. (1987) Metabolic flux and fitness. *Genetics*, **115**, 25–31.

Edelstein-Kashet, L. (1988) *Mathematical Models in Biology*, Random House, New York.

Edmunds, L.N. Jr (1984) Physiology of circadian rhythms in micro-organisms. *Adv. Microb. Physiol.*, **25**, 61–148.

Edmunds, L.N. Jr (1988) *Cellular and Molecular Basis of Biological Clocks. Models and Mechanisms for Circadian Timekeeping*, Springer-Verlag, New York.

Edwards, K.L. and Pickard, B.G. (1987) Detection and transduction of physical stimuli in plants, in *The Cell Surface in Signal Transduction*, NATO ASI Series, vol. H12 (ed. E. Wagner), Springer-Verlag, Berlin, Heidelberg, pp. 41–66.

Ehret, C.F. and Trucco, E. (1967) Molecular models for the circadian clock. I. The chronon concept. *J. Theor. Biol.*, **15**, 242–262.

Ehret, C.F., Wille, J.J. and Trucco, E. (1973) The circadian oscillation: an integral and undissociable property of eukaryotic gene-action systems, in *Biological and Biochemical Oscillators* (eds B. Chance, A.K. Ghosh, E.K. Pye and B. Hess), Academic Press, New York, pp. 503–512.

Eigen, M. and R. Winkler (1981) *Laws of the Game. How the Principles of Nature Govern Chance*, Penguin Books, New York.

Eigenbrodt, E., Fister, P. and Reinacher, M. (1985) New perspectives on carbohydrate metabolism in tumor cells, in *Regulation of Carbohydrate Metabolism*, vol. 2 (ed. R. Beitner), CRC Press, Boca Raton, pp. 141–179.

Elinson, R.P. and Houliston, E. (1990) Cytoskeleton in *Xenopus* oocytes and eggs. *Semin. in Cell Biol.*, **1**, 349–357.

Elinson, R.P. and Kao, K.R. (1989) The location of dorsal information in frog early development. *Dev. Growth Differ.*, **31**, 423–430.

Elinson, R.P. and Rowling, B. (1988) A transient array of parallel microtubules in frog eggs: Potential tracks for a cytoplasmic rotation that specifies the dorso-ventral axis. *Dev. Biol.*, **128**, 185–197.

Entian, K.D. (1986) Glucose repression: a complex regulatory system in yeast. *Microbiol. Sci.*, **3**, 366–371.

Epner, D.E., Partin, A.W., Schalken, J.A. *et al.* (1993) Association of glyceraldehyde-3-phosphate dehydrogenase expression with cell motility and metastatic potential of rat prostatic adenocarcinoma. *Cancer Res.*, **53**, 1995–1997.

Erickson, H.P. and O'Brien, E.T. (1992) Microtubule dynamic instability and GTP hydrolysis. *Annu. Rev. Biophys. Biomol. Struct.*, **21**, 145–166.

Ermentrout, G.B. and Edelstein-Keshet, L. (1993) Cellular automata approaches to biological modeling. *J. Theor. Biol.*, **160**, 97–133.

Espeso, E.A., Tilburn, J., Arst, H.N. and Peñalva, M.A. (1993) pH regulation is a major determinant in expression of a fungal penicillin biosynthetic gene. *EMBO J.*, **12**, 3947–3956.

Esposito, R.E. and Klapholz, S. (1981) Meiosis and ascospore development, in *The Molecular Biology of the Yeast* Saccharomyces. *Life Cycle and Inheritance* (eds. J.N. Strathern, E.W. Jones and J.R. Broach), Cold Spring Harbor Laboratory Press, Cold Spring Harbor, NY, pp. 211–287

Eubank, S. and Farmer, D. (1989) An introduction to chaos and prediction, in *Lectures in the Sciences of Complexity*, vol. 2 (ed. E. Jen), Addison-Wesley, Santa Fe Institute, pp. 75–190.

Evtodienko, Y.V., Teplova, V.V., Duszynski, J. *et al.* (1994) The role of cytoplasmic [Ca^{2+}] in glucose-induced inhibition of respiration and oxidative phosphorylation in Ehrlich ascites tumor cells: a novel mechanism of the Crabtree effect. *Cell Calcium*, **15**, 439–446.

Ewer, J., Frisch, B., Hamblen-Coyle, M.J. *et al.* (1992) Expression of the period clock gene within different cell types in the brain of *Drosophila* adults and mosaic analysis of these cells' influence on circadian behavioral rhythms. *J. Neurosci.*, **12**, 3321–3349.

Falke, L.C., Edwards, K.L., Pickard, B.G. and Misler, S. (1988) A stretch-activated anion channel in tobacco protoplasts. *FEBS Lett.*, **237**, 141–144.

Feder, J. (1988) *Fractals*, Plenum Press, New York.

Fell, D. (1992) Metabolic control analysis: a survey of its theoretical and experimental development. *Biochem. J.*, **286**, 313.

Felle, H., Brummer, B., Bertl, A. and Parish, R.W. (1986) Indole 3-acetic acid and fusicoccin cause cytosolic acidification of corn coleoptile cells. *Proc. Natl. Acad. Sci. USA*, **83**, 8992–8995.

Ferreira, A. and Cáceres, A. (1989) The expression of acetylated microtubules during axonal and dendritic growth in cerebellar macroneurons which develop *in vitro*. *Dev. Brain Res.*, **49**, 205–213.

Fiechter, A. and Gmunder, F.K. (1989) Metabolic control of glucose degradation in yeast and tumor cells. *Adv. Biochem. Eng. Biotechnol.*, **39**, 2–28.

Fiechter, A., Fuhrmann, G.F. and Kappeli, O. (1981) Regulation of glucose metabolism in growing yeast cells. *Adv. Microb. Physiol.*, **22**, 123–183.

Fishan, J., Mikschl, E. and Hansen, U.P. (1986) Separate oscillations of the electrogenic pump and of a K⁺-channel in *Nitella* as revealed by simultaneous measurement of membrane potential and of resistance. *J. Exp. Bot.*, **37**, 34–47.

Fluck, R.A., Miller A.L. and Jaffe, L.F. (1991) Slow calcium waves accompany cytokinesis in medaka fish eggs. *J. Cell Biol.*, **115**, 1259–1265.

Forgacs, G. (1995) On the possible role of cytoskeletal filamentous networks in intracellular signaling: an approach based on percolation. *J. Cell Sci.*, **108**, 2131–2143.

Forgacs, G. and Newman, S.A. (1994) Phase transitions, interfaces, and morphogenesis in a network of protein fibers. *Int. Rev. Cytol.*, **150**, 139–148.

Forsburg, S.L., and Nurse, P. (1991) Cell cycle regulation in the yeasts *Saccharomyces cerevisiae* and *Schizosaccharomyces pombe*. *Annu. Rev. Cell Biol.*, **7**, 227–256.

Fraenkel, D.G. (1982) Carbohydrate metabolism, in *The Molecular Biology of the Yeast* Saccharomyces (eds J.N. Strathern, E.W. Jones and J.R. Broach), Cold Spring Harbor Laboratory Press, Cold Spring Harbor, NY, pp. 1–37

Fraenkel, D.G. (1992) Genetics and intermediary metabolism. *Annu. Rev. Genet.*, **26**, 159–177.

Franco, C.M.M., Smith, J.E. and Berry, D.R. (1984) Effect of nitrogen and phosphate on the levels of intermediates in bakers' yeast grown in continuous cultures. *J. Gen. Microbiol.*, **130**, 2465–2472.

Freese, E.B., Chu, M.I. and Freese, E. (1982) Initiation of yeast sporulation by partial carbon, nitrogen, or phosphate deprivation. *J. Bacteriol.*, **149**, 840–851.

Friesen, W.O. and Block, G.D. (1984) What is a biological oscillator? *Am. J. Physiol.*, **246**, R847–R851.

Friesen, W.O., Block, G.D. and Hocker, C.G. (1993) Formal approaches to understanding biological oscillators. *Annu. Rev. Physiol.*, **55**, 661–681.

Frölich, H. (1975) The extraordinary dielectric properties of biological materials and the action of enzymes. *Proc. Natl. Acad. Sci. USA*, **72**, 4211–4215.

Frolich, K.U., Entian, K.D. and Mecke, D. (1984) Cloning and restriction analysis of the hexokinase PII gene of the yeast *Saccharomyces cerevisiae*. *Mol. Gen. Genet.*, **194**, 144.

Fulton, A.B. (1982) How crowded is the cytoplasm? *Cell*, **30**, 345–347.

Fulton, A.B. (1993) Spatial organization of the synthesis of cytoskeletal proteins. *J. Cell. Biochem.*, **52**, 148–152.

Fulton, A.B. and L'Ecuyer, T. (1993) Cotranslational assembly of some cytoskeletal proteins: implications and prospects. *J. Cell Sci.*, **105**, 867–871.

Fushimi, K. and Verkman, A.S. (1991) Low viscosity in the aqueous domain of cell cytoplasm measured by picosecond polarization microfluorimetry. *J. Cell Biol.*, **112**, 719–725.

Galazzo, J.L. and Bailey, J.E. (1990) Fermentation pathways kinetics and metabolic flux control in suspended and inmobilized *Saccharomyces cerevisiae*. *Enzyme Microb. Technol.*, **12**, 162–172.

Gancedo, C. and Serrano, R. (1989) Energy-yielding metabolism, in *The Yeasts*, vol. 3 (eds A.H. Rose and J.S. Harrison), Academic Press, London, pp. 205–259.

Gancedo, J.M. (1992) Carbon catabolite repression in yeast. *Eur. J. Biochem.*, **206**, 297–313.

Gancedo, J.M. an Gancedo, C. (1986) Catabolite repression mutants of yeasts. *FEMS Microbiol. Rev.*, **32**, 179–187.

Garay, A.S. (1987) Why is chirality so important? *BioSystems*, **20**, 1–6.

Garner, M.M. and Burg, M.B. (1994) Macromolecular crowding and confinement in cells exposed to hypertonicity. *Am. J. Physiol.*, **266** (*Cell Physiol.* 35) C877–C892.

Gause, G.F.(1966) in *Microbial Models of Cancer Cells* (eds A. Neuberger and E.L. Tatum), North Holland, Amsterdam, pp. 17–28.

Gbelska, Y., Subik, J., Svoboda, A. *et al.* (1983) Intramitochondrial ATP and cell functions: yeast cells depleted of intramitochondrial ATP lose the ability to grow and multiply. *Eur. J. Biochem.*, **130**, 281–286.

Gelfand, V.I. and Bershadsky, A.D. (1991) Microtubule dynamics: mechanism, regulation and function. *Annu. Rev. Cell Biol.*, **7**, 93–116.

Gellert, M., Mizuuchi, K., O'Dea, M.H. and Nash, H.A. (1976) DNA gyrase: an enzyme that introduces superhelical turns into DNA. *Proc. Natl. Acad. Sci. USA*, **73**, 3872–3876.

Gellert, M., Menzel, R., Mizuuchi, K. *et al.* (1983) Regulation of DNA supercoiling in *Escherichia coli*. *Cold Spring Harbor Symp. Quant. Biol.*, **47**, 763–767.

Georgiev, G.P., Vassetzky, Y.S. Jr, Luchnik, A.N. *et al.* (1991) Nuclear skeleton, DNA domains and control of replication and transcription. *Eur. J. Biochem.*, **200**, 613–624.

Gerhart, J. and Keller, R. (1986) Region-specific cell activities in amphibian gastrulation. *Annu. Rev. Cell Biol.*, **2**, 201–229.

Gerhart, J., Black, S. and Scharf, S. (1983) Cellular and pancellular organization of the amphibian embryo. *Modern Cell Biol.*, **2**, 483–507.

Gerisch, G. and Hess, B. (1974) Cyclic AMP controlled oscillations in suspended *Dictyostelium* cells: their relation to morphogenetic cell interactions. *Proc. Natl. Acad. Sci. USA*, **71**, 2118–2122.

German, M.S. (1993) Glucose sensing in pancreatic islet beta cells: the key role of glucokinase and the glycolytic intermediates. *Proc. Natl. Acad. Sci. USA*, **90**, 1781–1785.

German, M.S., Moss, L.G. and Rutter, W.J. (1990) Regulation of insulin gene expression by glucose and calcium in transfected primary islet cultures. *J. Biol. Chem.*, **265**, 22063–22066.

Getzenberg, R.H., Pienta, K.J., Ward, W.S. and Coffey, D.S. (1991) Nuclear structure and the three-dimensional organization of DNA. *J. Cell. Biochem.*, **47**, 289–299.

Ghosh, A.K., Chance, B. and Pye, E.K. (1971) Metabolic coupling and synchronization of NADH oscillations in yeast cell populations. *Arch. Biochem. Biophys.*, **145**, 319–331.

Gibbon, B.C. and Kropf, D.L. (1994) Cytosolic pH gradients associated with tip growth. *Science*, **263**, 1419–1421.

Giddings, T.H., Brower, D.L. and Staehelin, L.A. (1980) Visualization of particle complexes in the plasma membrane of *Micrasterias denticulata* associated with the formation of cellulose fibrils in primary and secondary cell walls. *J. Cell Biol.*, **84**, 327–339.

Gillies, R.J. (1981) Intracellular pH and growth control in eukaryotic cells, in *The Transformed Cell* (eds I.L. Cameron and T.B. Pool), Academic Press, New York, pp. 348–395.

Gillies, R.J. (1982) Intracellular pH and proliferation in yeast, Tetrahymena and sea urchin eggs, in *Intracellular pH: its Measurement, Regulation, and Utilization in Cellular Functions* (eds R. Nuccitelli and D.W. Deamer), Alan R. Liss, New York, pp. 341–359.

Gillies, R.J. and Benoit, A.G. (1983) *Biochim. Biophys. Acta*, **762**, 466–470.

Gillies, R.J., Ugurbil, K., den Hollander, J.A. and Shulman, R.G. (1981) ^{31}P NMR studies of intracellular pH and phosphate metabolism during cell division cycle of *Saccharomyces cerevisiae*. *Proc. Natl. Acad. Sci. USA*, **78**, 2125–2129.

Gimeno, C.J., Ljungdahl, P.O., Styles, C.A. and Fink, G.R. (1992) Unipolar cell divisions in the yeast *S. cerevisiae* lead to filamentous growth: regulation by starvation and *RAS*. *Cell*, **68**, 1077.

Glansdorff, P. and Prigogine, I. (1971) *Thermodynamic Theory of Structure, Stability and Fluctuations*, Wiley, London.

Glass, L. and Mackey, M.C. (1979) Pathological conditions resulting from instabilities in physiological control systems. *Ann. N.Y. Acad. Sci.*, **316**, 214–235.

Glass, L. and Mackey, M.C. (1988) *From Clocks to Chaos. The Rhythms of Life*, Princeton University Press, Princeton, NJ.

Go, N. (1983) Theoretical studies of protein folding. *Annu. Rev. Biophys. Bioeng.*, **12**, 183–210.

Gober, J.W. and Kashket, E.R. (1989) Role of DNA superhelicity in regulation of bacteroid-associated functions of *Bradyrhyzobium* sp. strain 32H1. *Appl. Environ. Microbiol.*, **55**, 1420–1425.

Goldbeter, A. and Caplan, S.R. (1976) Oscillatory enzymes. *Annu. Rev. Biophys. Bioeng.*, **5**, 449–476.

Goldbeter, A. and Dupont, G. (1990) Allosteric regulation, cooperativity, and biochemical oscillations. *Biophys. Chem.*, **37**, 341–353.

Goldbeter, A. and Martiel, J.L. (1987) Periodic behaviour and chaos in the mechanism of intercellular communication governing aggregation of *Dictyostelium* amoebae, in *Chaos in Biological Systems* (eds H. Degn, A.V. Holden and L.F. Olsen), Plenum Press, New York, pp. 79–89.

Goldbeter, A. and Segel, L.A. (1977) Unified mechanism for relay and oscillation of cyclic AMP in *Dictyostelium discoideum*. *Proc. Natl. Acad. Sci. USA*, **74**, 1543–1547.

Goldbeter, A. and Segel, L.A. (1980) Control of developmental transitions in the cyclic AMP signalling system of *Dictyostelium discoideum*. *Differentiation*, **17**, 127–135.

Goldstein, B., Hird, S.N. and White, J.G. (1993) Cell polarity in early *C. elegans* development. *Development*, Suppl., 279–287.

Goldstein, R.A. (1991) *Genetic Ecology: Working With Nature to Clean Up Wastes*, Electric Power Research Institute, Palo Alto.

Goldstein, R.A., Bourquin, A.W., Federle, T.W. *et al.* (1991) *Environmental Biotechnology for Waste Treatment*, Plenum Press, New York, pp. 271–278.

Gommers, P.J.F., van Schie, B.J., van Dijken, J.P. and Kuenen, J. G. (1988) Biochemical limits to microbial growth yields: An analysis of mixed substrate utilization. *Biotechnol. Bioeng.*, **32**, 86–94.

Goodner, B. and Quatrano, R.S. (1993) *Fucus* embryogenesis: a model to study the establishment of polarity. *Plant Cell*, **5**, 1471–1481.

Goodsell, D.S. (1991) Inside a living cell. *TIBS*, **16**, 203–206.

Goodwin, B.C. (1963) *Temporal Organization in Cells. A Dynamic Theory of Cellular Control Processes*, Academic Press, London.

Goodwin, B.C. (1986) What are the causes of morphogenesis? *BioEssays*, **3**, 32–36.

Gow, N.A.R. (1989) Circulating ionic currents in micro-organisms. *Adv. Microb. Physiol.*, **30**, 89–123.

Graham, I.A., Denby, K.J. and Leaver, Ch.J. (1994) Carbon catabolite repression regulates glyoxylate cycle gene expression in cucumber. *Plant Cell*, **6**, 761–772.

Graham, J.H., Freeman, D.C. and Emlen, J.M. (1993) Antisymmetry, directional asymmetry and dynamic morphogenesis. *Genetica*, **89**, 121–137.

Grandin, N. and Charbonneau, M. (1991a) Intracellular free calcium oscillations during cell division of *Xenopus* embryos. *J. Cell Biol.*, **112**, 711–718.

Grandin, N. and Charbonneau, M. (1991b) Cycling of intracellular free calcium and intracellular pH in *Xenopus* embryos: a possible role in the control of the cell cycle. *J. Cell Sci.*, **99**, 5–11.

Greaney, G.S. and Somero, G.N. (1979) Effects of anions on the activation thermodynamics and fluorescence emission spectrum of alkaline phosphatase: evidence for enzyme hydration changes during catalysis. *Biochemistry*, **18**, 5322–5332.

Gross, J.D. (1994) Developmental decisions in *Dictyostelium discoideum*. *Microbiol. Rev.*, **58**, 330–351.

Gross, J.D., Bradbury, J., Kay, R.R. and Peacey, M.J. (1983) Intracellular pH and the control of cell differentiation in *Dictyostelium discoideum*. *Nature*, **303**, 244–245.

Gubb, D. (1986) Intron-delay and the precision of expression of homeotic gene products in *Drosophila*. *Dev. Genet.*, **7**, 119–131.

Gubb, D. (1993) Genes controlling cellular polarity in *Drosophila*. *J. Cell Sci.*, Suppl., 269–273.

Guckenheimer, J. and Holmes, P. (1983) *Nonlinear Oscillations, Dynamical Systems and Bifurcation Vector Fields*, Springer Verlag, New York.

Gumaa, K.A. and McLean, P. (1969) A possible interrelationship between binding of hexokinase and the site of ATP formation in Krebs ascites cells. *Biochem. Biophys. Res. Commun.*, **36**, 771–779.

Gunsalus, I.C. and Shuster, C.W. (1961) *The Bacteria. Vol. II: Metabolism*, Academic Press, New York, pp. 1–57.

Haken, H. (1978) *Synergetics*, Springer-Verlag, Berlin, Heidelberg, p. 351.

Haken, H. (1987) Self-organizing systems. The emergence of order, in *Synergetics: an Approach to Self-organization* (eds F.E.Yates, A. Garfinkel, D.O. Walter and G.B. Yates), Plenum Press, New York, London, pp. 417–434.

Haken, H. (1991) Synergetics – can it help physiology?, in *Rhythms in Physiological Systems* (eds H. Haken and H.P. Koepchen), Springer-Verlag, Berlin, Heidelberg.

Hall, B.G. (1990) Spontaneous point mutations that occur more often when advantageous than when neutral. *Genetics*, **126**, 5–16.

Hall, J.L. and Baker, D.A. (1982) *Membranas Celulares y Transporte de Iones*, CECSA, Mexico.

Hara, K., Tydeman, P. and Kirschner, M. (1980) A cytoplasmic clock with the same period as the division cycle in *Xenopus* eggs. *Proc. Natl. Acad. Sci. USA*, **77**, 462–466.

Harder, W. and Dijkhuizen, L. (1983) Physiological responses to nutrient limitation. *Annu. Rev. Microbiol.*, **37**, 1–23.

Hardin, P.E., Hall, J.C. and Rosbash, M. (1990) Feedback of the *Drosophila period* gene product on circadian cycling of its messenger RNA levels. *Nature*, **343**, 536–540.

Harman, T.R. and Pace, G.W. (1984) Energy requirements for microbial exopolysaccharide synthesis. *Arch. Microbiol.*, **137**, 231–235.

Harold, F.M. (1982) Pumps and currents: A biological perspective. *Curr. Top. Membr. Transp.*, **16**, 485–516.

Harold, F.M. (1990) To shape a cell: an inquiry into the causes of morphogenesis of microorganisms. *Microbiol. Rev.*, **54**, 381–431.

Harold, F.M. (1991) Biochemical topology: from vectorial metabolism to morphogenesis. *Biosci. Rep.*, **11**, 347–385.

Harrison, L.G. (1982) An overview of kinetic theory in developmental modeling, in *Developmental Order: Its Origin and Regulation* (eds S. Subtelny and P.B. Green), Alan R. Liss, New York, pp. 3–33.

Harrison, L.G. (1987) What is the status of reaction–diffusion theory thirty-four years after Turing? *J. Theor. Biol.*, **125**, 369–384.

Hartwell, L.H. (1967) Macromolecule synthesis in temperature-sensitive mutants of yeast. *J. Bacteriol.*, **93**, 1662–1670.

Hartwell, L.H. (1970) Periodic density fluctuation during the yeast cell cycle and the selection of synchronous cultures. *J. Bacteriol.*, **104**, 1280–1285.

Hartwell, L.H. (1973) Three additional genes required for DNA synthesis in *Saccharomyces cerevisiae. J. Bacteriol.*, **115**, 966–974.

Hartwell, L.H. (1974) *Saccharomyces cerevisiae* cell cycle. *Bacteriol. Rev.*, **38**, 164–198.

Hartwell, L.H. (1991) Twenty-five years of cell cycle genetics. *Genetics*, **129**, 975–980.

Hartwell, L.H. and Kastan, M.B. (1994) Cell cycle control and cancer. *Science*, **266**, 1821–1828.

Hartwell, L.H. and Weinert, T.A. (1989) Checkpoints: controls that ensure the order of cell cycle events. *Science*, **246**, 629–634.

Hartwell, L.H., Culotti, J. and Reid, B. (1970) Genetic control of the cell-division cycle in yeast, I. Detection of mutants. *Proc. Natl. Acad. Sci. USA*, **66**, 352–359.

Hartwell, L.H., Mortimer, R.K., Culotti, J. and Culotti, M. (1973) Genetic control of the cell division cycle in yeast: V. Genetic analysis of cdc mutants. *Genetics*, **74**, 267–286.

Hartwell, L.H., Culotti, J., Pringle, J.R. and Reid, B.J. (1974) Genetic control of the cell division cycle in yeast: a model. *Science*, **183**, 46–51.

Hassell, M.P., Comins, H.N. and May, R.M. (1991) Spatial structure and chaos in insect population dynamics. *Nature*, **353**, 500–503.

Haussinger, D., Stoll, B. vom Dahl, S. *et al.* (1994a) Effect of hepatocyte swelling on microtubule stability and tubulin mRNA levels. *Biochem. Cell Biol.*, **72**, 12–19.

Haussinger, D., Lang, F. and Gerok, W. (1994b) Regulation of cell function by the cellular hydration state. *Am. J. Physiol. (Endocrinol. Metab.)*, **30**, E343–E355.

Hedeskov, C.J. (1968) Early effects of phytohaemagglutinin on glucose metabolism of normal human lymphocytes. *Biochem. J.*, **110**, 373–380.

Heinisch, J. (1986) Isolation and characterization of the two sructual genes coding for phosphofructokinase in yeast. *Mol. Gen. Genet.*, **202**, 75–82.

Heinrich, R. and Rappoport, T. (1974) A linear steady-state treatment of enzymatic chains. General properties, control and effector strength. *Eur. J. Biochem.*, **42**, 89–95.

Heinrich, R., Rapoport, S.M. and Rapoport, T.A. (1977) Metabolic regulation and mathematical models. *Prog. Biophys. Mol. Biol.*, **32**, 1–82.

Heins, S. and Aebi, U. (1994) Making heads and tails of intermediate filament assembly, dynamics and networks. *Curr. Opin. Cell Biol.*, **6**, 25–33.

Hellingwerf, K.J. and Konings, W.N. (1985) The energy flow in bacteria: the main free energy intermediates and their regulatory role. *Adv. Microb. Physiol.*, **26**, 125–154.

Hempfling, W.P. and Mainzer, S.E. (1975) Effects of varying the carbon source limiting growth on yield and maintenance characteristics of *Escherichia coli* in continuous cultures. *J. Bacteriol.*, **123**, 1078–1087.

Herbert, D. and Kornberg, H.L. (1976) Glucose transport as rate-limiting step in the growth of *Escherichia coli* in glucose. *Biochem. J.*, **156**, 477–480.

Hereford, L.M. and Hartwell, L.H. (1974) Sequential gene function in the initiation of *Saccharomyces cerevisiae* DNA synthesis. *J. Mol. Biol.*, **84**, 445–461.

Hernandez-Cruz, A., Sala, F. and Adams, P.R. (1990) Subcellular calcium transients visualized by confocal microscopy in a voltage-clamped vertebrate neuron. *Science*, **247**, 858–862.

Herschlag, D. (1988) The role of induced fit and conformational changes of enzymes in specificity and catalysis. *Bioorg. Chem.*, **16**, 62–96.

Herth, W. (1985) Plasma membrane rosettes involved in localized wall thickening during xylem vessel formation of *Lepidium sativum* L. *Planta*, **164**, 12–21.

Hervagault, J.F. and Canu, S. (1987) Bistability and irreversible transitions in a simple substrate cycle. *J. Theor. Biol.*, **127**, 439–449.

Hervagault, J.F. and Thomas, D. (1987) Oscillatory phenomena in immobilized enzyme systems. *Methods Enzymol.*, **135**, 554–569.

Hess, B. (1977) Oscillating reactions. *Trends Biochem. Sci.*, 193–195.

Hess, B. and Boiteux, A. (1968) Control of glycolysis, in *Regulatory Functions of Biological Membranes* (ed. J. Jarnefelt), Elsevier, Amsterdam, pp. 148–162.

Hess, B. and Boiteux, A. (1971) Oscillatory phenomena in biochemistry. *Annu. Rev. Biochem.*, **40**, 237–258.

Hess, B. and Boiteux, A. (1973) Substrate control of glycolytic oscillations, in *Biological and Biochemical Oscillators* (eds B. Chance, K. Pye, A.K. Ghosh and B. Hess), Academic Press, New York, pp. 229–242.

Hess, B. and B. Chance (1978) Oscillating enzyme reactions, in *Theoretical Chemistry. Periodicities in Chemistry and Biology*, vol. 4 (eds H. Eyring and D. Henderson), Academic Press, New York, pp. 159–179.

Hess, B. and Markus, M. (1987) Order and chaos in biochemistry. *TIBS*, **12**, 45–48.

Hess, B. and Mikhailov, A. (1994) Self-organization in living cells. *Science*, **264**, 223–224.

Higgins, Ch.F., Cairney, J., Stirling, D.A. *et al.* (1987) Osmotic regulation of gene expression: ionic strength as an intracellular signal? *TIBS*, **12**, 339–344.

Higgins, Ch.F., Dorman, Ch.J., Stirling, D.A. *et al.* (1988) A physiological role of DNA supercoiling in the osmotic regulation of gene expression in *S. typhimurium* and *E. coli*. *Cell*, **52**, 569–584.

Higgins, J. (1965) Dynamics and control in cellular reactions, in *Control of Energy Metabolism* (eds B. Chance, R.W. Estabrook and J.R. Williamson), Academic Press, New York, pp. 13–46.

Higgins, J., Frenkel, R., Hulme, E. *et al.* (1973) The control theoretic approach to the analysis of glycolytic oscillators, in *Biological and Biochemical Oscillators* (eds B. Chance, K. Pye, A.K.Ghosh and B. Hess), Academic Press, New York, pp. 127–175.

Hill, A. (1990) Entropy production as the selection rule between different growth morphologies. *Nature*, **348**, 426–428.

Hill, D.P. and Strome, S. (1988) An analysis of the role of microfilaments in the establishment and maintenance of asymmetry in *Caenorhabditis elegans* zygotes. *Dev. Biol.*, **125**, 75–84.

Hirokawa, N. (1991) Molecular architecture and dynamics of the neuronal cytoskeleton, in *The Neuronal Cytoskeleton*, Wiley-Liss, New York, pp. 5–74.

Hirsch, M.W. and Smale, S. (1974) *Differential Equations, Dynamical Systems and Linear Algebra*, Academic Press, London.

Hitt, A., Cross, A. and Williams, R.C. (1990) Microtubule solutions display nematic liquid crystalline structure. *J. Biol. Chem.*, **265**, 1639–1647.

Hochachka, P.W. (1988) Metabolic-, channel-, and pump-coupled functions: constraints and compromises of coadaptation. *Can. J. Zool.*, **66**, 1015–1027.

Holliday, R. (1988) Successes and limitations of molecular biology. *J. Theor. Biol.*, **132**, 253–262.

Holms, W.H. (1986) The central metabolic pathways of *Escherichia coli*: Relationships between flux and control at a branch point, efficiency of conversion to biomass and excretion of acetate. *Curr. Top. Cell. Regul.*, **28**, 69–105.

Honigberg, S.M. and Esposito, R.E. (1994) Reversal of cell determination in yeast meiosis: Postcommitment arrest allows return to mitotic growth. *Proc. Natl. Acad. Sci. USA*, **91**, 6559–6563.

Hoosein, M.A. and Lewin, A.S. (1984) Derepression of citrate synthase in *Saccharomyces cerevisiae* may occur at the level of transcription. *Mol. Cell. Biol.*, **4**, 247–253.

Hopf, E. (1942) Abzweigung einer periodischen Lssung von einer stationSren Lssung eines Differentialsystems. *Ber. Math. Kl. Sch. Phys. Akad. Wiss. Leipzig*, **94**, 3–22

Horio, T. and Hotani, H. (1986) Visualization of the dynamic instability of individual microtubules by dark-field microscopy. *Nature*, **321**, 605–607.

Houliston, E. and Elinson, R.P. (1991) Patterns of microtubule polymerization relating to cortical rotation in *Xenopus laevis* eggs. *Development*, **112**, 107–117.

Hsieh, L.S., Burger, R.M. and Drlica, K. (1991a) Bacterial DNA supercoiling and ATP/ADP changes associated to transition to anaerobic growth. *J. Mol. Biol.*, **199**, 443–450.

Hsieh, L.S., Rouviere-Yaniv, J. and Drlica, K. (1991b) Bacterial DNA supercoiling and ATP/ADP ratio: changes associated with salt shock. *J. Bacteriol.*, **173**, 3914–3917.

Hugo, F., Mazurek, S., Zander, U. and Eigenbrodt, E. (1992) In vitro effect of extracellular AMP on MCF-7 breast cancer cells: inhibition of glycolysis and cell proliferation. *J. Cell. Physiol.*, **153**, 539–549.

Ingber, D.E. (1993) The riddle of morphogenesis: a question of solution chemistry or molecular cell engineering? *Cell*, **75**, 1249–1252.

Ingber, D.E., Dike, L., Hansen, L. *et al.* (1994) Cellular tensegrity: Exploring how mechanical changes in the cytoskeleton regulate cell growth, migration, and tissue pattern during morphogenesis. *Int. Rev. Cytol.*, **150**, 173–224.

Ingham, P.W. (1988) The molecular genetics of embryonic pattern formation in *Drosophila*. *Nature*, **335**, 25–34.

Ingraham, J.L., Maaloe, O. and Neidhardt, F.C. (1983) Chemical synthesis of the bacterial cell: Polymerization, biosynthesis, fueling reactions, and transport, in *Growth of the Bacterial Cell*, Sinauer Associates, Sunderland, MA, pp. 87–174.

Ishide, N., Miura, M., Sakurai, M. and Takishima, T. (1992) Initiation and development of calcium waves in rat myocytes. *Am. J. Physiol.*, **263**, H327–332.

Jackson, D.A. (1991) Structure–function relationship in eukaryotic nuclei. *Bioessays*, **13**, 1–10.

Jackson, S.L. and Heath, I.B. (1993) Roles of calcium ions in hyphal tip growth. *Microbiol. Rev.*, **57**, 367–382.

Jacobs, Ch.W., Adams, A.E.M., Szaniszlo, P.J. and Pringle, J.R. (1988) Functions of microtubules in the *Saccharomyces cerevisiae* cell cycle. *J. Cell Biol.*, **107**, 1409–1426.

Jaffe, L. (1991) The path of calcium in cytosolic calcium oscillations: A unifying hypothesis. *Proc. Natl. Acad. Sci. USA*, **88**, 9883–9887.

Jaffe, L.A. (1980) Calcium explosions as triggers of development. *Ann. New York Acad. Sci.*, **339**, 86–101.

Jaffe, L.F. (1977) Electrophoresis along cell membranes. *Nature*, **265**, 600–602.

Jaffe, L.F. (1979) Control of development by ionic currents, in *Membrane Traduction Mechanism* (eds R.A. Cone and J.E. Prowling), Raven Press, New York, pp. 119–231.

Janmey, P.A. (1994) Phosphoinositides and calcium as regulators of cellular actin assembly and disassembly. *Annu. Rev. Physiol.*, **56**, 169–191.

Jarman, T.R. and Pace, G.W. (1984) Energy requirements for microbial exopolysaccharide synthesis. *Arch. Microbiol.*, **137**, 231–235.

Jeffrey, A. (1990) *Linear Algebra and Ordinary Differential Equations*, Blackwell, Oxford.

Jeffries, T.W. (1985) Emerging technology for fermenting D-xylose. *TIBTECH* , **3**, 208–212.

Jensen, P.R, Loman, L., Petra, B. *et al.* (1995) Energy buffering of DNA structure fails when *Escherichia coli* runs out of substrate. *J. Bacteriol.*, **177**, 3420–4326.

Johnston, G.C. and Singer, R.A. (1978) RNA synthesis and control of cell division in the yeast *S. cerevisiae*. *Cell*, **14**, 951–958.

Johnston, G.C., Pringle, J.R. and Hartwell, L.H. (1977a) Coordination of growth with cell division in the yeast *Saccharomyces cerevisiae*. *Exp. Cell Res.*, **105**, 79–98.

Johnston, G.C., Singer, R.A. and McFarlane, E.S. (1977b) Growth and cell division during nitrogen starvation of the yeast *Saccharomyces cerevisiae*. *J. Bacteriol.*, **132**, 723–730.

Johnston, G.C., Singer, R.A., Sharrow, S.O. and Slater, M.L. (1980) Cell division in the yeast *Saccharomyces cerevisiae* at different rates. *J. Gen. Microbiol.*, **118**, 479–484.

Johnston, M. and Carlson, M. (1992) Regulation of carbon and phosphate utilization, in *The Molecular and Cellular Biology of the Yeast* Saccharomyces (eds E.W. Jones, J.R. Pringle and J.R. Broach), Cold Spring Harbor Laboratory Press, Cold Spring Harbor, NY, pp. 193–281.

Jones, D.P., Shan, X. and Park, Y. (1992) Coordinated multisite regulation of cellular energy metabolism. *Annu. Rev. Nutr.*, **12**, 327–343.

Jones, E.W. and Fink, G.R. (1982) Metabolism and gene expression, in *The Molecular Biology of the Yeast* Saccharomyces (eds J.N. Strathern, E.W. Jones and J.R. Broach), Cold Spring Harbor Laboratory Press, Cold Spring Harbor, NY, pp. 181–299.

Jones, R.P. (1988) Intracellular ethanol – accumulation and exit from yeast and other cells. *FEMS Microbiol. Rev.*, **54**, 239–258.

Jordan, M.A., Himes, R.H. and Wilson, L. (1985) Comparison of the effect of vinblastine, vincristine, vindesine and vinepidine on microtubule dynamics and cell proliferation *in vitro*. *Cancer Res.*, **45**, 2741–2747.

Jordan, M.A., Thrower, D. and Wilson, L. (1991) Mechanism of inhibition of cell proliferation by Vinca alkaloids. *Cancer Res.*, **51**, 2212–2222.

Jordan, M.A., Toso, R.J., Thrower, D. and Wilson, L. (1993) Mechanism of mitotic block and inhibition of cell proliferation by taxol at low concentrations. *Proc. Natl. Acad. Sci. USA*, **90**, 9552–9556.

Kacser, H. and Burns, J.A. (1973) The control of flux. *Symp. Soc. Exp. Biol.*, **27**, 65–104.

Kacser, H. and Burns, J.A. (1981) The molecular basis of dominance. *Genetics*, **97**, 639–666.

Kacser, H. and Porteus, J.W. (1987) Control of metabolism: What do we have to measure? *TIBS*, **12**, 5–14.

Kagan, M.L., Kosloff, R., Citri, O. and Avnir, D. (1989) Chemical formation of spatial patterns induced by non linearity in a concentration-dependent diffusion coefficient. *J. Phys. Chem.*, **93**, 2728–2731.

Kai, S., Müller, S.C. and Ross, J. (1982) Measurements of temporal and spatial sequences of events in periodic precipitation processes, *J. Chem. Phys.*, **76**, 1392–1406.

Kappeli, O. (1986) Regulation of carbohydrate metabolism in *Saccharomyces cerevisiae* and related yeast. *Adv. Microb. Physiol.*, **28**, 181–209.

Kappeli, O. and Sonnleitner, B. (1986) Regulation of sugar metabolism in *Saccharomyces*-type yeast: Experimental and conceptual considerations. *CRC Crit. Rev. Biotechnol.*, **4**, 299–325.

Kappeli, O., Arreguin, M. and Rieger, M. (1985) The respirative breakdown of glucose by *Saccharomyces cerevisiae*: an assessment of a physiological state. *J. Gen. Microbiol.*, **131**, 1411–1416.

Kaprelyants, A.S. (1988) Dynamic spatial distribution of proteins in the cell. *TIBS*, **13**, 43–46.

Karadsheh, N.S. and Uyeda, K. (1977) Changes in allosteric properties of PFK bound to erythrocyte membranes. *J. Biol. Chem.*, **252**, 7418–7420.

Karplus, M. and McCammon (1983) Protein dynamics. *Annu. Rev. Biochem.*, **52**, 263–300.

Kashket, E.R. (1981) Proton motive force in growing *Streptococcus lactis* and *Staphylococcus aureus* cells under aerobic and anaerobic conditions. *J. Bacteriol.*, **146**, 369–376.

Kashket, E.R. (1982) Stoichiometry of the H^+-ATPase of growing and resting, aerobic *Escherichia coli*. *Biochemistry*, **21**, 5534–5538.

Kassas, M. and Polunin, N. (1989) The three systems of man. *Environ. Conservation*, **16**, 7–11.

Kassir, Y., Granto, D. and Simchem, G. (1988) *IME1*, a positive regulator gene of meiosis in *Saccharomyces cerevisiae*. *Cell*, **52**, 853–862.

Katchalsky, A. and Spangler, R. (1968) Dynamics of membrane processes. *Q. Rev. Biophys.*, **2**, 127–175.

Kauffman, S.A. (1989) Adaptation on rugged fitness landscapes, in *Lectures in the Sciences of Complexity*, Vol. 1 (ed. D.L. Stein), Addison-Wesley, Santa Fe Institute, New Mexico, pp. 527–712.

Kauffman, S.A., Shymko, R.M. and Trabert, K. (1978) Control of sequential compartment formation in *Drosophila*. *Science*, **199**, 259–270.

Keleti, T. and Ovádi, J. (1988) Control of metabolism by dynamic macromolecular interactions. *Curr. Top. Cell. Regul.*, **29**, 1–33.

Keleti, T., Ovádi, J. and Batke, J. (1989) Kinetic and physico-chemical analysis of enzyme complexes and their possible role in the control of metabolism. *Prog. Biophys. Mol. Biol.*, **53**, 105–152.

Kell, D.B. (1987) Forces, fluxes and the control of microbial growth and metabolism. *J. Gen. Microbiol.*, **133**, 1651–1665.

Kell, D.B., van Dam, K. and Westerhoff, H.V. (1989) Control analysis of microbial growth and productivity, in *Soc. Gen. Microbiol. Symp.*, Cambridge University Press, Cambridge, pp. 44–93.

Kelly, P.J., Kelleher, J.K. and Wright, B.E. (1979) The tricarboxylic acid cyle in *Dictyostelium discoideum*. *Biochem. J.*, **184**, 581–588.

Kernevez, J.P. (1980) *Enzyme Mathematics*, North Holland Publishing Company, Amsterdam.

Kernevez, J.P., Joly, G., Duban, M.C. *et al.* (1979) Hysteresis, oscillations and pattern formation in realistic immobilized enzyme systems. *J. Math. Biol.*, **7**, 41–56

Kilmartin, J.V. and Adams, A.E.M. (1984) Structural rearrangements of tubulin and actin during the cell cycle of the yeast *Saccharomyces*. *J. Cell Biol.*, **98**, 922–933.

Kimelman, D., Kirschner, M. and Scherson, T. (1987) The events of the midblastula transition in *Xenopus* are regulated by changes in the cell cycle. *Cell*, **48**, 399–407.

Kinraide, T.B. and Wisse, R.E. (1986) Electrical evidence for turgor inhibition of proton extrusion of sugar beet taproot. *Plant Physiol.*, **82**, 1148–1150.

Kirschner, M. (1982) Microtubules and their role in cell, tissue, and organismal polarity, In *Developmental Order: its Origin and Regulation* (eds S. Subtelny and P.B. Green), Alan R. Liss, New York, pp. 117–132.

Kirschner, M. (1992) The cell cycle then and now.*TIBS*, **17**, 281–285.

Kirschner, M. and Mitchinson, T. (1986) Beyond self-assembly: from microtubules to morphogenesis. *Cell*, **45**, 329–342.

Knull, H.R. and Walsh, J.L. (1992) Association of glycolytic enzymes with the cytoskeleton. *Curr. Top. Cell. Regul.*, **33**, 15–30.

Kobayashi, H. (1985) A proton-translocating ATPase regulates the pH of the bacterial cytoplasm. *J. Biol. Chem.*, **260**, 72–76.

Koerner, T.J., Hill, J. and Tzagoloff, A. (1985) Cloning and characterization of the yeast nuclear gene for subunits of cytochrome oxidase. *J. Biol. Chem.*, **260**, 9513.

Konings, W.N. (1985) Generation of metabolic energy by end-product flux. *TIBS*, **10**, 317–319.

Kooten, O.V., Snell, J.F.H. and Vredenberg, W.J. (1986) in *Photosynthesis Research* (eds M. Nijhoff and W. Junk), Kluwer, Dordrecht, pp. 211–227.

Kopelman, R. (1988) Fractal reaction kinetics, *Science*, **241**, 1620–1626.

Korn, E.D., Carlier, M.F. and Pantaloni, D. (1987) Actin polymerization and ATP hydrolysis. *Science*, **238**, 638–644.

Koshland, D.E. Jr (1987) Evolution of catalytic function. *Cold Spring Harbor Symp. Quant. Biol.*, **52**, 1–7

Kretschner, M., Tempst, P. and Fraenkel, D.G. (1991) Identification and cloning of yeast phosphofructokinase 2. *Eur. J. Biochem.*, **197**, 367.

Kristofferson, D., Mitchison, T. and Kirschner, M. (1986) Direct observation of steady-state microtubule dynamics. *J. Cell Biol.*, **102**, 1007–1019.

Kropf, D.L. (1986) Electrophysiological properties of *Achlya* hyphae: Ionic currents studied by intracellular potential recording. *J. Cell. Biol.*, **102**, 1209–1216.

Kropf, D.L. (1992) Establishment and expression of cellular polarity in fucoid zygotes. *Microbiol. Rev.*, **56**, 316–339.

Kropf, D.L. (1994) Cytoskeletal control of cell polarity in a plant zygote. *Dev. Biol.* **165**, 361–371.

Kropf, D.L., Coffman, H.R., Kloareg, B. *et al.* (1993) Cell wall and rhizoid polarity in *Pelvetia* embryos. *Dev. Biol.*, **160**, 303–314.

Kubicek, M. and Marek, M. (1983) *Computational Methods in Bifurcation Theory and Dissipative Structures*, Springer Verlag, New York.

Kuenzi, M.T. and Fiechter, A. (1972) Regulation of carbohydrate composition of *Saccharomyces cerevisiae* under growth limitation. *Arch. Mikrobiol.*, **84**, 254–261.

Kunau, W.-H. and Hartig, A. (1992) Peroxisome biogenesis in *Saccharomyces cerevisiae*. *Antonie van Leeuwenhoek*, **62**, 63.

Kutschera, U. and Briggs, W.R. (1987) Rapid auxin-induced stimulation of cell wall synthesis in pea internodes. *Proc. Natl. Acad. Sci. USA*, **84**, 2747–2751.

Lagunas, R. (1993) Sugar transport in *Saccharomyces cerevisiae*. *FEMS Microbiol. Rev.*, **104**, 229–242.

Lahoz-Beltra, R., Hameroff, S.R. and Dayhoff, J.E. (1993) Cytoskeletal logic: a model for molecular computation via Boolean operations in microtubules and microtubule-associated proteins. *Biosystems*, **29**, 1–23.

Lawrence, P.A. (1988) Background to *bicoid*. *Cell*, **54**, 1–2.

Lechleiter, J., Girard, S., Peralta, E. and Clapham, D. (1991) Spiral calcium wave propagation and annihilation in *Xenopus* laevis oocytes. *Science*, **252**, 123–126.

Lee, M.-S. and Garrard, W.T. (1991) Positive DNA supercoiling generates a chromatin conformation characteristic of highly transcribed genes. *Proc. Natl. Acad. Sci. USA*, **88**, 9675–9679.

Lehninger, A.L. (1977) *Biochemistry*, Worth Publishers, New York.

Lehotzky, A., Telegdi, M., Liliom, K. and Ovadi, J. (1993) Interaction of phosphofructokinase with tubulin and microtubules. *J. Biol. Chem.*, **268**, 10888–10894.

Lengsfeld, A.M., Dietrich, J. and Schultze-Maurer, B. (1982) Accumulation and release of vinblastine and vincristine by HeLa cells: light microscopic, cinematographic, and biochemical study. *Cancer Res.*, **42**, 3798–3805.

Leverve, X.M., Fontaine, E., Putod-Paramelle, F. and Rigoulet, M. (1994) Decrease in cytosolic ATP/ADP ratio and activation of pyruvate kinase after in vitro addition of almitrine in hepatocytes isolated from fasted rats. *Eur. J. Biochem.*, **224**, 967–974.

Lew, D.J. and Reed, S.I. (1993) Morphogenesis in the yeast cell cycle: regulation by Cdc28 and cyclins. *J. Cell Biol.*, **120**, 1305–1320.

Lew, D.J. and Reed, S.I. (1995) Cell cycle control of morphogenesis in budding yeast. *Curr Opin. Genet Dev.*, **5**, 17–23

Lewis, J., Slack, J.M. and Wolpert, L. (1977) Thresholds in development. *J. Theor. Biol.*, **65**, 579–590.

Liao, J.C. and Delgado, J. (1993) Advances in metabolic control analysis. *Biotechnol. Prog.*, **9**, 221–233.

Liao, J.C., Chao, Y.P. and Patnaik, R. (1994) Alteration of the biochemical valves in the central metabolism of *Escherichia coli*. *Ann. NY Acad. Sci.*, **745**, 21–34

Lieber, C.S., Decarli, L.M., Matsuzaki, S. *et al.* (1978) The microsomal ethanol oxidizing system (MEOS). *Methods Enzymol.*, **52**, 355–367.

Lillie, S.H. and Pringle, J.R. (1980) Reserve carbohydrate metabolism in *Saccharomyces cerevisiae*: responses to nutrient limitation. *J. Bacteriol.*, **143**, 1384–1394.

Lin-Liu, S., Adey, W.R. and Poo, M.M. (1984) Migration of cell surface concanavalin A receptors in pulsed electric fields. *Biophys. J.*, **45**, 1211–1218.

Linton, J.D. (1989) Potential of dual substrates for biomass and metabolite production, in *Proceedings of the Discussion Meeting Organized by the Microbial Physiology Working Party of the European Federation of Biotechnology*, Weinfelden, Switzerland.

Linton, J.D. (1990) The relationship between metabolite production and the growth efficiency of the producing organism. *FEMS Microbiol. Rev.*, **75**, 1–18.

Linton, J.D. and R.J. Stephenson (1978) A preliminary study on growth yields in relation to the carbon and energy content of various organic growth substrates. *FEMS Microbiol. Lett.*, **3**, 95–98.

Lipsitz, L.A. and Goldberger, A.L. (1992) Loss of 'complexity' and aging. Potential applications of fractals and chaos theory to senescence. *JAMA*, **267**, 1806–1809.

Litchfield, C.D. (1991) *Environmental Biotechnology for Waste Treatment*, Plenum Press, New York, pp. 147–157.

Liu, X., Zwiebel, L.J., Hinton, D. *et al.* (1992) The period gene encodes a predominantly nuclear protein in adult *Drosophila*. *J. Neurosci.*, **12**, 2735–2744.

Llinás, R. (1990) Intrinsic electrical properties of nerve cells and their role in network oscillations. *Cold Spring Harbor Symp. Quant. Biol.*, **60**, 933–938.

Lloyd, A.L. and Lloyd, D. (1993) Hypothesis: the central oscillator of the circadian clock is a controlled chaotic attractor. *BioSystems*, **29**, 77–85.

Lloyd, D. and Rossi, E.L. (1993) Biological rhythms as organization and information. *Biol. Rev.*, **68**, 563–577.

Lloyd, D. and Stupfel, M. (1991) The occurrence and functions of ultradian rhythms. *Biol. Rev.*, **66**, 275–299.

Lloyd, D., Poole, R.K. and Edwards, S.W. (1982) *The Cell Division Cycle. Temporal Organization and Control of Cellular Growth and Reproduction*, Academic Press, London.

Lombardo, A. and Scheffler, I.E. (1989) Isolation and characterization of a *Saccharomyces cerevisiae* mutant with a disrupted gene for the IP subunit of succinate dehydrogenase. *J. Biol. Chem.*, **264**, 18874.

Lombardo, A., Cereghino, G.P. and Scheffler, I.E. (1992) Control of mRNA turnover as a mechanism of glucose repression in *Saccharomyces cerevisiae*. *Mol. Cell. Biol.*, **12**, 2941–2948.

Lorenz, E.N. (1963) Deterministic nonperiodic flow. *J. Atmos. Sci.*, **20**, 130–141.

Lorincz, A.T., Miller, M.J., Xuong, N. and Geiduschek, E.P. (1982) Identification of proteins whose synthesis is modulated during the cell cycle of *Saccharomyces cerevisiae. Mol. Cell. Biol.*, **2**, 1532–1549.

Lowe, S.E., Jain, M.K. and Zeikus, J.G. (1993) Biology, ecology, and biotechnological applications of anaerobic bacteria adapted to environmental stresses in temperature, pH, salinity, or substrates. *Microbiol. Rev.*, **57**, 451–509

Luby-Phelps, K., Taylor, D.L. and Lanni, F. (1986) Probing the structure of cytoplasm. *J. Cell Biol.*, **102**, 2015–2022.

Luby-Phelps, K., Lanni, F. and Taylor, D.L. (1988) The submicroscopic properties of cytoplasm as a determinant of cellular function. *Annu. Rev. Biophys. Biophys. Chem.*, **17**, 369–396.

Lumbach, G.W.M., Cox, H.C. and Berends, W. (1970) Elucidation of the chemical structure of bongkrekic acid I. Isolation, purification and properties of bongkrekic acid. *Tetrahedron*, **26**, 5993–5999.

Lumry, R. and Biltonen, R. (1969) In *Structure and Stability of Biological Macromolecules* (eds S.N. Timasheff and A.G. Fasman), Marcel Dekker, New York, p. 65.

Luther, M.A. and Lee, J.C. (1986) The role of phosphorylation in the interaction of rabbit muscle phosphofructokinase with F-actin. *J. Biol. Chem.*, **261**, 1753–1759.

Lynch, J.M. and Hobbie, J.E. (1988) *Micro-organisms in Action: Concepts and Applications in Microbial Ecology*, Blackwell Scientific Publications, Oxford.

Lynch, R.M., Fogarty, K.E. and Fay, F.S. (1991) Modulation of hexokinase association with mitochondria analyzed with quantitative three-dimensional confocal microscopy. *J. Cell Biol.*, **112**, 385–395.

Lynen, F. (1963) Regulatory mechanisms in aerobic carbohydrate metabolism of yeasts, in *Control Mechanisms in Respiration and Fermentation* (ed. B. Wright), The Ronald Press Company, New York, pp. 289–306.

Macdonald, P.M. and Struhl, G. (1986) A molecular gradient in early *Drosophila* embryos and its role in specifying the body pattern. *Nature*, **324**, 537–545.

Macías, A., Casanova, J. and Morata, G. (1990) Expression and regulation of the *abd-A* gene of *Drosophila. Development*, **110**, 1197–1207.

Mackey, M.C. and Glass, L. (1977) Oscillations and chaos in physiological control systems. *Science*, **197**, 287–289.

Madden, T.L. and Herzfeld, J. (1993) Crowding-induced organization of cytoskeletal elements: I. Spontaneous demixing of cytosolic proteins and model filaments to form filament bundles. *Biophys. J.*, **65**, 1147–1154.

Maggio, B. (1985) Geometric and thermodynamic restrictions for the self-assembly of glycosphingolipid–phospholipid systems. *Biochim. Biophys. Acta*, **815**, 245–258.

Maggio, B. (1994) The surface behavior of glycosphingolipids in biomembranes: A new frontier of molecular ecology. *Prog. Biophys. Mol. Biol.*, **62**, 55–117.

Malchow, H. (1988) Dissipative pattern formation in ternary non-linear reaction–electrodiffusion systems with concentration dependent diffusivities. *J. Theor. Biol.*, **135**, 371–381.

Maloney, P.C. (1977) Obligatory coupling between proton entry and the synthesis of adenosine 5′-triphosphate in *Streptococcus lactis. J. Bacteriol.*, **132**, 564–575.

Maloney, P.C. (1983) Relationship between phosphorylation potential and electrochemical H^+ gradient during glycolysis in *Streptococcus lactis. J. Bacteriol.*, **153**, 1461–1470.

Maloney, P.C. (1987) Coupling to an energized membrane: role of ion-motive gradients in the transduction of metabolic energy, in Escherichia coli and Salmonella typhimurium, in *Cellular and Molecular Biology*, Vol. 1 (eds J.L. Ingraham, K. Brooks Low, B. Magasanik *et al.*), American Society for Microbiology, Washington, pp. 222–243.

Mandelbrot, B.B. (1982) *The Fractal Geometry of Nature*, W.H. Freeman, New York.

Mandelbrot, B.B. (1993) *Los Objetos Fractales*, Tusquets Editores, Barcelona.

Mandelkow, E.M. and Mandelkow, E. (1992) Microtubule oscillations. *Cell Motil. Cytoskel.*, **22**, 235–244.

Mandelkow, E., Mandelkow, E.M., Hotani, H. *et al.* (1989) Spatial patterns from oscillating microtubules. *Science*, **246**, 1291–1293.

Manes, M.E. and Elinson, R.P. (1980) Ultraviolet light inhibits grey crescent formation on the frog egg. *Wilhelm Roux's Arch.*, **189**, 73–76.

Manes, M.E., Elinson, R.P. and Barbieri, F.D. (1978) Formation of the amphibian grey crescent: Effects of colchicine and cytochalasin B. *Wilhelm Roux's Arch.*, **185**, 99–104.

Marie, S., Diaz-Guerra, M.J., Miquerol, L. *et al.* (1993) The pyruvate kinase gene as a model for studies of glucose-dependent regulation of gene expression in the endocrine pancreatic β-cell type. *J. Biol. Chem.*, **268**, 23881–23890.

Markus, M., Kuschmitz, D. and Hess, B. (1984) Chaotic dynamics in yeast glycolysis under periodic substrate input flux. *FEBS Lett.*, **172**, 235–238.

Markus, M., Kuschmitz, D. and Hess, B. (1985) Properties of strange attractors in yeast glycolysis. *Biophys. Chem.*, **22**, 95–105.

Marmillot, P., Keith, T., Srivastava, D.K. and Knull, H.R. (1994) Effect of tubulin on the activity of the muscle isoenzyme of lactate dehydrogenase. *Arch. Biochem. Biophys.*, **315**, 467–472.

Marr, A.G. (1991) Growth rate of *Escherichia coli*. *Microbiol. Rev.*, **55**, 316–333.

Marre, E. and Denti-Ballarin, A. (1985) The proton pumps of the plasmalemma and the tonoplast of higher plants. *J. Bioenerg. Biomemb.*, **17**, 1–21.

Mason, S.F. (1984) Origins of biomolecular handedness. *Nature*, **311**, 19–23.

Masters, C. (1984) Interactions between glycolytic enzymes and components of the cytomatrix. *J. Cell Biol.*, **99**, 222–225.

Mastro, A.M. and Keith, A.C. (1984) Diffusion in the aqueous compartment. *J. Cell Biol.*, **99**, 180–187.

Mastro, A.M., Babich, M.A., Taylor, W.D. and Keith, A.C. (1984) Diffusion of small molecules in the cytoplasm of mammalian cells. *Proc. Natl. Acad. Sci. USA*, **81**, 3414–3418.

Matsuno, T. (1987) Bioenergetic of tumor cells: Glutamine metabolism in tumor cell mitochondria. *Int. J. Biochem.*, **19**, 303–307.

May, R. (1976) Simple mathematical models with very complicated dynamics. *Nature*, **261**, 459–467.

McAlister-Henm, L. and Thompson, L.M. (1987) Isolation and expression of the gene encoding yeast mitochondrial malate dehydrogenase. *J. Bacteriol.*, **169**, 5157.

McCloskey, M.A. and Poo, M. (1986) Rates of membrane-associated reactions: reduction of dimensionality revisited. *J. Cell Biol.*, **102**, 88–96.

McClure, B.A. and Guilfoyle, T. (1989) Rapid redistribution of auxin-regulated RNAs during gravitropism. *Science*, **243**, 91–93.

Meinhardt, H. (1982) *Models of Biological Pattern Formation*, Academic Press, London.

Meinhardt, H. (1984) Digits, segments, somites – The superposition of sequential and periodic structures, in *Synergetics. From Microscopic to Macroscopic Order* (ed. E. Frehland) Springer-Verlag, Berlin, pp. 228–245.

Meinhardt, H. (1986) Hierarchical inductions of cell states: a model for segmentation in Drosophila. *J. Cell Sci.*, Suppl., **4**, 357–381.

Meinhardt, H. (1988) Models for maternally supplied positional information and the activation of segmentation genes in *Drosophila* embryogenesis. *Development*, **104**, 95–110.

Mendes, P., Kell, D.B. and Westerhoff, H.V. (1992) Channelling can decrease pool size. *Eur. J. Biochem.*, **204**, 257–266.

Menzel, R. and Gellert, M. (1983) Regulation of the genes for *E. coli* DNA gyrase: Homeostatic control of DNA supercoiling. *Cell*, **34**, 105–113.

Meury, J. and Kohiyama, M. (1992) Potassium ions and changes in bacterial DNA supercoiling under osmotic stress. *FEMS Microbiol. Lett.*, **99**, 159–164.

Meyer, T. (1991) Cell signaling by second messenger waves. *Cell*, **64**, 675–678.

Michel, S., Geusz, M.E., Zaritsky, J.J. and Block, G.D. (1993) Circadian rhythm in membrane conductance expressed in isolated neurons. *Science*, **259**, 239–241.

Michels, P.A.M., Michels, J.P.J., Boonstra, J. and W.N. Konings (1979) Generation of an electrochemical proton gradient in bacteria by the excretion of metabolic end products. *FEMS Microbiol. Lett.*, **5**, 357–364.

Miller, J.J. (1989) Sporulation in *Saccharomyces cerevisiae*, in *The Yeast*, 2nd edn, Vol. 2 (eds A.H. Rose and J.S. Harrison), Academic Press, London, pp. 489–550.

Mimura, T. and Tazawa, M. (1986) Light-induced membrane hyperpolarization and adenine nucleotide levels in perfused Characean cells. *Plant Cell Physiol.*, **27**, 319–330.

Minaschek, G., Groschel-Stewart, U., Blum, S. and Bereiter-Hahn, J. (1992) Microcompartmentation of glycolytic enzymes in cultured cells. *Eur. J. Cell Biol.*, **58**, 418–428.

Minton, A.P. (1983) The effect of volume occupancy upon the thermodynamic activity of proteins: some biochemical consequences. *Mol. Cell. Biochem.*, **55**, 119–140.

Minton, A.P. (1992) Confinement as a determinant of macromolecular structure and reactivity. *Biophys. J.*, **63**, 1090–1100.

Minton, A.P. and J. Wilf (1981) Effect of macromolecular crowding upon the structure and function of an enzyme: glyceraldehyde-3-phosphate dehydrogenase. *Biochemistry*, **20**, 4821–4826.

Mitchell, A.P. and Herskowitz, I. (1986) Activation of meiosis and sporulation by repression of the *RME1* product in yeast. *Nature*, **319**, 738–742.

Mitchell, P. (1961) Coupling of phosphorylation to electron and hydrogen transfer by a chemiosmotic type of mechanism. *Nature*, **191**, 144–148.

Mitchell, P. (1968) Biological transport phenomena and the spatially anisotropic characteristics of enzyme systems causing a vector component of metabolism, in *Membrane Transport and Metabolism* (eds A. Kleinzeller and A. Kotyk), Academic Press, London, pp. 22–34.

Mitchison, J.M. (1989) Cell cycle growth and periodicities, in *Molecular Biology of the Fission Yeast*, Academic Press, London, pp. 205–242.

Mitchison, T. and Kirschner, M. (1984a) Microtubule assembly nucleated by isolated centrosomes. *Nature*, **312**, 232–237.

Mitchison, T. and Kirschner, M. (1984b) Dynamic instability of microtubules growth. *Nature*, **312**, 237–242.

Mitchison, T. and Kirschner, M. (1988) Cytoskeletal dynamics and nerve growth. *Neuron*, **1**, 761–772.

Mitton, J.B. (1993) Enzyme heterozygosity, metabolism, and developmental stability, *Genetica*, **89**, 47–65

Mlodzik, M. and Gehring, W.J. (1987) Expression of the caudal gene in the germ line of *Drosophila* during early embryogenesis. *Cell*, **48**, 465–478.

Mónaco, M.E., Valdecantos, P. and Aon, M.A. (1995) Carbon and energetic uncoupling associated with cell cycle arrest of cdc mutants of *Saccharomyces cerevisiae* may be linked to glucose-induced catabolite repression. *Exp. Cell Res.*, **217**, 52–56.

Montero, F. and Morán, F. (1992) *Biofísica. Procesos de Autoorganización en Biología*, Eudema, Madrid.

Montgomery, D.L., Leung, D.W., Smith, M. *et al.* (1980) Isolation and sequence of the gene for iso-2-cytochrome c in *Saccharomyces cerevisiae*. *Proc. Natl. Acad. Sci. USA*, **77**, 541.

Moore, P.A., Sagliocco, F.A., Wood, R.M.C. and Brown, A.J.P. (1991) Yeast glycolytic mRNAs are differentially regulated. *Mol. Cell. Biol.*, **11**, 5330–5337.

Mora, J. (1990) Glutamine metabolism and cycling in *Neurospora crassa*. *Microbiol. Rev.*, **54**, 293–304.

Morfini, G., DiTella, M.C., Feiguin, F. *et al.* (1994) Neurotrophin-3 enhances neurite outgrowth in cultured hippocampal pyramidal neurons. *J. Neurosci. Res.*, **39**, 219–232.

Morita, R.Y. (1988) Bioavailability of energy and its relationship to growth and starvation survival in nature. *Can. J. Microbiol.*, **34**, 436–441.

Morris, C.E. and Sigurdson, W.J. (1989) Stretch-inactivated ion channels coexist with stretch-activated ion channels. *Science*, **243**, 807–809.

Mulholland, J., Preuss, D., Moon, A. *et al.* (1994) Ultrastructure of the yeast actin cytoskeleton and its association with the plasma membrane. *J. Cell Biol.*, **125**, 381–391.

Munch, T., Sonnleitner, B. and Fiechter, A. (1992a) New insights into the synchronization mechanism with forced synchronous cultures of *Saccharomyces cerevisiae*. *J. Biotechnol.*, **24**, 299–314.

Munch, T., Sonnleitner, B. and Fiechter, A. (1992b) The decisive role of the *Saccharomyces cerevisiae* cell cycle behavior for dynamic growth characterization. *J. Biotechnol.*, **22**, 329–352.

Murray, A.W. and Kirschner, M.W. (1989) Dominoes and clocks: the union of the two views of the cell cycle. *Science*, **246**, 614–621.

Murray, J.D. (1989) *Mathematical Biology*, Springer-Verlag, Berlin, Heidelberg.

Nagorcka, B.N. (1989) Wavelike isomorphic prepatterns in development. *J. Theor. Biol.*, **137**, 127–162.

Needham, J. (1934) Chemical heterogony and the ground-plan of animal growth. *Biol. Rev.*, **9**, 79–109.

Neidhardt, F.C. (1987) Multigene systems and regulons, in Escherichia coli *and* Salmonella typhimurium. *Cellular and Molecular Biology* (eds F.C. Neidhardt, J.L. Ingraham, K.B. Low *et al.*), ASM Press, Washington, DC, pp. 1313–1318.

Nelson, N. and Taiz, L. (1989) The evolution of H^+-ATPases. *TIBS*, **14**, 113–116.

Neumann, E. (1973) Molecular hysteresis and its cybernetic significance. *Angew. Chem. Int. Ed. Eng.*, **12**, 356–369.

Neumann, E. and Katchalsky, A. (1972) Long-lived conformation changes induced by electric impulses in biopolymers. *Proc. Natl. Acad. Sci. USA*, **69**, 993–997.

Newell, P.C. and Ross, F.M. (1982) Inhibition by adenosine of aggregation center initiation and cyclic AMP binding in *Dictyostelium. J. Gen. Microbiol.*, **128**, 2715–2724.

Ni Bhriain, N., Dorman, C.J. and Higgins, C.F. (1989) An overlap between osmotic and anaerobic stress responses: a potential role for DNA supercoiling in the coordinate regulation of gene expression. *Mol. Microbiol.*, **3**, 933–942.

Nicolis, G. and Prigogine, I. (1977) *Self-organization in Non-equilibrium Systems*, Wiley-Interscience, London.

Niederberger, P., Prasad, R., Miozzari, G. and Kacser, H. (1992) A strategy for increasing an *in vivo* flux by genetic manipulations. The tryptophan system of yeast. *Biochem. J.*, **287**, 473–479.

Nobrega, F.G. and Tzagoloff, A. (1980) Assembly of the mitochondrial membrane system. DNA sequence and organization of the cytochrome b gene in *Saccharomyces cerevisiae* D273-10B. *J. Biol. Chem.*, **255**, 9828.

Nooden, L.D. and Thimann, K.V. (1963) Evidence for a requirement for protein synthesis for auxin-induced cell enlargement. *Proc. Natl. Acad. Sci. USA*, **60**, 194–200.

Nossal, R. (1988) On the elasticity of cytoskeletal networks. *Biophys. J.*, **53**, 349–359.

Novak, B. and Tyson, J.J. (1993) Numerical analysis of a comprehensive model of M-phase control in *Xenopus* oocyte extracts and intact embryos. *J. Cell Sci.*, **106**, 1153–1168.

Novick, P. and Botstein, D. (1985) Phenotypic analysis of temperature-sensitive yeast actin mutants. *Cell*, **40**, 405–416.

Nowell, P.C. (1976) The clonal evolution of tumor cell populations. *Science*, **194**, 23–28.

Nuccitelli, R. (1984) Ionic currents in development, in *Pattern Formation* (ed. M. Malacinsky), Macmillan-Collier Macmillan, New York, London, pp. 23–35.

Nuccitelli. R. and Heiple, J.M. (1982) Summary of the evidence and discussion concerning the involvement of pH_i in the control of cellular functions, in *Intracellular pH: its Measurement, Regulation and Utilization in Cellular Functions* (eds R. Nuccitelli and D.W. Deamer), Alan R. Liss, New York, pp. 567–586.

Nüsslein-Volhard, Ch. (1991) Determination of the embryonic axes of *Drosophila. Development*, Suppl., **1**, 1–10.

Odum, E.P. (1989) Input management of production systems. *Science*, **241**, 177–182

Oliver, G., Wright, C.V.E., Hardwicke, J. and De Robertis, E.M. (1988) A gradient of homeodomain protein in developing forelimbs of *Xenopus* and mouse embryos. *Cell*, **55**, 1017–1024.

Olsen, L.F. and Degn, H. (1985) Chaos in biological systems. *Q. Rev. Biophys.*, **18**, 165–225.

Onsager, L. (1931a) Reciprocal relations in irreversible processes I. *Physiol. Rev.*, **37**, 405–426

Onsager, L. (1931b) Reciprocal relations in irreversible processes II. *Physiol. Rev.*, **38**, 2265–2279

Oohata, A.A. (1992) Induction of cell differentiation of isolated cells in *Dictyostelium discoideum* by low extracellular pH. *Dev. Growth Differ.*, **34**, 693–697.

Oosawa, F. and Asakura, S. (1975) *Thermodynamics of the Polymerization of Protein*, Academic Press, New York.

O'Rourke, B., Ramza, B.M. and Marban, E. (1994) Oscillations of membrane current and excitability driven by metabolic oscillations in heart cells. *Science*, **265**, 962–966.

Oster, G.F., Murray, J.D. and Harris, A.K. (1983) Mechanical aspects of mesenchymal morphogenesis. *J. Embryol. Exp. Morphol.*, **78**, 83–125.

Ott, E., Grebogi, C. and Yorke, J.A. (1990) Controlling chaos. *Phys. Rev. Lett.*, **64**, 1196–1199.

Otto, R., Sonnenberg, A.S.M., Veldkamp, H. and Konings, W.N. (1980) Generation of an electrochemical proton gradient in *Streptococcus cremoris* by lactate efflux. *Proc. Natl. Acad. Sci. USA*, **77**, 5502–5506.

Ovádi, J. and Orosz, F. (1992) Calmodulin and dynamics of interactions of cytosolic enzymes. *Curr. Top. Cell. Regul.*, **33**, 105–126.

Ovádi, J., Tompa, P., Vértessy, B. *et al.* (1989) Transient time analysis of substrate channelling in interacting enzyme systems. *Biochem. J.*, **257**, 187–192.

Pagel, M.D. and Harvey, P.H. (1989) Taxonomic differences in the scaling of brain on body weight among mammals. *Science*, **244**, 1589–1593.

Pardee, A.B. (1974) A restriction point for control of normal animal cell proliferation. *Proc. Natl. Acad. Sci. USA*, **71**, 1286–1290.

Pardee, A.B., Dubrow, R., Hamlin, J.L. and Kletzien, R.F. (1978) Animal cell cycle. *Annu. Rev. Biochem.*, **47**, 715–750.

Parker, J.C. (1993) In defense of cell volume? *Am. J. Physiol.*, **265**, C1191–C1200.

Pasteur, L. (1884) La dissymétrie moléculaire. *Rev. Sci. Bull. Soc. Chim. France*, **7**, 2–6.

Paton, R.C. (1993) Some computational models at the cellular level. *BioSystems*, **29**, 63–75.

Pavlidis, T. (1969) Populations of interacting oscillators and circadian rhythms. *J. Theor. Biol.*, **22**, 418–436.

Peacocke, A.R. (1983) *The Physical Chemistry of Biological Organization*, Clarendon Press, Oxford.

Pearson, J.E. (1993) Complex patterns in a simple system. *Science*, **261**, 189–192.

Pedersen, P.L. (1978) Tumor mitochondria and the biogenesis of cancer cells. *Prog. Exp. Tumor Res.*, **22**, 190–274.

Peng, B., Petrov, V. and Showalter, K. (1991) Controlling chemical chaos. *J. Phys. Chem.*, **95**, 4957–4959.

Penman, S., Fulton, A., Capco, D. *et al.* (1981) Cytoplasmic and nuclear architecture in cells and tissue: form, functions, and mode of assembly. *Cold Spring Harbor Symp. Quant. Biol.*, **45**, 1013–1028.

Petrov, V., Gáspár, V., Masere, J. and Showalter, K. (1993) Controlling chaos in the Belousov–Zhabotinsky reaction. *Nature*, **361**, 240–243.

Pienta, K.J. and Hoover, C.N. (1994) Coupling of cell structure to cell metabolism and function. *J. Cell. Biochem.*, **55**, 16–21.

Pienta, K.J., Partin, A.W. and Coffey, D.S. (1989) Cancer as a disease of DNA organization and dynamic cell structure. *Cancer Res.*, **49**, 2525–2532.

Pirt, J.S. (1975) *Principles of Microbe and Cell Cultivation*, Blackwell, Oxford.

Pokrywka, N.J. and Stephenson, E.C. (1991) Microtubules mediate the localization of *bicoid* RNA during *Drosophila* oogenesis. *Development*, **113**, 55–66.

Polakis, E.S., Bartley, W. and Meek, G.A. (1965) Changes in the activities of respiratory enzymes during the aerobic growth of yeast on different carbon sources. *Biochem. J.*, **97**, 298–302.

Poo, M. (1981) In situ electrophoresis of membrane components. *Annu. Rev. Biophys. Bioeng.*, **10**, 245–276.

Poolman, B., Driessen, A.J.M. and Konings, W.N. (1985) Regulation of solute transport in Streptococci by external and internal pH values. *Microbiol. Rev.*, **51**, 495–508.

Porter, K.R. (1984) The cytomatrix: a short history of its study. *J. Cell Biol.*, **99**, 3–12.

Porter, K.R. and Palade, G.E. (1957) Studies of the endoplasmic reticulum. III. Its form and distribution in striated muscle cells. *J. Biophys. Biochem. Cytol.*, **3**, 269–299.

Postma, P.W. (1986) The bacterial phosphoenolpyruvate:sugar phosphotransferase system of *Escherichia coli* and *Salmonella typhimurium*, in *Carbohydrate Metabolism in Cultured Cells* (ed. M.J. Morgan), Plenum Press, New York, pp. 357–408.

Postma, E., Verduyn, C., Scheffers, W.A. and van Dijken, J.P. (1989) Enzymic analysis of the Crabtree effect in glucose-limited chemostat cultures of *Saccharomyces cerevisiae*. *Appl. Environ. Microbiol.*, **55**, 468–477.

Postma, P.W., Lengeler, J.L. and Jacobson, G.R. (1993) Phosphoenolpyruvate: carbohydrate phosphotransferase systems of bacteria. *Microbiol. Rev.*, **57**, 543–594.

Pratt, O. (1974) *Ecuaciones Diferenciales Ordinarias*, Ed. Reverte, Barcelona.

Prescott, D. and Bender, M. (1962) Synthesis of RNA and protein during mitosis in mammalian tissue culture cells. *Exp. Cell Res.*, **26**, 260–268.

Prieto, S., Bouillaud, F., Ricquier, D. and Rial, E. (1992) Activation by ATP of a proton-conducting pathway in yeast mitochondria. *Eur. J. Biochem.*, **208**, 487–491.

Prigogine, I. (1967) *Introduction to Thermodynamics of Irreversible Processes*, Interscience, New York.

Prigogine, I. and Nicolis, G. (1971) Biological order, structure and instabilities. *Q. Rev. Biophys.*, **4**, 107–148.

Prigogine, I. and Stengers, I. (1983) *La Nueva Alianza. Metamorfosis de la Ciencia*, Alianza Editorial, Madrid.

Pringle, J.R. and Hartwell, L.H. (1981) The *Saccharomyces cerevisiae* cell cycle, in *The Molecular Biology of the Yeast* Saccharomyces: *Life Cycle and Inheritance* (eds J.N. Strathern, E.W. Jones and J.R. Broach), Cold Spring Harbor Laboratory Press, Cold Spring Harbor, NY, pp. 97–142.

Procaccia, I. (1988) Universal properties of dynamically complex systems: the organization of chaos. *Nature*, **333**, 618–623.

Pruss, G.J. and Drlica, K. (1989) DNA supercoiling and prokaryotic transcription. *Cell*, **56**, 521–523.

Przywara, D.A., Bhave, S.V., Bhave, A. *et al.* (1991) Stimulated rise in neuronal calcium is faster and greater in the nucleus than the cytosol. *FASEB J.*, **5**, 217–222.

Puck, T.T. and Krystosek, A. (1992) Role of the cytoskeleton in genome regulation and cancer. *Int. Rev. Cytol.*, **132**, 75–108.

Pye, K.E. (1973) Glycolytic oscillations in cells and extracts of yeast. Some unsolved problems, in *Biological and Biochemical Oscillators* (eds B. Chance, K. Pye, A.K. Ghosh and B. Hess), Academic Press, New York, pp. 269–284.

Rabinovitz, M. (1992) The pleiotypic response to amino acid deprivation is the result of interactions between components of the glycolysis and protein synthesis pathways. *FEBS Lett.*, **302**, 113–116.

Rabouille, C. (1988) Thèse de Doctorat, Université de Technologie de Compiègne.

Rabouille C., Aon, M.A. and Thomas, D. (1989) Interactions involved in ovomucin gel-forming properties: A rheological–biochemical approach. *Arch. Biochem. Biophys.*, **270**, 495–503.

Rabouille, C., Aon, M.A., Muller, G. *et al.* (1990) The supramolecular structure of ovomucin. Biophysical and morphological studies. *Biochem. J.*, **266**, 697–706.

Rabouille, C., Cortassa, S. and Aon, M.A. (1992) Fractal organization in biological macromolecular lattices. *J. Biomol. Struct. Dyn.*, **9**, 1013–1024.

Racker, E. (1976) *A New Look at Mechanisms in Bioenergetics*, Academic Press, New York.

Rapp, P.E. (1979) An atlas of cellular oscillators. *J. Exp. Biol.*, **81**, 281–306.

Rapp, P.E., Mees, A.I. and Sparrow, C.T. (1981) Frequency encoded biochemical regulation is more accurate than amplitude dependent control. *J. Theor. Biol.*, **90**, 531–544.

Rasmussen, L.J., Moller, P.L. and Atlung, T. (1991) Carbon metabolism regulates expression of the pfl (pyruvate formate-lyase) gene in *Escherichia coli*. *J. Bacteriol.*, **173**, 6390–6397.

Raven, J.A. (1987) Electrochemistry of plant membranes and signal transduction, in *NATO Series*, Vol. H12 (ed. E. Wagner), Springer-Verlag, Berlin, Heidelberg, pp. 205–235.

Ray, P.M. (1985) Auxin and fusicoccin enhancement of β-glucan synthase in peas. *Plant Physiol.*, **78**, 466–472.

Reed, S. (1980) The selection of *Saccharomyces cerevisiae* mutants defective in the start event of cell division. *Genetics*, **95**, 561–577.

Reed, S.I., Madwiger, J.A. and Lorincz, A.T. (1985) Protein kinase activity associated with the product of the yeast cell division cycle gene CDC28. *Proc. Natl. Acad. Sci. USA*, **82**, 4055–4059.

Rehmus, P., Termonia, Y. and Ross, J. (1981) Dissipation in driven oscillatory chemical systems with application to glycolysis. *Kinam*, **3**, 123–155.

Reibstein, D., den Hollander, J.A., Pilkis, S.J. and Shulmann, R.G. (1986) Studies on the regulation of yeast phosphofructo-1-kinase: its role in aerobic and anaerobic glycolysis. *Biochemistry*, **25**, 219–227

Reich, J.G. and Sel'kov, E.E. (1981) *Energy Metabolism of the Cell*, Academic Press, London.

Reiss, H.D., Schnepf, E. and Herth, W. (1984) The plasma membrane of the *Funaria* caulonema tip cell: morphology and distribution of particle rosettes and the kinetics of cellulose synthesis. *Planta*, **160**, 428–435

Reiss, M.J. (1989) *The Allometry of Growth and Reproduction*, Cambridge University Press, Cambridge.

Ricard, J. (1987) Dynamics of multi-enzyme reactions, cell growth and perception of ionic signals from the external milieu. *J. Theor. Biol.*, **128**, 253–278.

Ricard, J. and Soulié, J.M. (1982) Self-organization and dynamics of an open futile cycle. *J. Theor. Biol.*, **95**, 105–121.

Ricard, J., Meunier, J.C. and Buc, J. (1974) Regulatory behavior of monomeric enzymes, I. The Mnemonical enzyme concept. *Eur. J. Biochem.*, **49**, 195–208.

Richard, P., Teusink, B., Westerhoff, H.V. and van Dam, K. (1993) Around the growth phase transition *S. cerevisiae*'s make-up favours sustained oscillations of intracellular metabolites. *FEBS Lett.*, **318**, 80–82.

Richard, P., Diderich, J.A., Bakker, B.M. *et al.* (1994) Yeast cells with a specific cellular make-up and an environment that removes acetaldehyde are prone to sustained glycolytic oscillations. *FEBS Lett.*, **341**, 223–226.

Rieger, M., Kappeli, O. and Fiechter, A. (1983) The role of limited respiration in the incomplete oxidation of glucose by *Saccharomyces cerevisiae*. *J. Gen. Microbiol.*, **129**, 653–661.

Rietman, E. (1989) Exploring the geometry of nature, in *Computer Modeling of Chaos, Fractals, Cellular Automata and Neural Networks*, Windcrest Books, pp. 52–57.

Rivin, C.J. and Fangman, W.L. (1980) Cell cycle phase expansion in nitrogen-limited cultures of *Saccharomyces cerevisiae*. *J. Cell Biol.*, **85**, 96–107.

Roberts, S.J. and Somero, G.N. (1987) Binding of phosphofructokinase to filamentous actin. *Biochemistry*, **26**, 3437–3442.

Robinson, K.R. (1985) The responses of cells to electrical fields: A review. *J. Cell Biol.*, **101**, 2023–2027.

Rodicio, R. and Heinisch, J. (1987) Isolation of the yeast phosphoglyceromutase gene and construction of deletion mutants. *Mol. Gen. Genet.*, **206**, 133.

Roels, J.A. (1983) *Energetics and Kinetics in Biotechnology*, Elsevier Biomedical Press, Amsterdam.

Roggenkamp, R., Didion, T. and Kowallik, K.V. (1989) Formation of irregular giant peroxisomes by overproduction of the crystalloid core protein methanol oxidase in the methylotrophic yeast *Hansenula polymorpha*. *Mol. Cell. Biol.*, **9**, 988–994.

Roland, J.C. (1973) The relationship between the plasmalemma and plant cell wall. *Int. Rev. Cytol.*, **36**, 45–92.

Roland, J.C., Reis, D., Mosiniak, M. and Vian, B. (1982) Cell wall texture along the growth gradient of the mung bean *Vigna radiata* hypocotyl ordered assembly and dissipative processes. *J. Cell Sci.*, **56**, 303–318.

Romano, A.H. (1986) Microbial sugar transport systems and their importance in biotechnology. *TIBTECH*, **4**, 207–213.

Romer, A.S. (1967) *The Vertebrate Body*, W.B. Saunders Company, London, pp. 1–13.

Rose, I.A. and Warms, J.V.B. (1967) Mitochondrial hexokinase. Release, rebinding, and location. *J. Biol. Chem.*, **242**, 1635–1645.

Rose, M., Albig, W. and Entian, K.-D. (1991) Glucose repression in *Saccharomyces cerevisiae* is directly associated with hexose phosphorylation by hexokinase PI and PII. *Eur. J. Biochem.*, **199**, 511–518.

Rosen, R. (1967) *Optimality Principles in Biology*, Butterworths, London.

Rosen, R. (1970) *Dynamical System Theory in Biology*, John Wiley & Sons, New York.

Ross, J. and Schell, M. (1987) Thermodynamic efficiency in nonlinear biochemical reactions. *Annu. Rev. Biophys. Biophys. Chem.*, **16**, 401–422.

Ross, J., Pugh, S. and Schell, M. (1988) Spectral kinetics and the efficiency of (bio)chemical reactions, in *From Chemical to Biological Organization* (eds M. Markus, S.C. Muller and G. Nicolis), Springer-Verlag, Berlin, Heidelberg, pp. 34–46.

Roth, R. and Lusnak, K. (1970) DNA synthesis during yeast sporulation: genetic control of an early developmental event. *Science*, **168**, 493–494.

Rothschild, K.J., Ellias, S.A., Essig, A. and Stanley, H.E. (1980) Nonequilibrium linear behavior of biological systems. Existence of enzyme mediated multidimensional inflection points. *Biophys J.*, **30**, 209–230.

Rottenberg, H. (1973) The thermodynamic description of enzyme catalyzed reactions. The linear relation between the reaction rate and the affinity. *Biophys. J.*, **13**, 503–511.

Ruijter, G. (1992) Enzyme II^{glc} of the bacterial phosphotransferase system. Coupling of glucose transport to phosphorylation and control of glucose metabolism. PhD Thesis, University of Amsterdam.

Ruijter, G., Postma, P.W. and van Dam, K. (1992) Control of glucose metabolism by enzyme II^{Glc} of the phosphoenolpyruvate-dependent phosphotransferase system in *Escherichia coli. J. Bacteriol.*, **173**, 6184–6191.

Russell, D.W., Smith, M., Williamson, V.M. and Young, E.T. (1983) Nucleotide sequence of the yeast alcohol dehydrogenase II gene. *J. Biol. Chem.*, **258**, 2674.

Ryser, U. (1985) Cell wall biosynthesis in differentiating cotton fibres. *Eur. J. Cell Biol.*, **39**, 236–256.

Sá-Correia, I., Darzins, A., Wang, S.K. *et al.* (1987) Alginate biosynthetic enzymes in mucoid and nonmucoid *Pseudomonas aeruginosa*: overproduction of phosphomannose isomerase, phosphomannomutase, and GDP-mannose pyruvophosphorylase by overexpression of the phosphomannose isomerase (pmi) gene. *J. Bacteriol.*, **169**, 3224–3231.

Saigal, D., Cunningham, S.J., Farrés, J. and Weiner, H. (1991) Molecular cloning of the mitochondrial aldehyde dehydrogenase gene of *Saccharomyces cerevisiae* by genetic complementation. *J. Bacteriol.*, **173**, 3199.

Salmon, E.D., Saxton, W.M., Leslie, R.J. *et al.* (1984a) Diffusion coefficient of fluorescein-labeled tubulin in the cytoplasm of embryonic cells of a sea urchin: video image analysis of fluorescence redistribution after photobleaching. *J. Cell Biol.*, **99**, 2157–2194.

Salmon, E.D., Leslie, R.J., Saxton, W.M. *et al.* (1984b) Spindle microtubule dynamics in sea urchin embryos. Analysis using a fluorescein-labelled tubulin and measurements of fluorescence redistribution after laser photobleaching. *J. Cell Biol.*, **99**, 2165–2174.

Sampath, P. and Pollard, T.D. (1991) Effects of cytochalasin, phalloidin, and pH on the elongation of actin filaments. *Biochemistry*, **30**, 1973–1980.

Sander, L.M. (1985) Viscous fingers and fractal growth. *Nature*, **314**, 405–406.

Saunders, P.T. and Kubal, C. (1989) Bifurcations and the epigenetic landscape, in *Theoretical Biology. Epigenetic and Evolutionary Order from Complex Systems* (eds B. Goodwin and P. Saunders), Edinburgh University Press, Edinburgh, pp. 16–30.

Sauro, H.M., Small, J.R. and Fell, D.A. (1987) Metabolic control and its analysis. Extensions to the theory and matrix method. *Eur. J. Biochem.*, **165**, 215–221.

Savageau, M.A. (1976) *Biochemical Systems Analysis*, Addison-Wesley, Reading, MA.

Savageau, M.A. (1979) Allometric morphogenesis of complex systems: Derivation of the basis equations from first principles. *Proc. Natl. Acad. Sci. USA*, **76**, 6023–6025.

Savageau, M.A. (1985) Mathematics of organizationally complex systems. *Biomed. Biochim. Acta*, **44**, 839–844.

Savageau, M.A. (1995) Michaelis–Menten mechanism reconsidered: Implications of fractal kinetics. *J. Theor. Biol.*, **176**, 115–124.

Schaaf, I., Heinisch, J. and Zimmermann, K. (1989) Overproduction of glycolytic enzymes in yeast. *Yeast*, **5**, 285–290.

Scharf, S.R. and Gerhart, J. (1980) Determination of the dorsal–ventral axis in eggs of *Xenopus laevis*. Rescue of UV-impaired eggs by oblique orientation. *Dev. Biol.*, **79**, 181–198.

Scharf, S.R., and Gerhart, J. (1983) Axis determination in eggs of *Xenopus laevis*: A critical period before first cleavage, identified by the common effects of cold, pressure and ultraviolet irradiation. *Dev. Biol.*, **99**, 75–87.

Schellenberger, W. and Hervagault, J.F. (1991) Irreversible transitions in the 6-phosphofructokinase/fructose 1,6-bisphosphatase cycle. *Eur. J. Biochem.*, **195**, 109–113

Schierenberg, E. (1987) Reversal of cellular polarity and early cell–cell interactions in the embryo of *Caenorhabditis elegans*. *Dev. Biol.*, **122**, 452–463.

Schiffmann, Y. (1982) Chemical triggering and hysteresis. *Physica*, **114A**, 74–83.

Schlegel, R. and Craig, R.W. (1991) Mitosis: normal control mechanisms and consequences of aberrant regulation in mammalian cells, in *Perspectives on Cellular Regulation: From Bacteria to Cancer*, Wiley-Liss, Inc., New York, pp. 235–249.

Schnepf, E. (1986) Cellular polarity. *Annu. Rev. Plant Physiol.*, **37**, 23–47.

Schnepf, E., Witte, D., Rudolph, U. *et al.* (1985) Tip cell growth and the frequency and distribution of particle rosettes in the plasmalemma: Experimental studies in *Funaria* protonema cells. *Protoplasma*, **127**, 222–229.

Schoppink, P.J., Hemrika, W., Reynen, J.M. *et al.* (1988) Yeast ubiquinol:cytochrome c oxidoreductase is still active after inactivation of the gene encoding the 17 kDa subunit VI. *Eur. J. Biochem.*, **173**, 115.

Schoppink, P.J., Berden, J.A. and Grivell, L.A. (1989a) Irradiation of the gene encoding the 14 kDa subunit VII of yeast ubiquinol:cytochrome c oxidoreductase and analysis of the resulting mutant. PhD Thesis, University of Amsterdam.

Schoppink, P.J., de Jong, M., Berden, J.A. and Grivell, L.A. (1989b) The C-terminal half of the 11 kDa subunit VIII is not necessary for the enzymic activity of yeast ubiquinol:cytochrome c oxidoreductase. PhD Thesis, University of Amsterdam.

Segel, I.H. (1975) *Enzyme Kinetics. Behavior and Analysis of Rapid Equilibrium and Steady State Enzyme Systems*, Wiley-Interscience Publications, New York.

Segel, L.A. (ed.) (1980) *Models in Molecular and Cellular Biology*, Cambridge University Press, Cambridge.

Segel, L.A. (1984) *Modeling Dynamic Phenomena in Molecular and Cellular Biology*, Cambridge University Press, New York.

Sejnowski, T.J., Koch, Ch. and Churchland, P.S. (1988) Computational neuroscience. *Science*, **241**, 1299–1306.

Sel'kov, E.E. (1975) Stabilization of energy charge, generation of oscillations and multiple steady state in energy metabolism as a result of purely stoichiometric regulation. *Eur. J. Biochem.*, **59**, 151–157.

Senior, P.J., Beech, G.A., Ritchie, G.A.F. and Dawes, E.A.(1972) The role of oxygen limitation in the formation of poly-b-hydroxybutyrate during batch and continuous culture of *Azotobacter beijerinckii. Biochem. J.*, **128**, 1193–1201.

Sernetz, M., Gélleri, B. and Hofmann, J. (1985) The organism as bioreactor. Interpretation of the reduction law of metabolism in terms of heterogeneous catalysis and fractal structure. *J. Theor. Biol.*, **117**, 209–230.

Serrano, R., Gancedo, J.M. and Gancedo, C. (1973) Assay of yeast enzymes *in situ*. A potential tool in regulation studies. *Eur. J. Biochem.*, **34**, 479–482.

Severina, I.I., Skulachev, V.P. and Zorov, D.B. (1988) Coupling membranes as energy-transmitting cables. II. Cyanobacterial trichomes. *J. Cell Biol.*, **107**, 497–501.

Shah, J.C. and Clancy, M.J. (1992) *IME4*, a gene that mediates *MAT* and nutritional control of meiosis in *Saccharomyces cerevisiae. Mol. Cell. Biol.*, **12**, 1078–1086.

Shaw, G. and Kamen, R. (1986) A conserved AU sequence from the 3' untranslated region of GM-CSF mRNA mediates selective mRNA degradation. *Cell*, **46**, 659–667.

Sheen, J. (1990) Metabolic repression of transcription in higher plants. *Plant Cell*, **2**, 1027–1038.

Sheen, J. (1994) Feedback control of gene expression. *Photosynth. Res.*, **39**, 427–438.

Shen, J.-J., Jan, S.-P., Lee, H.-T. and Yu, S.-M. (1994) Control of transcription and mRNA turnover as mechanisms of metabolic repression of α-amylase gene expression. *Plant Cell*, **5**, 655–664.

Shermoen, A.W. and O'Farrell, P.H. (1991) Progression of the cell cycle through mitosis leads to abortion of nascent transcripts. *Cell*, **67**, 303–310.

Shilo, V., Simchen, G. and Shilo, B. (1978) Initiation of meiosis in cell cycle initiation mutants of *Saccharomyces cerevisiae. Exp. Cell Res.*, **112**, 241–248.

Shinbrot, T., Grebogi, C., Ott, E. and Yorke, J.A. (1993) Using small perturbations to control chaos. *Nature*, **363**, 411–417.

Shinohara, Y., Nishida, E. and Sakai, H. (1989) Initiation of DNA synthesis by microtubule disruption in quiescent rat 3Y1 cells. *Eur. J. Biochem.*, **183**, 275–280.

Shulman, R.G. (1988). High resolution NMR *in vivo. TIBS*, **13**, 37–39.

Shulman, R.W., Conjeevaram, E.S. and Warner, J.R. (1977) Noncoordinated transcription in the absence of protein synthesis in yeast. *J. Biol. Chem.*, **252**, 1344–1349.

Siegele, D.A. and Kolter, R. (1992) Life after log. *J. Bacteriol.*, **174**, 345–348.

Sierkstra, L.N., Verbakel, J.M.A. and Verrips, C.T. (1992) Analysis of transcription and translation of glycolytic enzymes in glucose-limited continuous cultures of *Saccharomyces cerevisiae. J. Gen. Microbiol.*, **138**, 2559–2566.

Sierkstra, L.N., Nouwen, N.P., Verbakel, J.M.A. and Verrips, C.T. (1993) Regulation of glycolytic enzymes and the Crabtree effect in galactose-limited continuous cultures of *Saccharomyces cerevisiae. Yeast*, **9**, 787–795.

Simchen, G. and Kassir, Y. (1989) Genetic regulation of differentiation towards meiosis in the yeast *Saccharomyces cerevisiae*. *Genome*, **31**, 95–99.

Simchem, G., Piñon, R. and Salts, Y. (1972) Sporulation in *Saccharomyces cerevisiae*: premeiotic DNA synthesis, readiness and commitment. *Exp. Cell Res.*, **75**, 207–218.

Skulachev, V.P. (1990) Power transmission along biological membranes. *J. Membrane Biol.*, **114**, 97–112.

Skulachev, V.P. (1991) Chemiosmotic systems in bioenergetics: H^+-cycles and Na^+-cycles. *Biosci. Rep.*, **11**, 387–444.

Slack, J.M.W. (1987) Morphogenetic gradients – past and present. *TIBS*, **12**, 200–204.

Slater, J.H. (1988) Microbial population and community dynamics, in *Micro-organisms in Action: Concepts and Applications in Microbial Ecology* (eds J.M. Lynch and J.E. Hobbie), Blackwell Scientific Publications, Oxford, pp. 51–74.

Slater, M.L., Sharrow, S.O. and Gart, J.J. (1977) Cell cycle of *Saccharomyces cerevisiae* in populations growing at different rates. *Proc. Natl. Acad. Sci. USA*, **74**, 3850–3854.

Smarr, L.L. (1985) An approach to complexity: Numerical computations. *Science*, **228**, 403–408.

Smith, A.P. and Segall, J. (1986) Characterization and mutational analysis of a cluster of three genes expressed preferentially during sporulation of *Saccharomyces cerevisiae*. *Mol. Cell. Biol.*, **6**, 2443–2451.

Smith, M.W. and Neidhardt, F.C. (1983a) Proteins induced by anaerobiosis in *Escherichia coli*. *J. Bacteriol.*, **154**, 336–343.

Smith, M.W. and Neidhardt, F.C. (1983b) Proteins induced by aerobiosis in *Escherichia coli*. *J. Bacteriol.*, **154**, 344–350.

Sneyd, J. and Kalachev, L.V. (1994) A profile analysis of propagating calcium waves. *Calcium*, **15**, 289–296.

Solomon, F. (1981) Specification of cell morphology by endogenous determinants. *J. Cell Biol.*, **90**, 547–553.

Solomon, F. (1991) Analyses of the cytoskeleton in *Saccharomyces cerevisiae*. *Annu. Rev. Cell Biol.*, **7**, 633–662.

Sols, A. (1981) Multimodulation of enzyme activity. *Curr. Top. Cell. Regul.*, **19**, 77–101.

Spanswick, R.M. (1981) Electrogenic ion pumps. *Annu. Rev. Plant Physiol.*, **32**, 267–289.

Srere, P. (1987) Complexes of sequential metabolic enzymes. *Annu. Rev. Biochem.*, **56**, 205–223.

Srivastava, D.K. and Bernhard, S.A. (1987) Biophysical chemistry of metabolic reaction sequences in concentrated enzyme solution and in the cell. *Annu. Rev. Biophys. Biophys. Chem.*, **16**, 175–204.

Stauffer, D. and Aharony, A. (1994) *Introduction to Percolation Theory*, Taylor & Francis, London.

Stein, D.L. (1989) Preface, in *Lectures in the Sciences of Complexity* (ed. D.L. Stein), Addison-Wesley, Santa Fe Institute, pp. XIII–XXII.

Steinert, P.M., Jones, J.C.R. and Goldman, R.D. (1984) Intermediate filaments. *J. Cell Biol.*, **99**, 22–27.

Stephanopoulos, G. and Vallino, J.J. (1991) Network rigidity and metabolic engineering in metabolite overproduction. *Science*, **252**, 1675–1681.

Stewart, I. and Golubitsky, M. (1992) *Fearful Symmetry. Is God a Geometer?*, Blackwell, Oxford.

Stewart-Savage, J., Grey, R.D. and Elinson, R.P. (1991) Polarity of the surface and cortex of the amphibian egg from fertilization to first cleavage. *J. Electron Microsc. Techn.*, **17**, 369–383.

Stouthamer, A.H. (1979) The search for correlation between theoretical and experimental growth yields. *Int. Rev. Biochem.*, **21**, 1–47.

Stouthamer, A.H. and van Verseveld, H.W. (1987) Microbial energetics should be considered in manipulating metabolism for biotechnological purposes. *TIBTECH*, **5**, 149–155.

Strauss, B.S. (1992) The origin of point mutations. *Cancer Res.*, **52**, 249–253.

Stricker, S.A., Centonze, V.E., Paddock, S.W. and Schatten, G. (1992) Confocal microscopy of fertilization-induced calcium dynamics in sea urchin eggs. *Dev. Biol.*, **149**, 370–380.

Strome, S. and Wood, W.B. (1983) Generation of asymmetry and segregation of germline granules in early *Caenorhabditis elegans*. *Cell*, **35**, 15–25.

Stryer, L. (1988) *Biochemistry*, W.H. Freeman and Co., New York.

Stucki, J.W. (1978) Stability analysis of biochemical systems. A practical guide. *Prog. Biophys. Mol. Biol.*, **33**, 99–187.

Stucki, J.W. (1982) Thermodynamic optimizing principles in mitochondrial energy conversions, in *Metabolic Compartmentation* (ed. H. Sies), Academic Press, London, pp. 39–69.

Suihko, M.L. and Drazic, M. (1983) Pentose fermentation by yeasts. *Biotechnol. Lett.*, **5**, 107–112.

Sumegy, B., Sherry, A.D., Malloy, C.R. *et al.* (1991) Is there tight channelling in the tricarboxylic acid cycle metabolon? *Biochem. Soc. Trans.*, **19**, 1002–1005.

Supply, P., Wach, A., Thinés-Sempoux, D. and A. Goffeau (1993) Proliferation of intracellular structures upon overexpression of the PMA2 ATPase in *Saccharomyces cerevisiae*. *J. Biol. Chem.*, **268**, 19744–19752.

Suprenant, K.A. (1991) Unidirectional microtubule assembly in cell-free extracts of *Spisula solidissima* oocytes is regulated by subtle changes in pH. *Cell Motil. Cytoskel.*, **19**, 207–220.

Suprenant, K.A. (1993) Microtubules, ribosomes, and RNA: evidence for cytoplasmic localization and translational regulation. *Cell Motil. Cytoskel.*, **25**, 1–9.

Suzuki, A., Yamazaki, M. and Ito, T. (1989) Osmoelastic coupling in biological structures: formation of parallel bundles of actin filaments in a crystalline-like structure caused by osmotic stress. *Biochemistry*, **28**, 6513–6518.

Tabony, J. (1994) Morphological bifurcations involving reaction–diffusion processes during microtubule formation. *Science*, **264**, 245–248.

Tabony, J. and Job, D. (1992) Gravitational symmetry breaking in microtubular dissipative structures. *Proc. Natl. Acad. Sci. USA*, **89**, 6948–6952.

Taiz, L. (1984) Plant cell expansion: regulation of cell wall mechanical properties. *Annu. Rev. Plant Physiol.*, **35**, 585–657.

Takahashi, J.S. (1993) Circadian-clock regulation of gene expression. *Curr. Opin. Gen. Devel.*, **3**, 301–309.

Takashima, S. and Schwan, H.P. (1985) Alignment of microscopic particles in electric fields and its biological implications. *Biophys. J.*, **47**, 513–518.

Takeda, M., Vassarotti, A. and Douglas, M.G. (1985) Nuclear genes coding the yeast mitochondrial adenosine triphosphatase complex. Primary sequence analysis of ATP_2 encoding the F1-ATPase b-subunit precursor. *J. Biol. Chem.*, **260**, 15458.

Takeda, M., Chen, W.J., Saltzgaber, J. and Douglas, M.G. (1986) Nuclear genes encoding the yeast mitochondrial ATPase complex. Analysis of ATP_1 coding the F1-ATPase a-subunit and its assembly. *J. Biol. Chem.*, **261**, 15126.

Tanaka, E.M. and Kirschner, M.W. (1991) Microtubule behavior in the growth cones of living neurons during axon elongation. *J. Cell Biol.*, **115**, 345–363.

Taylor, J. (1960) Nucleic acid synthesis in relation to the cell division cycle. *Ann. NY Acad. Sci.*, **90**, 409–421.

Tempest, D.W. and Neijssel, O.M. (1984) The status of Y_{ATP} and maintenance energy as biologically interpretable phenomena. *Annu. Rev. Microbiol.*, **38**, 459–486.

Tempest, D.W. and Neijssel, O.M. (1992) Physiological and energetic aspects of bacterial metabolite overproduction. *FEMS Microbiol. Lett.*, **100**, 169–176.

Ten Brink, B. and Konings, W.N. (1982a) Electrochemical proton gradient and lactate concentration gradient in *Streptococcus cremoris* cells grown in batch culture. *J. Bacteriol.*, **152**, 682–686.

Ten Brink, B. and Konings, W.N. (1982b) Generation of a protonmotive force in anaerobic bacteria by end-product efflux. *Methods Enzymol.*, **125**, 492–510.

Teplova, V.V., Bogucka, K., Czyz, A. *et al.* (1993) Effect of glucose and deoxyglucose on cytoplasmic $[Ca^{2+}]$ in Ehrlich ascites tumor cells. *Biochem. Biophys. Res. Commun.*, **196**, 1148–1154.

Termonia, Y. and Ross, J. (1981) Oscillations and control features in glycolysis in numerical analysis of a comprehensive model. *Proc. Natl. Acad. Sci. USA*, **78**, 2952–2956.

Thaller, C. and Eichele, G. (1987) Identification and spatial distribution of retinoids in the developing chick bud. *Nature*, **327**, 625–628.

Thaller, C. and Eichele, G. (1990) Isolation of 3,4-didehydroretinoic acid, a novel morphogenetic signal in the chick wing bud. *Nature*, **345**, 815–819.

Theologis, A. (1985) Rapid induction of specific mRNAs by auxin in pea epicotyl tissue. *J. Mol. Biol.*, **183**, 53–68.

Theologis, A. and Ray, P.M. (1982) Early auxin-regulated polyadenylated mRNA sequences in pea stem tissue. *Proc. Natl. Acad. Sci. USA*, **79**, 418–421.

Thevelein, J.M. (1992) The RAS-adenylate cyclase pathway and cell cycle control in *Saccharomyces cerevisiae*. *Antonie van Leeuwenhoek*, **62**, 109–130.

Thom, R. (1972) *Stabilité Structurelle et Morphogenese*, Intereditions, Paris.

Thompson, D'A.W. (1980) *Sobre el Crecimiento y la Forma*, H. Blume Edic., Madrid.

Thuren, T., Tulkki, A.-P., Virtanen, J.A. and Kinnunen, P.K.J. (1987) Triggering of the activity of phospholipase A2 by an electric field. *Biochemistry*, **26**, 4907–4910

Toivola, A., Yarrow, D., van den Bosch, E. *et al.* (1984) Alcoholic fermentation of D-xylose by yeasts. *Appl. Environ. Microbiol.*, **47**, 1221–1223.

Toso, R.J., Jordan, M.A., Farrell, K.W. *et al.* (1993) Kinetic stabilization of microtubule dynamic instability *in vitro* by vinblastine. *Biochemistry*, **32**, 1285–1293.

Town, C.D., Dominov, J.A., Karpinski, B.A. and Jentoft, J.E. (1987) Relationships between extracellular pH, intracellular pH, and gene expression in *Dictyostelium discoideum*. *Dev. Biol.*, **122**, 354–362.

Traub, R.D., Miles, R. and Wong, R.K.S. (1989) Model of the origin of rhythmic population oscillations in the hippocampal slice. *Science*, **243**, 1319–1325.

Tseng, Ch.P., Hansen, A.K., Cotter, P. and Gunsalus, R.P. (1994) Effect of cell growth rate on expression of the anaerobic respiratory pathway operons *frdABCD*, *dmsABC*, and *narGHJI* of *Escherichia coli*. *J. Bacteriol.*, **176**, 6599–6605.

Tsong, T.Y. (1989) Deciphering the language of cells. *TIBS*, **14**, 89–92.

Tsong, T.Y. and Astumian, R.D. (1986) Absorption and conversion of electric field energy by membrane bound ATPases. *Bioelectrochem. Bioenerg.*, **15**, 457–476.

Tsong, T.Y. and Astumian, R.D. (1988) Electroconformational coupling: How membrane-bound ATPase transduces energy from dynamic electric fields. *Annu. Rev. Physiol.*, **50**, 273–290.

Tsonis, P.A. (1987) The nature of positional information. *TIBS*, **12**, 249.

Tucker, R.P. (1990) The roles of microtubule-associated proteins in brain morphogenesis: a review. *Brain Res. Rev.*, **15**, 101–120.

Turing, A.M. (1952) The chemical basis of morphogenesis. *Phil. Trans. Royal Soc. Lond.*, **B237**, 37–72.

Tuttle, J.P. and Wilson, J.E. (1970) Rat brain hexokinase: A kinetic comparison of soluble and particulate forms. *Biochim. Biophys. Acta*, **212**, 185–188.

Tyson, J.J. (1976) The Belousov–Zhabotinskii reaction, in *Lecture Notes in Biomathematics*, Vol. 10. (ed. S. Levin), Springer-Verlag, New York.

Tyson, J.J. (1991) Modeling of the cell division cycle: cdc2 and cyclin interactions. *Proc. Natl. Acad. Sci. USA*, **88**, 7328–7332.

Tyson, J.J. and Kagan, M.L. (1988) Spatiotemporal organization in biological and chemical systems: Historical review, in *From Chemical to Biological Organization* (eds M. Markus, S.C. Muller and G. Nicolis), Springer-Verlag, Berlin, pp. 14–21.

Tzagoloff, A., Wu, M. and Crivellone, M. (1986) Assembly of the mitochondrial membrane system. Characterization of COR1, the structural gene for the 44-kilodalton core protein of yeast coenzyme QH2–cytochrome c reductase. *J. Biol. Chem.*, **261**, 17163.

Unger, M.W. and Hartwell, L.H. (1976) Control of cell division in *Saccharomyces cerevisiae* by methionyl-tRNA. *Proc. Natl. Acad. Sci. USA*, **73**, 1664–1668.

Uyeda, K. (1992) Interactions of glycolytic enzymes with cellular membranes. *Curr. Top. Cell. Regul.*, **33**, 31–46.

Valberg, P.A. and Albertini, D.F. (1985) Cytoplasmic motions, rheology, and structure probed by a novel magnetic particle method. *J. Cell Biol.*, **101**, 130–140.

Vallari, R.C., Cook, W.J., Audino, D.C. *et al.* (1992) Glucose repression of the yeast *ADH2* gene occurs through multiple mechanisms, including control of the protein synthesis of its transcriptional activator, ADR1. *Mol. Cell. Biol.*, **12**, 1663–1673.

Vanden Broeck, J., De Loof, A. and Callaerts, P. (1992) Electrical–ionic control of gene expression. *Int. J. Biochem.*, **24**, 1907–1916.

Van der Aar, P. (1990) Consequences of 3-phosphoglycerate kinase overproduction for the growth and physiology of *Saccharomyces cerevisiae*. PhD Thesis, Vrije Universiteit, Amsterdam.

Van der Meer, R., Westerhoff, H.V. and van Dam, K. (1980) Linear relation between rate and thermodynamic force in enzyme catalyzed reactions. *Biochim. Biophys. Acta*, **591**, 488–493.

Van Dijken, J.P. and Scheffers, W.A. (1986) Redox balances in the metabolism of sugars by yeasts. *FEMS Microbiol. Rev.*, **32**, 199–224,

Van Dijken, J.P., Weusthuis, R.A. and Pronk, J.T. (1993) Kinetics of growth and sugar consumption in yeasts. *Antonie van Leeuwenhoek*, **63**, 343–352.

Van Driel, R., Humbel, B. and de Jong, L. (1991) The nucleus: a black box being opened. *J. Cell. Biochem.*, **47**, 311–316.

Van Urk, H., Schipper, D., Breedveld, G.J. *et al.* (1989) Localization and kinetics of pyruvate-metabolizing enzymes in relation to aerobic alcoholic fermentation in *Saccharomyces cerevisiae* CBS 8066 and *Candida utilis* CBS 621. *Biochim. Biophys. Acta*, **992**, 78–86.

Van Venrooij, W.J.W., Henshaw, E.C. and Hirsch, C.A. (1972) Effects of deprival of glucose or individual amino acids on polyribosome distribution and rate of protein synthesis in cultured mammalian cells. *Biochim. Biophys. Acta*, **259**, 127–137.

Varela, J.C.S., van Beekvelt, C., Planta, R.J. and Mager, W.H. (1992) Osmostress-induced changes in yeast gene expression. *Mol. Microbiol.*, **6**, 2183–2190.

Vastano, J.A., Pearson, J.E., Horsthemke, W. and Swinney, H.L. (1987) Chemical pattern formation with equal diffusion coefficients. *Phys. Lett.*, **A 124**, 320–325.

Vaulont, S. and Kahn, A. (1994) Transcriptional control of metabolic regulation genes by carbohydrates. *FASEB J.*, **8**, 28–35.

Vedeler, A., Pryme, I.F. and Hesketh, J.E. (1991) Compartmentalization of polysomes into free, cytoskeletal-bound and membrane-bound populations. *Biochem. Soc. Trans.*, **19**, 1108–1111.

Veenhuis, M., van der Klei, I.J., Titorenko, V. and Harder, W. (1992) *Hansenula polymorpha*: An attractive model organism for molecular studies of peroxisome biogenesis and function. *FEMS Microbiol. Lett.*, **100**, 393–404.

Verde, F., Labbe, J., Doree, M. and Karsenti, E. (1990) Regulation of microtubule dynamics by CDC2 kinase in cell free extracts of *Xenopus* eggs. *Nature*, **343**, 233–238.

Verdoni, N., Aon, M.A., Lebeault, J.M. and Thomas, D. (1990) Proton motive force, energy recycling by end product excretion and metabolic uncoupling during anaerobic growth of *Pseudomonas mendocina*. *J. Bacteriol.*, **172**, 6673–6681.

Verdoni, N., Aon, M.A. and Lebeault, J.M. (1992) Metabolic and energetic control of *Pseudomonas mendocina* growth during transitions from aerobic to oxygen-limited conditions in chemostat cultures. *Appl. Environ. Microbiol.*, **58**, 3150–3156.

Vincent, J.P., Oster, G.F. and Gerhart, J.C. (1986) Kinematics of gray crescent formation in *Xenopus* eggs: the displacement of subcortical cytoplasm relative to the egg surface. *Dev. Biol.*, **113**, 484–500.

Vincent, J.P., Scharf, S.R. and Gerhart, J.C. (1987) Subcortical rotation in *Xenopus* eggs: a preliminary study of its mechanochemical basis. *Cell Motil. Cytoskel.*, **8**, 143–154.

Von Bertalanffy, L. (1987) *Theorie Générale des Systèmes*, Dunod, Paris, pp. 168–176.

Von Meyenburg, K.H. (1969) Energetics of the budding cycle of *Saccharomyces cerevisiae* during glucose limited aerobic growth. *Arch. Mikrobiol.*, **66**, 289–303.

Von Meyenburg, K.H. (1973) Stable synchrony oscillations in continuous cultures of *Saccharomyces cerevisiae* under glucose limitation, in *Biological and Biochemical Oscillators* (eds B. Chance, K. Pye, A.K. Ghosh and B. Hess), Academic Press, New York, pp. 411–418.

Waddington, C.H. (1947) *Organisers and Genes*, Cambridge University Press, London.

Wadsworth, W.G. and Riddle, D.L. (1989) Developmental regulation of energy metabolism in *Caenorhabditis elegans*. *Dev. Biol.*, **132**, 167–173.

Wales, D.S., Cartledge, T.G. and Lloyd, D. (1980) Effects of glucose repression and anaerobiosis on the activities and subcellular distribution of tricarboxylic acid cycle and associated enzymes in *Saccharomyces carlsbergensis*. *J. Gen. Microbiol.*, **116**, 93–98.

Walker, M.E., Val, D.L., Rhode, M. *et al.* (1991) Yeast pyruvate decarboxylase: Identification of two genes encoding isoenzymes. *Biochem. Biophys. Res. Commun.*, **176**, 1210.

Walsh, K. and Koshland, D.E. Jr (1985) Characterization of rate-controlling steps *in vivo* by use of an adjustable expression vector. *Proc. Natl. Acad. Sci. USA*, **82**, 3577–3581.

Wang, J.C. (1985) DNA topoisomerases. *Annu. Rev. Biochem.*, **54**, 665–697.

Wang, N. and Ingber, D.E. (1994) Control of cytoskeletal mechanics by extracellular matrix, cell shape, and mechanical tension. *Biophys. J.*, **66**, 2181–2189.

Wang, N., Butler, J.P. and Ingber, D.E. (1993) Mechanotransduction across the cell surface and through the cytoskeleton. *Science*, **260**, 1124–1127.

Warburg, O. (1930) *Metabolism of Tumors*, Arnold Constable, London.

Waygood, E.B. (1987) Kinetic aspects of the PEP-dependent phosphotransferase system, in *Sugar Transport and Metabolism in Gram-positive Bacteria* (eds J. Reizen and H. Peterkofsky), Plenum Press, New York, pp. 235–254.

Weber, G. (1983) Biochemical strategy of cancer cells and the design of chemotherapy: G.H.A. Clowes memorial lecture. *Cancer Res.*, **43**, 3466–3492.

Weber, G. (1990) Metabolic strategies in cancer chemotherapy. *Biochem. Soc. Trans.*, **18**, 74–78.

Weibel, E.R. (1991) Fractal geometry: a design principle for living organisms. *Am. J. Physiol.*, **261** (*Lung Cell. Mol. Physiol. 5*), L361–L369.

Weiner, J.H., Lemire, B.D., Elmes, M.L. *et al.* (1984) Overproduction of fumarate reductase in *Escherichia coli* induces a novel intracellular lipid–protein organelle. *J. Bacteriol.*, **158**, 590–596.

Weisenberg, R.C. and Cianci, Ch. (1984) ATP-induced gelation-contraction of microtubules assembled *in vitro*. *J. Cell Biol.*, **99**, 1527–1533.

Weiss, J.N, Garfinkel, A., Spano, M.L. and Ditto, W.L. (1994) Chaos and chaos control in biology. *J. Clin. Invest.*, **93**, 1355–1360.

Welch, G.R. (1977) On the role of organized multienzyme systems in cellular metabolism: A general synthesis. *Prog. Biophys. Mol. Biol.*, **32**, 103–191.

Welch, G.R. (1984) Biochemical dynamics in organized states: a holistic approach, in *Dynamics of Biochemical Systems. NATO Series* (eds J. Ricard and A. Cornish-Bowden), Plenum Press, New York, pp. 85–101.

Welch, M.D., Holtzman, D.A. and Drubin, D.G. (1994) The yeast actin cytoskeleton. *Curr. Opin. Cell Biol.*, **6**, 110–119.

Werner-Washburne, M., Braun, E., Johnston, G.C. and Singer, R.A. (1993) Stationary phase in the yeast *Saccharomyces cerevisiae*. *Microbiol. Rev.*, **57**, 383–401.

Wessels, J.G.H. (1986) Cell wall synthesis in apical hyphal growth. *Int. Rev. Cytol.*, **104**, 37–79.

Westerhoff, H.V. and van Dam, K. (1987) *Thermodynamics and Control of Biological Free-energy Transduction*, Elsevier, Amsterdam.

Westerhoff, H.V. and Kell, D.B. (1987) Matrix method for determining steps most rate-limiting to metabolic fluxes in biotechnological processes. *Biotechnol. Bioeng.*, **30**, 101–107.

Westerhoff, H.V., Lolkema, J.S., Otto, R. and Hellingwerf, K.J. (1982) Thermodynamics of growth. Non-equilibrium thermodynamics of bacterial growth. The phenomenological and the mosaic approach. *Biochim. Biophys. Acta*, **683**, 181–220.

Westerhoff, H.V., Hellingwerf, K.J. and van Dam, K. (1983) Thermodynamic efficiency of microbial growth is low but optimal for maximal growth rate. *Proc. Natl. Acad. Sci. USA*, **80**, 305–309.

Westerhoff, H.V., Tsong, T.Y., Chock, P.B. *et al.* (1986) How enzymes can capture and transmit free energy from an oscillating electric field. *Proc. Natl. Acad. Sci. USA*, **83**, 4734–4738.

Westerhoff, H.V., O'Dea, M.H., Maxwell, A. and Gellert, M. (1988) *Cell Biophys.*, **12**, 167–181

Westerhoff, H.V., van Heeswijk, W., Kahn, D. and Kell, D.B. (1991) Quantitative approaches to the analysis of the control and regulation of microbial metabolism. *Antonie van Leeuwenhoek*, **60**, 193–207.

Weusthuis, R.A., Pronk, J.T., van den Broek, P.J.A. and van Dijken, J.P. (1994) Chemostat cultivation as a tool for studies on sugar transport in yeasts. *Microbiol. Rev.*, **58**, 616–630.

Wheals, A.E. (1987) Biology of the cell cycle in yeasts, in *The Yeasts*, Vol. 1 (eds A.H. Rose and J.S. Harrison), Academic Press, London, pp. 284–390.

Wheatley, D.N. and Malone, P.C. (1993) Heat conductance, diffusion theory and intracellular metabolic regulation. *Biol. Cell*, **79**, 1–5.

Williams, R.P.J. (1990) The mineral elements in homeostasis and morphogenesis. *Biochem. Soc. Trans.*, **18**, 689–705.

Wilson, J.E. (1980) Brain hexokinase, the prototype ambiquitous enzyme. *Curr. Top. Cell. Regul.*, **16**, 1–54.

Winfree, A.T. (1967) Biological rhythms and the behavior of populations of coupled oscillators. *J. Theor. Biol.*, **16**, 15–42.

Winfree, A.T. (1974) Rotating chemical reactions. *Sci. Am.*, 82–95.

Winfree, A.T. (1980) *The Geometry of Biological Time*, Springer Verlag, New York, pp. 285–299.

Winfree, A.T. (1984) The prehistory of the Belousov–Zhabotinskii oscillator. *J. Chem. Educ.*, **61**, 661–663.

Winfree, A.T. (1987) *When Time Breaks Down*, Princeton University Press, Princeton, NJ.

Winfree, A.T. (1991) Crystals from dreams. *Science*, **352**, 568–569.

Winfree, A.T. and Strogatz, S.H. (1984) Organizing centres for three-dimensional chemical waves, *Nature*, **311**, 611–615.

Winston, D., Arora, M., Maselko, J. *et al.* (1991) Cross-membrane coupling of chemical spatiotemporal patterns. *Nature*, **351**, 132–135.

Wittenberg, C. and Reed, S.S. (1988) Control of yeast cell cycle is associated with assembly–disassembly of the Cdc28 protein kinase complex. *Cell*, **54**, 1061–1072.

Wittenberg, C., Richardson, S.L. and Reed, S.S. (1987) Subcellular localization of a protein kinase required for cell cycle initiation in *Saccharomyces cerevisiae*;

evidence for an association between the CDC28 gene product and the insoluble cytoplasmic matrix. *J. Cell Biol.*, **105**, 1527–1538.

Wojcieszyn, J.W., Schlegel, R.A., Wu, E.-S. and Jacobson, K.A. (1981) Diffusion of injected macromolecules within the cytoplasm of living cells. *Proc. Natl. Acad. Sci. USA*, **78**, 4407–4410.

Wolfner, M., Yep, D., Messenguy, F. and Fink, G.R. (1975) Integration of amino acid biosynthesis into the cell cycle of *Saccharomyces cerevisiae. J. Mol. Biol.*, **96**, 273–290.

Wolpert, L (1969) Positional information and the spatial pattern of cellular differentiation. *J. Theor. Biol.*, **25**, 1–47.

Wright, B.E. and Albe, K.R. (1994a) Carbohydrate metabolism in *Dictyostelium discoideum*: I. Model construction. *J. Theor. Biol.*, **169**, 231–241.

Wright, B.E. and Albe, K.R. (1994b) Carbohydrate metabolism in *Dictyostelium discoideum*: II. Systems' analysis. *J. Theor. Biol.*, **169**, 243–251.

Wright, B.E. and Kelly, P.J. (1981) Kinetic models of metabolism in intact cells, tissues and organisms. *Curr. Top. Cell. Regul.*, **19**, 103–158.

Wright, B.E. and Reimers, J.M. (1988) Steady-state models of glucose-perturbed *Dictyostelium discoideum. J. Biol. Chem.*, **263**, 14906–14912.

Wright, R.M., Ro, Ch., Cumsky, M.G. and Poyton, R.O. (1984) Isolation and sequence of the structural gene for cytochrome c oxidase subunit VI from *Saccharomyces cerevisiae. J. Biol. Chem.*, **259**, 15401.

Wu, M. and Tzagoloff, A. (1987) Mitochondrial and cytoplasmic fumarases in *Saccharomyces cerevisiae* are encoded by a single nuclear gene FUM1. *J. Biol. Chem.*, **262**, 12275.

Wu, R. and Racker, E. (1963) Control of rate-limiting factors of glycolysis in tumor cells, in *Control Mechanisms in Respiration and Fermentation* (ed. B. Wright), The Ronald Press Company, New York, pp. 265–288.

Wylie, C.C., Heasman, J., Parke, J.M. *et al.* (1986) Cytoskeletal changes during oogenesis and early development of *Xenopus laevis. J. Cell Sci.*, Suppl. **5**, 329–341.

Yamashita, I. and Fukui, S. (1985) Transcriptional control of the sporulation-specific glucoamylase gene in the yeast *Saccharomyces cerevisiae. Mol. Cell. Biol.*, **5**, 3069.

Yanishevsky, R.M. and Stein, G.H. (1981) Regulation of the cell cycle in eukaryotic cells. *Int. Rev. Cytol.*, **69**, 223–259.

Yates, E.F. (1992) Fractal applications in biology: scaling time in biochemical networks. *Methods Enzymol.*, **210**, 636–675.

Yuan, Z., Medina, M.A., Boiteux, A. *et al.* (1990) The role of fructose 2,6-bisphosphate in glycolytic oscillations in extracts and cells of *Saccharomyces cerevisiae. Eur. J. Biochem.*, **192**, 791–795.

Zhai, Y., Kronebusch, P.J. and Borisy, G.G. (1995) Kinetochore microtubule dynamics and the metaphase–anaphase transition. *J. Cell Biol.*, **131**, 721–734.

Zimmermann, U. (1978) Physics of turgor- and osmoregulation. *Annu. Rev. Plant Physiol.*, **29**, 121–148.

Zitomer, R.S., and Lowry, Ch.V. (1992) Regulation of gene expression by oxygen in *Saccharomyces cerevisiae. Microbiol. Rev.*, **56**, 1–11.

Zurfluh, L. L. and Guilfoyle, T. J. (1980) Auxin-induced changes in the patterns of protein synthesis in soyabean hypocotyl. *Proc. Natl. Acad. Sci. USA*, **77**, 357–361.

Index

Page references appearing in *italics* refer to tables and those in **bold** refer to figures.

Nuclear magnetic resonance 234, 235
   proton 202
Nuclear matrix 184, 359, 364

Onsager, 9, 109
Oocytes 86
Order, macroscopic 164, 168, 171–2, 229
Orthophenantroline 332
Organization
   cytoplasmic, 45, 112
   genome 404
   levels of 45, *47–50*, 50, **51**, **55**, 62
      coherence 324
      in development and
         morphogenesis 116, 143
      functional self-similarity 285
      in physiological processes 53, 201,
         430
      transitions between 47
   spatio-temporal
      of cytostructure 363
   stability of 115
   transition point 63–4, 116, 163
Oscillations 31, 35, 64, 73, 83
   calcium 87
   co-substrate model 461, **463**, 464, 480
   complex 74
   coupled 79
   cyanide-induced 250
   damped 249
   energy-driven 444–5
   in glycolysis 59, 87–9, 93, 246
   high frequency 80
   macroscopic damping 460, 464
   membrane potential 82, 472, 490
   pH_i 87
   sustained 249
   ultradian 89
Ouabain 100
Ovomucin, 197
   gel 59, 198
Oxidative phosphorylation 27, 29, 89
   electron carrier transcript length *410*
   generation of ATP by 296–7
   inhibition of 249
   P/O ratio 290–1, *292*
   in tumour cells 234

Pachysolen tannophilus 337

Pancreatic islets of Langerhans 449–50
Parameter
   control 147
   optimization 238
   space 14, 16, 29, 282
   see also Bifurcation, parameter
Pasteur effect 91, 244, 349–51
   reverse 434,
Patch-clamp 142, 235
Pattern
   asymmetric 168
   banding 160, 170, 171
   chaotic 154
   formation 124, 146, 152, 160, 162
   spatial pH 157
   symmetric 160
PEG, *see* Poly(ethylene glycol)
Pelvetia 132
   P. fastigiata 142
   zygote polarity 469, **470**
Pentose phosphate pathway 236, 245,
   291, *292*
Per gene 74, 77, 100
Percolation
   cluster 70–1, 123, 204, 213–14, 220,
     261
      dynamics of biochemical reaction
        in 272–3
   lattice 214, 224
      connectedness and dynamics in
        272–3
      geometry of cytoskeleton 274–5
      of microtubular protein 232
      probability of occupation 228
   probability 70
   spanning cluster 214, 232, 273
   threshold 70, 137, 214, 217, 227
      in development 123–4
      spatio-temporal scaling at 224
Period doubling 98, 246
Peroxidase 98
Peroxisome 414–16
Perturbation 50, 52
Perturbation, methods 53
pH dependence
   autocatalysis 156–7
   of conformational changes 149
   enzyme activity 149, 173
Phase portrait 15, 393